Progress in Probability
Volume 28

Random Walks, Brownian Motion, and Interacting Particle Systems

A Festschrift in Honor of Frank Spitzer

Rick Durrett
Harry Kesten
Editors

1991 Birkhäuser
Boston · Basel · Berlin

Rick Durrett
Department of Mathematics
Cornell University
White Hall
Ithaca, NY 14853

Harry Kesten
Department of Mathematics
Cornell University
White Hall
Ithaca, NY 14853

Library of Congress cataloging-in-publication data

Random walks, Brownian motion, and interacting particle systems /
Rick Durrett, Harry Kesten, editors.
p. cm. -- (Progress in probability : v. 28)
Festschrift in honor of Frank Spitzer.
Includes bibliographical references and index.
ISBN 0-8176-3509-2 (acid free). -- ISBN 3-7643-3509-2 (Basel : acid-free)
1. Random walks (Mathematics) 2. Brownian motion processes.
3. Probabilities. 4. Statistical physics. 5. Spitzer, Frank, 1926-
I. Durrett, Richard, 1951- II. Kesten, Harry, 1931-
III. Spitzer, Frank, 1926- . IV. Series: Progress in probability
: 28.
QC174.85.R37R37 1991 91-31173
519.2--dc20 CIP

Printed on acid-free paper.

ISBN 0-8176-3509-2
ISBN 3-7643-3509-2

Typset by authors in TeX.
Printed and bound by Edwards Brothers, Inc., Ann Arbor, Michigan.
Printed in the U.S.A.

9 8 7 6 5 4 3 2 1

ABOUT THE COVER

The editors wish to thank Robert Fisch and David Griffeath for the graphic on the cover. The graphic depicts convergence to equilibrium in a system of particles which perform random walks under the "exclusion" restriction that there may be at most one particle per lattice site. Initially the particles are packed in a tight ball and successive "still frames" show the state of the system at times 10, 100, 1000 and 10000 (on a 256×256 lattice with wrap-around).

The picture was produced using the hardware and software of T. Toffoli and N. Margolus. A Matrix film recorder, funded by the National Science Foundation, made the transparency used in the printing process.

STUDENTS OF FRANK SPITZER

1957 J. W. Lamperti, *On the asymptotic behavior of recurrent and almost-recurrent events.*

1964 W. W. Whitman, *Some strong laws for random walks and Brownian motion.*

1965 J. C. Mineka, *The existence and uniqueness of positive solutions to the Wiener-Hopf equation with positive kernel.*

1969 R. A. Holley, *The motion of a heavy particle in an infinite one-dimensional gas of hard spheres*

1972 F. C. Solomon, *Random walks in a random environment.*

1974 K. G. Logan, *Time reversible solutions in statistical mechanics.*

R. L. Thompson, *Equilibrium states on thin energy shells.*

1976 J. T. Cox, *Entrance laws for Markov chains and one-dimensional diffusions.*

D. S. Griffeath, *Coupling methods for Markov processes.*

1977 L. F. Gray, *Controlled spin-flip systems.*

A. Kleinerman, *Limit theorems for infinitely divisible random fields.*

1979 C.-T. Hsiao, *The stochastic time evolution of Gaussian interacting systems.*

1984 N. N. Madras, *A process in a randomly fluctuating environment.*

PREFACE

This collection of articles is dedicated to Frank Spitzer on the occasion of his 65th birthday. The articles, written by a group of his friends, colleagues, former students and coauthors, are intended to demonstrate the major influence Frank has had on probability theory for the last 30 years and most likely will have for many years to come. Frank has always liked new phenomena, clean formulations and elegant proofs. He has created or opened up several research areas and it is not surprising that many people are still working out the consequences of his inventions.

By way of introduction we have reprinted some of Frank's seminal articles so that the reader can easily see for himself the point of origin for much of the research presented here. These articles of Frank's deal with properties of Brownian motion, fluctuation theory and potential theory for random walks, and, of course, interacting particle systems.

The last area was started by Frank as part of the general resurgence of treating problems of statistical mechanics with rigorous probabilistic tools. It is presently one of the most exciting and active areas in probability, and is one of the reasons for the fruitful interaction between statistical physicists and probabilists nowadays. About one-third of the papers in this collection are in this area. The variety of topics of these papers illustrates how far the subject has developed over the last fifteen years. Among the subjects of the other articles directly related to Frank's work are potential theory of Brownian motion, winding of one or more two-dimensional Brownian motions and aspects of random walk. Despite the age of these topics, they keep generating new questions.

We regret that limitations of space prevented us from inviting more of Frank's friends to contribute to this Festschrift. Frank always enjoyed explaining his ideas to others and has been generous in sharing them and attracting coauthors. We are sure we speak for all mathematicians who have worked with Frank or who have been influenced by his work, when we thank him for introducing us to beautiful mathematics.

Rick Durrett
Harry Kesten

Table of Contents

PUBLICATIONS OF FRANK SPITZER
(THE ARTICLES MARKED WITH AN ASTERISK ARE REPRINTED IN THIS VOLUME.)

1. *On a class of random variables*, Proc. Amer. Math. Soc. **6** (1955), 494–505.

2. *On interval recurrent sums of independent random variables*, Proc. Amer. Math. Soc. **7** (1956), 164–171.

*3. *A combinatorial lemma and its application to probability theory*, Trans. Amer. Math. Soc. **82** (1956), 323–339.

4. *The Wiener–Hopf equation whose kernel is a probability density*, Duke Math. J. **24** (1957), 327–343.

*5. *Some theorems concerning 2–dimensional Brownian motion*, Trans. Amer. Math. Soc. **87** (1958), 187–197.

6. (with A. Calderon and H. Widom), *Inversion of Toeplitz matrices*, Illinois J. Math. **3** (1959), 490–498.

7. *Some probability limit theorems*, Bull. Amer. Math. Soc. **65** (1959), 117–119.

8. *A Tauberian theorem and its probability interpretation*, Trans. Amer. Math. Soc. **94** (1960), 150–169.

9. *The Wiener–Hopf equation whose kernel is a probability density, II*, Duke Math. J. **27** (1960), 363–372.

10. (with C. Stone), *A class of Toeplitz forms and their application to probability*, Illinois J. Math. **4** (1960), 253–277.

11. (with P. Schmidt), *The Toeplitz matrices of an arbitrary Laurent polynomial*, Math. Scand. **8** (1960), 15–38.

12. *Some properties of recurrent random walk*, Illinois J. Math. **5** (1961), 234–245.

13. (with H. Widom), *The circumference of a convex polygon*, Proc. Amer. Math. Soc. **12** (1961), 506–509.

*14. *Recurrent random walk and logarithmic potential*, Proc. 4th Berkeley Symposium on Probability and Statistics **II** (1961), 515–534.

15. (with H. Kesten and D. Ornstein), *A general property of random walk*, Bull. Amer. Math. Soc. **68** (1962), 526–528.

16. *Hitting probabilities*, J. Math. Mech. **11** (1962), 593–614.

17. (with H. Kesten), *Ratio theorems for random walks, I*, J. Analyse Math. **XI** (1963), 285–322.

18. *Principles of random walk*, D. Van Nostrand Co., 1964; 2nd edition, Springer Verlag, 1976; Russian translation Mir, Moscow 1969; French translation Dunod, Paris,1970.

*19. *Electrostatic capacity, heat flow, and Brownian motion*, Z. Wahrsch. verw. Gebiete **3** (1964),110–121.

20. (with H. Kesten), *Random walk on countably infinite abelian groups*, Acta Math. **114** (1965), 237–265.

21. (with P. Ney), *The Martin boundary for random walk*, Trans. Amer. Math. Soc. **121** (1966), 116–132.

22. (with H. Kesten and P. Ney), *The Galton Watson process with mean one and finite variance*, Teor. Veroyatnost i Primenen **11** (1966), 579–611.

23. *Promenades aléatoires recurrentes sur les groupes discrets*, Université de Strasbourg, 1966.

24. *Two explicit Martin boundary constructions*, Symposium on Probability Methods in Analysis (Loutraki, 1966), Springer Lecture Notes in Math. **31** (1967), 296–298.

25. *Renewal theorems for Markov chains*, Proc. 5th Berkeley Symposium on Probability and Statistics **II, part 2** (1967), 311–320.

26. (with A. Joffe), *On multitype branching processes with $\rho \leq 1$*, J. Math. Anal. and Appl. **19** (1967), 409–430.

27. *Uniform motion with elastic collision of an infinite particle system*, J. Math. Mech. **18** (1969), 973–989.

28. *Random processes defined through the interaction of an infinite particle system*, Probability and Information Theory (Proc. International Symposium, McMaster University, Hamilton, Ontario, 1968), Springer Lecture Notes in Math. **89** (1969), 201–223.

*29. *Interaction of Markov processes*, Adv. Math. **5** (1970), 246–290.

30. *Markov random fields and Gibbs ensembles*, Amer. Math. Monthly **78** (1971), 142–154.

31. *Random fields and interacting particle systems*, Notes on lectures given at the 1971 MAA Summer Session, Williams College, Williamstown, Mass., Mathematical Association of America, Washington, D.C..

32. *A variational characterization of finite Markov chains*, Ann. Math. Statist. **43** (1972), 303–307.

33. *Contribution to discussion of "Subadditive ergodic theory" by J. F. C. Kingman*, Ann. Probab. **1** (1973), 883–909.

34. *Recurrent random walk of an infinite particle system*, Trans. Amer. Math. Soc. **198** (1974), 191–199.

35. *Introduction aux processus de Markov a parametre dans $\mathbb{Z}_\nu$*, Lectures given at Saint Flour Summer School 1973, Springer Lecture Notes in Math **390** (1974), 114–189.

36. *Random time evolution of infinite particle systems*, Proc. Intern. Congress of Math. (Vancouver 1974) **2**, 169–171.

37. *Markov random fields on an infinite tree*, Ann. Probab. **3** (1975), 387–398.

38. (with H. Kesten), *Controlled Markov chains*, Ann. Probab. **3** (1975), 32–40.

39. *Random time evolution of infinite particle systems*, Adv. Math. **16** (1975), 139–143 ; Also appeared in Surveys in Applied Mathematics, (N. Metropolis, S. Orszag and G.-C. Rota eds.), Academic Press, 1976.

40. *Phase transition in one-dimensional nearest-neighbor systems*, J. Funct. Anal. **20** (1975), 240–255.

41. (with H. Kesten and M. V. Kozlov), *A limit law for random walk in a random environment*, Compositio Math. **30** (1975), 145–168.

42. (with H. Wan, Jr.), *The characterization of optimal saving programs in a quadratic model*, J. Math. Econom. **3** (1976), 43–79.

43. *Stochastic time evolution of one dimensional infinite particle systems*, Bull. Amer. Math. Soc. **83** (1977), 880–890.

44. *Contribution to discussion of "Abhangige Verschiebungen, III"*, Serdica **4** (1978), 135–163.

45. (with H. Kesten), *A limit theorem related to a new class of self-similar processes*, Z. Wahrsch. verw. Gebiete **50** (1979), 5–25.

46. *Infinite systems with locally interacting components*, Ann. Probab. **9** (1981), 349–364.

47. (with T. M. Liggett), *Ergodic theorems for coupled random walks and other systems with locally interacting components*, Z. Wahrsch. verw. Gebiete **56** (1981), 443–468.

48. (with H. Kesten), *Convergence in distribution of products of random matrices*, Z. Wahrsch. verw. Gebiete **67** (1984), 363–386.

49. *A multidimensional renewal theorem*, Probability, statistical mechanics, and number theory (G.-C. Rota, ed.), Academic Press, 1986, pp. 147–155.

REPRINTS OF FRANK SPITZER

Reprinted from the
TRANSACTIONS OF THE AMERICAN MATHEMATICAL SOCIETY
Vol. 82, No. 2, pp. 323-339
July, 1956

with permission of the American Mathematics Society

A COMBINATORIAL LEMMA AND ITS APPLICATION TO PROBABILITY THEORY

BY
FRANK SPITZER[1]

1. **Introduction.** To explain the idea behind the present paper the following fundamental principle is emphasized. Let $X=(X_1, \cdots, X_n)$ be an n-dimensional vector valued random variable, and let $\mu(x)=\mu(x_1, \cdots, x_n)$ be its probability measure (defined on euclidean n-space E_n). Suppose that X has the property that $\mu(x)=\mu(gx)$ for every element g of a group G of order h of transformations of E_n into itself. Let $f(x)=f(x_1, \cdots, x_n)$ be a μ-integrable complex valued function on E_n. Then the expected value of $f(x)$ is

$$(1.1) \qquad Ef(X) = \int f(x)d\mu(x) = \int \bar{f}(x)d\mu(x),$$

where

$$(1.2) \qquad \bar{f}(x) = \frac{1}{h} \sum_{g \in G} f(gx).$$

This principle will be fruitful when it is possible to write $\bar{f}(x)$ in a form which is simpler to integrate than $f(x)$.

We shall consider only the case when G is the symmetric group of permulations σ on n symbols, so that

$$\sigma x = \begin{pmatrix} 1 & 2 & \cdots & n \\ \sigma_1 & \sigma_2 & \cdots & \sigma_n \end{pmatrix} x = (x_{\sigma_1}, x_{\sigma_2}, \cdots, x_{\sigma_n}).$$

Vector random variables $X=(X_1, \cdots, X_n)$ with the property that $\mu(x)=\mu(\sigma x)$ for every permutation are called symmetrically dependent and have been treated at length by E. Sparre Andersen [1; 2; 3]. For the most interesting applications of the present theory it will be necessary to require in addition that $X_1, \cdots, X_n$ be identically distributed and independent.

For the function $f(x)$ in equation (1.1) we shall take

$$f(x) = \max [0, x_1, x_1 + x_2, \cdots, x_1 + x_2 + \cdots + x_n].$$

The fundamental combinatorial result of this paper (Theorem 2.2) identifies

Presented to the Society, September 2, 1955 under the title *On the maximum of partial sums of random variables*; received by the editors September 2, 1955.

(1) The research of this author was supported in part by the United States Air Force under Contract No. AF18(600)-685 monitored by the Office of Scientific Research.

the set of numbers $[f(\sigma x)]$, as σ ranges over G, with a set of numbers $[g(\sigma x)]$ which have very desirable properties with regard to integration. The author is indebted to H. F. Bohnenblust for valuable discussions. The proof of Theorem 2.2 is due to him.

When applied to the case of independent identically distributed random variables $X_1, X_2, \cdots$ and their partial sums $S_k = X_1 + \cdots + X_k$, Theorem 2.2 and the principle of equation (1.1) produce Theorem 3.1:

For $|t| < 1$,

$$\sum_{n=0}^{\infty} \phi_n(\lambda) t^n = \exp\left[\sum_{n=1}^{\infty} \frac{\psi_n(\lambda)}{n} t^n \right], \tag{1.3}$$

where $\phi_n(\lambda)$ *is the characteristic function of* $\max [0, S_1, S_2, \cdots, S_n]$, *and* $\psi_n(\lambda)$ *the characteristic function of* $\max [0, S_n]$. Equation (1.3) generalizes results of M. Kac and G. A. Hunt (Theorem 4.1 in [7]), and of E. Sparre Andersen (Theorem 1 in [3]). In §6 equation (1.3) is generalized in Theorem 6.1 which gives the joint characteristic function of S_n and $\max [0, S_1, \cdots, S_n]$.

§§4, 5, and 7 contain a number of new results concerning the limiting behavior of the random variables $\max [0, S_1, \cdots, S_n]$ and N_n = the number of positive S_k, $k = 1, 2, \cdots, n$. These results are chosen to illustrate the power of the combinatorial method, without in any way exhausting its possibilities. The case of sums of nonidentically distributed independent random variables clearly is beyond the scope of this method. The same remark applies to continuous parameter stochastic processes and consequently the present method is a natural one for the class of processes with stationary independent increments.

2. **Combinatorial considerations.** The following proposition will be very useful.

THEOREM 2.1. *Let* $x = (x_1, \cdots, x_n)$ *be a vector such that* $x_1 + x_2 + \cdots + x_n = 0$, *but no other partial sum of distinct components vanishes. Let* $x_{k+n} = x_k$, *and* $x(k) = (x_k, x_{k+1}, \cdots, x_{k+n})$, $k = 1, 2, \cdots, n$. *Then, for each* $r = 0, 1, \cdots, n-1$, *exactly one of the cyclic permutations* $x(k)$ *of* x *is such that exactly* r *of its successive partial sums are positive.*

Proof. Let $s_k = x_1 + x_2 + \cdots + x_k$, $s_0 = 0$, $s_{k+n} = s_k$. Then the successive partial sums of the components of $x(k)$ are $s_k - s_{k-1}, s_{k+1} - s_{k-1}, \cdots, s_{k+n-1} - s_{k-1}$. By assumption the s_k are all distinct for $k = 1, 2, \cdots, n$, so that the number r of positive terms in the above sequence equals the number of positive terms among $s_j - s_{k-1}$, $j = 1, 2, \cdots, n$. Hence r may be given any value between 0 and $n-1$ in one and only one way, i.e. by choosing k so that s_{k-1} is $(r+1)$st from the top in order of magnitude among the s_k, $k = 1, 2, \cdots, n$.

One can obtain an interesting version of Theorem 2.1 which does not require the assumption that $s_k = 0$. We call the polygon connecting the points $(0, 0), (1, s_1), \cdots, (k, s_k), \cdots, (n, s_n)$ the sum polygon of the vector x. The sum polygon for the cyclically permuted vector $x(k)$ is defined the same way. Then the numbers $s_k - (k/n)s_n$ represent the vertical distances from the vertices of the sum polygon to its chord, the line connecting $(0, 0)$ with (n, s_n). Now we may apply Theorem 2.1 to the numbers $x_k - s_n/n$, whose sum vanishes, if they satisfy the incompatibility assumption. (It is clearly sufficient that the x_k be rationally independent.) The result is that if we consider a sum polygon and its n cyclic permutations, and prescribe an integer r between 0 and $n-1$, then exactly one of the cyclic permutations of the sum polygon will have the property that exactly r of its vertices lie strictly above its chord.

The geometric meaning of this theorem will provide the clue to Theorem 2.2. Its proof depends on the possibility of finding a unique cyclic permutation of certain subsets of components, such that their sum polygon (the polygon connecting the points (k, s_k)), lies entirely below its chord. Rational independence of the components will be assumed there.

Now we introduce certain notations. For any real a

$$a^+ = (|a| + a)/2 = \max [0, a],$$

so that

$$\max [0, a_1, \cdots, a_n] = \max_{1 \leq k \leq n} \overset{+}{a_k}.$$

$$\sigma x = \begin{pmatrix} 1 & 2 & \cdots & n \\ \sigma_1 & \sigma_2 & \cdots & \sigma_n \end{pmatrix} x = (x_{\sigma_1}, x_{\sigma_2}, \cdots, x_{\sigma_n}), \tag{2.1}$$

$$s_k(\sigma x) = x_{\sigma_1} + x_{\sigma_2} + \cdots + x_{\sigma_k},$$

$$S(\sigma x) = \max_{1 \leq k \leq n} \overset{+}{s_k}(\sigma x) = \max_{1 \leq k \leq n} \left(\sum_{i=1}^{k} x_{\sigma_i} \right)^+. \tag{2.2}$$

Consider the permutation τ represented as a product of cycles, including the one-cycles, and with no index contained in more than one cycle. For example suppose that $n = 7$, and that

$$\tau = (14)(2)(3756). \tag{2.3}$$

Then we define

$$T(\tau x) = (x_1 + x_4)^+ + \overset{+}{x_2} + (x_3 + x_7 + x_5 + x_6)^+. \tag{2.4}$$

In formal notation, let

$$\tau = (\alpha_1(\tau))(\alpha_2(\tau)) \cdots (\alpha_{n(\tau)}(\tau)), \tag{2.3'}$$

where the $\alpha_i(\tau)$, $i = 1, 2, \cdots, n(\tau)$, are disjoint sets of integers whose union is the set $[1, 2, \cdots, n]$. Then define

$$T(\tau x) = \sum_{i=1}^{n(\tau)} \Big(\sum_{k \in \alpha_i(\tau)} (\tau) x_k \Big)^+. \tag{2.4'}$$

Now it is claimed that

THEOREM 2.2. *For an arbitrary fixed vector $x=(x_1, x_2, \cdots, x_n)$ the sets $[S(\sigma x)]$ and $[T(\tau x)]$, which are generated by letting σ and τ run through all of the $n!$ permutations, are identical sets.*

Proof. If the theorem is proved for a set of x which is dense in E_n, then its truth follows for arbitrary x, since the numbers $S(\sigma x)$ and $T(\tau x)$ are continuous functions of x. Therefore the proof is given for an arbitrary fixed x with rationally independent components. (For rational r_i, $r_1x_1+r_2x_2+\cdots+r_nx_n=0$ if and only if each $r_i=0$.)

It is planned to exhibit between permutations of the form σ, defined in (2.1), and permutations of the form τ, defined in (2.3′) a one to one correspondence $\sigma_x(\tau)$. This mapping will depend on x and will have the property that, for each τ and for each x with rationally independent components,

$$T(\tau x) = S(\sigma_x(\tau)x). \tag{2.5}$$

The proof will then be complete.

Suppose a permutation τ is given in the form of (2.3′). The order of the indices in each set $\alpha_i(\tau)$ is then prescribed up to an arbitrary cyclic permutation. In accordance with the remarks following the proof of Theorem 2.1 we choose that unique cyclic permutation of the indices which ensures that the sum polygon lies below its chord. In terms of the example of equation (2.3),

$$\tau = (14)(2)(3756)$$

is rewritten as

$$\tau = (14)(2)(5637),$$

if it turns out that

$$x_1 < \frac{x_1+x_4}{2}; \quad \max\left[x_5, \frac{x_5+x_6}{2}, \frac{x_5+x_6+x_3}{3}\right] < \frac{x_5+x_6+x_3+x_7}{4}. \tag{2.6}$$

In formal notation, suppose that the cycle $(\alpha_\nu(\tau))=(j_1, j_2, \cdots, j_k)$. Then without changing that cycle as a permutation we can rewrite it in one and only one way, as $(a_\nu(\tau))=(i_1, i_2, \cdots, i_k)$ so that

$$\max\left[x_{i_1}, \frac{x_{i_1}+x_{i_2}}{2}, \cdots, \frac{x_{i_1}+\cdots+x_{i_{k-1}}}{k-1}\right] < \frac{x_{i_1}+\cdots+x_{i_k}}{k}, \tag{2.6'}$$

where $(i_1, i_2, \cdots, i_k)$ is a cyclic permutation of $(j_1, \cdots, j_k)$. This is done for each cycle.

Finally, the cycles $\alpha_i(\tau)$ may be permuted among themselves. Again using the rational independence of the components of x, there is a unique permutation (relabeling) of the $\alpha_i(\tau)$, after which

$$(2.7')\qquad \frac{\sum_{k\in\alpha_1(\tau)} x_k}{\sum_{k\in\alpha_1(\tau)} 1} > \frac{\sum_{k\in\alpha_2(\tau)} x_k}{\sum_{k\in\alpha_2(\tau)} 1} > \cdots > \frac{\sum_{k\in\alpha_{n(\tau)}(\tau)} x_k}{\sum_{k\in\alpha_{n(\tau)}(\tau)} 1}.$$

Thus in our example τ should now be rewritten as

$$\tau = (5637)(2)(14),$$

if it turns out that

$$(2.7)\qquad \frac{x_5 + x_6 + x_3 + x_7}{4} > x_2 > \frac{x_1 + x_4}{2}.$$

Finally we define

$$\sigma_x(\tau) = \begin{pmatrix} 1 & 2 & 3 & 4 & 5 & 6 & 7 \\ 5 & 6 & 3 & 7 & 2 & 1 & 4 \end{pmatrix},$$

or in general

$$(2.8)\qquad \sigma_x(\tau) = \begin{pmatrix} 1 & 2 \cdots n \\ i_1 & i_2 \cdots i_n \end{pmatrix},$$

where the indices $i_1, \cdots, i_n$ are the integers $[1, 2, \cdots, n]$ in that unique order in which they now appear in the successive sets $\alpha_i(\tau)$.

It remains to define $\tau_x(\sigma)$ as a function of σ, and to show that $\sigma_x[\tau_x(\sigma)] = \sigma$, and finally that equation (2.5) holds. Given

$$\sigma = \begin{pmatrix} 1, & 2, & \cdots, n \\ i_1, & i_2, & \cdots, i_n \end{pmatrix},$$

consider the sum polygon through the points $(0, 0)$, $(1, x_{i_1})$, $\cdots$, $(k, x_{i_1} + \cdots + x_{i_k})$, $\cdots$, (n, s_n). Now we define the lowest convex majorant of the sum polygon as that unique polygon which goes through $(0, 0)$ and (n, s_n) in such a way that all its vertices are also vertices of the sum polygon and that it always lies above or coincides with the sum polygon. (Uniqueness follows from the fact that the x_i are rationally independent.) Suppose now that the lowest convex majorant constructed for the permutation σ has the vertices $(0, 0)$, $(k_1, x_{i_1} + \cdots + x_{i_{k_1}})$, $\cdots$, $(k_\nu, x_{i_1} + \cdots + x_{i_{k_\nu}})$, (n, s_n), where $0 < k_1 < \cdots < k_\nu < n$. Then we define

$$(2.9)\qquad \tau_x(\sigma) = (i_1, \cdots, i_{k_1})(i_{k_1}, \cdots, i_{k_2}) \cdots (i_{k_\nu+1}, \cdots, i_n).$$

It is geometrically obvious that $\tau_x(\sigma)$ is left unchanged by the transformations used in (2.6′) and (2.7′) to define $\sigma_x(\tau)$, so that $\sigma_x[\tau_x(\sigma)] = \sigma$, establishing the desired one to one correspondence.

Finally $S(\sigma x)$ is the maximum ordinate among those of the vertices of the sum polygon, but this maximum is clearly also attained at a vertex of the lowest convex majorant corresponding to σ. It follows from (2.9) that

$$S(\sigma x) = (x_{i_1} + \cdots + x_{i_{k_1}})^+ + \cdots + (x_{i_{k_\nu}+1} + \cdots + x_{i_n})^+ = T(\tau_x(\sigma)x),$$

or

$$T(\tau x) = S(\sigma_x(\tau)x),$$

which completes the proof.

In §5, we shall give a simple proof of a very surprising theorem of E. Sparre Andersen, which is stated in Equation (5.3). That theorem will be seen to follow from a combinatorial fact rather similar to Theorem 2.1.

Let $\theta(a)=1$ if $a>0$ and 0 otherwise. Let $x=(x_1, \cdots, x_n)$ be a given n-tuple as before, and let $\sigma x=(x_{\sigma_1}, \cdots, x_{\sigma_n})$ be a permutation of x. Let $\tau=(\alpha_1)(\alpha_2)\cdots(\alpha_{n(\tau)})$ be a permutation decomposed into cycles, and define

$$A(\sigma x) = \sum_{k=1}^{n} \theta(x_{\sigma_1} + \cdots + x_{\sigma_k}),$$

$$B(\tau x) = \sum_{i=1}^{n(\tau)} \Big(\sum_{j\in\alpha_i} 1 \Big) \theta\Big[\sum_{j\in\alpha_i} x_j \Big].$$

Then we have

THEOREM 2.3. *The sets* $\{A(\sigma x)\}$ *and* $\{B(\tau x)\}$, *which are obtained by letting* σ *and* τ *run through all permutations on* n *objects, are identical sets.*

The proof is omitted, since it has recently been shown by H. F. Bohnenblust that far more general theorems than those considered here are easier to prove than the elegant but somewhat too special theorems considered here.

3. **The distribution of** max $[0, S_1, S_2, \cdots, S_n]$. Before proceeding to the applications of Theorem 2.2, it should be mentioned that even the simple Theorem 2.1 is not without probabilistic interest. When applied to equation (1.1) it immediately yields the following result of E. Sparre Andersen (Theorem 3 in [2]):

"*Let* $X_1, \cdots, X_{n+1}$ *be symmetrically dependent random variables and let* C *be an event which is symmetric with respect to* $X_1, \cdots, X_{n+1}$. *Let* N_n^* *be the number of points* (j, S_j), $j=1, \cdots, n$, *which lie above the straight line from* $(0, 0)$ *to* $(n+1, S_{n+1})$. *Then for* $\Pr[C]>0$,

$$\Pr[N_n^* = m \mid C] = (n+1)^{-1}, \qquad m = 0, 1, \cdots, n,$$

if and only if

$$\Pr\{[i^{-1}S_i = (n+1)^{-1}S_{n+1}] \cap C\} = 0, \qquad i = 1, 2, \cdots, n."$$

From now on $X_1, X_2, \cdots$ will be an infinite sequence of indentically distributed independent random variables, with $S_k=X_1+\cdots+X_k$ and $S_k^+=\max[0, S_k]$. If τ is the permutation

$$\tau = (\alpha_1)(\alpha_2)\cdots(\alpha_{n(\tau)}),$$

then

$$T(\tau X) = \sum_{i=1}^{n(\tau)}\Bigl(\sum_{k\in\alpha_i} X_k\Bigr)^+,$$

and if

$$\sigma = \begin{pmatrix} 1 & 2 & \cdots & n \\ \sigma_1 & \sigma_2 & \cdots & \sigma_n \end{pmatrix},$$

then

$$S(\sigma X) = \max_{1\leq k\leq n}\Bigl(\sum_{i=1}^{k} X_{\sigma_i}\Bigr)^+ .$$

For every complex-valued $f(x)$ it follows from equation (1.1) and from Theorem 2.2 that

$$Ef\Bigl[\max_{1\leq k\leq n} S_k^+\Bigr] = \frac{1}{n!}\sum_{\sigma} Ef[S(\sigma X)] = \frac{1}{n!}\sum_{\tau} Ef[T(\tau X)]. \tag{3.1}$$

Equation (3.1) must be interpreted in the sense that each member is finite and equal to the other two, provided that one of them is finite.

Everything said so far is valid for symmetrically dependent random variables. To take advantage of the independence of the X_i it is convenient to take

$$f(x) = \exp(i\lambda x), \qquad \mathrm{Im}(\lambda) \geqq 0,$$

$$\phi_n(\lambda) = E\exp(i\lambda \max_{1\leq k\leq n} S_k^+), \qquad \phi_0(\lambda) = 1, \tag{3.2}$$

$$\psi_k(\lambda) = E\exp(i\lambda S_k^+). \tag{3.3}$$

Consider the last member of equation (3.1). If a permutation τ consists of k_ν cycles of length ν, $\nu=1, 2, \cdots, n$, with $k_1+2k_2+\cdots+nk_n=n$, then

$$Ef[T(\tau X)] = \prod_{\nu=1}^{n}[\psi_\nu(\lambda)]^{k_\nu}.$$

The number of permutations on n objects, which, when decomposed into disjoint cycles, exhibit the above structure is exactly

$$n!\prod_{\nu=1}^{n}\nu^{-k_\nu}(k_\nu!)^{-1}.$$

Hence

$$\phi_n(\lambda) = \sum{}^* \prod_{\nu=1}^{n} \left(\frac{\psi_\nu(\lambda)}{\nu}\right)^{k_\nu} \frac{1}{k_\nu!}, \qquad n \geqq 0, \tag{3.4}$$

where the summation $\sum^*$ extends over all n-tuples $(k_1, k_2, \cdots, k_n)$ of non-negative integers with the property $k_1+2k_2+\cdots+nk_n=n$. It is easy to verify that the identity between generating functions in equation (1.3) is equivalent to equation (3.4). Hence we have proved

THEOREM 3.1. *For* $|t|<1$, $\operatorname{Im}(\lambda)\geqq 0$,

$$\sum_{n=0}^{\infty} \phi_n(\lambda)t^n = \exp\left[\sum_{k=1}^{\infty} \frac{\psi_k(\lambda)}{k} t^k\right]. \tag{3.5}$$

In §6, Theorem 3.1 will be found to be a special case of Theorem 6.1.

The results of M. Kac, G. A. Hunt, and E. Sparre Andersen are simple consequences of Theorem 3.1.

COROLLARY 1 (*Theorem* 4.1 *in* [7]).

$$E \max_{1\leqq k\leqq n} S_k^+ = \sum_{k=1}^{n} \frac{1}{k} ES_k^+. \tag{3.6}$$

This formula is obtained by differentiating (3.5) with respect to λ and setting $\lambda=0$. Alternatively, it can be proved directly from equation (3.1) with the choice of $f(x)=x$.

COROLLARY 2.

$$\sum_{n=0}^{\infty} \Pr\left\{\bigcap_{k=1}^{n} [S_k \geqq 0]\right\} t^n = \exp\left\{\sum_{k=1}^{\infty} \frac{t^k}{k} \Pr [S_k \geqq 0]\right\}. \tag{3.7}$$

This result is obtained by applying equation (3.5) to the random variables $-X_i$, setting $\lambda=iu$, and letting $u\to\infty$. Actually (3.7) remains correct if both inequalities in (3.7) are modified to be strict inequalities. In that form (3.7) was discovered by E. Sparre Andersen. It is equation (3.6) of Theorem 1 in [3]. To obtain it directly, it is easiest to use the following weak form of Theorem 2.3. The probability that the first n partial sums $S_1, S_2, \cdots, S_n$ are all positive is the same as the probability of success in the following experiment. One selects a permutation at random (i.e. with equal probability) from the $n!$ permutations of size n, and observes the lengths $\nu_1, \nu_2, \cdots, \nu_r$ of its successive disjoint cycles. One then observes r independent random variables with the same distribution as $S_{\nu_1}, S_{\nu_2}, \cdots, S_{\nu_r}$, and success is defined as the event that they are all positive. This fact then leads to a direct proof of equation (3.7) with strict inequalities quite similar to the proof of (3.5).

The analogue of Theorem 3.1 for continuous parameter separable stochastic processes with stationary independent increments will be given elsewhere.

4. **The limit behavior of** max $[0, S_1, \cdots, S_n]$.

THEOREM 4.1. *Let* $a_k = \Pr[S_k > 0]$, $\psi_k(\lambda) = E\exp(i\lambda S_k^+)$.
(a) *If* $\sum_1^\infty a_k/k < \infty$, *then*

$$\max_{1 \leq k \leq n} S_k^+ \to \sup_{k \geq 1} S_k^+ = \max_{k \geq 1} S_k^+ < \infty \tag{4.1}$$

with probability one;

$$\limsup_{n \to \infty} S_n = -\infty \text{ with probability one,} \tag{4.2}$$

except in the trivial case when $\Pr[X_i = 0] = 1$. $\max_{k \geq 1} S_k^+$ *has the infinitely divisible characteristic function*

$$E\exp(i\lambda \max_{k \geq 1} S_k^+) = \prod_{k=1}^{\infty} \exp\left[\frac{\psi_k(\lambda) - 1}{k}\right]. \tag{4.3}$$

(b) *If* $\sum_1^\infty a_k/k = \infty$, *then*

$$\max_{1 \leq k \leq n} S_k^+ \to \sup_{k \geq 1} S_k^+ = \limsup_{n \to \infty} S_n = \infty \tag{4.4}$$

with probability one.
(c) *If* $E|X_i| < \infty$ *and* $\Pr[X_i = 0] < 1$, *then case* (a) *corresponds to* $EX_i < 0$, *while case* (b) *corresponds to* $EX_i \geq 0$.

Proof. To prove (4.1) and (4.2) it is sufficient to show that in case (a)

$$\Pr[S_n > x \text{ i.o.}] = 0, \qquad -\infty < x < \infty. \tag{4.5}$$

Let $q_n = \Pr[\max_{1 \leq k \leq n} S_k \leq 0]$, $q_0 = 1$. In Theorem 3.1 set $\lambda = iu$ and let $u \to \infty$. Then

$$\sum_0^\infty q_n t^n = \exp\left[\sum_1^\infty \left(\frac{1 - a_k}{k}\right) t^k\right], \qquad |t| < 1,$$

$$(1 - t)\sum_0^\infty q_n t^n = \exp\left[-\sum_1^\infty \frac{a_k}{k} t^k\right], \qquad |t| < 1. \tag{4.6}$$

Since the coefficients q_n are monotone nonincreasing, a simple Tauberian argument gives

$$\lim_{n \to \infty} q_n = \exp\left[-\sum_1^\infty \frac{a_k}{k}\right] > 0, \tag{4.7}$$

so that

$$\Pr[S_n > 0 \text{ i.o.}] \leq 1 - \lim_{n \to \infty} q_n < 1. \tag{4.8}$$

But it was shown by P. Lévy [8, p. 147], that when $\Pr[X_i=0]<1$,

$$\Pr[S_n > x \text{ i.o.}] = 0 \quad \text{or} \quad 1, \qquad -\infty < x < \infty, \tag{4.9}$$

and that the probability in (4.8) is the same for every x. Equations (4.8) and (4.9) now imply (4.5), and hence (4.1) and (4.2) are proved.

Equation (4.1) implies that the sequence

$$\phi_n(\lambda) = E \exp\left[i\lambda \max_{1 \leq k \leq n} S_k^+\right]$$

converges to a characteristic function. Therefore

$$\begin{aligned} E \exp\left[i\lambda \max_{k \geq 1} S_k^+\right] &= \lim_{n\to\infty} \phi_n(\lambda) = \lim_{t\to 1} (1-t) \sum_1^\infty \phi_n(\lambda)t^n \\ &= \lim_{t\to 1} (1-t) \exp\left[\sum_1^\infty \frac{\psi_k(\lambda)}{k} t^k\right] \\ &= \exp\left[\sum_1^\infty \frac{\psi_k(\lambda)-1}{k}\right]. \end{aligned}$$

This characteristic function is infinitely divisible since its factors

$$\exp\left(\frac{\psi_k(\lambda)-1}{k}\right) = \exp\left[\frac{1}{k}\left\{\int_0^\infty (e^{i\lambda x}-1)d\Pr[S_k \leq x]\right\}\right]$$

are infinitely divisible.

To prove equation (4.4), suppose that $\sum_1^\infty a_k/k = \infty$. Then it follows from (4.6) that

$$\lim_{n\to\infty} q_n = \Pr\left[\sup_{k\geq 1} S_k \leq 0\right] = 0.$$

Now

$$\begin{aligned} 1 - \Pr[S_k > 0 \text{ i.o.}] &= \Pr\left[\sup_{k\geq 1} S_k \leq 0\right] + \sum_{k=1}^\infty \Pr\left[S_k > 0;\ \sup_{m\geq 1} S_{k+m} \leq 0\right] \\ &\leq \lim_{n\to\infty} \sum_{k=1}^n \Pr\left[\sup_{m\geq 1} S_m \leq 0\right] = 0. \end{aligned}$$

Hence, by (4.9)

$$\Pr[S_k > 0 \text{ i.o.}] = \Pr[S_k > x \text{ i.o.}] = 1, \qquad -\infty < x < \infty,$$

which implies equation (4.4).

It remains to prove part (c). We assume that $E|X_i| < \infty$. Then, if $EX_i<0$, the strong law of large numbers implies that $\Pr[S_n>0 \text{ i.o.}]=0$, so that $\sum_1^\infty a_k/k < \infty$. Conversely, if $EX_i>0$, we get $\Pr[S_n<0 \text{ i.o.}]=0$, so that

$\Pr[S_n>0 \text{ i.o.}]=1$, and $\sum_1^\infty a_k/k=\infty$. Finally, if $EX_i=0$, then the sequence of partial sums S_n is recurrent in the sense of Chung and Fuchs [4], so that $\Pr[S_n>0 \text{ i.o.}]=1$ if $\Pr[X_i=0]<1$, and $\sum_1^\infty a_k/k=\infty$.

Theorem 4.1 has a curious corollary which constitutes a novel form of the strong law of large numbers for identically distributed random variables. It gains in interest by comparison to the following result of P. Erdös [6].

$EX_i=m$ and $EX_i^2<\infty$, if and only if

$$\sum_1^\infty \Pr\left[\left|\frac{S_k}{k}-m\right|>\epsilon\right]<\infty \text{ for every } \epsilon>0.$$

We shall prove

THEOREM 4.2. *$EX_i=m$ if and only if*

$$\sum_1^\infty \frac{1}{k}\Pr\left[\left|\frac{S_k}{k}-m\right|>\epsilon\right]<\infty \text{ for every } \epsilon>0. \tag{4.10}$$

Proof. It will suffice to consider the case when $m=0$. Suppose therefore that $EX_i=0$. Then $E(X_i-\epsilon)<0$, if $\epsilon>0$, and by Theorem 4.1 (c)

$$\sum_1^\infty \frac{1}{k}\Pr[S_k-k\epsilon>0]<\infty. \tag{4.11}$$

By the same argument

$$\sum_1^\infty \frac{1}{k}\Pr[S_k+k\epsilon<0]<\infty. \tag{4.12}$$

Adding (4.11) and (4.12), we have (4.10). Conversely, assume that (4.10) holds with $m=0$. Then (4.11) and (4.12) must hold for every $\epsilon>0$, which means that

$$\Pr\left[\frac{S_n}{n}\to 0\right]=1.$$

Now it follows from the converse of the strong law of large numbers that $EX_i=0$.

5. **The number of positive partial sums.** We summarize earlier definitions and make some new ones.

$$a_k=\Pr[S_k>0],$$

$$p_n=\Pr\left[\min_{1\le k\le n} S_k>0\right], \qquad p_0=1,$$

$$q_n=\Pr\left[\max_{1\le k\le n} S_k\le 0\right], \qquad q_0=1.$$

(5.1) N_n = the number of $S_k>0$, $k=1, 2, \cdots, n$.

(5.2) T_n = the index (time) k at which $\max_{1\leq k\leq n} S_k$ is first attained, with the provision that $T_n=0$ if $\max_{1\leq k\leq n} S_k\leqq 0$.

The results of this section will depend on the following important theorem of E. Sparre Andersen (Theorem 1 in [1]).

"*If* $p_k=\Pr[N_k=k]$, $q_k=\Pr[N_k=0]$ (as is the case in our notation), *then*

$$\Pr[N_n = k] = p_kq_{n-k} = \Pr[T_n = k], \quad k = 0, 1, \cdots, n.\text{"} \tag{5.3}$$

The second half of equation (5.3) is a simple consequence of the markovian nature of the process of successive partial sums S_n. To prove the first half, we let

$$\rho_n(\lambda) = E[e^{-\lambda n\theta(S_n)}] = a_ne^{-\lambda n} + 1 - a_n, \qquad \lambda \geqq 0. \tag{5.4}$$

$$\chi_n(\lambda) = E[e^{-\lambda N_n}], \qquad \lambda \geqq 0. \tag{5.5}$$

One sees that in view of Theorem 2.3 the random variable N_n has the same distribution as the random variable

$$\nu_1\theta(S_{\nu_1}) + \nu_2\theta(S_{\nu_2}) + \cdots + \nu_r\theta(S_{\nu_r}),$$

where $\nu_1, \cdots, \nu_r$ are random variables indicating the length of the disjoint cycles of a permutation chosen at random from among the permutations on n objects. The partial sums $S_{\nu_1}, \cdots, S_{\nu_r}$ are taken as independent random variables with the distribution indicated by their subscript. Hence the method used to prove Theorem 3.1 applies with only the change of $\phi_n(\lambda)$ into $\chi_n(\lambda)$, and $\psi_n(\lambda)$ into $\rho_n(\lambda)$, and yields the following result analogous to equation (3.5):

$$\sum_{n=0}^{\infty} \chi_n(\lambda)t^n = \exp\left[\sum_{k=1}^{\infty} \frac{\rho_k(\lambda)}{k} t^k\right], \qquad \lambda \geqq 0. \tag{5.6}$$

Using (5.4) this becomes

$$\sum_0^{\infty} \chi_n(\lambda)t^n = \exp\left[\sum_1^{\infty} \frac{a_k}{k}(te^{-\lambda})^k\right]\exp\left[-\sum_1^{\infty} \frac{a_k}{k} t^k\right],$$

and in view of equation (3.7) and the discussion following it,

$$\sum_0^{\infty} \chi_n(\lambda)t^n = \left(\sum_0^{\infty} e^{-\lambda k}t^k p_k\right)\left(\sum_0^{\infty} t^kq_k\right),$$

so that

$$\chi_n(\lambda) = \sum_{k=0}^{n} e^{-\lambda k}\Pr[N_n = k] = \sum_{k=0}^{n} e^{-\lambda k}p_kq_{n-k},$$

from which the first half of (5.3) follows.

Now we consider the limiting behavior of N_n and T_n.

THEOREM 5.2. (a) *If* $\sum_1^\infty a_k/k < \infty$, *then* $N_n \to N < \infty$ *and* $T_n \to T < \infty$, *with probability one, and*

$$Et^N = Et^T = \exp\left[\sum_1^\infty \frac{a_k}{k}(t^k - 1)\right], \qquad |t| \leqq 1. \tag{5.7}$$

(b) *If* $\sum_1^\infty a_k/k = \infty$, *then* $N_n \to \infty$ *and* $T_n \to \infty$ *with probability one.*

Proof. Let $\sum_1^\infty a_k/k < \infty$. Then equations (4.1) and (4.2) of Theorem 4.1 imply that

$$\Pr[N_{n+1} \neq N_n \text{ i.o.}] = \Pr[T_{n+1} \neq T_n \text{ i.o.}] = 0.$$

On the other hand, if $\sum_1^\infty a_k/k = \infty$, then it follows from equation (4.4) that

$$\Pr[N_{n+1} = N_n + 1 \text{ i.o.}] = \Pr[T_n = n \text{ i.o.}] = 1.$$

This proves the probability one statements in parts (a) and (b) of the present theorem.

To prove equation (5.7) it suffices to consider the random variables N_n. In case (a), when they converge to a random variable N with probability one, we have

$$Et^N = \sum_{k=0}^\infty t^k \lim_{n\to\infty} \Pr[N_n = k].$$

By (5.3)

$$\lim_{n\to\infty} \Pr[N_n = k] = p_k \lim_{n\to\infty} q_n = p_k q.$$

In fact, (4.7) states that

$$q = \exp\left[-\sum_1^\infty \frac{a_k}{k}\right],$$

and (3.7) that

$$\sum_0^\infty p_k t^k = \exp\left[\sum_1^\infty \frac{a_k}{k} t^k\right],$$

so that

$$Et^N = Et^T = \exp\left[\sum_1^\infty \frac{a_k}{k}(t^k - 1)\right], \qquad |t| \leqq 1.$$

6. **The joint distribution of S_n and** max $[0, S_1, \cdots, S_n]$. The results and methods of the preceding sections enable us to find the joint characteristic function of S_n and max $[0, S_1, \cdots, S_n]$. For reasons of symmetry we shall concentrate on the bivariate characteristic function

$$(6.1)\qquad \phi_n(\alpha,\beta) = E\exp\left[i\alpha \max_{1\le k\le n} S_k^+ + i\beta\left(\max_{1\le k\le n} S_k^+ - S_n\right)\right], \qquad \phi_0(\alpha,\beta) = 1.$$

The joint characteristic function of S_n and max $[0, S_1, \cdots, S_n]$ will then be

$$(6.2)\qquad E\exp\left[i\alpha S_n + i\beta \max_{1\le k\le n} S_k^+\right] = \phi_n(\alpha+\beta, -\alpha).$$

It will be seen that $\phi_n(\alpha, \beta)$ depends only on the characteristic functions

$$(6.3)\qquad u_k(\alpha) = E\exp[i\alpha S_k^+], \qquad (2S_k^+ = |S_k| + S_k),$$

$$(6.4)\qquad v_k(\beta) = E\exp[i\beta S_k^-], \qquad (2S_k^- = |S_k| - S_k),$$

for $k = 1, 2, \cdots, n$.

THEOREM 6.1. *For* $|t| < 1$, Im $(\alpha) \geqq 0$, Im $(\beta) \geqq 0$,

$$(6.5)\qquad \sum_{n=0}^{\infty} \phi_n(\alpha,\beta)t^n = \exp\left[\sum_{k=1}^{\infty} \frac{1}{k}(u_k(\alpha) + v_k(\beta) - 1)t^k\right].$$

The proof depends on equation (5.3)

$$\Pr[T_n = k] = p_k q_{n-k}.$$

If max $[0, S_1, \cdots, S_n]$ is assumed at $T_n = k$, then the random variables $S_i - S_k$, $i \leqq k$, and $S_j - S_k$, $j \geqq k$, are independent. By a simple argument one obtains

$$(6.6)\qquad \phi_n(\alpha,\beta) = \sum_{k=0}^{n} p_k q_{n-k} \frac{\int_{A_k} e^{i\alpha S_k} dP}{\Pr[A_k]} \cdot \frac{\int_{B_{n-k}} e^{-i\beta S_{n-k}} dP}{\Pr[B_{n-k}]},$$

where

$$A_k = \bigcap_{i=1}^{k} [S_i > 0], \quad B_k = \bigcap_{i=1}^{k} [S_i \leqq 0],$$

so that $\Pr[A_k] = p_k$, $\Pr[B_k] = q_k$. To evaluate the integrals in (6.6) one must go back to Theorem 2.2 and apply the method of equation (1.1) just as was done in the proof of Theorem 3.1. One obtains

$$(6.7)\qquad \sum_{n=0}^{\infty} t^n \int_{A_n} e^{i\alpha S_n} dP = \exp\left[\sum_{k=1}^{\infty} \frac{1}{k}(u_k(\alpha) - 1 + a_k)t^k\right],$$

$$(6.8)\qquad \sum_{n=0}^{\infty} t^n \int_{B_n} e^{i\beta S_n} dP = \exp\left[\sum_{k=1}^{\infty} \frac{1}{k}(v_k(\beta) - a_k)t^k\right],$$

where $a_k = \Pr[S_k > 0]$. But now it follows from equation (6.6) that the series on the left in (6.5) is the product of the generating function in (6.7) and (6.8). This product gives the right-hand side of (6.5) as was to be proved.

7. **The generalized arc-sine law.** By the generalized arc-sine laws we mean the following one parameter family of distributions $F_\alpha(x)$.

$$F_0(x) = 0 \text{ if } x < 0, \qquad 1 \text{ if } x \geqq 0,$$

$$F_\alpha(x) = \frac{\sin \pi\alpha}{\pi} \int_0^x s^{\alpha-1}(1-s)^{-\alpha} ds, \qquad 0 < \alpha < 1,$$

$$F_1(x) = 0 \text{ if } x < 1, \qquad 1 \text{ if } x \geqq 1.$$

E. Sparre Andersen was the first to prove that the sequence of random variables N_n/n, i.e. the fraction of times that $S_n > 0$, converges in distribution to the law $F_\alpha(x)$, if the sequence $a_k = \Pr[S_k > 0]$ converges to α (Theorem 3 in [3]). We generalize his result in the following way.

THEOREM 7.1. *If* $(a_1 + a_2 + \cdots + a_n)/n \to \alpha$, *then*

$$\Pr\left[\frac{N_n}{n} \leqq x\right] \to F_\alpha(x). \tag{7.1}$$

If $(a_1 + \cdots + a_n)/n$ *does not tend to a limit, then neither does* $\Pr[N_n/n \leqq x]$.

The second half of the theorem is trivial, since

$$\frac{a_1 + \cdots + a_n}{n} = E\frac{N_n}{n},$$

and a sequence of uniformly bounded random variables cannot converge in distribution unless their first moments converge. Moreover, this half of the theorem may also be vacuous, since it is not known whether there exists a sequence of identically distributed independent random variables with the property that the a_k fail to have a $(C, 1)$ limit.

However, the so called universal laws of Doeblin [5] show that the ordinary limit of a_k need not exist. Hence Theorem 7.1 is a generalization of E. Sparre Andersen's.

The proof of (7.1) will be based on the following Abelian theorem.

LEMMA 7.2. *If the sequence* $\{a_k\}$, $k = 1, 2, \cdots$, *is* $(C, 1)$ *summable to* α, *and* $\lambda \geqq 0$, *then as* $s \to 1$ *through real* $s < 1$,

$$\lim_{s\to 1} \sum_{k=1}^{\infty} \frac{a_k}{k} s^k [1 - e^{-\lambda k(1-s)}] = \alpha \log (1 + \lambda). \tag{7.2}$$

Proof. Let $A_n = (a_1 + a_2 + \cdots + a_n)/n$. Then after summation by parts, equation (7.2) becomes

$$\lim_{s\to 1} \sum_{k=1}^{\infty} A_k c_k(s) = \alpha \log (1 + \lambda),$$

where $c_k(s) = s^k[1 - e^{-\lambda k(1-s)}] - (ks^{k+1}/(k+1))[1 - e^{-\lambda(k+1)(1-s)}]$. By Toeplitz, Theorem we have to prove that

(a) $$\lim_{s\to 1} c_k(s) = 0,$$

(b) $$\lim_{s\to 1} \sum_{k=1}^{\infty} c_k(s) = \log (1 + \lambda),$$

(c) $$\sum_{k=1}^{\infty} | c_k(s) | < K \text{ in some interval } 1 - \delta \leqq s \leqq 1.$$

It is easily deduced that (a) and (b) hold. To verify (c), note that

$$c_k(s) = \lambda s^k \int_0^{k(1-s)} e^{-x\lambda} dx - \frac{\lambda k s^{k+1}}{k+1} \int_0^{(k+1)(1-s)} e^{-x\lambda} dx \geqq - \lambda s^k \int_{k(1-s)}^{(k+1)(1-s)} e^{-x\lambda} dx$$
$$\geqq - \lambda s^k (1 - s) e^{-\lambda k(1-s)}.$$

Hence $|c_k(s)| \leqq c_k(s) + 2\lambda s^k(1-s)e^{-\lambda k(1-s)}$, so that

$$\limsup_{s\to 1} \sum_{k=1}^{\infty} | c_k(s) | \leqq \log (1 + \lambda) + 2\lambda,$$

so that (c) is satisfied for some $\delta > 0$ with $K = 4\lambda$.

We proceed to prove Theorem 7.1. As in §5 let

$$\chi_n(\lambda) = E[e^{-\lambda N_n}],$$

and equation (5.6) can be written in the form

$$(7.3) \quad (1 - s) \sum_{n=0}^{\infty} \chi_n[\lambda(1 - s)]s^n = \exp \left\{ \sum_{n=1}^{\infty} \frac{a_n}{n} s^n [1 - e^{-\lambda n(1-s)}] \right\},$$

when $|s| < 1$ and $\lambda \geqq 0$. For $|s| < 1$ both members of (7.3) can be written as convergent power series in λ and we may equate the coefficients $c_k(s)$ of λ^k, $k = 0, 1, \cdots$. Lemma 7.2 simply states that

$$\lim_{s\to 1} \sum_{k=0}^{\infty} c_k(s)\lambda^k = (1 + \lambda)^{-\alpha},$$

so that

$$(7.4) \qquad \lim_{s\to 1} c_k(s) = \binom{-\alpha}{k}, \qquad k = 0, 1, \cdots.$$

Since $\chi_n(\lambda)$ is a moment generating function for N_n we can use equation (7.3) to express the coefficients $c_k(s)$ in terms of the moments of N_n/n. Let

$$\mu_k^{(n)} = E\left(\frac{N_n}{n}\right)^k.$$

Then

$$c_k(s) = (1-s)^{k+1}\frac{(-1)^k}{k!}\sum_{n=0}^{\infty} n^k\mu_k^{(n)}s^n, \qquad |s| < 1,$$

and equation (7.4) implies that

$$(7.5) \qquad (1-s)\sum_{n=0}^{\infty} n^k\mu_k^{(n)}s^n \sim (-1)^k\binom{-\alpha}{k}\frac{k!}{(1-s)^k}, \qquad s \to 1.$$

The left-hand member as a power series in s has the coefficients

$$(n+1)^k\mu_k^{(n+1)} - n^k\mu_k^{(n)} = E(N_{n+1}^k - N_n^k),$$

which are non-negative. Karamata's Tauberian theorem therefore applies, and yields

$$\lim_{n\to\infty}\mu_k^{(n)} = (-1)^k\binom{-\alpha}{k}, \qquad k = 0, 1, \cdots.$$

But it is easy to verify that

$$(-1)^k\binom{-\alpha}{k} = \int_{-\infty}^{\infty} x^k dF_\alpha(x),$$

and because the moment problem in this case has a unique solution, the proof of Theorem 7.1 is complete.

Bibliography

1. E. Sparre Andersen, *On sums of symmetrically dependent random variables*, Skandinavisk Aktuarietidskrift vol. 36 (1953) pp. 123–138.

2. ———, *On the fluctuations of sums of random variables*, Math. Scand. vol. 1 (1953) pp. 263–285.

3. ———, *On the fluctuations of sums of random variables* II, Math. Scand. vol. 2 (1954) pp. 195–223.

4. K. L. Chung and W. H. J. Fuchs, *On the distribution of values of sums of random variables*, Memoirs of the American Mathematical Society vol. 6 (1951) pp. 1–12.

5. W. Doeblin, *Sur l'ensemble de puissances d'une loi de probabilité*, Studia Mathematica vol. 9 (1941) pp. 71–96.

6. P. Erdös, *Remark on my paper "On a theorem of Hsu and Robbins,"* Ann. Math. Statist. vol. 21 (1950) p. 138.

7. M. Kac, *Toeplitz matrices, translation kernels and a related problem in probability theory*, Duke Math. J. vol. 21 (1954) pp. 501–509.

8. P. Lévy, *Théorie de l'addition des variables aléatoires*, Paris, 1937.

California Institute of Technology,
Pasadena, Calif.

Reprinted from the
TRANSACTIONS OF THE AMERICAN MATHEMATICAL SOCIETY
Vol. 87, No. 1, pp. 187-197
January, 1958
with permission of the American Mathematics Society

SOME THEOREMS CONCERNING 2-DIMENSIONAL BROWNIAN MOTION

BY
FRANK SPITZER

This paper consists of three separate parts(1) which are related mainly in that they treat different stochastic processes which arise in the study of plane brownian motion. §1 is concerned with the process $R(t) = |Z(t)|$, denoting the distance of the 2-dimensional separable Bachelier-Wiener process $Z(t) = X(t) + iY(t)$ from the origin. We shall derive a law of the so-called strong type concerning the frequency of small values of $R(t)$. This theorem disproves a conjecture of Paul Lévy. In the next section we study the process $\theta(t) = \arg Z(t)$. Results are obtained concerning the transition probabilities and absorption probabilities of $\theta(t)$. The limiting distribution of $(2^{-1} \log t)^{-1}\theta(t)$ is found to be the Cauchy distribution. This problem has also been considered by P. Lévy, who showed that the distribution of $\theta(t)$ must have infinite variance. The two-sided absorption time is shown to be a random variable which has a finite nth moment if and only if the wedge which constitutes the absorbing barrier has an interior angle $\beta < \pi/2n$. In §3 we point out how plane brownian motion can be used to represent the Cauchy process. A theorem on brownian motion due to P. Lévy is then used to gain information about the Cauchy process $C(t)$. If $-1 < C(0) = x < 1$ the probability that $C(t) \geqq 1$ before $C(t) \leqq -1$ is found to be $1/2 + \pi^{-1} \sin^{-1} x$.

1. In his recent book on brownian motion [4, pp. 59–60] P. Lévy quotes a result of Dvoretzky and Erdös [3, Theorem 5] concerning brownian motion in $n \geqq 3$ dimensions. He goes on to point out that the analogue of their theorem for the plane could be found if one had an asymptotic estimate for the probability

$$H(t_1, t_2; r) = \Pr\left[\min_{t_1 \leqq t \leqq t_2} R(t) < r \,\middle|\, R(0) = 0 \right], \qquad \text{as } r \to 0.$$

We shall find such an estimate and call it

LEMMA 1.

Received by the editors September 8, 1956.

(1) The results of §1 and part of §2 are taken from the author's 1953 Ph.D. dissertation, and were presented to the Mathematical Society in Abstracts 247 and 248, December 1952. The result of §1 was also obtained independently in 1954 by P. Erdös, A. Dvoretzky, and S. Kakutani, who kindly urged the author to publish it.

$$H(t_1, t_2; r) \sim \frac{\log t_2 - \log t_1}{\log r^{-2}} \qquad (r \to 0).$$

Using this lemma, we can prove that the correct analogue of the result of Dvoretzky and Erdös is

THEOREM 1. *For every positive nonincreasing* $g(t)$, $R(t) \geqq g(t)t^{1/2}$ *for all sufficiently large* t *with probability one or zero, according as the series* $\sum_{k=1}^{\infty} [k|\log g(k)|]^{-1}$ *converges or diverges.* $R(0)$ *is arbitrary.*

P. Lévy's conjecture that, for every $\epsilon > 0$, $R(t) > t^{-\epsilon}$ for all sufficiently large t with probability one, is therefore false. $t^{-\epsilon}$ does not decrease rapidly enough but for instance $t^{-\log t}$ does.

Lévy also points out in this context that, if $R(0) = 0$, the invariance of the process $t^{-1/2}R(t)$ under the transformation $t \to t^{-1}$ transforms results of the type discussed into interesting local theorems. By this device, Theorem 1 produces the

COROLLARY. *For positive, nonincreasing* $g(t)$, *and* $R(0) = 0$, $R(t) \geqq t^{1/2}g(1/t)$ *for all sufficiently small positive values of* t *with probability zero or one according as the series in Theorem* 1 *converges or diverges.*

To arrive at Lemma 1 we make the definitions

$$F(r, t; R) = \Pr[R(t) < r \mid R(0) = R],$$

$$G(r, t; R) = \Pr\left[\min_{0 \leqq \tau \leqq t} R(\tau) < r \mid R(0) = R\right] \text{ if } R \geqq r,$$

$$= 1 \text{ if } R \leqq r.$$

$$\hat{F}(r, \lambda; R) = \int_0^\infty e^{-\lambda t} F(r, t; R)\,dt,$$

$$\hat{G}(r, \lambda; R) = \int_0^\infty e^{-\lambda t} G(r, t; R)\,dt.$$

It will not be necessary to compute the probabilities F and G explicitly. $F(r, t; R)$ satisfies the backward diffusion equation

$$(1.1) \qquad 2\frac{\partial F}{\partial t} = \frac{\partial^2 F}{\partial R^2} + \frac{1}{R}\frac{\partial F}{\partial R}\cdot$$

Since $F(r, t; R) \to 0$ as $t \to 0$, it follows that

$$(1.2) \qquad \frac{\partial^2 \hat{F}}{\partial R^2} + \frac{1}{R}\frac{\partial \hat{F}}{\partial R} - 2\lambda \hat{F} = 0.$$

The general solution of (1.2) is

$$AI_0((2\lambda)^{1/2}R) + BK_0((2\lambda)^{1/2}R),$$

where A, B may depend on λ and on r, and $I_0(z)$, $K_0(z)$ are Bessel functions in standard notation. As $R \to \infty$, only K_0 is regular so that

$$\hat{F}(r, \lambda; R) = B(\lambda, r)K_0((2\lambda)^{1/2}R).$$

Next we make use of the continuity of the sample functions and of the Markovian nature of the process $R(t)$ to write down the following renewal-type integral equation

$$F(r, t; R) = \int_0^t F(r, t - s; r)d_sG(r, s; R), \qquad r \leqq R. \tag{1.3}$$

It follows by taking the Laplace transform that for $r \leqq R$

$$\hat{G}(r, \lambda; R) = \frac{1}{\lambda}\frac{\hat{F}(r, \lambda; R)}{\hat{F}(r, \lambda; r)} = \frac{1}{\lambda}\frac{K_0((2\lambda)^{1/2}R)}{K_0((2\lambda)^{1/2}r)} \cdot \tag{1.4}$$

But, as $z \to 0$, $K_0(z) = -\log z + C \log 2 + o(|z|)$, so that

$$\lim_{r \to 0} \log\left(\frac{1}{r}\right)\hat{G}(r, \lambda; R) = \frac{1}{\lambda} K_0((2\lambda)^{1/2}; R). \tag{1.5}$$

Now it can be verified by studying the inversion integrals of the Laplace transforms in (1.4) and (1.5), that

$$\lim_{r \to 0} \log\left(\frac{1}{r}\right)G(r, t; R) = \int_{R^2/2t}^{\infty} \frac{e^{-x}}{2x} dx. \tag{1.6}$$

Lemma 1 now follows, since

$$H(t_1, t_2; r) = \int_0^\infty G(r, t_1 - t_2; \rho)dF(\rho, t_1; 0).$$

We find

$$\lim_{r \to 0} \log\left(\frac{1}{r}\right)H(t_1, t_2; r) = \frac{1}{t_1}\int_0^\infty \rho \exp\left(\frac{\rho^2}{2t_1}\right)d\rho \int_{\rho^2/2(t_2 - t_1)}^{\infty} \frac{e^{-x}}{2x} dx = \frac{1}{2}\log\frac{t_2}{t_1} \cdot$$

The proof of Theorem 1 follows closely that given in [2] by Dvoretzky and Erdös for the case $n \geqq 3$. There is no loss in assuming $R(0) = 0$, since plane brownian motion is a recurrent process. We first assume that the series in Theorem 1 converges and define the events

$$A_n \equiv R(t) < 2^{n/2}g(2^{n-1}) \text{ for some } 2^{n-1} \leqq t \leqq 2^n, \qquad n = 1, 2, \cdots,$$

$$A = \limsup_{n \to \infty} A_n.$$

It is seen that

$$
\begin{aligned}
(1.7)\qquad & 1 - \Pr\left[R(t) \geqq t^{1/2}g(t) \text{ for all sufficiently large } t\right] \\
& \qquad = \Pr\left[R(t) < t^{1/2}g(t) \text{ for arbitrarily large } t\right] \leqq \Pr\left[A\right].
\end{aligned}
$$

It is well known that $F(r, t; R)$ and $G(r, t; R)$ can be expressed as functions of the variables $t^{-1/2}r$ and $t^{-1/2}R$. Hence

$$
(1.8)\qquad H(t_1, t_2; r) = H(c^2t_1, c^2t_2; cr), \qquad c > 0.
$$

Equation (1.8) and Lemma 1 imply

$$
\Pr\left[A_n\right] = H(1, 2; 2^{1/2}g(2^{n-1})) = -\frac{\log 2}{2 \log\left[2^{1/2}g(2^{n-1})\right]}(1 + \delta_n), \quad \lim_{n\to\infty} \delta_n = 0.
$$

Our assumption about $g(t)$ shows that $\sum_{n=1}^{\infty} \Pr\left[A_n\right]$ converges, so that $\Pr\left[A\right] = 0$, and by (1.7) the event of Theorem 1 is certain.

To prove the second part of Theorem 1 we assume that $g(t)$ has the property that the series diverges. Since $\log g(2^n)$ is a monotone function of n, there exists an increasing sequence $\{n_k\}$ of integers with the property that, as $k \to \infty$

$$
(1.9)\qquad n_{k+1} - n_k \to \infty, \qquad \sum \left| \log g(2^{n_k}) \right|^{-1} = \sum \left[n_k \left| \log g(n_k) \right| \right]^{-1} = \infty.
$$

Now the events B_k and B are defined as

$$
\begin{aligned}
B_k &\equiv R(t) < 2^{n_k/2}g(2^{n_k+1}) \text{ for some } 2^{n_k} \leqq t \leqq 2^{n_k+1}, \\
B &= \limsup_{k\to\infty} B_k,
\end{aligned}
$$

so that

$$
(1.10)\qquad \Pr\left[R(t) < t^{1/2}g(t) \text{ for arbitrarily large } t\right] \geqq \Pr\left[B\right].
$$

By the method used above, it follows from equation (1.8) and (1.9) that

$$
\sum_{1}^{\infty} \Pr\left[B_k\right] = \infty.
$$

If the events B_k were independent we would have $\Pr\left[B\right] = 1$ by the Borel-Contelli Lemma, so that (1.10) would show that the probability in Theorem 1 is zero. We can in fact conclude that $\Pr\left[B\right] = 1$, because the choice of the sequence n_k has made the events B_k almost independent. Rather than giving a detailed proof we refer to the extension of the Borel-Contelli lemma of Chung and Erdös [1], the conditions of which can be verified to hold for the sequence of events B_k.

2. Now we define the process $\theta_R(t)$, the angular part of the plane brownian motion. The subscript R denotes the radial initial condition, i.e. $\left| X(0) + iY(0) \right| = R > 0$. $\theta_R(t)$ is then the winding number of the continuous path $Z(\tau) = X(\tau) + iY(\tau)$, $0 \leqq \tau \leqq t$, about the origin. Or, in other words, one may imagine the original brownian motion to take place on the Riemann surface of

$\log z$, so that $\theta_R(t) = \text{Im} \log [Z(t)]$. The initial condition $\theta_R(0)$ is left unspecified for the moment. It is clear that $\theta_R(t)$ will be a continuous function of t with probability one (the probability is zero that $Z(t)=0$ in any t-interval). However a little intuition shows that $\theta_R(t)$, unlike $R(t)$, is not a Markov process. Nevertheless some interesting questions can be answered concerning this process. It was first considered by P. Lévy [3, p. 252] who deduced that the distribution of $\theta_R(t)$ must have infinite variance from the fact that $\theta_R(t)$ tends to assume very large values, when $R(t)$ is small. We return to this problem after studying the absorption problem which, in a way, is more general.

Let $\theta_R(0) = \alpha > 0$, and $\beta > \alpha$, and let

$$T = T_{R,\alpha,\beta} = \inf_\tau [\tau \mid \theta_R(\tau) \leqq 0, \quad \text{or} \quad \theta_R(\tau) \geqq \beta].$$

Then $u(R, \alpha, t) = \Pr [T > t] = \Pr [0 < \theta_R(\tau) < \beta, \text{ for all } 0 \leqq \tau \leqq t]$. This absorption distribution satisfies the diffusion equation in polar coordinates as its backward equation, and since the process $Z(t)$ is recurrent, it is determined by the boundary and initial value problem

$$2 \frac{\partial u}{\partial t} = \frac{\partial^2 u}{\partial R^2} + \frac{1}{R} \frac{\partial u}{\partial R} + \frac{1}{R^2} \frac{\partial^2 u}{\partial \alpha^2}, \qquad R > 0, t > 0, 0 < \alpha < \beta, \tag{2.1}$$

$$u(R, \alpha, 0) = 1, \qquad u(R, 0, t) = u(R, \beta, t) = 0.$$

This problem has been solved by many authors, but we prefer to use a new method which gives the solution in an elegant form. Let

$$v(s, \alpha, t) = \left(\frac{2}{\pi s}\right)^{1/2} \int_0^\infty u(r, \alpha, t) e^{-r^2/2s} dr, \qquad s > 0. \tag{2.2}$$

One verifies, through integration by parts, that v is the solution of the problem

$$\frac{\partial^2 v}{\partial \alpha^2} + 4s^2 \frac{\partial^2 v}{\partial s^2} + 4s \frac{\partial v}{\partial s} - 4s^2 \frac{\partial^2 v}{\partial s \partial t} - 2s \frac{\partial v}{\partial t} = 0, \tag{2.3}$$

$$v(s, \alpha, 0) = 1, \qquad v(s, 0, t) = v(s, \beta, t) = 0.$$

It remains to use the invariance of u under transformations of the state space leaving $Rt^{-1/2}$ invariant, as was done for the function $H(t_1, t_2; r)$ in the last section. Hence u is a function of α and $Rt^{-1/2}$, and v is a function of α and ts^{-1}. Now let

$$t^{1/2} = s^{1/2} \sinh y, \qquad v(s, \alpha, t) = w(\alpha, y). \tag{2.4}$$

By this change of variable (2.3) now reduces to the simple Dirichlet problem

$$\frac{\partial^2 w}{\partial \alpha^2} + \frac{\partial^2 w}{\partial y^2} = 0, \qquad y > 0, 0 < \alpha < \beta, \tag{2.5}$$

$$w(\alpha, 0) = 1, \qquad w(0, y) = w(\beta, 0) = 0.$$

The solution is

$$w(\alpha, y) = \frac{2}{\pi} \tan^{-1}\left[\sin\frac{\alpha\pi}{\beta} \Big/ \sinh\frac{y\pi}{\beta}\right], \tag{2.6}$$

and the distribution of $T_{R,\alpha,\beta}$ can be obtained through inversion of the integral transform in (2.2).

If $\beta=\pi$, then it is known that $T_{r,\alpha,\pi}$ is a positive stable random variable of index one-half, since it is the time before absorption in a half-plane. Therefore it has an infinite first moment. On the other hand, the time of absorption in a strip has also been studied. It is the same random variable as the absorption time for an interval in one-dimensional brownian motion and has moments of all orders. (This distribution is contained as a special case in the present model, from which it can be obtained by letting $\beta\to 0$, and $R\to\infty$ so that $R\sin\beta=\text{const.}$) The time to absorption in a wedge is now found to have the interesting behavior that the existence of moments depends on the angle of the wedge.

THEOREM 2. *For $\delta\geqq 0$, $(T_{R,\alpha,\beta})^\delta$ has finite expected value if and only if $2\delta\beta<\pi$. This criterion is independent of the initial position $R>0$ and of α with $0<\alpha<\beta$.*

Proof. In the terminology of Markov processes all states inside the wedge communicate. Hence the moment of order δ is either finite or infinite, independent of R and of α. For simplicity we therefore set $\alpha=\beta/2$, and equations (2.2) and (2.4) imply that

$$E[T^{\delta}_{R,\beta/2,\beta}] = -\int_0^\infty t^\delta d_t u\left(R, \frac{\beta}{2}, t\right)$$

is finite or infinite according as the integral

$$-\int_0^\infty t^\delta d_t w\left[\frac{\beta}{2}, \sinh^{-1}\left(\frac{t}{s}\right)^{1/2}\right]$$

converges or diverges. Making use of (2.6), the criterion of Theorem 2 is readily verified.

Next we let $\beta\to\infty$ to obtain one-sided absorption probabilities. Because of the evident symmetry of the process $\theta_R(t)$ about $\theta_R(0)=0$, we can write

$$\begin{aligned} u_0(R,\alpha,t) &= \lim_{\beta\to\infty} u(R,\alpha,t) = \Pr\left[\min_{0\leqq\tau\leqq t}\theta_R(\tau)>0 \,\middle|\, \theta_R(0)=\alpha\right] \\ &= \Pr\left[\max_{0\leqq\tau\leqq t}\theta_R(\tau)<\alpha \,\middle|\, \theta_R(0)=0\right]. \end{aligned} \tag{2.7}$$

Since

$$\lim_{\beta\to\infty} \omega(\alpha, y) = \frac{2}{\pi}\tan^{-1}\frac{\alpha}{y},$$

we have from (2.2) and (2.4)

$$(2.8)\quad \left(\frac{2}{\pi s}\right)^{1/2}\int_0^\infty u_0(r, \alpha, t)e^{-r^2/2s}dr = \frac{2}{\pi}\tan^{-1}\frac{\alpha}{\log\left[(1+t/s)^{1/2}+(t/s)^{1/2}\right]}.$$

Before inverting the integral transform in (2.8) we prove a lemma which shows how to obtain the transition probabilities of the process $\theta_R(t)$.

Lemma 2. *The transition probability* $P_R(\alpha, t) = \Pr\left[\theta_R(t) < \alpha \mid \theta_R(0) = 0\right] = 1/2 + u_0(R, \alpha, t)/2$, *for* $\alpha \geqq 0$. *For* $\alpha \leqq 0$ *it is defined by the requirement that* $\theta_R(t)$ *be a symmetric random variable.*

Proof. We make use of the continuity of the sample functions $\theta_R(t)$ as well as the above mentioned symmetry of $\theta_R(t)$. The argument used is nothing but the well known reflection principle, and it is interesting to note that it goes through for a process which is not Markovian.

Let A be the event that $\max_{0\le\tau\le t}\theta_R(\tau) > \alpha$, and B the event that $\theta_R(t) > \alpha$, both subject to the initial condition $\theta_R(0) = 0$. Ω is the space of all sample functions. Symmetry and continuity together imply $\Pr[B|A] = \Pr[\Omega - B|A]$. Hence $\Pr[B\cap A] = \Pr[(\Omega-B)\cap A]$. But $B\subset A$, so that $\Pr[B] = \Pr[A-B] = \Pr[A-B]$, or $\Pr[\Omega - B] = 1/2 + \Pr[\Omega - A]/2$. But that is the content of the lemma.

From the above Lemma and from equation (2.8) we obtain

$$\left(\frac{2}{\pi s}\right)^{1/2}\int_0^\infty P_R(\alpha, t)e^{-R^2/2s}dR = \frac{1}{2} + \frac{1}{\pi}\tan^{-1}\left[\frac{\alpha}{\log\left((1+t/s)^{1/2}+(t/s)^{1/2}\right)}\right].$$

If the characteristic function of $\theta_R(t)$ is

$$\phi(\lambda, t; R) = E\left[e^{i\lambda\theta_R(t)} \mid \theta_R(0) = 0\right] = \int_{-\infty}^{\infty} e^{i\lambda\alpha}dP_R(\alpha, t),$$

it follows that

$$(2.9)\quad \int_0^\infty \frac{\phi(\lambda, t; (2x)^{1/2})}{(2x)^{1/2}}e^{-px}dx = \left(\frac{\pi}{2p}\right)^{1/2}\left[(tp+1)^{1/2} - t^{1/2}p^{1/2}\right]^{|\lambda|}.$$

It is not difficult to verify that the (unique) inversion of the Laplace transform in (2.9) gives

$$(2.10)\quad \phi(\lambda, t; R) = \left(\frac{\pi}{2}\right)^{1/2}\left(\frac{R^2}{4t}\right)^{1/2}e^{-R^2/4t}\left[I_{(|\lambda|-1)/2}\left(\frac{R^2}{4t}\right) + I_{(|\lambda|+1)/2}\left(\frac{R^2}{4t}\right)\right],$$

where $I_\nu(z)$ is the modified Bessel function of the first kind of order.

Finally, (2.10) enables us to find the normalized limiting distribution of $\theta_R(t)$. Let

$$\delta = R(8t)^{-1/2}, \qquad \mu = -\frac{\lambda}{\log \delta}, \qquad \lambda > 0.$$

$$\lim_{t\to\infty} \phi\left(\frac{2\lambda}{\log (8t/R^2)}, t; R\right) = \lim_{\delta\to 0} \pi^{1/2}\delta e^{-2\delta^2}\left[I_{(\mu-1)/2}(2\delta^2) + I_{(\mu+1)/2}(2\delta^2)\right]$$

$$= \lim_{\delta\to 0} \pi^{1/2}\delta e^{-2\delta^2} \sum_{n=0}^{\infty}\left[\frac{\delta^{4n+\mu-1}}{n!\Gamma(n+\mu/2+1/2)} + \frac{\delta^{4n+\mu+1}}{n!\Gamma(n+\mu/2+3/2)}\right].$$

Only the first term of the first series contributes to the limit, which becomes

$$\lim_{\delta\to 0} \frac{\pi^{1/2}e^{-2\delta^2}\delta^{\mu}}{\Gamma(1/2+\mu/2)} = e^{-\lambda} = \lim_{t\to\infty} \phi\left(\frac{2\lambda}{\log t}, t; R\right).$$

Due to the continuity theorem for characteristic functions, we have indeed proved

THEOREM 3. *If* $\theta_R(0) = 0$, *then the characteristic function of* $\theta_R(t)$ *is given by equation* (2.10), *and*

$$\lim_{t\to\infty} \Pr\left[\theta_R(t) < \frac{\alpha}{2}\log t\right] = \frac{1}{\pi}\int_{-\infty}^{\alpha} \frac{dx}{1+x^2}.$$

The limit theorem could have been obtained with little effort directly from Equation (2.8) and from Lemma 2, or alternatively, directly from Equation (2.9).

3. The Cauchy process $C(t)$ is defined as the separable process with stationary independent increments, whose distribution is given by

$$\Pr\left[C(t+s) - C(s) < x\right] = \frac{t}{\pi}\int_{-\infty}^{x} \frac{d\xi}{t^2+\xi^2}. \tag{3.1}$$

This process has a very simple representation in terms of our 2-dimensional brownian motion $X(t)+iY(t)$. We assume throughout that $X(0) = x_0$, $Y(0) = 0$, and define for $y \geqq 0$ the random variable

$$T(y) = \min_{0\leqq t}\left[t \mid Y(t) \geqq y\right].$$

We shall leave aside the measure theoretical justification which should accompany each such definition. (Since $Y(t)$ is separable, a proof that $T(y)$ is a random variable is obtained by representing the set $[T(y) < \tau]$ as a countable union of simpler measurable subsets of the space of sample functions $Z(t)$.)

Now the Cauchy process is represented by $Z(t)$, i.e., the sets $[C(y_k) \leqq a_k,$ $k = 1, \cdots, m]$ are in the Borel field of sets generated by $[X(t_k) \leqq \alpha_k,\ Y(t_k) \leqq \beta_k,\ k = 1, \cdots, m]$, by virtue of

LEMMA 3.

$$C(y) = X[T(y)], \text{ if } C(0) = Z(0).$$

Proof. $X[T(y)]$ is the x-coordinate of $Z(t)$ at the first time when $\operatorname{Im} Z(t) = y$. Hence the fact that increments $X[T(y+\eta)] - X[T(\eta)]$ are independent and stationary (their distribution is independent of η). Equation (3.1) follows from

$$\Pr[T(y) \leqq s] = 2\Pr[Y(s) > t] = \left(\frac{2}{\pi}\right)^{1/2} \int_{y/s^{1/2}}^{\infty} e^{-x^2/2} dx.$$

Hence

$$\Pr[X(T(y)) \leqq x] = \int_0^{\infty} \Pr[X(s) \leqq x \mid X(0) = x_0] ds \Pr[T(y) \leqq s]$$

$$= \frac{y}{\pi} \int_{-\infty}^{x-x_0} \frac{y}{\xi^2 + y^2} d\xi.$$

The separability of $C(y)$ follows from that of $Z(t)$.

To exhibit the advantage of this representation we pose the simple but not trivial problem of computing the probability $P(x_0)$ that the Cauchy process $C(t)$, with $-1 < C(0) = x_0 < 1$ enters the region $x \geqq 1$ before entering $x \leqq -1$. That the corresponding problem for 1-dimensional brownian motion has the solution $(1+x_0)/2$ is an immediate consequence of the symmetry and continuity of the process. The answer to the present problem should be different due to the discontinuities of the sample functions $C(t)$.

We define a new process

$$M(t) = \max_{0 \leqq \tau \leqq t} Y(\tau), \qquad Y(0) = 0.$$

Now we have a picture, albeit not a simple one, of $P(x_0)$ as the probability of a certain set of paths of $Z(t)$ with $Z(0) = x_0 \cdot P(x_0)$ is the probability that, if we observe $X(t)$ only at the instants τ when $M(\tau) - Y(\tau) = 0$, the event $X(\tau) \geqq 1$ should occur before the event $X(\tau) \leqq -1$. This picture is rendered simple by a theorem of P. Lévy [3, Theorem 49.1]:

"*If* $Y(0) = 0$, *then* $M(t) - Y(t)$ *is the same stochastic process as* $|Y(t)|$."

A proof based on Doob's theory of separability would be similar to that of Lemma 3. It is necessary to show that each process is Markovian, that both processes have the same one-step transition probabilities, and that both are separable.

Since $X(t)$ and $Y(t)$ are independent processes, we can replace $M(t) - Y(t)$ by $|Y(t)|$ in the above description of $P(x_0)$. Let E_2 be the complex plane, A and B its subsets

$$A = [x \geqq 1; y = 0], \qquad B = [x \leqq -1; y = 0].$$

Let

$$u(x, y) = \Pr [Z(t) \in A \text{ before } Z(t) \in B \mid Z(0) = x + iy],$$

with $x+iy \in E_2 - A \cup B$. Then $P(x_0) = u(x_0, 0)$. But it is well known that $u(x, y)$ is the solution of the Dirichlet problem

$$\frac{\partial^2 u}{\partial x^2} + \frac{\partial^2 u}{\partial y^2} = 0, \qquad x + iy \in E_2 - A \cup B, \tag{3.2}$$
$$u(x, y) = 1 \text{ for } x + iy \in A,$$
$$u(x, y) = 0 \text{ for } x + iy \in B.$$

The solution is

$$u(x, y) = \frac{1}{\pi} \tan^{-1} \frac{2 \operatorname{Im}(w)}{w\bar{w} - 1},$$
$$x + iy = \frac{1 + w^2}{1 - w^2}, \qquad \operatorname{Im} w > 0,$$

so that

$$P(x) = \frac{2}{\pi} \tan^{-1} \left(\frac{1+x}{1-x}\right)^{1/2} = \frac{1}{2} + \frac{1}{\pi} \sin^{-1} x. \tag{3.3}$$

This is

THEOREM 4. *The Cauchy process starting at* $x \in (-1, 1)$ *is absorbed on the right with probability* $1/2 + \sin^{-1} x/\pi$.

It is possible to do more. By changing the boundary condition in (3.2) to $u(x, y) = 1$ for $x+iy$ in a measurable subset S of $A \cup B$, and $u(x, y) = 0$ for $x+iy$ in $A \cup B - S$, one can deduce that a Cauchy process $C(t)$ with $-1 < C(0) = x_0 < 1$, will first be absorbed in $A \cup B$ at a point y with the probability density

$$f(x_0; y) = \frac{1}{\pi} (1 - x_0^2)^{1/2} \cdot \frac{1}{|y - x_0| (y^2 - 1)^{1/2}}, \qquad |y| > 1.$$

Both $f(x; y)$ and $P(x)$ satisfy the equation

$$\text{P.V.} \int_{-1}^{1} \frac{\partial f(x; y)/\partial x}{x - t} dx = \text{P.V.} \int_{-1}^{1} \frac{P'(x)}{x - t} dx = 0, \qquad |t| < 1. \tag{3.4}$$

It is difficult to derive these equations directly, and to impose appropriate conditions. $P(x)$ for instance, as given by (3.3), happens to be the unique solution of (3.4) with $P(1)=1$, $P(-1)=0$, and $P'(x)\cdot(1-x^2)^{1/2}$ in $L_2(-1, 1)$.

References

1. K. L. Chung and P. Erdös, *On the application of the Borel-Contelli lemma*, Trans. Amer. Math. Soc. vol. 72 (1952) pp. 179–186.

2. A. Dvoretzky and P. Erdös, *Some problems on random walk in space*, Proceedings of the 2nd Berkeley Symposium, Univ. of California Press, 1951, pp. 353–368.

3. P. Lévy, *Processus stochastiques et mouvement brownien*, Paris, 1948.

4. ———, *Le mouvement browien*, Mémorial des Sciences Mathématiques, Fasc. 76, Paris, 1954.

California Institute of Technology,
Pasadena 4, Calif.

Reprinted with permission from
the University of California Press
from the Proceedings of the 4th Berkeley Symposium
on Math. Statistics and Probability
Vol. II, 515-534 (1961)

RECURRENT RANDOM WALK AND LOGARITHMIC POTENTIAL

FRANK SPITZER
UNIVERSITY OF MINNESOTA

1. Introduction

This is an attempt to show that a potential theory is associated with certain recurrent Markov processes in a natural way. For transient Markov processes this fact has been studied intensely. Thus Hunt [9] bases a general potential theory on transient continuous parameter processes, Doob [7] and Hunt [10] use the theory to construct boundaries for discrete parameter processes, Itô and McKean [11] solve the problem of characterizing the recurrent sets for simple random walk in three and higher dimension within the framework of the associated potential theory.

The class of recurrent Markov processes considered here are defined as follows. The state space will be E, the set of all ordered pairs (lattice points) $x = (m, n)$ where m and n are integers. X is a random variable with values in E. Its characteristic function $\phi(\theta)$ is

$$\phi(\theta) = E[e^{i\theta\cdot X}] = \sum_{x\in E} e^{i\theta\cdot x} P\{X = x\}, \tag{1.1}$$

where $\theta = (\theta_1, \theta_2)$ and $\theta\cdot x = \theta_1 m + \theta_2 n$ if $x = (m, n)$. Here $X_1, X_2, \cdots$ is an infinite sequence of independent random variables with the same distribution as X. Each characteristic function $\phi(\theta)$ defines a Markov process

$$S_n = S_0 + X_1 + \cdots + X_n, \qquad n \geqq 1, \tag{1.2}$$

where the starting point S_0 is an arbitrary point in E.

For simplicity we assume throughout that $\phi(\theta) = \phi(-\theta)$, or equivalently that $P\{X = x\} = P\{X = -x\}$. Two further assumptions are essential to the theory. Let S denote the square $|\theta_1| \leqq \pi$, $|\theta_2| \leqq \pi$ in the θ-plane and let $\int [\ ]\, d\theta$ denote integration over S where $d\theta$ is two-dimensional Lebesgue measure. For $\theta \in S$, it is assumed that $\phi(\theta)$ satisfies

$$\phi(\theta) = 1 \Rightarrow \theta = 0 \tag{1.3}$$

$$\int \frac{d\theta}{1 - \phi(\theta)} = \infty. \tag{1.4}$$

A number theoretical argument may be used to show that (1.3) is equivalent to the condition that every x in E is a possible value of the process S_n, that is, that, given S_0 and x, there is a positive integer n such that $P\{S_n = x\} > 0$. Condition (1.4) is equivalent to

$$\sum_{n=0}^{\infty} P\{S_n = S_0\} = \infty \tag{1.5}$$

and if (1.3) holds, this is equivalent to

$$P\left\{\bigcap_{n=1}^{\infty} \bigcup_{k=n}^{\infty} (S_k = x)\right\} = 1 \tag{1.6}$$

for every S_0, that is, every point in E is visited infinitely often with probability one. These things are discussed in [4].

Conditions (1.3) and (1.4) are obviously satisfied by the *simple random walk* with

$$\phi(\theta) = \frac{1}{2}\cos\theta_1 + \frac{1}{2}\cos\theta_2, \tag{1.7}$$

and, according to [4], a sufficient condition for (1.4) to hold is, using the assumption that $\phi(\theta) = \phi(-\theta)$,

$$\sum_{x\in E} |x|^2 P\{X = x\} = E(|X|^2) < \infty. \tag{1.8}$$

The potential theory will be based on three operators P, A, and K. P has the representation (kernel)

$$P(x, y) = P\{X = x - y\}, \tag{1.9}$$

and will act on the space $\mathcal{L}$ of real-valued functions on E for which $P|f| < \infty$, that is, $P\colon \mathcal{L} \to \mathcal{L}$, where

$$Pf(x) = \sum_{y\in E} P(x, y)f(y), \qquad f \in \mathcal{L} = \{f|P|f| < \infty\}. \tag{1.10}$$

If I is the identity then $P - I$ (being the second difference operator for simple random walk) is our version of the *Laplace operator*. We let Ω denote a finite subset of E and $\mathcal{L}(\Omega)$ the vector space of real-valued functions on Ω whose dimension $|\Omega|$ is the number of points in Ω. In section 2, lemma 2.1, it is shown that, for each pair x, y in E, the integral

$$A(x, y) = (2\pi)^{-2} \int \frac{1 - \cos (x - y)\cdot\theta}{1 - \phi(\theta)}\, d\theta \tag{1.11}$$

exists and we define A as an operator (*potential kernel*) $A\colon \mathcal{L}(\Omega) \to \mathcal{L}$ by

$$Af(x) = \sum_{a\in\Omega} A(x, a)f(a), \qquad x \in E, \quad f \in \mathcal{L}(\Omega). \tag{1.12}$$

The one-dimensional analogue of the kernel $A(x, y)$ was first studied by Kac [12] in a somewhat different context. In section 3, lemma 3.1 we use one of his results concerning $A(x, y)$. (This kernel is studied by W. Hoeffding in his paper "On sequences of sums of independent random vectors," given at this Symposium. In particular, many of the results in section 2 are also obtained by Hoeffding, and used in a context closely related to potential theory.)

Lemma 2.2 will show that $A\psi \in \mathcal{L}$ when $\psi \in \mathcal{L}(\Omega)$. It will show more, namely that $(P - I)A\psi(x) = 0$ when $x \in E - \Omega$ whereas $(P - I)A\psi(x) = \psi(x)$ when $x \in \Omega$. This is familiar from potential theory. If ψ is a nonnegative function (charge distribution on Ω) in $\mathcal{L}(\Omega)$, $f = A\psi$ satisfies Laplace's equation $(P - I)f = 0$ off Ω whereas f satisfies Poisson's equation $(P - I)f = \psi$ on Ω. In the classical case the sign is reversed in the last equation. This could be remedied by using $-A$ as the potential kernel instead of A. Our notation has the advantage of a more natural probability interpretation which is given in section 4.

If the range of the operator A is restricted to $\mathcal{L}(\Omega)$, A becomes a (matrix) operator mapping $\mathcal{L}(\Omega)$ into $\mathcal{L}(\Omega)$. In section 3, lemma 3.1, this mapping will be shown to be onto, that is, the restricted version of A has an inverse $K = K(x, y)$, $x, y \in \Omega$ defined by $K: \mathcal{L}(\Omega) \to \mathcal{L}(\Omega)$,

$$(1.13)\qquad \begin{aligned} Kf(x) &= \sum_{a\in\Omega} K(x, a)f(a), && \text{for } f \in \mathcal{L}(\Omega), \quad x \in \Omega, \\ AKf(x) &= f(x), && \text{when } f \in \mathcal{L}(\Omega), \quad x \in \Omega. \end{aligned}$$

The first of several equivalent definitions, in section 8, of the *capacity* of the set Ω will be in terms of the quadratic form (e, Ke) where $e = e(x) \equiv 1$ for $x \in \Omega$.

This paper relies to some extent on the *imbedded Markov chain* induced by the finite set Ω. This chain is simply the process S_n observed only at those (random) times n when S_n is in Ω. Thus its transition matrix is the (stochastic) matrix Π of size $|\Omega,|$ defined by

$$(1.14)\qquad \Pi(x, y) = \sum_{n=1}^{\infty} P\{S_\nu \notin \Omega \quad \text{for} \quad \nu = 1, \cdots, n - 1;\ S_n = y | S_0 = x\}, \qquad x, y \in \Omega.$$

In section 3, theorem 3.1, a formula for $\Pi(x, y)$ is obtained in terms of $A(x, y)$ and $K(x, y)$ by the method of the imbedded Markov chain, the foundation for which was laid in [14]. The method adds probabilistic interest to the potential theory and makes it more explicit, without being essential to its logical development. Thus, in section 5, theorem 5.2 gives an explicit formula for the *harmonic measure*. It was first discovered by the method of the imbedded Markov chain; nevertheless the proof which is given in section 5 consists of verifying that the formula obtained is the appropriate solution to the exterior Dirichlet problem, which has a unique solution in view of theorem 5.1. The last three sections, sections 6, 7, 8, imitate classical potential theory, giving explicit representations as well as the standard theorems for the *potential*, *Green function*, and *capacity* of finite sets of lattice points, corresponding to each process S_n, satisfying $\phi(\theta) = \phi(-\theta)$ as well as conditions (1.3) and (1.4).

The methods and results of this paper should be compared to those of Itô and McKean [11]. Our method will yield part of their results, namely those concerned with finite sets, in a different and more explicit form. Thus the matrix Π of the imbedded Markov chain can be expressed in terms of the inverse of Itô and McKean's potential kernel, restricted to the set in question. (Π is of course substochastic in their theory.) Just as we do in section 8, the capacity of a finite set can be expressed as a quadratic form, and the equilibrium charge on a finite set Ω turns out to be the limit of $p(x, a)$ as $x \to \infty$, $p(x, a)$ being the conditional probability that the set Ω is first hit at a, starting at x, given that it is hit in a finite time.

The fact that we get more formulas than Itô and McKean is of dubious merit. The hard calculations in [11] are based on maximum principles, such as we obtain in section 7, enabling one to use integral estimates from classical potential theory rather than "explicit" formulas, involving matrix inversion.

Typical new problems arising from this work are, in increasing order of difficulty: (1) Its extension to nonlattice random walk in the plane of the same sum of independent variable type as here. The theory of integral equations with completely continuous kernels should imitate the present matrix theory without serious difficulty. (2) To understand the connection of the limit theorems (as time tends to infinity) of Kac [12] which lead to the same potential kernel as the present (time-independent) theory. (3) To discover the domain of validity of present key theorems among general irreducible recurrent Markov chains. For example let x_k, $k = 1, 2, \cdots$ be the states of such a chain and Ω a finite set of states. What are necessary and sufficient conditions, that is what does it mean, for $\lim_{n\to\infty} p(x_n, a)$ to exist for all a in Ω, for every Ω, where $p(x, a)$ is the harmonic measure, defined as in (5.2), of Ω?

2. Some Fourier analysis

To justify the definition in (1.11) we first show

LEMMA 2.1. *For $x, y \in E$, $[1 - \cos (x - y)\cdot\theta][1 - \phi(\theta)]^{-1}$ is Lebesgue integrable on S and*

$$A(x, y) = (2\pi)^{-2} \int \frac{1 - \cos (x - y)\cdot\theta}{1 - \phi(\theta)}\, d\theta \tag{2.1}$$

$$= \lim_{N\to\infty} \sum_{n=0}^{N} [P\{S_n = 0|S_0 = 0\} - P\{S_n = x - y|S_0 = 0\}].$$

PROOF. Let $S_0 = 0$. In view of (1.3), $[1 - \phi(\theta)]^{-1}$ is continuous on S except at the origin. $\phi(\theta)$ is real. Suppose that

$$Q(\theta) = E[(\theta\cdot X)^2] = \theta_1^2\sigma_{11} + 2\theta_1\theta_2\sigma_{12} + \theta_2^2\sigma_{22} < \infty. \tag{2.2}$$

Then

$$\lim_{\theta\to 0} \frac{1 - \phi(\theta)}{Q(\theta)} = \frac{1}{2}. \tag{2.3}$$

Equation (1.3) implies that $Q(\theta)$ is a positive definite quadratic form. Consequently there is a constant $d > 0$ such that $[1 - \phi(\theta)]^{-1} \leqq d(\theta\cdot\theta)^{-1}$ for all θ in S. The integral defining $A(x, y)$ therefore exists in the sense of Lebesgue, since $[1 - \cos (x - y)\cdot\theta](\theta\cdot\theta)^{-1}$ is Lebesgue integrable on S. If on the other hand $Q(\theta) = +\infty$, for some $\theta \neq 0$ we have

$$E[|X|^2] = +\infty, \quad \lim_{\theta\to 0} \frac{1-\phi(\theta)}{\theta\cdot\theta} = +\infty, \qquad |X|^2 = X\cdot X, \tag{2.4}$$

and the previous argument may be resumed.

$$\begin{aligned}\sum_{n=0}^{N} [P\{S_n = 0\} - P\{S_n = x - y\}] &= (2\pi)^{-2} \int [1 - \cos (x-y)\cdot\theta][1 + \phi(\theta) + \cdots + \phi(\theta)^N]\, d\theta \\ &= A(x, y) - (2\pi)^{-2} \int \frac{1 - \cos (x-y)\cdot\theta}{1 - \phi(\theta)} [\phi(\theta)]^{N+1}\, d\theta.\end{aligned} \tag{2.5}$$

Since $|\phi(\theta)| < 1$ except at most at a finite number of points in S, the last term tends to zero as N tends to infinity.

$A(x, y)$ will play the role of a potential kernel because of

LEMMA 2.2. *For* $f \in \mathfrak{L}(\Omega)$, $Af \in \mathfrak{L}$, *and*

$$(P - I)Af(x) = \begin{cases} f(x), & x \in \Omega \\ 0, & x \in E - \Omega. \end{cases} \tag{2.6}$$

(For the case of simple random walk this fact is due to Stöhr [15], as is the result of lemma 2.4.)

PROOF. Since $A(x, y)$ is a difference kernel we assume without loss of generality that Ω contains 0 (the origin). Let $f(0) = 1$, and $f(x) \equiv 0$ when $x \in \Omega$, $x \neq 0$. It remains to show that, for $x \in E$,

$$\sum_{t\in E} P(x, t)A(t, 0) - A(x, 0) = \begin{cases} 1, & x = 0, \\ 0, & x \neq 0. \end{cases} \tag{2.7}$$

The left side is

$$\begin{aligned}&(2\pi)^{-2} \int [1 - \phi(\theta)]^{-1}[1 - \sum_{t\in E} P(x, t) \cos t\cdot\theta - 1 + \cos x\cdot\theta]\, d\theta \\ &\qquad = (2\pi)^{-2} \int [1 - \phi(\theta)]^{-1}[-(\cos x\cdot\theta)\phi(\theta) + \cos x\cdot\theta]\, d\theta \\ &\qquad = (2\pi)^{-2} \int \cos x\cdot\theta\, d\theta = 1\end{aligned} \tag{2.8}$$

when $x = 0$ and 0 otherwise.

By linearity (2.6) holds for all $f \in \mathfrak{L}(\Omega)$ and by the triangle inequality $P|Af| < \infty$ so that $Af \in \mathfrak{L}$.

The next lemma, concerning the asymptotic behavior of $A(x, y)$ is responsible for the asymptotic results in our potential theory. All we assume is that $\phi(\theta) = \phi(-\theta)$ and that $\phi(\theta)$ satisfies (1.3) and (1.4). Let

$$a(x) = A(x, 0) = (2\pi)^{-2} \int \frac{1 - \cos x\cdot\theta}{1 - \phi(\theta)}\, d\theta. \tag{2.9}$$

LEMMA 2.3.

$$\lim_{x\to\infty} a(x) = +\infty. \tag{2.10}$$

For x_1, x_2 *in* E

$$\lim_{x\to\infty} [A(x, x_1) - A(x, x_2)] = 0. \tag{2.11}$$

PROOF. Let S_ϵ be the subset of S defined by $\theta\cdot\theta < \epsilon^2$, with $0 < \epsilon < \pi$, and decompose

$$\begin{aligned}(2\pi)^2 a(x) &= \int_{S_\epsilon} \frac{1 - \cos x\cdot\theta}{1 - \phi(\theta)}\, d\theta + \int_{S-S_\epsilon} \frac{1 - \cos x\cdot\theta}{1 - \phi(\theta)}\, d\theta \\ &\geqq \int_{S-S_\epsilon} \frac{d\theta}{1 - \phi(\theta)} - \int_{S-S_\epsilon} \frac{\cos x\cdot\theta}{1 - \phi(\theta)}\, d\theta.\end{aligned} \tag{2.12}$$

With ϵ fixed let $x\cdot x = |x|^2$ tend to infinity, and the two-dimensional Riemann Lebesgue lemma [1] gives $\lim_{x\to\infty} (2\pi)^2 a(x) \geqq \int_{S-S_\epsilon} d\theta/[1 - \phi(\theta)]$ for every $0 < \epsilon < \pi$. Because of (1.4), the integral on the right tends to infinity as ϵ tends to zero, and this proves (2.10).

Since $A(x, y)$ is a difference kernel it clearly suffices to prove (2.11) for $x_2 = 0$, with x_1 arbitrary in E. Letting $u_1 = (1, 0)$, $u_2 = (0, 1)$, $x = mu_1 + nu_2$, it is clear that

$$\begin{aligned}&\lim_{x\to\infty} [A(x_1, x) - A(0, x)] \\ &\qquad = m \lim_{x\to\infty} [A(u_1, x) - A(0, x)] + n \lim_{x\to\infty} [A(u_2, x) - A(0, x)],\end{aligned} \tag{2.13}$$

provided one can show that the last two limits exist. Since they are of the same type it suffices to consider only the first one and the proof of (2.11) is completed by showing that it is zero.

$$\begin{aligned}(2\pi)^2[A(u_1, x) - A(0, x)] &= \int \frac{\cos x\cdot\theta - \cos (x - u_1)\cdot\theta}{1 - \phi(\theta)}\, d\theta \\ &= -2 \int \frac{\sin \frac{\theta_1}{2} \sin \left(x - \frac{u_1}{2}\right)\cdot\theta}{1 - \phi(\theta)}\, d\theta.\end{aligned} \tag{2.14}$$

But $f(\theta) = -2 \sin (\theta_1/2)[1 - \phi(\theta)]^{-1}$ is integrable on S since $|\theta_1|(\theta\cdot\theta)^{-1}$ is integrable. Letting $x - u_1/2 = y$ and noting that y tends to infinity with x, we have

$$\lim_{x\to\infty} (2\pi)^2[A(u_1, x) - A(0, x)] = \lim_{y\to\infty} \int f(\theta) \sin y\cdot\theta\, d\theta = 0 \tag{2.15}$$

by the two-dimensional Riemann-Lebesgue lemma.

Sharper results concerning the asymptotic behavior of $A(x, y)$ require further assumptions concerning the random walk S_n, or its characteristic function $\phi(\theta)$. Under the assumption that $E(|X|^2) < \infty$ it can be shown that $a(x) \sim c \log |x|$ as $|x| \to \infty$, where c is a positive constant. (Theorem 5.1 of Hoeffding's Symposium paper.) However, the behavior of the difference $a(x) - c \log |x|$ depends on the direction in which x tends to infinity, unless we assume the random walk to be isotropic, that is, that the quadratic form $Q(\theta)$ has no preferred principal axes. To get a result as strong as in classical potential theory it also seems necessary to assume that $|X|$ has a finite moment of order greater than two.

LEMMA 2.4. *If* $Q(\theta) = E[(\theta\cdot X)^2] = \sigma^2\theta\cdot\theta < \infty$, *and if* $E(|X|^{2+\delta}) < \infty$ *for some* $\delta > 0$, *then*

$$\lim_{x\to\infty}\left[a(x) - \frac{1}{\pi\sigma^2}\log|x|\right] = c + \frac{1}{\pi\sigma^2}d = k, \tag{2.16}$$

$$0 < c = (2\pi)^{-2}\int\left[\frac{1}{1-\phi(\theta)} - \frac{2}{Q(\theta)}\right]d\theta < \infty, \tag{2.17}$$

where $\gamma = .5772\cdots$ *is Euler's constant,* $\lambda = \sum_{n=0}^{\infty}(-1)^n(2n+1)^{-2}$ *is Catalan's constant, and* $d = \gamma + \log\pi - 2\lambda/\pi$. *For the simple random walk* $\sigma^2 = 1/2$, $k = (1/\pi)\log 8 + 2\gamma/\pi$.

PROOF. First we check that the integrand

$$\psi(\theta) = [1-\phi(\theta)]^{-1}Q^{-1}(\theta)\{Q(\theta) - 2[1-\phi(\theta)]\} \tag{2.18}$$

in (2.17) is Lebesgue integrable. Since $[1-\phi(\theta)]^{-1}Q^{-1}(\theta)$ has a positive constant times $(\theta\cdot\theta)^{-2}$ as a lower bound, it suffices to prove integrability for

$$\begin{aligned}\chi(\theta) &= (\theta\cdot\theta)^{-2}\left[\frac{\sigma^2}{2}(\theta\cdot\theta) - 1 + \phi(\theta)\right] \\ &= (\theta\cdot\theta)^{-2}\,E\left[\frac{1}{2}(\theta\cdot X)^2 - 1 + \cos\theta\cdot X\right].\end{aligned} \tag{2.19}$$

For some constant $h > 0$, by the Schwarz inequality,

$$\begin{aligned}|\chi(\theta)| &\leqq h(\theta\cdot\theta)^{-2}\,E(|\theta\cdot X|^{2+\delta}) \\ &\leqq h(\theta\cdot\theta)^{-2}\,E\{[(\theta\cdot\theta)(X\cdot X)]^{1+\delta/2}\} \\ &\leqq h(\theta\cdot\theta)^{-1+\delta/2}E(|X|^{2+\delta}),\end{aligned} \tag{2.20}$$

which is integrable.

Now we decompose

$$\begin{aligned}a(x) &= (2\pi)^{-2}\int\frac{1-\cos x\cdot\theta}{1-\phi(\theta)}\,d\theta \\ &= \frac{2}{\sigma^2}(2\pi)^{-2}\int\frac{1-\cos\theta\cdot x}{\theta\cdot\theta}\,d\theta + (2\pi)^{-2}\int(1-\cos\cdot x\,\theta)\psi(\theta)\,d\theta.\end{aligned} \tag{2.21}$$

The last integral tends to c by the two-dimensional Riemann Lebesgue lemma and it only remains to verify that

$$\begin{aligned}&\lim_{x\to\infty}\left(\frac{1}{2\pi}\int\frac{1-\cos x\cdot\theta}{\theta\cdot\theta}\,d\theta - \log|x|\right) \\ &\qquad = \gamma + \log\pi - \frac{2}{\pi}\lambda\,\frac{1}{2\pi}\int\frac{1-\cos x\cdot\theta}{\theta\cdot\theta}\,d\theta = I_1(x) + I_2(x),\end{aligned} \tag{2.22}$$

where

$$I_1(x) = \frac{1}{2\pi} \int_{0 \leq \sqrt{\theta\cdot\theta} < \pi} \frac{1 - \cos x\cdot\theta}{\theta\cdot\theta}\, d\theta, \tag{2.23}$$

and $I_2(x)$ is the integral over the remaining part of the square S
By introducing polar coordinates it is shown that $I_1(x) - \log|x| \to \gamma + \log(\pi/2)$, using the formula

$$\gamma = \int_0^1 \frac{1 - \cos t}{t}\, dt - \int_1^\infty \frac{\cos t}{t}\, dt. \tag{2.24}$$

The second integral $I_2(x)$, by the Riemann-Lebesgue lemma, tends to a limit as $x \to \infty$ which may be expressed in terms of Catalan's constant λ. The details are omitted as is the calculation of k for simple random walk.

3. The imbedded Markov chain

For the purpose of this section only let $A = A(x, y)$ be restricted to x, y in $\Omega \subset E$, and let $2 \leq |\Omega| < \infty$. The case when $|\Omega| = 1$ is trivial. Imitating an argument of Kac [12], we obtain more than is needed to assert that A has an inverse K according to (1.13). A real number λ is an eigenvalue of A if there is a function f in $\mathfrak{L}(\Omega)$ such that $Af = \lambda f$ on Ω. Thus the inverse K exists if $\lambda = 0$ is not an eigenvalue. Also $e = e(x) \equiv 1$ for x in Ω and $(f, g) = \sum_{x\in\Omega} f(x)g(x)$.

Lemma 3.1. *A has one simple positive eigenvalue, its other eigenvalues are negative and the quadratic form* $(e, Ke) = \sum_{x\in\Omega}\sum_{y\in\Omega} K(x, y) \neq 0$.

Proof. Suppose that λ_1 and λ_2 are two distinct positive eigenvalues of A with eigenvectors u_1 and u_2 in $\mathfrak{L}(\Omega)$. Since A is symmetric the functions u_1 and u_2 may be taken real and such that $(u_1, u_1) = (u_2, u_2) = 1$, $(u_1, u_2) = 0$. One may choose real constants α_1, α_2, not both zero, such that $(v, e) = 0$ where $v = \alpha_1 u_1 + \alpha_2 u_2$. Direct computation yields

$$(v, Av) = \alpha_1^2\lambda_1 + \alpha_2^2\lambda_2 > 0. \tag{3.1}$$

On the other hand, for every v such that $(v, e) = 0$ the definition of $A(x, y)$ in (1.11) gives

$$(v, Av) = -(2\pi)^{-2} \int \Big|\sum_{x\in\Omega} v(x)\, e^{ix\cdot\theta}\Big|^2\, |1 - \phi(\theta)|^{-1}\, d\theta \leq 0 \tag{3.2}$$

with strict inequality if and only if $v \equiv 0$ on Ω. Thus (3.2) contradicts (3.1) so that A has at most one positive eigenvalue, and there must be one, since the trace of A is zero. This eigenvalue is simple by the theory of positive matrices of Frobenius (for every $\epsilon > 0$, $A + \epsilon I$ is a strictly positive matrix whose largest eigenvalue is simple).

A is nonsingular, for suppose $Au = 0$, $(u, u) = 1$. Then either $(u, e) = 0$ in which case (3.2) applied to u gives $(u, Au) < 0$ which is impossible, or $(u, e) \neq 0$. In the latter case let $\mu > 0$, $Av = \mu v$, $(v, v) = 1$. By (3.2) $(v, e) \neq 0$. Hence we can choose real $\alpha \neq 0$, $\beta \neq 0$, such that $(\alpha u + \beta v, e) = 0$ and calculate, for $w = \alpha u + \beta v$, that $(w, Aw) = \beta^2 \mu < 0$ which is impossible.

Finally let K be the inverse of A which has just been shown to exist, and take $v = Ke$. Since K has an inverse, v is not the zero vector. If $(v, e) = 0$, then (3.2) implies $(v, Av) < 0$ which is impossible since $(v, Av) = (e, Ke) = (v, e)$. Hence $(v, e) = (e, Ke) \neq 0$.

Unfortunately the eigenvalues and eigenfunctions of A do not seem to have a natural place in the potential theory. An additional fact which does, to be proved in section 5, is that

$$Ke(x) \geqq 0 \quad \text{for} \quad x \in \Omega, \quad \text{and} \quad (e, Ke) > 0. \tag{3.3}$$

Lemma 3.1 is used to derive a representation for the transition matrix $\Pi = \Pi(x, y)$, defined in (1.14), of the imbedded Markov chain. For $f, g \in \mathfrak{L}(\Omega)$ we write $f \otimes g$ for the matrix $f(x)\, g(y)$, with $x, y \in \Omega$.

THEOREM 3.1. *If* $2 \leqq |\Omega| < \infty$,

$$\Pi = I + K - \frac{Ke \otimes Ke}{(e, Ke)}. \tag{3.4}$$

PROOF. Define $\Pi_t = \Pi_t(x, y)$ for $0 \leqq t < 1$, by

(3.5)

$$\Pi_t(x, y) = \sum_{n=1}^{\infty} t^n P\{S_\nu \notin \Omega \quad \text{for} \quad \nu = 1, \cdots, n-1; S_n = y | S_0 = x\},$$

$$x, y \in \Omega.$$

Then $\Pi(x, y) = \lim_{t \to 1^-} \Pi_t(x, y)$ and a simple renewal argument, given in detail in [14], shows that

$$(I - \Pi_t)^{-1}(x, y) = \sum_{n=0}^{\infty} t^n P\{S_n = y | S_0 = x\}, \quad 0 \leqq t < 1; x, y \in \Omega. \tag{3.6}$$

This renewal argument gives $I + \Pi_t + \Pi_t^2 + \cdots$ on the left, but by (3.5) Π_t has all row sums less than one when $0 \leqq t < 1$ so that one obtains (3.6). Let

$$s(t) = (2\pi)^{-2} \int \frac{d\theta}{1 - t\phi(\theta)}, \qquad 0 \leqq t < 1. \tag{3.7}$$

Using lemma 2.1, equation (3.6) may be written for $0 \leqq t < 1$,

$$(I - \Pi_t)^{-1} = s(t) e \otimes e - A + B_t, \tag{3.8}$$

where

$$\begin{aligned} B_t &= B_t(x, y) \\ &= (2\pi)^{-2} \int [1 - \cos (x - y) \cdot \theta] \left[\frac{1}{1 - \phi(\theta)} - \frac{1}{1 - t\phi(\theta)} \right] d\theta. \end{aligned} \tag{3.9}$$

A matrix M of the form $M = f \otimes f + C$, where C is nonsingular and $(f, C^{-1}f) + 1 \neq 0$, has the elegant property [8] that ($\tilde{f}$ being the transpose of the vector f)

$$M^{-1} = C^{-1} - \frac{C^{-1}f\tilde{f}C^{-1}}{1 + \tilde{f}C^{-1}f}. \tag{3.10}$$

Thanks are due to I. Olkin, who pointed out this fact to me. (3.10) is easily verified, and it may be used to invert the matrix $(I - \Pi_t)^{-1}$ in (3.8) if one can verify that, for t close enough to one, the conditions for (3.10) are satisfied. We shall take $\sqrt{s(t)}\, e = f$ and $B_t - A = C$. Since A is nonsingular and since B_t tends to the zero matrix as t tends to one, by (3.9) and lemma 2.1, it is clear that $B_t - A$ is nonsingular for t sufficiently close to one. Also

$$1 + s(t)[e, (B_t - A)^{-1}e] \neq 0 \tag{3.11}$$

for t close enough to one by the above argument and lemma 3.1.

Observing that $B_t - A$ is symmetric, one obtains from (3.8) and (3.10) for some $\delta > 0$

$$I - \Pi_t = (B_t - A)^{-1} - \frac{(B_t - A)^{-1}e \otimes (B_t - A)^{-1}e}{[s(t)]^{-1} + (e, (B_t - A)^{-1}e)} \tag{3.12}$$

for $1 - \delta \leqq t < 1$, and passing to the limit

$$\lim_{t\to 1^-} [I - \Pi_t] = I - \Pi = -A^{-1} + \frac{A^{-1}e \otimes A^{-1}e}{(e, A^{-1}e)} = -K + \frac{Ke \otimes Ke}{(e, Ke)}, \tag{3.13}$$

proving theorem 3.1.

4. Probability interpretation of the potential kernel

Taking x, y, z in E, not necessarily distinct, we let ${}_zE_{xy}$ denote the expected number of visits of the process S_n, with $S_0 = x$, to the point y, before the first visit to z. Formally

$${}_zE_{xy} = \sum_{n=1}^{\infty} P\{S_\nu \neq z, \quad \text{for} \quad \nu = 1, \cdots, n-1;\ S_n = y \mid S_0 = x\}. \tag{4.1}$$

It turns out that

$${}_zE_{xy} = A(z, y) + A(z, x) - A(x, y) + \delta(z, y) + \delta(z, x) - \delta(x, y), \tag{4.2}$$

where $\delta(a, b) = 1$ if $a = b$ and 0 otherwise. In section 7, theorem 7.1, the Green function $g_\Omega(x, y)$ is obtained explicitly and that result will generalize (4.2) as ${}_zE_{xy} = g_\Omega(x, y)$ with Ω consisting of the single point z, when x, y, z are distinct. Equation (4.2) will not be proved there as its proof requires only trivial modifications of that of theorem 2 in [14] which concerns one-dimensional random walk. A simple direct proof may be based on theorem 3.1 for arbitrary sets Ω with $|\Omega| = 2$ and $|\Omega| = 3$ and requires only certain calculations based on the geometric distribution of the number of returns to a point before visiting a different, prescribed, point.

$A(x, y)$ has the probability interpretation given by

THEOREM 4.1. *If* $x \neq y$,

$$ {}_yE_{xx} = {}_xE_{yy} = 2A(x, y) - 1 \tag{4.3}$$

$$ \lim_{z\to\infty} {}_xE_{yz} = \lim_{z\to\infty} {}_yE_{xz} = A(x, y). \tag{4.4}$$

PROOF. (4.3) is just a special case of (4.2) and (4.4) follows from lemma 2.3 applied to (4.2).

5. The harmonic measure

Given a finite subset Ω of E, the harmonic measure of Ω is defined as a function $p(x, a)$ on the product space $E \times \Omega$ by $x \in E$, $a \in \Omega$,

$$ p(x, a) = \delta(x, a) \qquad \text{when} \quad x \in \Omega \tag{5.1}$$

$$ p(x, a) = \sum_{n=1}^{\infty} P\{S_\nu \notin \Omega \quad \text{for} \quad \nu = 1, \cdots, n-1; S_n = a | S_0 = x\} \qquad \text{when} \quad x \in E - \Omega. \tag{5.2}$$

Thus the harmonic measure defines, for each x, a probability measure on Ω which assigns to each point a the probability that Ω is first visited at a, if $S_0 = x$.

This is well known, as is the fact that the *harmonic measure* yields the solution to the *exterior Dirichlet problem*, that is, the boundary value problem

$$ (P - I)f = 0 \quad \text{on } E - \Omega, \tag{5.3'}$$

$$ f = \phi \quad \text{on } \Omega, \tag{5.3''}$$

$$ f \in \mathcal{L} \quad \text{is bounded on } E. \tag{5.3'''}$$

Here $\phi \in \mathcal{L}(\Omega)$ is a known function and (5.3) turns out to have a unique solution $f \in \mathcal{L}$. It is given by

THEOREM 5.1. *The exterior Dirichlet problem has the unique solution*

$$ f(x) = \sum_{a\in\Omega} p(x, a)\phi(a). \tag{5.4}$$

PROOF. The definition of $p(x, a)$ in (5.1) and (5.2) and a simple renewal argument (based on the strong Markov property of the process S_n) imply that (5.4) is indeed a solution of (5.3). To prove uniqueness, suppose f_1 and f_2 are two solutions of (5.3) and let $h = f_1 - f_2$. Then h satisfies (5.3) with $\phi \equiv 0$ on Ω and we have to show that $h \equiv 0$ on E. Define

$$ \begin{aligned} P_\Omega^{(1)}(x, y) &= P(x, y), \\ P_\Omega^{(n+1)}(x, y) &= \sum_{t\in E-\Omega} P(x, t)P_\Omega^{(n)}(t, y) \qquad \text{for} \quad n \geqq 1. \end{aligned} \tag{5.5}$$

As $h \equiv 0$ on Ω, and $Ph = h$ on $E - \Omega$,

$$ h(x) = \sum_{y\in E-\Omega} P_\Omega^{(1)}(x, y)h(y) = \cdots = \sum_{y\in E-\Omega} P_\Omega^{(n)}(x, y)h(y), \tag{5.6}$$

for all $n \geqq 1$ and every x in $E - \Omega$. Choosing M such that $|h(x)| \leqq M$ on E, we have for every $x \in E - \Omega$,

$$(5.7) \qquad |h(x)| \leqq M \sum_{y \in E - \Omega} P_{\Omega}^{(n)}(x, y)$$
$$= M\, P\{S_\nu \in E - \Omega \quad \text{for} \quad \nu = 1, \cdots, n | S_0 = x\}$$

for all $n \geqq 1$. As the process S_n is recurrent the right side tends to zero as n approaches infinity, so that $h(x) \equiv 0$.

An equally short proof comes from the following observation of Doob. If $S_0 = x$ and if f satisfies (5.3) and if S_n' is the process S_n, stopped when it first reaches Ω, then $f(S_n')$ is a bounded martingale [6]. (5.4) is the martingale theorem:

$$(5.8) \qquad f(x) = E[f(S_n')] = E[\lim_{n \to \infty} f(S_n')] = E[\phi(\lim_{n \to \infty} S_n')] = \sum_{a \in \Omega} p(x, a)\phi(a).$$

The next theorem gives an explicit formula for the harmonic measure $p(x, a)$ which will be quite indispensable for the further development of the potential theory. It need be given only for $2 \leqq |\Omega| < \infty$. When $|\Omega| = 1$, $p(x, a) \equiv 1$ for all $x \in E$, and when Ω is the empty set the only bounded (harmonic) functions satisfying $Pf = f$ on E are the constants by Derman's theorem, giving the uniqueness of invariant measure [5].

THEOREM 5.2. *When* $2 \leqq |\Omega| < \infty$,

$$(5.9) \qquad p(x, a) = \frac{Ke(a)}{(e, Ke)} + \sum_{b \in \Omega} A(x, b)[\Pi(b, a) - \delta(b, a)].$$

PROOF. As mentioned in section 1 this result was first derived probabilistically as a consequence of theorem 3.1. (It is not hard to see how $p(x, a)$ is related to the imbedded Markov chain corresponding to the set Ω' obtained by adjoining the point x to the set Ω.) We shall be content to verify that (5.9) is correct. In view of theorem 5.1 it suffices to verify that the right side in (5.9) satisfies (5.3) with $\phi(x) = 1$ when $x = a$ and $\phi(x) = 0$ when $x \in \Omega - \{a\}$. As a is fixed we write $h(\cdot a)$ for the function $h(x, a)$ on the right in (5.9), and $(P - I)h(\cdot a)(x)$ for the image of $h(\cdot a)$ after applying $P - I$ to it. The first term on the right in (5.9) is independent of x so that $P - I$ maps it into zero. Using lemma 2.2 we get

$$(5.10) \qquad (P - I)h(\cdot a)(x) = \sum_{b \in \Omega} [(P - I)A](x, b)[\Pi(b, a) - \delta(b, a)] = 0$$

when $x \notin \Omega$, so that (5.3′) holds. Using theorem 4.1

$$(5.11) \qquad h(x, a) = \frac{Ke(a)}{(e, Ke)} + \sum_{b \in \Omega} A(x, b)\left[K(b, a) - \frac{Ke(b)Ke(a)}{(e, Ke)}\right].$$

The definition of K in (1.13) applied to (5.11) shows that $h(x, a) = \delta(x, a)$ for x in Ω so that (5.3″) holds. To verify that $h(x, a)$ is bounded in x, let $c(b) = \Pi(b, a) - \delta(b, a)$. Note that for each fixed a, $\sum_{b \in \Omega} c(b) = 0$, as Π is a symmetric stochastic matrix. (Even if S_n were unsymmetric but recurrent, with $\phi(\theta)$ complex this conclusion would hold as Π would still be doubly stochastic without being symmetric.) A trivial extension of equation (2.11) in lemma 2.3 implies that

$$\lim_{x \to \infty} \sum_{b \in \Omega} A(x, b)[\Pi(b, a) - \delta(b, a)] = 0 \tag{5.12}$$

so that

$$\lim_{x \to \infty} h(x, a) = \frac{Ke(a)}{(e, Ke)} \tag{5.13}$$

which is independent of x, so that (5.3‴) is verified. Summarizing the observations which led to equations (5.10), (5.11) and (5.12) and introducing the notation

$$p_\infty = p_\infty(a) = \frac{Ke(a)}{(e, Ke)} \tag{5.14}$$

for the harmonic measure of Ω "seen from infinity" which will play an important role in the sequel, we can state

THEOREM 5.3. *The unique solution of* 5.4 *is*

$$f = (p_\infty, \phi) + A(\Pi - I)\phi \tag{5.15}$$

$$\lim_{x \to \infty} f(x) = (p_\infty, \phi) \tag{5.16}$$

$$(P - I)f = (\Pi - I)\phi \quad \text{on} \quad \Omega. \tag{5.17}$$

Here (5.15) follows from (5.4) and (5.9), (5.16) from (5.12), and (5.17) from (5.10) with x in Ω.

A final observation is the

COROLLARY. *Equation* (3.3) *holds for* $2 \leqq |\Omega| < \infty$.

PROOF. The probability interpretation of $p(x, a)$ yields that $p_\infty(a) \geqq 0$. To get (3.3) we have to rule out the possibility that $Ke(a) \leqq 0$ and $(e, Ke) < 0$. That is impossible since $\sum_{a \in \Omega} A(b, a)Ke(a) = e(b) = 1$ for $b \in \Omega$ and $A(b, a) \geqq 0$. Therefore $Ke(a) \geqq 0$ on Ω and $(e, Ke) > 0$.

6. Logarithmic potential

When $1 \leqq |\Omega| < \infty$ and when $0 \leqq \psi \in \mathcal{L}(\Omega)$, we call the function $f(x) = \sum_{a \in \Omega} A(x, a)\psi(a)]$ the *potential* due to the *charge distribution* ψ. The total charge on Ω is (ψ, e). The potential $f = A\psi$ has the properties (1) $(P - I)f = 0$ on $E - \Omega$, (2) $(P - I)f = \psi$ on Ω, (3) $f(x) - (\psi, e)a(x) \to 0$ as $x \to \infty$. Properties (1) and (2) are immediate consequences of lemma 2.2, and, recalling that $a(x) = A(x, 0)$, property (3) follows from lemma 2.3. Thus $f = A\psi$ is similar (except for a change in sign) to classical *logarithmic potential* [13], if the additional conditions of lemma 2.4 are met, so that

$$a(x) - \frac{1}{\pi\sigma^2} \log |x| \to k \tag{6.1}$$

with k given in lemma 2.4, as x tends to infinity. Note that in any case the potential $f(x) = A\psi(x)$ of a $\psi \geqq 0$ tends to infinity with x, unless $\psi \equiv 0$ on Ω. That follows from (2.10) in lemma 2.3.

It is also true that properties (1), (2), (3) characterize our potentials:

THEOREM 6.1. *If* $0 \leqq \psi \in \mathcal{L}(\Omega)$ *then* $f = A\psi$ *satisfies* (1), (2), (3). *Conversely if* f *satisfies* (1), (2), (3) *with* $0 \leqq \psi \in \mathcal{L}(\Omega)$, *then* $f = A\psi$.

PROOF. Having already proved the first part, we assume that a function $g \in \mathcal{L}$, as well as $f = A\psi$ satisfies (1), (2), (3) with $0 \leqq \psi \in \mathcal{L}(\Omega)$. Letting $h = f - g$, (1) and (2) imply that $(P - I)h \equiv 0$ on E and (3) implies that h is bounded, in fact that $\lim_{x\to\infty} h(x) = 0$. From the uniqueness theorem for invariant measure [5] it follows that $h \equiv 0$. That completes the proof.

Much more than theorem 6.1 could be desired. The second (converse) part could probably be altered by assuming that $f \geqq 0$ and that f satisfies (1) and (2) with $0 \leqq \psi \in \mathcal{L}(\Omega)$. One would then expect the conclusion that $f =$ constant $+ A\psi$.

Our methods break down, as well as the simple argument in section 3 of [11] where it is shown for simple random walk in three and higher dimension that $f \geqq 0$ and $(P - I)f \leqq 0$ on E implies that f is a constant plus a (Newtonian) potential. *Note added in proof.* It will be shown elsewhere that the above conjecture is correct for the random walks considered here. For one-dimensional random walks the situation turns out to be more complicated.

In the next section we shall require the quite obvious extension of theorem 6.1

THEOREM 6.2. *If* $(P - I)u = 0$ *on* $E - \Omega$, $(P - I)u = \psi \geqq 0$ *on* Ω, $(\psi, e) = 1$, $|u(x) - a(x)| \leqq M$ *on* E *for some* $M > 0$, *then* $u =$ constant $+ A\psi$.

A number of inequalities which imitate maximum principles from classical potential theory will be proved in the next section (theorem 7.3).

7. The Green function

The Green function for the exterior domain $E - \Omega$ is first defined probabilistically as the function $g_\Omega(x, y)$ on $E \times E$

When $y \in \Omega$, $g_\Omega(x, y) = 0$;

When $y \in E - \Omega$ and $x \in \Omega$, $g_\Omega(x, y) = 0$;

When $y \in E - \Omega$, $x \in E - \Omega$, $x \neq y$;

$$g_\Omega(x, y) = \sum_{n=1}^{\infty} P\{S_\nu \in E - \Omega \text{ for } \nu = 1, \cdots, n-1; S_n = y | S_0 = x\}. \tag{7.1}$$

When $x = y \in E - \Omega$

$$g_\Omega(x, y) = 1 + \sum_{n=1}^{\infty} P\{S_\nu \in E - \Omega \text{ for } \nu = 1, \cdots, n-1; S_n = x | S_0 = x\}. \tag{7.1'}$$

When $x \neq y$ and both x and y are outside Ω, then $g_\Omega(x, y)$ is the expected number of visits to y of the process S_n with $S_0 = x$, before the first visit to Ω. In spite of the unsymmetric definition we shall show that $g_\Omega(x, y) = g_\Omega(y, x)$ in the course of finding an explicit formula for the Green function. This could be done probabilistically, using theorem 3.1 and the method of the imbedded Markov chain, but we shall appeal instead to the potential theory developed in section 6.

First we need a definition.

$$f(x) = \sum_{a\in\Omega} A(x, a)p_\infty(a) \tag{7.2}$$

will denote the potential due to the (nonnegative unit charge) distribution p_∞ on Ω. In terms of f we shall prove

THEOREM 7.1. *For* $2 \leqq |\Omega| < \infty$ *and* $x, y \in E$

$$g_\Omega(x, y) = -A(x, y) + f(x) + f(y) - \frac{1}{(e, Ke)} + \sum_{a\in\Omega}\sum_{b\in\Omega} A(x, a)[\Pi(a, b) - \delta(a, b)]A(b, y). \tag{7.3}$$

When $|\Omega| = 1$, theorem 7.1 is meaningless since K is undefined, but in this case (suppose Ω consists of the single point $\{a\}$) equation (4.2) and the definition (7.1) gives

$$g_\Omega(x, y) = -A(x, y) + A(x, a) + A(a, y). \tag{7.4}$$

This agrees with (7.3) if we define $(e, Ke)^{-1} = 0$ when $|\Omega| = 1$.

Before proceeding to the proof of theorem 7.1 we need an analogue of the classical theorem [13] that the harmonic measure of Ω is the normal derivative of the Green function at the boundary of Ω. Here the normal derivative of a function $u(t)$ evaluated at a in Ω becomes $\sum_{t\in E-\Omega} P(a, t)u(t) - u(a)$ and applied to the Green function, which vanishes on Ω, this is

THEOREM 7.2. *For* $a \in \Omega$, $x \in E - \Omega$,

$$p(x, a) = \sum_{t\in E} P(a, t)g_\Omega(t, x). \tag{7.5}$$

One can also obtain Π from g_Ω, since for $a, b \in \Omega$,

$$\Pi(a, b) = \sum_{s\in E} P(a, s)p(s, b) = \sum_{s\in E}\sum_{t\in E} P(a, t)g_\Omega(t, s)P(s, b). \tag{7.6}$$

PROOF of (7.5). In view of (7.1) the right side in (7.5) is

$$\sum_{n=1}^{\infty} P\{S_\nu = E - \Omega \quad \text{for} \quad \nu = 1, \cdots, n-1; S_n = x | S_0 = a\}. \tag{7.7}$$

The symmetry of the process S_n, $\phi(\theta) = \phi(-\theta)$, enables one to set up a probability preserving one to one correspondence between paths going from a to x in time n without hitting Ω and paths going from x to a in time n without hitting Ω. Therefore the above sum is

$$\sum_{n=1}^{\infty} P\{S_\nu \in E - \Omega \quad \text{for} \quad \nu = 1, \cdots, n-1; S_n = a | S_0 = x\} \tag{7.8}$$

and this is the harmonic measure $p(x, a)$ according to (5.2).

PROOF of (7.3). Take a fixed y in $E - \Omega$. A simple renewal argument (based on the strong Markov property of S_n) and the definition (7.1) gives

$$(P - I)g_\Omega(\cdot y)(x) = \begin{cases} 0, & x \neq y, x \in E - \Omega, \\ -1, & x = y \in E - \Omega. \end{cases} \tag{7.9}$$

Let $u(x) = u_y(x) = g_\Omega(x, y) + A(x, y)$. By lemma 2.2 and (7.9)

$$(P - I)u = 0 \qquad \text{on} \quad E - \Omega, \tag{7.10}$$

and by theorem 7.2 and lemma 2.2

$$(P - I)u(x) = \sum_{t \in E} P(x, t) g_\Omega(t, y) = p(y, x), \qquad x \in \Omega. \tag{7.11}$$

We wish to solve the system consisting of (7.10) and (7.11) to find an explicit formula for $u = u_y$. To do so one must first check that u satisfies the conditions in theorem 6.2. The first two conditions in theorem 6.2 are equivalent to (7.10) and (7.11) with $\psi(x) = p(y, x) \geqq 0$ for $x \in \Omega$. Clearly $(\psi, e) = 1$. For each fixed y, $A(x, y) - a(x)$ is bounded, as it tends to a finite limit when x tends to infinity by lemma 2.2. Thus the result of theorem 6.2 is applicable if we show that $g_\Omega(x, y)$ is bounded in x for each fixed y. To do so let Ω' be a subset of Ω consisting of a single point. Then the probability interpretation of the definition (7.1) of the Green function implies that $0 \leqq g_\Omega(x, y) \leqq g_{\Omega'}(x, y)$, and either equation (4.2) or theorem 4.1 proves that $g_{\Omega'}(x, y)$ is bounded in x for each fixed y.

Applying the result of theorem 6.2 we have $u = \text{constant} + A\psi$, where $\psi(x) = p(y, x)$ for $x \in \Omega$ and we let $h(y)$ be the constant, which may of course depend on y. Then theorem 5.2 together with (7.2) yields

$$\begin{aligned} u(x) &= u_y(x) = g_\Omega(x, y) + A(x, y) \\ &= h(y) + f(x) + \sum_{a \in \Omega} \sum_{b \in \Omega} A(x, a)[\Pi(a, b) - \delta(a, b)] A(b, y), \\ &\qquad x \in E,\ y \in E - \Omega. \end{aligned} \tag{7.12}$$

Setting $x = c \in \Omega$ in (7.12) gives for $y \in E - \Omega$,

$$\begin{aligned} A(c, y) &= h(y) + \frac{1}{(e, Ke)} + \sum_{a \in \Omega} \sum_{b \in \Omega} A(c, a)[K(a, b) - (e, Ke) p_\infty(a) p_\infty(b)] A(b, y) \\ &= h(y) + \frac{1}{(e, Ke)} + A(c, y) - f(y), \end{aligned} \tag{7.13}$$

or

$$h(y) = f(y) - \frac{1}{(e, Ke)}. \tag{7.14}$$

Equations (7.12) and (7.14) together complete the proof of theorem 7.1 for the case when $y \in E - \Omega$. For the case when $y \in \Omega$ the proof is completed by straightforward calculation which shows that the right side in (7.3) vanishes.

The function $h(x)$ defined by equation (7.14) is important as it follows from theorem 7.1 and lemma 2.3 that $g_\Omega(x, y)$ has a limit as y tends to infinity, and this limit is $h(x)$,

$$\begin{aligned} \lim_{y \to \infty} g_\Omega(x, y) &= g_\Omega(x, \infty) = g_\Omega(\infty, x) = h(x) = f(x) - \frac{1}{(e, Ke)}, \\ f(x) &= A p_\infty(x) = \sum_{a \in \Omega} A(x, a) p_\infty(a). \end{aligned} \tag{7.15}$$

We use the resulting inequality

$$h(x) = f(x) - \frac{1}{(e, Ke)} \geqq 0, \qquad x \in E, \tag{7.16}$$

to establish a maximum (or rather minimum) principle for potentials.

THEOREM 7.3. *For ψ_1 and ψ_2 in $\mathcal{L}(\Omega)$, $A\psi_1 \geqq A\psi_2$ on Ω implies that $A\psi_1 \geqq A\psi_2$ on E, and that $(\psi_1, e) \geqq (\psi_2, e)$. If $0 \leqq \psi \in \mathcal{L}(\Omega)$, then*

$$A\psi(x) \geqq \min_{a \in \Omega} A\psi(a) \qquad \text{for all } x \text{ in } E. \tag{7.17}$$

PROOF. For $a, t \in \Omega$, theorem 3.1 gives

$$K(t, a) = \frac{Ke(t)Ke(a)}{(e, Ke)} + \Pi(t, a) - \delta(t, a). \tag{7.18}$$

For $x \in E - \Omega$ and $a \in \Omega$, therefore, using theorem 5.2,

$$\begin{aligned} \sum_{t \in \Omega} A(x, t)K(t, a) &= f(x)Ke(a) + p(x, a) - \frac{Ke(a)}{(e, Ke)} \\ &= \left[f(x) - \frac{1}{(e, Ke)}\right] Ke(a) + p(x, a). \end{aligned} \tag{7.19}$$

In view of (7.16), and (1.13)

$$\sum_{t \in \Omega} A(x, t)K(t, a) \geqq p(x, a) \geqq 0 \quad \text{when} \quad x \in E, a \in \Omega. \tag{7.20}$$

Hence, if $0 \leqq g \in \mathcal{L}(\Omega)$,

$$\begin{aligned} AK\, g(x) \\ = \sum_{t \in \Omega} \sum_{a \in \Omega} A(x, t)K(t, a)g(a) \geqq \sum_{a \in \Omega} p(x, a)g(a) \geqq 0, \qquad x \in E. \end{aligned} \tag{7.21}$$

If $A\psi_1 \geqq A\psi_2$ on Ω, let $g = A(\psi_1 - \psi_2)$ restricted to Ω, giving, for all $x \in E$,

$$A\psi_1(x) - A\psi_2(x) \geqq \sum_{a \in \Omega} \sum_{b \in \Omega} p(x, a)A(a, b)[\psi_1(b) - \psi_2(b)] \geqq 0. \tag{7.22}$$

Since $g \geqq 0$ and $Ke \geqq 0$ on Ω we also have $(g, Ke) \geqq 0$, but $(g, Ke) = (A(\psi_1 - \psi_2), Ke) = (\psi_1 - \psi_2, e) = (\psi_1, e) - (\psi_2, e) \geqq 0$. This proves the first part of theorem 7.3. For the second part let $\psi_2(a) = cp_\infty(a)$ for $a \in \Omega$ where $c > 0$ is chosen so that

$$\min_{a \in \Omega} A\psi(a) = A\psi_2(t) = \frac{c}{(e, Ke)} \tag{7.23}$$

for all t in Ω. (It follows from the definition of p_∞ that its potential $f(x) = Ap_\infty(x)$ is constant on Ω.) Since $A\psi_2$ is constant on Ω we have $A\psi \geqq A\psi_2$ on Ω and by the first part of the theorem $A\psi \geqq A\psi_2$ on E. But

$$A\psi_2(x) = cf(x) \geqq \frac{c}{(e, Ke)} = \min_{a \in \Omega} A\psi(a) \tag{7.24}$$

because of (7.16) and (7.23).

8. Capacity

For a discussion of *capacity* in classical logarithmic potential theory see Nevanlinna [13] and for the modern theory of capacity (of the Newtonian type) Choquet [2] and Choquet and Deny [3]. To preserve the analogy of the present to the classical theory, the capacity of a finite set Ω should be defined as $\exp[(e, Ke)^{-1}]$ where K is the (matrix) operator given by (1.13). The quantity $-(e, Ke)^{-1}$ then is our analogue of Robin's constant. But for the sake of convenience we call $(e, Ke)^{-1}$ the capacity of Ω.

DEFINITION. *If $|\Omega| = 1$ the capacity of Ω is $C(\Omega) = 0$. If $2 \leqq |\Omega| < \infty$ the capacity of Ω is*

$$C(\Omega) = \frac{1}{(e, Ke)}. \tag{8.1}$$

The first theorem has the nature of a maximum principle. If $0 \leqq \psi \in \mathcal{L}(\Omega)$ and $(\psi, e) = 1$ then the potential $A\psi$ of ψ is greater than or equal to $C(\Omega)$ somewhere on Ω and less than or equal to $C(\Omega)$ somewhere on Ω. But there is a unique ψ (namely p_∞) such that $0 \leqq \psi \in \mathcal{L}(\Omega)$, $(\psi, e) = 1$, and such that $A\psi$ is constant on Ω. This constant is $C(\Omega)$ and p_∞ is called the *equilibrium unit charge.* A finite set Ω has the same capacity as its boundary,

$$\partial\Omega = [x | x \in \Omega, p_\infty(x) > 0]. \tag{8.2}$$

THEOREM 8.1. *If $0 \leqq \psi \in \mathcal{L}(\Omega)$ and $(\psi, e) = 1$, then*

$$\min_{a \in \Omega} A\psi(a) \leqq C(\Omega) \leqq \max_{a \in \Omega} A\psi(a) \tag{8.3}$$

$$C(\Omega) = Ap_\infty(x) = f(x) \qquad \textit{for} \quad x \in \Omega \tag{8.4}$$

$$C(\Omega) = C(\partial\Omega). \tag{8.5}$$

PROOF. When $|\Omega| = 1$ theorem 8.1 is obvious, therefore let $2 \leqq |\Omega| < \infty$. Suppose (8.3) fails, in such a way that the second inequality is violated. Then

$$A\psi < \frac{e}{(e, Ke)} \quad \text{on} \quad \Omega. \tag{8.6}$$

This inequality is preserved under the inner product with Ke, so that

$$(Ke, A\psi) = (e, KA\psi) = (e, \psi) > \frac{(Ke, e)}{(e, Ke)} = 1. \tag{8.7}$$

This contradicts the hypothesis $(e, \psi) = 1$. If the first inequality fails the same argument works.

Equation (8.4) follows immediately from (5.14), (7.2) and (8.1). Equation (8.5) is true because p_∞, restricted to $\partial\Omega$ is a unit charge distribution on $\partial\Omega$; the potential due to p_∞ is constant on $\partial\Omega$ and has the value

$$\sum_{a \in \partial\Omega} A(x, c) \frac{Ke(c)}{(e, Ke)} = \frac{1}{(e, Ke)} = C(\Omega), \qquad x \in \Omega, \tag{8.8}$$

where K is the inverse of A restricted to Ω, but by (8.3) $C(\Omega)$ is the capacity of $\partial\Omega$.

Equation (8.3) may be expressed in the language of *game theory*. Let u, v range over $\mathcal{L}(\Omega)$ with $(u, e) = (v, e) = 1, u \geqq 0, v \geqq 0$, and let $(u, Av) = \sum_{a\in\Omega}\sum_{b\in\Omega} u(a)A(a, b)v(b)$. Then (8.3) becomes

$$\max_v \min_u (u, Av) = C(\Omega) = \min_u \max_v (u, Av). \tag{8.9}$$

Thus $C(\Omega)$ is the *value* of the game whose matrix is A.

As in classical potential theory the capacity of Ω may also be defined through the asymptotic behavior of the Green function.

THEOREM 8.2. *Let* $g_\Omega(x, \infty) = \lim_{y\to\infty} g_\Omega(x, y)$, $a(x) = A(x, 0)$. *Then*

$$C(\Omega) = \lim_{x\to\infty} [a(x) - g_\Omega(x, \infty)]. \tag{8.10}$$

PROOF. When $|\Omega| = 1$, $\Omega = \{a\}$, equation (7.5) and lemma (2.2) give $g_\Omega(x, \infty) = A(x, a)$. Lemma (2.3) gives $C(\Omega) = 0$ as we defined it. When $2 \leqq |\Omega| < \infty$, (7.15) and (8.1) give $g_\Omega(x, \infty) = f(x) - C(\Omega)$, $a(x) - g_\Omega(x, \infty) = C(\Omega) + a(x) - f(x)$ and by lemma (2.3) and (7.2)

$$\lim_{x\to\infty} [a(x) - f(x)] = 0. \tag{8.11}$$

This characterization of capacity furnishes simple proofs of its behavior under set theoretical operations. Typical results are

THEOREM 8.3. *Let F, G be two finite subsets of E, and $C(F \cap G) = -\infty$ if $F \cap G$ is the empty set. Then*

$$C(F \cap G) + C(F \cup G) \leqq C(F) + C(G). \tag{8.12}$$

If $F \subset G \subset E$,

$$C(F) \leqq C(G). \tag{8.13}$$

If $\{x\}$ is the set consisting of the single point $x \in E - \Omega$, $|\Omega| < \infty$,

$$C(\Omega) \leqq C(\Omega \cup \{x\}) \leqq C(\Omega) + \frac{1}{2} \min_{y\in\Omega} A(x, y). \tag{8.14}$$

PROOF. If F and G have nonempty intersection, the definition (7.1) of $g_\Omega(x, y)$ implies that

$$g_{F\cap G}(x, y) - g_G(x, y) \geqq g_F(x, y) - g_{F\cup G}(x, y), \tag{8.15}$$

for all x, y in E. Letting first y tend to infinity, and then x, one obtains in view of theorem 8.2

$$\begin{aligned} &\lim_{x\to\infty} [g_{F\cap G}(x, \infty) - g_G(x, \infty)] \\ &\qquad = C(G) - C(F \cap G) \geqq \lim_{x\to\infty} [g_F(x, \infty) - g_{F\cup G}(x, \infty)] \\ &\qquad = C(F \cup G) - C(F). \end{aligned} \tag{8.16}$$

The proof of (8.13) is based on the simpler inequality $g_F(x, y) \geqq g_G(x, y)$ when $F \subset G$. The first inequality in (8.14) follows from (8.13). The second one comes from (8.12) with $F = \Omega$ and $G = \{x\} \cup \{y\}$ where $y \in \Omega$, $x \in E - \Omega$. Equation (8.12) yields

$$C(\{y\}) + C(\Omega \cup \{x\}) = C(\Omega \cup \{x\}) \leqq C(\Omega) + C(\{x\} \cup \{y\}). \tag{8.17}$$

For every pair of distinct points x, y in E, (1.13) and (8.1) give $C(\{x\} \cup \{y\}) = A(x, y)/2$. Since (8.17) holds for all $y \in \Omega$, this proves (8.14).

Still another definition of capacity follows from theorem 7.1 and equation (4.2). Let Ω be an arbitrary finite subset of E, and a an arbitrary fixed point in E. Then

$$\lim_{x \to \infty} [g_{\{a\}}(x, x) - g_{\Omega}(x, x)] = C(\Omega). \tag{8.18}$$

This formula has a comparatively pleasant probability interpretation, as $g_{\Omega}(x, x)$ (or $g_{\{a\}}(x, x)$) is the expected number of returns to x before the first visit to the set Ω (or the set $\{a\}$). Thus $C(\Omega)$ is a measure of the difference in "size" of the sets Ω and $\{a\}$, seen from infinity.

REFERENCES

[1] S. BOCHNER and K. CHANDRASEKHARAN, *Fourier Transforms*, Princeton, Princeton University Press, 1949.

[2] G. CHOQUET, "Theory of capacities," *Ann. Inst. Fourier, Grenoble*, Vol. 5 (1953), pp. 131–295.

[3] G. CHOQUET and J. DENY, "Modèles finis en théorie du potentiel," *J. Analyse Math.*, Vol. 5 (1956), pp. 77–135.

[4] K. L. CHUNG and W. FUCHS, "On the distribution of values of sums of random variables," *Mem. Amer. Math. Soc.*, No. 6 (1951), pp. 1–12.

[5] C. DERMAN, "A solution to a set of fundamental equations in Markov chains," *Proc. Amer. Math. Soc.*, Vol. 5 (1954), pp. 332–334.

[6] J. L. DOOB, *Stochastic Processes*, New York, Wiley, 1953.

[7] ———, "Discrete potential theory and boundaries," *J. Math. Mech.*, Vol. 8 (1959), pp. 433–458.

[8] A. S. HOUSEHOLDER, *Principles of Numerical Analysis*, New York, McGraw-Hill, 1953.

[9] G. A. HUNT, "Markoff processes and potentials," Parts I, II, and III, *Illinois J. Math.*, Vol. 1 (1957), pp. 44–93 and pp. 316–369, Vol. 2 (1958), pp. 151–213.

[10] ———, "Markoff chains and Martin boundaries," *Illinois J. Math.*, Vol. 4 (1960), pp. 313–340.

[11] K. ITÔ and H. J. MCKEAN, JR., "Potentials and the random walk," *Illinois J. Math.*, Vol. 4 (1960), pp. 119–132.

[12] M. KAC, "A class of limit theorems," *Trans. Amer. Math. Soc.*, Vol. 84 (1957), pp. 459–471.

[13] R. H. NEVANLINNA, *Eindeutige Analytische Funktionen*, Berlin, Springer, 1953 (2nd ed.).

[14] F. SPITZER, "Some properties of recurrent random walk," *Illinois J. Math.*, Vol. 5 (1961), pp. 234–245.

[15] A. STÖHR, "Über einige lineare partielle Differenzengleichungen mit konstant en Koeffi-cienten," Parts I, II, and III, *Math. Nachr.*, Vol. 3 (1949–50), pp. 208–242, pp. 295–315 and pp. 330–357.

Z. Wahrscheinlichkeitstheorie 3, 110—121 (1964)

Electrostatic Capacity, Heat Flow, and Brownian Motion

By

FRANK SPITZER

1. Introduction

We shall point out how the electrostatic capacity of a compact set in Euclidean space governs the rate of heat flow from such a set to a surrounding conducting medium. The answers will of course be quite different in dimensions two and three. In both cases, however, we have to rely heavily on the careful probabilistic formulation of classical diffusion theory by DOOB [2] and HUNT [4]. A systematic account of these matters will also be found in the book of ITÔ and McKEAN [5].

The standard Brownian motion process in dimension $s = 2$ or 3 will be denoted $x(t)$ ($= x(t, \omega)$ for ω in the appropriate probability space as defined in [5], Ch. 1 and 7). Measurable sets (events) in this space have measure $P_x[\cdot]$ when $x(0) = x$, and measurable functions (random variables) have expectation $E_x[\cdot]$. We shall always work with *compact* subsets A of Euclidean space R of dimension $s = 2$ or 3. When $s = 3$ it will be assumed that the *Newtonian capacity* $C(A)$ of A is positive, and when $s = 2$ that the *logarithmic capacity* of A is positive. In the latter case, however, we shall be concerned with the set function $K(A) = -\gamma(A)$, where $\gamma(A) =$ *Robin's constant*. Thus $K(A)$ is the natural logarithm of the logarithmic capacity of A (see [8], Ch. V).

For such a set A we define

$$T_A = \inf[t \mid t > 0,\ x(t) \in A] \leqq \infty .$$

A point $x \in A$ is called *regular* if

$$P_x[T_A = 0] = 1 .$$

It is known [4] that under our assumptions concerning A

$$f_A(x, t) = P_x[T_A \leqq t] > 0 \qquad \text{for all } t > 0$$

and that $f_A(x, t)$ is the unique solution of the heat conduction problem

$$\frac{\partial f}{\partial t} = \frac{1}{2} \Delta f \qquad \text{for } t > 0,\ x \in R - A$$

subject to the initial condition

$$f(x, 0) = 0 \qquad \text{for } x \in R - A$$

and boundary condition

$$\lim_{x \to y} f(x, t) = 1 \quad \text{for} \quad t > 0, y = \text{regular point} \in A .$$

Thus $f_A(x,t)$ may be interpreted as the temperature at time t at the point $x \in R - A$, and we proceed to investigate the integral of this temperature over $R - A$, i. e.

the *total energy flow in time t from the set A into the surrounding medium R − A*. Let us call it

(1.1) $$E_A(t) = \int_{R-A} P_x[T_A \leqq t]\,dx = \int_{R-A} f_A(x,t)\,dx\,.$$

The asymptotic behavior of $E_A(t)$ will be shown to depend on the capacity of the set A. In dimension $s = 3$ the capacity quite naturally determines the principal term, but when $s = 2$ it enters only the second term of the asymptotic expansion of $E_A(t)$.

Theorem 1. *In dimension three*

$$E_A(t) = t\,C(A) + 4(2\pi)^{-3/2}[C(A)]^2 t^{1/2} + o(t^{1/2})\,,$$

as $t \to \infty$. *When A is a sphere,* $o(t^{1/2}) \equiv 0$, $t > 0$*.

Theorem 2. *In dimension two, let A and B be two compact sets of positive (logarithmic) capacity. Then*

$$E_A(t) - E_B(t) \sim \frac{2\pi t}{(\ln t)^2}[K(A) - K(B)]\,, \quad as \quad t \to \infty^{**}.$$

Alternatively (specializing to the case when B is the unit disc)

$$E_A(t) = \frac{2\pi t}{\ln t} + \frac{2\pi t}{(\ln t)^2}[K(A) + 1 + \gamma - \ln 2] + o(t \ln^{-2} t)$$

as $t \to \infty$. *Here* $\gamma = .5772\ldots$ *is Euler's constant.*

Remark: There are many equivalent definitions of capacity and the normalization is rather arbitrary. For the above theorems to hold the proper normalization is that which gives Newtonian capacity $2\pi\varrho$ to a sphere of radius ϱ and logarithmic capacity ϱ^2 to a disc of radius ϱ (i. e. if $A = [z|\ |z| \leqq \varrho]$, then $K(A) = 2\ln\varrho$).

2. Proof of theorem 1

For arbitrary Borel sets $B \subset R$

$$P_x[x(t) \in B] = \int_B p(x,y,t)\,dy\,, \qquad x \in R\,,$$

where p is the transition function

$$p(x,y,t) = (2\pi t)^{-s/2} e^{-|x-y|^2/2t}\,, \qquad x, y \in R\,, t > 0\,.$$

Furthermore

(2.1) $$P_x[T_A > t, x(t) \in B] = \int_B q_A(x,y,t)\,dy\,, \qquad x \in R, t > 0\,,$$

* M. Kac has investigated the asymptotic expansion of the Laplace transform of $E_A(t)$ (private communication). His expansion is valid under slight regularity conditions concering the set A (which must have positive volume $|A|$). Formal term by term inversion yields the same terms of order t and $t^{1/2}$ as in Theorem 1. The rest, which we denoted $o(t^{1/2})$ becomes

$$o(t)^{1/2} = \frac{1}{2\pi^2}[C(A)]^3 - |A| - \frac{1}{2\pi}\int_A\int_A e(dx)\,|x-y|\,e(dy) + o(1)\,.$$

Here $e(dx)$ is the equilibrium charge of the set A, to be defined below, in the proof of Theorem 1.

** $f(t) \sim g(t)$ means $f(t)/g(t) \to 1$ as $t \to \infty$.

defines the density $q_A(x, y, t)$ which exists according to HUNT ([4] section 4) and is a *symmetric* function of x and y. Note that the symmetry of q_A which is not the trivial matter it might seem at first glance, is the key to the entire proof.

To study the asymptotic behavior of $E_A(t)$ we use (1.1) to write, for $0 < h < t$,

$$E_A(t) - E_A(t-h) = \int_{R-A} \{P_x[T_A > t-h] - P_x[T_A > t]\}\, dx\,. \tag{2.2}$$

Using the strong Markov property of Brownian motion (see [4] section 2; it will be taken for granted from now on) the integrand on the right in (2.2) becomes

$$\int_R \int_R q_A(x, u, t-h)\, p(u, y, h)\, du\, dy - \int_R q_A(x, y, t)\, dy\,. \tag{2.3}$$

The integrations are over R, since $q_A(x, y, t) = 0$ when $x \in A$ or $y \in A$. The symmetry of each of the densities p and q_A in the space variables permits the transformation of (2.3) into

$$\int_R \int_R p(y, u, h)\, q_A(u, x, t-h)\, du\, dy - \int_R q_A(y, x, t)\, dy\,. \tag{2.4}$$

Now we substitute (2.4) into (2.2) and carry out the integration of x over $R - A$. Thus

$$\begin{aligned} & E_A(t) - E_A(t-h) \\ &= \int_R \int_R p(y, u, h)\, P_u[T_A > t-h]\, du\, dy - \int_R P_y[T_A > t]\, dy \\ &= \int_R \{P_y[x(\tau) \in R - A \quad \text{for} \quad h < \tau \leqq t] - P_y[T_A > t]\}\, dy \\ &= \int_R P_y[T_A \leqq h; x(\tau) \in R - A \quad \text{for} \quad h < \tau \leqq t]\, dy\,. \end{aligned} \tag{2.5}$$

Equation (2.5) shows that $E_A(t) - E_A(t-h)$ is a monotone function of t, so that

$$\lim_{t\to\infty} [E_A(t) - E_A(t-h)] = \int_R P_y[T_A \leqq h; x(\tau) \in R - A \quad \text{for} \quad \tau > h]\, dy\,. \tag{2.6}$$

This limit is an additive function of h. It is obviously bounded and measurable so that it is a linear function of h. Hence we may write

$$C'(A) = \frac{1}{h} \int_R P_y[T_A \leqq h, x(\tau) \in R - A \quad \text{for} \quad \tau > h]\, dy, \quad h > 0\,. \tag{2.7}$$

In dimension $s \leqq 2$ clearly $C'(A) = 0$. Theorem 1 however is concerned with $s = 3$, and in this case we have to identify the set function $C'(A)$ with the Newtonian capacity $C(A)$. That is done by ITÔ and MCKEAN ([5] section 7.7). They define the family of measures (charges)

$$e_h(dy) = \frac{1}{h} P_y[T_A \leqq h; x(\tau) \in R - A \quad \text{for} \quad \tau > h]\, dy\,,$$

and prove by a short calculation that the corresponding potentials

$$p_h(x) = \frac{1}{2\pi} \int_R |x - y|^{-1}\, e_h(dy)$$

increase to $P_x[T_A < \infty]$ as $h \searrow 0$. They further show that a subsequence of these charges therefore converges (weakly) to an equilibrium charge $e(dy)$, supported by A. The capacitory potential corresponding to $e(dy)$ is $P_x[T_A < \infty] \leqq 1$, which, as pointed out in section 1, has boundary value one at the regular points of A. Thus the total charge

$$\int_R e_h(dy) = \int_R e(dy) = \int_A e(dy) = C'(A)$$

is the usual capacity $C(A)$ associated with the Green function $(2\pi)^{-1}|x-y|^{-1}$. If A is a sphere of radius ϱ, then

$$C(A) = \sup \int_A e(dy)$$

over all charges e such that

$$\frac{1}{2\pi}\int_A \frac{e(dy)}{|x-y|} \leqq 1, \; x \in R,$$

which gives $C(A) = 2\pi\varrho$. Equations (2.6) and (2.7) therefore yield the principal term in Theorem 1.

To obtain the second term in Theorem 1 let $|A|$ denote the volume of A. In view of equation (2.7)

$$\begin{aligned} &E_A(t) - tC(A) \\ &= \int_{R-A} P_x[T_A \leqq t]\,dx - \int_R P_x[T_A \leqq t; \quad x(\tau) \in R-A \quad \text{for} \quad \tau > t]\,dx \\ &= \int_R P_x[T_A \leqq t; \quad x(\tau) \in A \quad \text{for some} \quad \tau > t]\,dx - |A| \\ &= \int_R \int_R [p(x,y,t) - q_A(x,y,t)]\,P_y[T_A < \infty]\,dy\,dx - |A|\,. \end{aligned}$$

Using the symmetry of p and q_A,

$$\begin{aligned} E_A(t) &- tC(A) \\ &= \int_R \{\int_R [p(y,x,t) - q_A(y,x,t)]\,dx\}\,P_y[T_A < \infty]\,dy - |A| \\ &= \int_R P_y[T_A \leqq t]\,P_y[T_A < \infty]\,dy - |A| \\ &= \int_{R-A} P_y[T_A \leqq t]\,P_y[T_A < \infty]\,dy\,. \end{aligned} \tag{2.8}$$

By equation (2.8)

$$\begin{aligned} D_A(t,h) &= E_A(t) - E_A(t-h) - hC(A) \\ &= \int_{R-A} P_x[t-h < T_A \leqq t]\,P_x[T_A < \infty]\,dx \\ &= \int_{x \in R-A} \int_{y \in R} P_x[t-h < T_A \leqq t; x(t) \in dy]\,P_x[T_A < \infty]\,dx\,. \end{aligned}$$

Just as in going from (2.3) to (2.4) we now use the symmetry of $q_A(x, y, t)$ to reverse the direction of time. The result is

$$D_A(t,h) = \int_{x\in R-A} \int_{y\in R} P_y[T_A \leqq h; x(\tau) \in R - A \quad \text{for} \quad h < \tau \leqq t; x(t) \in dx]\, P_x[T_A < \infty]\, dx$$

$$(2.9) \qquad = \int_R P_y[T_A \leqq h; x(\tau) \in R - A \quad \text{for} \quad h < \tau \leqq t; x(\tau) \in A \quad \text{for some} \quad \tau > t]\, dy$$

$$= \int_{y\in R} \int_{z\in R} P_y[T_A \leqq h; x(h) \in dz]\, P_z[t - h < T_A < \infty]\, dy\,.$$

Now we shall apply the following

Lemma. *In dimension three, for* $x \in R - A$,

$$f_A(x,\infty) - f_A(x,t) = P_x[t < T_A < \infty]$$
$$\sim f_A(x,\infty)\, C(A)\, 2(2\pi)^{-3/2} t^{-1/2}, \quad \text{as} \quad t \to \infty\,,$$

uniformly for x in a compact set.

This lemma is due to JOFFE [6] who proved it for sets A with positive volume $|A|$, using the theory of additive functionals of the Wiener process of M. KAC. We shall first complete the proof of Theorem 1, and then sketch a proof of the lemma which does not require Tauberian arguments, and which applies to any compact set A with positive capacity.

The lemma applied to (2.9) gives

$$D_A(t,h) \sim 2\, C(A)\, (2\pi)^{-3/2}\, (t-h)^{-1/2} \cdot$$
$$\cdot \int_{y\in R} \int_{z\in R-A} P_y[T_A \leqq h; x(h) \in dz]\, P_z[T_A = \infty]\, dy$$
$$\sim 2\, C(A)\, (2\pi)^{-3/2}\, t^{-1/2} \int_R P_y[T_A \leqq h; x(\tau) \in R - A \quad \text{for} \quad \tau > h]\, dy\,.$$

Comparison with equation (2.7) shows that

$$D_A(t,h) \sim 2h\, C(A)\, (2\pi)^{-3/2}\, t^{-1/2} \int_R e_h(dy)\,,$$

where the total mass of e_h is $C(A)$. That proves Theorem 1.

To derive the Lemma observe that

$$(2.10) \qquad \lim_{|x|\to\infty} |x|\, f_A(x,\infty) = \lim_{|x|\to\infty} |x|\, P_x[T_A < \infty] = \frac{C(A)}{2\pi}\,,$$

since $f_A(x, \infty)$ is the capacitory potential

$$f_A(x,\infty) = \frac{1}{2\pi} \int_A \frac{e(dy)}{|x-y|}\,,$$

where $e(dy)$ is the weak limit of the measures $e_h(dy)$ in (2.7). Now we write

$$P_x[t < T_A < \infty] = \int_{R-A} q_A(x,y,t)\, P_y[T_A < \infty]\, dy\,,$$

and use (2.10) to conclude

$$P_x[t < T_A < \infty] \sim \frac{C(A)}{2\pi} \int_{R-A} \frac{q_A(x,y,t)}{|y|} dy .$$

Decomposing further

$$P_x[t < T_A < \infty] \sim \frac{C(A)}{2\pi} \int_{R-A} \frac{p(x,y,t)}{|y|} dy$$

$$- \frac{C(A)}{2\pi} \int_{R-A} \frac{P_x[x(t) \in dy; T_A \le t]}{|y|} = \frac{C(A)}{2\pi} [A_x(t) - B_x(t)] .$$

It is easy to check that

$$A_x(t) = \int_{R-A} \frac{p(x,y,t)}{|y|} dy \sim \sqrt{\frac{2}{\pi}} t^{-1/2}, \quad \text{as} \quad t \to \infty ,$$

and therefore the Lemma will be proved if we show that

$$B_x(t) = \int_{R-A} \frac{P_x[x(t) \in dy; T_A \le t]}{|y|} \sim \sqrt{\frac{2}{\pi}} P_x[T_A < \infty] t^{-1/2} \quad \text{as} \quad t \to \infty .$$

That is the only delicate part of the proof. Decompose

$$B_x(t) = \int_{y \in R-A} \int_{z \in R} \int_0^t \frac{1}{|y|} d_\tau \, P_x[T_A \leqq \tau; x(T_A) \in dz] p(z,y,t-\tau) dy .$$

Interchanging the order of integration, and calculating the integral on y (in polar coordinates) gives

$$B_x(t) \sim \int_{z \in R} \int_0^t d_\tau \, P_x[T_A \leqq \tau; x(T_A) \in dz] \left\{ \int_R \frac{p(0,y,t-\tau)}{|y|} dy \right\}$$

$$= \sqrt{\frac{2}{\pi}} \int_0^t \frac{d_\tau P_x[T_A \le \tau]}{\sqrt{t-\tau}} .$$

Finally, decompose

$$B_x(t) \sim \sqrt{\frac{2}{\pi}} \frac{P_x[T_A \le t]}{\sqrt{t}} + C_x(t) ,$$

where $C_x(t)$ is the error term

$$C_x(t) = \sqrt{\frac{2}{\pi}} \int_0^t \left[\frac{1}{\sqrt{t-\tau}} - \frac{1}{\sqrt{t}} \right] d_\tau \, P_x[T_A \leqq \tau] .$$

If we can show that $\sqrt{t} C_x(t) \to 0$ as $t \to \infty$, then

$$B_x(t) \sim \sqrt{\frac{2}{\pi}} P_x(T_A < \infty] t^{-1/2}$$

and the proof is complete. Since the integrand in the formula for $C_x(t)$ is non-negative, that is readily accomplished by probability arguments which give the estimate

$$\int_t^{t+1} d_\tau P_x[T_A \leqq \tau] = P_x[t < T_A \leqq t+1]$$
$$\leqq \int_R p(x, y, t)\, P_y[T_S \leqq 1]\, dy \leqq k t^{-3/2}.$$

Here S is a large sphere containing A and $k > 0$, depending on x and on A but not on t. A look back over the proof shows that uniformity on compact sets is automatic.

It only remains to verify theat the error term $o(t^{-1/2})$ vanishes when the set A is a sphere. That is immediate from an explicit calculation of $f_A(x, t)$, (see [1], p. 247).

Remark: Only for diffusion in dimension $s \geqq 5$ is the limit

$$\lim_{t\to\infty} [E_A(t) - t\, C(A)] = \int_{R-A} (P_y[T_A < \infty])^2\, dy$$

finite for sets of positive capacity.

3. Proof of theorem 2

Here it is essential that the dimension be $s = 2$, and our proof is entirely based on Hunt's investigation of the asymptotic behavior of $P_x[T_A > t]$. Following HUNT ([4] section 5) we define, for any compact $A \subset R$ with positive capacity, the Green function

$$G_A(x, y) = \int_0^\infty q_A(x, y, t)\, dt, \qquad x, y \in R.$$

Here q_A was defined in (2.1). HUNT showed that

$$H_A(x) = \lim_{|y|\to\infty} G_A(x, y) < \infty, \qquad x \in R$$

exists, and that also

$$\lim_{|x|\to\infty} [H_A(x) - \frac{1}{\pi} \ln |x|] = -(2\pi)^{-1} K(A) \tag{3.1}$$

exists and is finite. For the identification of this limit with $-(2\pi)^{-1}K(A)$ we refer to Nevanlinna's characterization of Robin's constant $-K(A)$ in terms of the asymptotic behavior of the Green function with pole at infinity ([8] Ch. 5, section 3). We further record from [2] or [4] that H_A is harmonic on $R - A$, so that for large enough $\varrho > 0$ the mean value theorem yields

$$\int_{|x|=\varrho} [H_A(x) - \frac{1}{\pi} \ln |x|]\, d\zeta(x) = -(2\pi)^{-1} K(A). \tag{3.2}$$

Here ζ denotes the constant measure of mass one on the circumference $|x| = \varrho$, and we shall use the symbol ζ for this purpose exclusively.

We need also some preliminary information concerning $E_A(t)$ when A is a disc. To simplify the notation write

$$E_{[z||z|\leq r]}(t) = E_r(t), \quad T_{[z||z|\leq r]} = T_r.$$

It is simplest to solve the heat conduction problem in section 1 by the method of Laplace transforms to obtain

$$\int_0^\infty P_x[T_r \leq t]\, e^{-\lambda t}\, dt = \lambda^{-1} \frac{K_0(\sqrt{2\lambda}\,|x|)}{K_0(\sqrt{2\lambda}\, r)} \tag{3.3}$$

when $\lambda > 0$, $|x| > r$. Here $K_0(z)$ is the Bessel function of the second kind with imaginary argument, of order 0, and $K_1(z)$ is the corresponding function of order 1. It is well known that

$$z^{-1} \frac{d}{dz}[z^\nu K_\nu(z)] = -z^{\nu-1} K_{\nu-1}(z),$$

which gives

$$\int_0^\infty E_r(t)\, e^{-\lambda t}\, dt = \int_{|x|>r} \int_0^\infty P_x[T_r \leq t]\, e^{-\lambda t}\, dt\, dx$$

$$= 2\pi r \lambda^{-3/2} \frac{K_1(r\sqrt{2\lambda})}{K_0(r\sqrt{2\lambda})}.$$

For small positive z

$$K_0(z) = -\log\frac{z}{2} - \gamma + O(z),$$

$$K_1(z) = z^{-1} + O(1),$$

so that

$$\int_0^\infty E_r(t)\, e^{-\lambda t}\, dt \sim \frac{2\pi}{\lambda^2 \ln(1/\lambda)}, \quad \text{as } \lambda \to 0.$$

As $E_r(t)$ is monotone Karamata's theorem yields the conclusion

$$E_r(t) \sim \frac{2\pi t}{\ln t}, \quad \text{as } t \to \infty. \tag{3.4}$$

Observe that the first order approximation to $E_r(t)$ in (3.4) is independent of r so that we are still far from the goal of obtaining an approximation which does depend on the set A. The crucial step is again due to Hunt who used (3.1), (3.2), and a Tauberian argument similar to that which gave (3.4) to verify the conjecture of Kac that for $x \in R - A$

$$P_x[T_A > t] \sim \frac{2\pi}{\ln t} H_A(x), \quad \text{as } t \to \infty.$$

Hunt's proof of this result ([4], section 6) requires no essential modification to yield

$$\int_{|x|=\varrho} P_x[T_A > t]\, d\zeta(x) \sim \frac{2\pi}{\ln t} \int_{|x|=\varrho} H_A(x)\, d\zeta(x), \quad t \to \infty \tag{3.5}$$

for every $\varrho > 0$, large enough so that A lies in the open disc $|z| < \varrho$.

The rest of our proof depends on the observation that a Brownian motion starting at a point outside a large disc has to hit the circumference before visiting a set A in the interior. Given A we choose r so large that $A \subset [z \,|\, |z| < r]$ and write

$$E_r(t) - E_A(t) = \int_{|x|>r} \{P_x[T_r \leqq t] - P_x[T_A \leqq t]\}\, dx$$

$$- \int_{[x|x \in R-A, |x| \leqq r]} P_x[T_A \leqq t]\, dx = \int_{|x|>r} P_x[T_r \leqq t < T_A]\, dx + O(1),$$

$O(1)$ being bounded as $t \to \infty$. Using the strong Markov property

$$\begin{aligned} E_r(t) - E_A(t) \\ = \int_{|x|>r} \Big\{ \int_{|y|=r} \int_0^t d_\tau P_x[T_r \leqq \tau;\ x(T_r) \in dy]\, P_y[T_A > t - \tau]\Big\}\, dx + O(1). \end{aligned} \tag{3.6}$$

Now let

$$g_r(t) = \int_{|y|=r} P_y[T_A > t]\, d\zeta(y).$$

In view of the rotation invariance of Brownian motion

$$\int_{|x|>r} P_x[T_r \leqq t; x(T_r) \in dy]\, dx = \int_{|x|>r} P_x[T_r \leqq t]\, dx\, d\zeta(y) = E_r(t)\, d\zeta(y),$$

so that (3.6) becomes

$$E_r(t) - E_A(t) = \int_0^t g_r(t-\tau)\, d_\tau E_r(\tau) + O(1), \quad t > 0. \tag{3.7}$$

The asymptotic behavior of $E_r(t)$ was given in (3.4) so that we proceed to estimate $g_r(t)$. By use of (3.5)

$$\begin{aligned} g_r(t) &\sim \frac{2\pi}{\ln t} \int_{|x|=r} H_A(x)\, d\zeta(x) \\ &= \frac{2\pi}{\ln t} \int_{|x|=r} \Big[H_A(x) - \frac{1}{\pi} \ln |x|\Big]\, d\zeta(x) + \frac{2 \ln r}{\ln t}, \end{aligned}$$

and equation (3.2) now yields

$$g_r(t) \sim (\ln t)^{-1} [2 \ln r - K(A)], \quad t \to \infty. \tag{3.8}$$

It only remains to substitute (3.4) and (3.8) into (3.7) to conclude that as $t \to \infty$

$$\begin{aligned} &E_r(t) - E_A(t) \\ &\sim 2\pi [2 \ln r - K(A)] \int_2^{t-2} \frac{1}{\ln(t-\tau)}\, d\Big(\frac{\tau}{\ln \tau}\Big) \\ &\sim \frac{2\pi t}{(\ln t)^2} [2 \ln r - K(A)]. \end{aligned} \tag{3.9}$$

Given 2 compacet sets A and B, each with positive capacity, we may choose r so large that (3.9) holds both for A and for B. The difference of the resulting versions of (3.9) reads

$$E_A(t) - E_B(t) \sim \frac{2\pi t}{(\ln t)^2}[K(A) - K(B)], \quad t \to \infty, \tag{3.10}$$

which proves the first half of Theorem 2. The proof would be complete if we knew that the unit disc dissipates the energy

$$E_1(t) = \frac{2\pi t}{\ln t} + \frac{2\pi t}{(\ln t)^2}(1 + \gamma - \ln 2) + o(t \ln^{-2} t). \tag{3.11}$$

For this fact see CARSLAW and JAEGER ([1] pp. 335—336), who invert the Laplace transform in (3.3) to obtain a complicated expression for $f(x, t) = P_x[T_1 \leqq t]$, and a simpler one for the temperature gradient at the circumference of the unit disc, namely

$$\Phi(t) = \left[\frac{\partial}{\partial |x|} f(x,t)\right]_{|x|=1} = -\frac{4}{\pi^2}\int_0^\infty \frac{1}{u} e^{-u^2 t/2} \frac{du}{J_0^2(u) + Y_0^2(u)}. \tag{3.12}$$

But

$$\frac{\partial E_1(t)}{\partial t} = \frac{\partial}{\partial t}\int_{|x|\geqq 1} f(x,t)\,dx = \frac{1}{2}\int_{|x|\geqq 1} \Delta f(x,t)\,dx,$$

and integration by parts yields

$$\frac{\partial E_1}{\partial t} = -\pi\Phi(t). \tag{3.13}$$

JAEGER ([1] p. 336) states that

$$-\Phi(t) = \frac{2}{\ln 2t - 2\gamma} - \frac{2\gamma}{(\ln 2t - 2\gamma)^2} + o(\ln^{-2} t), \quad t \to \infty. \tag{3.14}$$

To sketch an elementary proof (all the arguments in the literature seem to infer the behavior of $\Phi(t)$ as $t \to \infty$ from that of its Laplace transform near zero) let $\approx$ denote equality up to terms of smaller order than $(\ln t)^{-2}$, as $t \to \infty$. Then, by (3.12)

$$-\Phi(t) \approx \frac{4}{\pi^2}\int_0^A \frac{e^{-\frac{u^2}{2}t}}{u[J_0^2(u) + Y_0^2(u)]}\,du$$

for every $A > 0$, since $J_0^2(u) + Y_0^2(u)$ is bounded away from 0. From the asymptotic behavior of these Bessel functions

$$J_0^2(u) + Y_0^2(u) = 1 + \frac{4}{\pi^2}\left(\ln\frac{u}{2} + \gamma\right)^2 + o(u), \quad u \to \infty,$$

so that

$$-\Phi(t) \approx \int_0^A \frac{e^{-\frac{u^2}{2}t}}{u\left[\frac{\pi^2}{4} + \left(\ln\frac{u}{2} + \gamma\right)^2\right]}\,du = 2\int_0^{\frac{A}{2}t^2} \frac{e^{-x}}{x[\pi^2 + (\ln x - \ln \tau)^2]}\,dx,$$

where we have defined τ by

$$\ln\tau = \ln 2t - 2\gamma.$$

It follows that for every $B > 0$

$$-\Phi(t) \approx 2\int_0^{B\tau} \frac{e^{-x}}{x[\pi^2 + (\ln x - \ln\tau)^2]}\,dx$$

$$\approx 2\int_0^{B\tau} \frac{e^{-x}}{x(\ln x - \ln\tau)^2}\,dx, \quad \tau\to\infty.$$

Now an integration by parts yields

$$-\Phi(t) \approx 2\int_0^{B\tau} \frac{e^{-x}}{\ln\tau - \ln x}\,dx.$$

Finally

$$-\Phi(t) \approx \frac{2}{\ln\tau} + 2\int_0^{B\tau} e^{-x}\left[\frac{1}{\ln\tau - \ln x} - \frac{1}{\ln\tau}\right]dx$$

$$\approx \frac{2}{\ln\tau} + \frac{2}{(\ln\tau)^2}\int_0^{B\tau} e^{-x}\ln x\,dx$$

$$\approx \frac{2}{\ln\tau} + \frac{2}{(\ln\tau)^2}\int_0^{\infty} e^{-x}\ln x\,dx$$

$$= \frac{2}{\ln\tau} - \frac{2\gamma}{(\ln\tau)^2}, \quad \tau\to\infty.$$

Thus (3.14) is proved, and by (3.13) we may integrate (3.14) to obtain (3.11). That completes the proof of Theorem 2.

4. Related probability problems

Both theorems are valid in a more general setting, roughly speaking for stochastic processes whose potential theory resembles the classical one. In particular the result of KAC and HUNT concerning the asymptotic behavior of $P_x[T_A > t]$ has been extended to arbitrary recurrent random walk in [7]. (See also [9], Ch. 7, problem 10 for an incomplete generalization of Theorem 2.) However, Theorem 1 has the more interesting history of having suggested improvements even in the case of Brownian motion. In 1951 DVORETZKY and ERDÖS [3] considered a related problem for simple random walk on the lattice points of s dimensional Euclidean space. Let $L_n =$ the number of distinct points visited by the random walk in time n. If T_x denotes the time of the first visit to the point x, then

$$E[L_n] = \sum_{x\neq 0} P[T_x \leq n].$$

Thus $E[L_n]$ is the discrete analogue of $E_A(t)$ with the set $A = \{0\}$ consisting of a single point. Theorem 1 then has the natural analogue that $E[L_n]/n$ tends to a limit, but DVORETZKY and ERDÖS proved more, i. e., they showed that L_n/n *converges with probability one.* Now replace L_n by $L_A(n)$ = the number of lattice points x such that the random walk $x_k \in x + A$ for some k in the interval $[1, n]$. Again one finds that $E[L_A(n)]$ is the analogue of the energy $E_A(t)$ and that the corresponding strong (probability one) statement holds. (For a proof valid for arbitrary random walk, and for related results see [*10*].) The corresponding strong law for Brownian motion (a proof based on Theorem 1 and Birkhoff's ergodic theorem is given by WHITMAN [*10*]) reads as follows:

For Brownian motion $x(t)$ in 3 space R (in any dimension s if $C(A)$ below is interpreted correctly) and for any compact set $A \subset R$ let $A_t = [x \mid x \in R, x(\tau) \in x + A$ for some $0 \leqq \tau \leqq t]$. Let $A_0 = A$, and denote the Lebesgue measure (volume) of A_t by $|A_t|$. Then

$$E[|A_t|] = E_A(t) + |A|$$

in the notation of (1.1) and

$$\lim_{t \to \infty} \frac{|A_t|}{t} = C(A)$$

with probability one on the probability space of the Brownian motion.

References

[*1*] CARSLAW, H. S., and J. C. JAEGER: Conduction of heat in solids. 2nd ed. Oxford: Clarendon Press, 1959.

[*2*] DOOB, J. L.: A probability approach to the heat equation. Trans. Amer. math. Soc. **80**, 216–280 (1955).

[*3*] DVORETZKY, A., and P. ERDÖS: Some problems on random walk in spac. Proc. 2nd Berkeley Sympos. math. Statist. Probability. 353–368 (1951).

[*4*] HUNT, G. A.: Some theorems concerning Brownian motion. Trans. Amer. math. Soc. **81**, 294–319 (1956).

[*5*] ITÔ, K., and H. P. MCKEAN: Diffusion processes and their sample paths. Berlin: Springer 1964.

[*6*] JOFFE, A.: Sojourn time for stable processes. Thesis, Cornell U., 1959.

[*7*] KESTEN, H., and F. SPITZER: Ratio theorems for random walk (I). J. Analyse math. **11**, 285–321 (1963).

[*8*] NEVANLINNA, R. H.: Eindeutige analytische Funktionen. 2nd ed. Berlin: Springer, 1953.

[*9*] SPITZER, F.: Principles of random walk. Princeton: Van Nostrand 1964.

[*10*] WHITMAN, W.: Thesis. Cornell U. 1964.

Department of Mathematics
Cornell University
Ithaca, New York, USA

(Received August 20, 1963)

Reprinted from ADVANCES IN MATHEMATICS
Vol. 5, No. 2, October 1970

Interaction of Markov Processes

FRANK SPITZER*

Department of Mathematics, Cornell University, Ithaca, New York

INTRODUCTION

For a gentle approach to the problems connected with interacting Markov processes we review first what is known in the absence of interaction. For simplicity let S be a countable (or finite) set, and consider a countable (or finite if S is finite) collection of independent Markov processes on S, with common transition function $P_t(x, y)$, x, $y \in S$. It is natural to assume constant invariant measure, i.e.,

$$\sum_{x \in S} P_t(x, y) = \sum_{y \in S} P_t(x, y) = 1, \qquad x, y \in S, \quad t \geqslant 0. \tag{0.1}$$

Let Σ denote the set of all integer-valued measures on S; since S is countable $\sigma \in \Sigma$ means $\sigma : S \to \mathbf{N} = \{0, 1,...\}$, so that each σ represents a configuration of nondistinguishable particles on S, with $\sigma(x)$ equal to the number of particles at $x \in S$. Let $\mathscr{B}$ denote the Borel field of subsets of Σ generated by the cylinder sets of the form $B = [\sigma : \sigma(x_i) = k_i$ for $i = 1,..., n]$, $x_i \in S$. For each $\lambda > 0$ define the Poisson point process (P.p.p.) on S of density λ as the probability measure μ_λ on $(\Sigma, \mathscr{B})$, which assigns mass

$$\mu_\lambda(B) = e^{-n\lambda} \prod_{i=1}^{n} \frac{\lambda^{k_i}}{k_i!} \tag{0.2}$$

to the cyclinder sets B described above. A familiar theorem of Doob (Ref. [5], pp. 404–407), formulated in the present setting by Derman [2], states that for each $\lambda > 0$ the P.p.p. with density λ is invariant under Markovian translation according to (0.1). More intuitively, place particles on S at time $t = 0$ according to μ_λ defined by (0.2). Let these particles move independently (without interaction) as Markov processes

* Research supported by the National Science Foundation at Cornell University.

with the common transition function P_t, satisfying (0.1), and let $\sigma_t = \{\sigma_t(x)\}_{x \in S}$ describe the particle configuration at time $t \geqslant 0$. Then for each $t > 0$, σ_t is a random variable on the product of the space of $(\Sigma, \mathscr{B}, \mu_\lambda)$ with the probability space of the Markovian motions, and the probability law of σ_t is the same as that assigned to $\sigma_0 = \sigma$ by (0.2). Moreover it can be shown (just as in Ref. [2] where the time parameter is discrete) that $\{\sigma_t\}_{t \geqslant 0}$ is a Markov process with state space Σ. In other words, for each $\lambda > 0$, μ_λ *is an invariant probability measure*[1] *for the Markov process* $\{\sigma_t\}_{t \geqslant 0}$.

The purpose of this paper is to break away from the sphere of influence of the P.p.p., by exhibiting and studying *Markovian particle motions* (*of necessity with interaction between particles*) *under which point processes other than the* P.p.p. *are invariant*, in the sense just described. The literature of stochastic processes contains several recent exploratory studies in this direction [4, 6, 7, 14, 15]. In some of these [6, 14, 15], the P.p.p. is again shown to be the invariant measure, even in the presence of interaction. Harris, in Ref. [7], was first in giving a systematic treatment of interactions in discrete time, including criteria for a point process to be invariant. An example due to Dobrushin, Pyatetski-Shapiro, and Vasilev [4] is much closer in spirit to ours in that it concerns interactions in continuous time which are motivated by problems of statistical mechanics. The invariant point process (or *random field* in the attractive terminology of Dobrushin) is the classical Gibbs ensemble, which was studied systematically in Ref. [3]. Since the example of Dobrushin, Pyatetski-Shapiro, and Vasilev has the drawback of not conserving the total number of particles, we shall use a different example to illustrate the way in which the classical Gibbs ensemble (cf. Ruelle, Ref. [11], Chapters 1–5) serves as a rich and intuitively appealing source of inspiration.

Let S be $\mathbf{Z}_n$ (the lattice points of n-dimensional Euclidean space) or a finite subset thereof, and define $(\Sigma, \mathscr{B})$ as above. The equilibrium state of a *lattice gas* is then defined as a certain probability measure μ on $(\Sigma, \mathscr{B})$. Suppose for a simple illustration that S is finite, that all particles of the gas are of the same type (no spin), that each lattice site is occupied by at most one particle, that there are no external forces acting, and,

[1] Other invariant probability measures can be obtained by taking convex combinations of μ_λ with respect to λ. The converse, to the effect that these are the only invariant probability measures, has not yet been established in sufficient generality. See Thedéen in Ref. [17] for the best results when S is Euclidean space.

finally, that the particles interact through a pair potential U, satisfying

$$U(x, y) = U(y, x) \quad \text{for} \quad x, y \in S, \qquad U(x, x) = \text{constant on} \quad S. \tag{0.3}$$

Then the equilibrium state for N particles ($N < |S|$ = cardinality of S) is described by the probability measure

$$\mu[\sigma \in \textstyle\sum : \sigma(x) = 1 \text{ for } x \in A, \sigma(x) = 0 \text{ for } x \in S - A]$$
$$= Z^{-1} \exp - \tfrac{1}{2} \sum_{x \in A} \sum_{y \in A} U(x, y), \tag{0.4}$$
$$Z = \sum_{[A \subset S : |A| = N]} \exp - \tfrac{1}{2} \sum_{x \in A} \sum_{y \in A} U(x, y).$$

The constant $Z = Z(S)$ is called the partition function of the canonical ensemble.

We shall now describe a class of Markovian motions of N particles with an interaction such that the above probability measure μ in (0.4) is invariant, i.e., such that μ is the invariant measure of the Markov process $\{A_t\}$, $t \geq 0$, whose state $A_t = A$ at any time $t \geq 0$ is that subset $A \subset S$ which is occupied by particles. Thus $A_t = [x \in S : \sigma_t(x) = 1]$. Unfortunately it seems impossible to derive these motions from statistical mechanics. The partition function of the Gibbs ensemble in (0.4) describes an approximation to a physical gas which perhaps cannot be derived on the basis of Hamiltonian mechanics alone (cf. Khinchin in Ref. [10], Chapter V.) Thus there is nothing, neither in the method (trial and error) by which the motions below were found, nor in the theory leading to (0.4), on which to base any claim to physical significance of these motions. Perhaps this is not too surprising since we shall see that the measure μ in (0.4) is in fact invariant under essentially all possible Markovian motions with the same interaction, just as (0.2) is invariant under all Markovian motions without any interaction. Thus *it is the interaction which determines the invariant measure, not the random motions on which the interaction is, so to speak, superimposed.* This state of affairs is typical of all interaction models we consider.

Let $P(x, y)$ be a *symmetric* Markov kernel (stochastic matrix) on $S \times S$, and let

$$P_t(x, y) = \exp t[P - I](x, y) = \exp(-t) \sum_{n=0}^{\infty} \frac{t^n}{n!} P^n(x, y), \qquad x, y \in S, \quad t \geq 0. \tag{0.5}$$

Thus P_t defines a continuous time Markov chain on S (jumps according to P after random holding times which are independent and exponential with mean one). We now describe the interaction to be imposed on N particles, each moving according to (0.5), which will force the measure μ in (0.4) to become the invariant measure. *First* we introduce a *change of speed.* At an instant when the N particles occupy the set $A \subset S$, of cardinality $|A| = N$, we shall give them different speeds, depending on their position relative to each other, as determined by the interaction potential U in (0.3). Thus a particle at x has speed

$$c(x, A) = \exp \sum_{y \in A} U(x, y). \tag{0.6}$$

This means it shall obey the transition law P_{ct} instead of P_t, where c varies in accord with (0.6). Equivalently, the probability for a particle at x to make a transition to y in time dt is $c(x, A)\, P(x, y)\, dt + O(dt^2)$, when A is the set of all particles. The *second* part of the interaction consists in the *exclusion of multiple occupancy*. This is achieved by simply suppressing any transition which would take a particle to an already occupied site of S. It should be clear that the resulting motion of N particles on S can be viewed as a Markov process on S^N, or, equivalently, as a Markov process $\{\sigma_t\}_{t \geqslant 0}$ with state space $\Sigma' = \Sigma \cap [\sigma : \sigma(x) = 0$ or $1, \sum_S \sigma(x) = N]$. As will be shown in Section 4a, the measure μ *in* (0.4) *is an invariant measure for* σ_t. Indeed, since S is finite, so is Σ, and one can conclude that the corresponding ergodic theorem holds: *If the matrix P is irreducible, then for arbitrary $\sigma_0 \in \Sigma'$, the probability measure of σ_t must converge to μ as $t \to \infty$.*

The above example is quite typical of the interactions which we shall study. The paper is divided into ten sections, in such a way that Sections 1a–5a will exhibit five different interaction models on a finite set S, while Sections 1b–5b will contain partial results for the far more interesting and delicate extensions of these models to an infinite set S (countable or a continuum)[2]. Thus Section 1 will deal with pure speed change interaction (multiple occupancy allowed); Section 2 concerns zero-range interaction (the speed change of a particle at x depends only on $\sigma(x)$, the number of particles which occupy its site); in Section 3 we study pure exclusion of multiple occupancy, without any speed change; Section 4 deals with an interaction containing the lattice gas described

[2] The reader interested in the infinite case, say in Section 3b, can proceed directly from Section 3a to Section 3b.

above as a special case; finally, in Section 5 we study a type of nearest neighbor interaction which is only meaningful when S is a finite cyclic group, or the circle group T, or $\mathbf{Z}$ or $\mathbf{R}$.

In general it is difficult to extend an interaction model from the case $|S| < \infty$ to $|S| = \infty$. The first difficulty when $|S| = \infty$ is to prove that a specified transition operator P together with prescribed speed change and/or exclusion rules determine a unique Markov process σ_t on Σ. We did not succeed in solving this problem and proceed as though the existence of σ_t had been proved. Next will come the problem of verifying that a given probability measure μ on $(\Sigma, \mathscr{B})$ is an invariant measure for σ_t. This problem is of the same order of difficulty as that of proving the existence of $\{\sigma_t\}$, and it is closely related to it. We shall be content to establish conjectures based on the formal passage from $|S| < \infty$ to $|S| = \infty$. This will be easiest in those cases when the invariant measure μ permits a simple description. (It is product measure in certain cases. Thus the occupation numbers are independent in Sections 2b and 3b, and we have independent spacings between successive particles in Section 5b.) In the cases of Sections 1b and 4b, however, the invariant measure μ is extremely complicated. When $S = \mathbf{Z}_n$ or $\mathbf{R}_n$ there is a method (the so-called passage to the thermodynamic limit, cf. Ref. [11], Chapters 2, 3) used in statistical mechanics to obtain the finite dimensional distributions (correlation functions) of μ. It consists of letting a sequence of cubes S_k expand to S in such a way that the expected density of particles in S_k tends to a finite positive limit. Our treatment in Section 4b is based on a theorem of Ruelle [12] which asserts that the invariant measures on μ_k on $(\Sigma_k, \mathscr{B}_k)$ converge to a measure μ on $(\Sigma, \mathscr{B})$.

The conjectures established in Sections 2b–5b will be proved in a companion paper to this one by Richard Holley [9] in a class of physically interesting cases. The set S has to be the group $\mathbf{Z}$ of one-dimensional integers, and the transition function $P(x, y)$ (and the law of speed change in Section 2b) have to be subjected to somewhat more stringent conditions than seems necessary for the conjectured state of affairs. It is in particular because Holley's method is limited to $S = \mathbf{Z}$ that we have developed the conjectures for arbitrary S in sufficient detail to emphasize the need for further work.

Finally, there are two problems which are trivial when $|S| < \infty$ but present considerable interest when $|S| = \infty$. One is how to prove ergodic theorems which assert that for "reasonable" initial σ_0, the probability measure of σ_t converges to the invariant measure μ as $t \to \infty$.

This is easy in the absence of interactions when S is denumerable [2], much harder when S is a continuum [16]. We have partial results only in case 3b. The other problem concerns the properties of the random, non-Markovian, motion $x_t \in S$, $t \geqslant 0$, of a single *tagged particle* in its interaction with all other particles. When the system is started with the stationary probability measure μ, and when S is an Abelian group, then it will be seen (Sections 2b, 3b, 5b) that the motion of x_t has an important stationarity property: The probability law of the entire system of particles, as seen from the trajectory x_t, is the same for each $t \geqslant 0$. This implies, for example, that $\{x_t\}_{t\geqslant 0}$ is a process with stationary (although dependent) increments. That in turn yields information concerning the behavior of x_t for large t (as indicated in Sections 2b and 3b).

1a. Speed Change Interaction

Let S be an arbitrary finite set and consider the following motion of n particles on S. We assume given an arbitrary *irreducible*, *doubly stochastic* matrix P on S, i.e.,

$$P(x, y) \geqslant 0, \qquad \sum_{x\in S} P(x, y) = \sum_{y\in S} P(x, y) = 1,$$
$$P^{n_1}(x, y) \cdot P^{n_2}(y, x) > 0, \qquad x, y \in S, \tag{1.1}$$

for suitable $n_1 \geqslant 0$, $n_2 \geqslant 0$, which may depend on x and y. The n particles will move on the trajectories of independent Markov processes with transition function $\exp t(P - I)$, except that their speed is changed in a manner now to be described. Let Φ denote a *strictly positive symmetric* function from S^n to $\mathbf{R}^+$, normalized to be a probability density on S^n, i.e.,

$$\sum_{x\in S^n} \Phi(x) = 1, \qquad \Phi(x) = \Phi(\tau x), \tag{1.2}$$

for every permutation $\tau x = (x_{\tau_1}, \ldots, x_{\tau_n})$ *of* $x = (x_1, \ldots, x_n) \in S^n$. Similarly, let ψ denote a *strictly positive symmetric* function from S^{n-1} to $\mathbf{R}$, and define π_k as the projection from S^n on S^{n-1} obtained by deleting the k-th coordinate. Thus $\pi_1 x = (x_2, x_3, \ldots, x_n)$, $\pi_2 x = (x_1, x_3, x_4, \ldots, x_n)$, etc. We define

$$c_k(x) = \frac{\psi(\pi_k x)}{\Phi(x)}, \qquad x \in S^n, \quad 1 \leqslant k \leqslant n, \tag{1.3}$$

as the speed of the k-th particle at an instant when the position of the n (labeled) particles in phase space S^n is $x = (x_1, ..., x_n)$. Thus the k-th particle jumps from x_k to y_k in time dt with probability $c_k(x) P(x_k, y_k) dt + O(dt^2)$. It is not hard to believe that our recipe defines a unique Markov process $x(t) = (x_1(t), ..., x_n(t))$ with state space S^n. Denoting by $P_a[\cdot]$ its probability measure corresponding to the starting point $a \in S^n$, the main result concerning this interaction is

$$\lim_{t\to\infty} P_a[x(t) = x] = \Phi(x), \qquad x \in S^n, \tag{1.4}$$

for arbitrary $a \in S^n$, *and* Φ *is the probability density of the unique probability measure invariant under the transition function* $Q_t(a, b) = P_a[x(t) = b]$, $a, b \in S^n$, $t \geqslant 0$.

To prove (1.4) we first assume the existence of the Markov process $x(t)$ on S^n, and use (1.3) to determine its infinitesimal generator Ω (see (1.5) below). Then inspection of Ω will suffice to show that it is the generator of a unique Markov process on S^n. Finally it will be obvious, although we omit the details, that this process indeed possesses the speed described by (1.3) which will have led to the formula for Ω, etc.

To compute $\Omega(x, y)$ when $x \neq y$, we introduce the substitution operator $s_k(\xi) : S^n \to S^n$ in such a way that $s_k(\xi)\, x$ is the vector x except for the k-th component which has been changed from x_k to ξ. With this notation (1.3) implies that

$$\Omega(x, y) = \lim_{t\downarrow 0} t^{-1} Q_t(x, y) = c_k(x)\, P(x_k, \eta)$$

when $y = s_k(\eta)\, x$ and $\eta \neq x_k$. When x and y differ in more than one component, however, then $Q_t(x, y) = O(t^2)$ as $t \downarrow 0$ and $\Omega(x, y) = 0$. From the fact that $\sum_y Q_t(x, y) = 1$ we obtain $\sum_y \Omega(x, y) = 0$. Summarizing,

$$\begin{aligned} \Omega(x, y) &= c_k(x)\, P(x_k, \eta) \qquad \text{when} \quad y = s_k(\eta)\, x, \\ &= -\sum_{k=1}^{n} \sum_{\eta \in S - \{x_k\}} c_k(x)\, P(x_k, \eta) \qquad \text{when} \quad y = x, \\ &= 0 \qquad \text{otherwise.} \end{aligned} \tag{1.5}$$

To prove that Φ is invariant under Q_t observe that $Q_t = \exp t\Omega$, since Ω is a bounded operator (matrix). The invariance of Φ is therefore equivalent to

$$\sum_{x \in S^n} \Phi(x)\, \Omega(x, y) = 0, \qquad y \in S^n. \tag{1.6}$$

Using (1.5), the left side of (1.6) becomes

$$\sum_{k=1}^{n} \sum_{\xi \in S - \{y_k\}} \Phi[s_k(\xi) y] \, \Omega(s_k(\xi) y, y) + \Phi(y) \, \Omega(y, y)$$

$$= \sum_{k=1}^{n} \sum_{\xi \in S - \{y_k\}} \Phi[s_k(\xi) y] \, c_k[s_k(\xi) y] \, P(\xi, y_k)$$

$$- \sum_{k=1}^{n} \sum_{\xi \in S - \{y_k\}} \Phi(y) \, c_k(y) \, P(y_k, \xi).$$

Substitution of the definition of $c_k(x)$ from (1.3) gives

$$\sum_{x \in S^n} \Phi(x) \, \Omega(x, y) = \sum_{k=1}^{n} \sum_{\xi \in S - \{y_k\}} \{\psi[\pi_k(s_k(\xi) y)] \, P(\xi, y_k) - \psi(\pi_k y) \, P(y_k, \xi)\}.$$

But by definition, $\pi_k[s_k(\xi) \, y] = \pi_k \, y$, so that

$$\sum_{x \in S^n} \Phi(x) \, \Omega(x, y) = \sum_{k=1}^{n} \psi(\pi_k y) \sum_{\xi \in S - \{y_k\}} [P(\xi, y_k) - P(y_k, \xi)] = 0,$$

where the last step of the proof used the hypothesis (1.1) that P is doubly stochastic.

Thus Φ is invariant. Finally, the irreducibility of the matrix P, together with the fact that $c_k(x) > 0$, implies that $Q_t(x, y) > 0$ for all $t > 0$ and all $x, y \in S$. From this it follows (from the ergodic theory of Markov chains [5]) that (1.4) holds, and that Φ is the unique invariant measure.

Observe that (1.4) contains a complete description of the ergodic behavior of the Markov process σ_t mentioned in the Introduction. The states $\sigma \in \Sigma$ of σ_t are simply the equivalence classes of points of S^n under permutation of coordinates. Thus the probability that $\sigma_t = \sigma$ tends, as $t \to \infty$, to the sum of $\Phi(x)$ over all x in the equivalence class represented by σ (see (2.5) in Section 2a).

For a specific example of an interaction defined by (1.3) let U be a pair potential satisfying (0.3). Let

$$\Phi(x) = Z^{-1} \exp - \tfrac{1}{2} \sum_{k=1}^{n} \sum_{j=1}^{n} U(x_k, x_j), \qquad x \in S^n, \tag{1.7}$$

where Z is a normalizing constant ensuring (1.2), and let

$$\psi(\pi_n x) = \psi(x_1, x_2, \ldots, x_{n-1}) = c \cdot \exp - \tfrac{1}{2} \sum_{k=1}^{n-1} \sum_{j=1}^{n-1} U(x_k, x_j). \tag{1.8}$$

Then (1.3) reduces (since $U(x, x) \equiv \alpha =$ constant) to

$$c_k(x) = c' \cdot \exp \sum_{j=1}^{n} U(x_k, x_j), \qquad x \in S^n, \quad 1 \leqslant k \leqslant n, \tag{1.9}$$

where c' is a constant depending on c, Z, n, and α.

In the simplest special case ($U \equiv$ constant) the invariant measure μ has density $\Phi \equiv$ constant, so that μ is the P.p.p. conditioned on the fact that there are exactly n particles. On the other hand it can be shown that (1.9) has a natural generalization to higher order interactions, i.e., to Gibbs distributions of the form

$$\Phi(x) = \text{constant} \cdot \exp - \sum_{\nu=1}^{\infty} V_\nu(x), \qquad x \in S^n,$$

where each V_ν is determined by a symmetric map $U_\nu : S^\nu \to \mathbf{R}$ via

$$V_\nu(x) = \sum_{\{i_1, i_2, \ldots, i_\nu\} \subset \{1, 2, \ldots, n\}} U_\nu(x_{i_1}, x_{i_2}, \ldots, x_{i_\nu}).$$

2a. Zero Range Interaction

This model is in fact a special case of the one in Section 1, but this will not be immediately obvious. Just as in Section 1 we consider the motion of n particles on a finite set S and take a stochastic matrix P satisfying (1.1). The interaction will have *zero range* in the sense that the speed $c_k(x)$ of the k-th particle at time t will depend only on the number of particles which occupy the same site at time t. Let

$$\nu(\xi, x) = \sum_{j=1}^{n} \delta(\xi, x_j), \qquad \xi \in S, \quad x = (x_1, \ldots, x_n) \in S^n, \tag{2.1}$$

denote the number of components of x which equal ξ, and let φ be a *strictly positive* function from $\mathbf{N}^+ = \{1, 2, 3, \ldots\}$ to $\mathbf{R}^+$. We define

$$c_k(x) = \varphi[\nu(x_k, x)], \qquad x \in S, \qquad 1 \leqslant k \leqslant n. \tag{2.2}$$

Next we define the symmetric, strictly positive function $\Phi : S^n \to \mathbf{R}^+$ which will ultimately be the density of the invariant measure defined by

$$\Phi(x) = Z^{-1} \prod_{\xi \in S} \{p[\nu(\xi, x)]\}^{-1}, \tag{2.3}$$

where Z is the normalizing constant which ensures that $\sum_{x \in S^n} \Phi(x) = 1$, and

$$p(0) = 1, \qquad p(k) = \varphi(1)\,\varphi(2) \cdots \varphi(k), \qquad k \geqslant 1. \tag{2.4}$$

Then, just as in Section 1, there is a unique Markov process $x(t)$, $t \geqslant 0$ on S^n which has the speed interaction given by (2.2), and the main result concerning its probability measure $P_a[\cdot]$, $a \in S^n$, reads

$$\lim_{t \to \infty} P_a[x(t) = x] = \Phi(x), \qquad x \in S^n, \tag{2.5}$$

for arbitrary $a \in S^n$, *and* Φ *is the density of the unique invariant probability measure of the Markov process* $x(t)$, $t \geqslant 0$.

The proof consists in showing that we have a special case of the model in Section 1, so that (2.5) is a special case of (1.4). This will be true if (1.3) is satisfied, i.e., if there exists a symmetric strictly positive function $\psi : S^{n-1} \to \mathbf{R}^+$ such that

$$c_k(x)\,\Phi(x) = \psi(\pi_k x), \qquad x \in S^n. \tag{2.6}$$

It is obvious from the definitions of c_k and Φ in (2.2) and (2.3) that $c_k(x)$ $\Phi(x)$ is positive and a symmetric function of the $n - 1$ components of $\pi_k x$. To see that it is indeed independent of x_k we take two points x and x' in S^n, such that $x_k = \alpha$, $x_k' = \beta$, with $\alpha \neq \beta$, and show that $c_k(x)\,\Phi(x) = c_k(x')\,\Phi(x')$. We obtain from (2.3)

$$\frac{\Phi(x)}{\Phi(x')} = \prod_{\xi \in S} \frac{p[\nu(\xi, x')]}{p[\nu(\xi, x)]} = \frac{p[\nu(\alpha, x')]\,p[\nu(\beta, x')]}{p[\nu(\alpha, x)]\,p[\nu(\beta, x)]},$$

and note that

$$\nu(\alpha, x') + 1 = \nu(\alpha, x), \qquad \nu(\beta, x) + 1 = \nu(\beta, x').$$

Hence, using (2.4),

$$\frac{\Phi(x)}{\Phi(x')} = \frac{\varphi[\nu(\beta, x')]}{\varphi[\nu(\alpha, x)]} = \frac{\varphi[\nu(x_k', x')]}{\varphi[\nu(x_k, x)]} = \frac{c_k(x')}{c_k(x)},$$

which concludes the proof.

Here it is very natural to rephrase (2.5) as an ergodic theorem for the Markov process σ_t, $t \geqslant 0$, whose state space is Σ_n, the set of all configurations of n indistinguishable particles on S. Thus $\sigma_t(\xi)$ is the number of particles at $\xi \in S$ at time t. We have

$$P_a[\sigma_t = \sigma] = \sum_{x \in A_\sigma} P_a[x(t) = x],$$

where A_σ is the set of $x \in S^n$ such that

$$\nu(\xi, x) = \sum_{k=1}^{n} \delta(x_k, \xi) = \sigma(\xi), \qquad \text{for all} \quad \xi \in S.$$

In the limit, as $t \to \infty$, it follows from (2.5) that

$$\lim_{t\to\infty} P_a[\sigma_t = \sigma] = \sum_{x \in A_\sigma} \Phi(x), \qquad a \in S^n, \tag{2.7}$$

and since Φ is symmetric (constant on A_σ for each $\sigma \in \Sigma$)

$$\begin{aligned} \mu(\sigma) &= \lim_{t\to\infty} P_a[\sigma_t = \sigma] = \Phi(x)\,|A_\sigma| \\ &= \Phi(x)\, n! \prod_{\xi \in S} \frac{1}{\sigma(\xi)!} = Z^{-1} n! \prod_{\xi \in S} \frac{1}{p[\sigma(\xi)]\,\sigma(\xi)!}, \qquad \sigma \in \Sigma_n. \end{aligned} \tag{2.8}$$

Thus μ as given in (2.8) is the unique invariant probability measure of the Markov process σ_t. The normalizing constant Z^{-1} is easily computed in special cases. We consider

EXAMPLE 2.1 (The Maxwell–Boltzmann Distribution). When $\varphi(k) =$ constant on $k \geqslant 1$ (*no interaction*)

$$\mu(\sigma) = \frac{n!}{|S|^n} \prod_{\xi \in S} \frac{1}{\sigma(\xi)!}, \qquad \sigma \in \Sigma_n. \tag{2.9}$$

EXAMPLE 2.2 (The Bose–Einstein Distribution). Let $\varphi(k) =$ constant$/k$ on $k \geqslant 1$ (a type of *attraction*). Then

$$\mu(\sigma) = \frac{1}{\binom{|S| - 1 + n}{n}}, \qquad \sigma \in \Sigma_n. \tag{2.10}$$

EXAMPLE 2.3 (The Fermi–Dirac Distribution). Let $\varphi(1) = \tau > 0$, while we set (formally) $\varphi(k) = +\infty$ in (2.8) when $k \geqslant 2$. This produces a ***repulsion*** so strong that no particle remains a positive time at a site occupied by another particle. Thus we must assume that $0 \leqslant n \leqslant |S|$, and obtain from (2.8)

$$\mu(\sigma) = \frac{1}{\binom{|S|}{n}} \quad \text{for } \sigma \in \Sigma_n \text{ such that all } \sigma(\xi) = 0 \text{ or } 1, \qquad (2.11)$$
$$= 0 \quad \text{otherwise.}$$

The rigorous derivation of an interaction leading to (2.11) is given in the next section.

3a. Simple Exclusion Interaction

In the case of exclusion of multiple occupancy the properties of the process σ_t are much simpler than those of $x(t)$ which describes identifiable particles. Since $\sigma_t(\xi)$ will take only the values 0 and 1, σ_t is determined by the subset $A_t \subset S$ which is occupied at time t. Thus we shall study the Markov process A_t and since we consider a fixed number $n < |S|$ of particles the state space will be X_n = the subsets of S of cardinality n. We shall describe two different interactions, both with the same property of yielding ***convergence to a constant invariant probability measure***, viz.,

$$\lim_{t \to \infty} P_A[A_t = B] = \frac{1}{\binom{|S|}{n}}, \quad \text{for } A, B \in X_n. \qquad (3.1)$$

In both cases we shall assume that the basic process is defined by the transition function $\exp t(P - I)$, with P satisfying (1.1). In the first model (Case I) we assume that *transitions to an occupied site are simply suppressed.* The speed remains unchanged. This yields the generator[3]

$$\Omega(A, B) = P(x, y) \quad \text{if } x \in A, \ y \in A^c, \ B = A \backslash x \cup y$$
$$= -\sum_{x \in A} \sum_{y \in A^c} P(x, y) \quad \text{if } A = B \qquad (3.2)$$
$$= 0 \quad \text{otherwise, } A, B \in X_n.$$

[3] A^c is the complement of A, $A \backslash x \cup y = [A - \{x\}] \cup \{y\}$.

Case II is more complicated, but arose in a natural way as the limiting case (2.11) of (2.8). An interpretation of this limit suggests that a particle which is in motion is so strongly repelled by occupied sites that it ***keeps on going*** (according to P) ***until, for the first time, it reaches an unoccupied site*** (which may be the site it started from). To describe this motion in terms of familiar concepts let $B \subset S$, and define $\Pi_B(x, y)$, $x, y \in B$ as the transition matrix of the imbedded Markov chain induced by the transition matrix P. (Thus the original discrete time P chain is observed only when it visits B. Π_B is the transition matrix of this new chain.) The interaction may now be described: The occupied set at time t is $A_t = A$, and a particle at $x \in A$ is ready to make a transition. It makes this transition instantaneously to a point y in $(A\backslash x)^c$, according to the probability law $\Pi_{(A\backslash x)^c}(x, y)$, since it must keep on going until it attains an unoccupied site. Thus the generator is

$$
\begin{aligned}
\Omega(A, B) &= \Pi_{(A\backslash x)^c}(x, y) \qquad \text{if} \quad x \in A, \quad y \in A^c, \quad B = A\backslash x \cup y \\
&= -\sum_{x\in A}\sum_{y\in A^c} \Pi_{(A\backslash x)^c}(x, y), \qquad \text{if} \quad A = B \qquad (3.3) \\
&= 0 \qquad \text{otherwise,} \quad A, B \in X_n .
\end{aligned}
$$

The ergodic theorem (3.1) in both cases is an immediate consequence of the fact that the constant measure $\mu(A) \equiv 1$ is invariant, or

$$\sum_A \Omega(A, B) = 0, \qquad B \subset S, \qquad |A| = |B| = n. \qquad (3.4)$$

In Case I one gets from (3.2)

$$
\begin{aligned}
\sum_A \Omega(A, B) &= \sum_{y\in B}\sum_{x\in B^c} \Omega(B\backslash y \cup x, B) + \Omega(B, B) \\
&= \sum_{y\in B}\sum_{x\in B^c} P(x, y) - \sum_{x\in B}\sum_{y\in B^c} P(x, y),
\end{aligned}
$$

which may be seen to vanish since P is doubly stochastic.

In Case II the result depends on the well-known fact that the matrix Π_B inherits from P the property of being doubly stochastic.

Therefore one finds, using first (3.3),

$$\sum_A \Omega(A, B) = \sum_{y\in B}\sum_{x\in B^c} \Omega(B\backslash y \cup x, B) + \Omega(B, B)$$

$$= \sum_{y\in B}\sum_{x\in B^c} \Pi_{(B\backslash y)^c}(x, y) - \sum_{x\in B}\sum_{y\in B^c} \Pi_{(B\backslash x)^c}(x, y)$$

$$= \sum_{y\in B}\sum_{x\in B^c} \Pi_{B^c\cup y}(x, y) - \sum_{x\in B}\sum_{y\in B^c} \Pi_{B^c\cup x}(x, y)$$

$$= \sum_{y\in B} [1 - \Pi_{B^c\cup y}(y, y)] - \sum_{x\in B} [1 - \Pi_{B^c\cup x}(x, x)] = 0.$$

4a. Speed Change with Exclusion

Here we combine the interactions of Section 1 and of Case I in Section 3. As a small price for the more elaborate interaction we have to require the transition matrix $P(x, y)$ to be *symmetric* ($P(x, y) = P(y, x)$) in addition to satisfying (1.1). We require as in Section 3a, Case I, that every transition to an occupied site be suppressed, and finally define the speed just as in Section 1. But since it is more convenient to study the Markov process

$$A_t = \{x \in S : \sigma_t(x) = 1\}, \qquad t \geqslant 0,$$

we denote the speed of a particle at x by $c(x, A)$, when A is the state of A_t. We consider the motion of n particles with $n < | S |$, so that the state space of A_t is $X_n = \{A : A \subset S, | A | = n\}$, assume given a strictly positive map $\Phi : X_n \to \mathbf{R}^+$, such that

$$\sum_{A\in X_n} \Phi(A) = 1,$$

and also choose a strictly positive map $\psi : X_{n-1} \to \mathbf{R}^+$. We now define the speed as

$$c(x, A) = \frac{\psi(A\backslash x)}{\Phi(A)}, \qquad x \in A, \quad A \subset S, \quad A \,| = n. \tag{4.1}$$

Condition (4.1) together with the exclusion rule determines the infinitesimal generator

$$\begin{aligned}\Omega(A, B) &= c(x, A)\, P(x, y) \qquad \text{if} \quad B = A\backslash x \cup y,\\ &= -\sum_{x\in A}\sum_{y\in A^c} c(x, A)\, P(x, y) \qquad \text{if} \quad B = A,\\ &= 0 \qquad \text{otherwise.}\end{aligned} \tag{4.2}$$

The definition of the speed will now be shown to be exactly such that Φ is the invariant measure of the Markov process A_t. We compute

$$\begin{aligned}\sum_A \Phi(A)\,\Omega(A, B) &= \sum_{x\in B^c}\sum_{y\in B} \Phi(B\backslash y \cup x)\,\Omega(B\backslash y \cup x, B) + \Phi(B)\,\Omega(B, B)\\ &= \sum_{x\in B^c}\sum_{y\in B} \Phi(B\backslash y \cup x)\, c(x, B\backslash y \cup x)\, P(x, y)\\ &\quad - \Phi(B) \sum_{x\in B}\sum_{y\in B^c} c(x, B)\, P(x, y)\\ &= \sum_{x\in B^c}\sum_{y\in B} \psi(B\backslash y)\, P(x, y) - \sum_{x\in B}\sum_{y\in B^c} \psi(B\backslash x)\, P(x, y)\\ &= \sum_{x\in B}\sum_{y\in B} \psi(B\backslash y)\, [P(x,y) - P(y,x)] = 0.\end{aligned}$$

Since P is irreducible it is clear that the transition function $Q_t(A, B)$ of the Markov process A_t, $t \geqslant 0$, is strictly positive for all A, B with $|A| = |B|$ and all $t > 0$. Thus it is seen that the invariant measure Φ is unique, and we have proved

$$\lim_{t\to\infty} Q_t(A, B) = \Phi(B), \qquad \textit{for all} \quad A \subset S, \quad B \subset S, \tag{4.3}$$

such that $|A| = |B| = n$, *and* Φ *is the unique invariant probability measure of the Markov process* $\{A_t\}$ *with transition function* Q_t, *determined by the generator* Ω *in* (4.2).

To obtain the results for the lattice gas discussed in the Introduction, let U be a pair potential satisfying (0.3) and define

$$\Phi(A) = Z^{-1} \exp - \tfrac{1}{2} \sum_{x\in A}\sum_{y\in A} U(x, y), \quad A \subset S, \quad |A| = n, \tag{4.4}$$

Z being a normalizing constant. Then the choice of

$$\psi(A) = \exp - \tfrac{1}{2} \sum_{x \in A} \sum_{y \in A} U(x, y), \qquad A \subset S, \qquad |A| = n - 1, \tag{4.5}$$

produces the interaction defined by exclusion and the speed

$$c(x, A) = \text{constant} \cdot \exp \sum_{y \in A} U(x, y). \tag{4.6}$$

5a. Nearest Neighbor Interaction

Here we require S to be the cyclic group of order N, written additively as $\{0, 1,..., N - 1\}$ with addition mod(N). On S we shall construct a motion with both speed change and exclusion interaction of n particles, $1 < n < N$. The *exclusion will be such that the particles will remain in a fixed order* (viewed as points on the circle obtained by identifying k with $\exp(2\pi ik/N)$). We label the position vector of the particles $x = (x_1, x_2, ..., x_n)$, $x_i \in S$, and denote the spacings between successive particles by

$$\delta = (\delta_1, \delta_2, ..., \delta_n) \qquad \text{where} \quad \delta_1 = x_2 - x_1, \quad \delta_2 = x_3 - x_2, ...,$$

$$\delta_{n-1} = x_n - x_{n-1}, \qquad \delta_n = x_1 - x_n, \qquad \text{and all} \quad 1 \leqslant \delta_i < N.$$

We use an even function U on $S(U(x) = U(-x)$ for $x \in S)$ to define the *nearest neighbor interaction potential* of the k-th particle as $U(\delta_{k-1}) + U(\delta_k)$, where $\delta_{-1} = \delta_n$. (Note that δ_{k-1} *and* δ_k *are the spacings between* x_k *and its two nearest neighbors.*)

On S we assume given the transition function P of a random walk satisfying

$$P(x, y) = P(y, x) = P(0, y - x), \qquad x, y \in S, \tag{5.1}$$

and in addition $P(0, 1) > 0$. This will ensure (irreducibility of P is not enough) that the interaction process to be defined below satisfies (5.6). Without interaction we would define the Markov process with transition function $\exp t(P - I)$. The interaction will consist first of all in changing the speed of the k-th particle to

$$c_k(\delta) = \exp[U(\delta_{k-1}) + U(\delta_k)]. \tag{5.2}$$

Thus it depends only on the spacings between x_k and its neighbors. Secondly the interaction will *suppress all transitions which would upset the order of the particles*. Given $\delta = (\delta_1, \delta_2, \ldots, \delta_n)$ at any time, with $\delta_i > 0$, and $\delta_1 + \delta_2 + \cdots + \delta_n = N$, the k-th particle is only allowed transitions which will take it from x_k to some point x_k' in the interval $(x_k - \delta_{k-1}, x_k + \delta_k)$. Thus in particular double occupancy is ruled out.

Let D be the space of vectors $\delta = (\delta_1, \ldots, \delta_n)$ with positive integer components, and $\delta_1 + \cdots + \delta_n = N$. On D define

$$\Phi(\delta) = Z^{-1} \exp\left[-\sum_{i=1}^{n} U(\delta_i)\right], \qquad \delta \in D, \tag{5.3}$$

where Z is the normalizing constant

$$Z = \sum_{\delta \in D} \exp\left[-\sum_{i=1}^{n} U(\delta_i)\right].$$

We shall show that the integration determines a Markov process $\delta(t)$ on D, such that its probability measure $P_\alpha[\ ]$ satisfies

$$\lim_{t\to\infty} P_\alpha[\delta(t) = \beta] = \Phi(\beta), \qquad \textit{for} \quad \alpha, \beta \in D, \tag{5.4}$$

and Φ is the unique invariant probability measure of the process $\delta(t)$.

The proof is just a matter of showing that the generator Ω satisfies

$$\sum_{\alpha \in D} \Phi(\alpha)\, \Omega(\alpha, \beta) = 0, \qquad \beta \in D. \tag{5.5}$$

and that

$$P_\alpha[\delta(t) = \beta] > 0 \qquad \text{for all} \quad t > 0, \quad \alpha, \beta \in D. \tag{5.6}$$

It is clear that (5.6) follows from $P(0, 1) > 0$.

Let e_k denote the vector (n-tuple) with k-th component $+1$, $(k - 1)$-st component -1, and all other components 0. If $k = 1$, then the $(k - 1)$-st component is interpreted as the n-th, so that $e_1 = (1, 0, 0, \ldots, 0, -1)$, $e_2 = (-1, 1, 0, 0, \ldots, 0)$, etc. When the state is $\delta \in D$, and when the k-th particle makes a transition from x_k to $x_k + \nu$, then the resulting change in δ is from δ to $\delta - \nu e_k$. This transition is permitted if $\delta - \nu e_k \in D$ and is suppressed otherwise.

Therefore one obtains

$$\Omega(\alpha, \beta) = c_k(\alpha)\, P(0, \nu) \qquad \text{if} \quad \beta = \alpha - \nu e_k \quad \text{for some} \quad 1 \leqslant k \leqslant n,$$

$$\text{and if} \quad \beta \in D - \{\alpha\},$$

$$= -\sum_{k=1}^{n} c_k(\alpha) \sum_{\nu:\alpha-\nu e_k \in D} P(0, \nu) \qquad \text{if} \quad \alpha = \beta,$$

$$= 0 \qquad \text{otherwise.}$$

We get

$$\sum_{\alpha \in D} \Phi(\alpha)\, \Omega(\alpha, \beta)$$

$$= \sum_{k=1}^{n} \sum_{\substack{\nu:\nu \neq 0 \\ \beta+\nu e_k \in D}} \Phi(\beta + \nu e_k)\, \Omega(\beta + \nu e_k, \beta) + \Phi(\beta)\, \Omega(\beta, \beta)$$

$$= \sum_{k=1}^{n} \sum_{\substack{\nu:\nu \neq 0 \\ \beta+\nu e_k \in D}} \Phi(\beta + \nu e_k)\, c_k(\beta + \nu e_k)\, P(0, \nu) - \sum_{k=1}^{n} \sum_{\substack{\nu:\nu \neq 0 \\ \beta-\nu e_k \in D}} \Phi(\beta)\, c_k(\beta)\, P(0, \nu).$$

Since $P(0, \nu) = P(0, -\nu)$ this becomes

$$\sum_{k=1}^{n} \sum_{\substack{\nu:\nu \neq 0 \\ \beta+\nu e_k \in D}} P(0, \nu)[\Phi(\beta + \nu e_k)\, c_k(\beta + \nu e_k) - \Phi(\beta)\, c_k(\beta)].$$

But in view of (5.2) and (5.3) this is zero, since

$$\log \frac{\Phi(\beta + \nu e_k)}{\Phi(\beta)} = -U(\beta_k + \nu) - U(\beta_{k-1} - \nu) + U(\beta_k) + U(\beta_{k-1}),$$

while

$$\log \frac{c_k(\beta)}{c_k(\beta + \nu e_k)} = U(\beta_k) + U(\beta_{k-1}) - U(\beta_k + \nu) - U(\beta_{k-1} - \nu).$$

Therefore Φ is the invariant measure and (5.4) holds.

The principal interest of (5.4) lies in the extension from the finite group S to the groups $\mathbf{Z}$ and $\mathbf{R}$ which will be given in Section 5b.

1b. Speed Change Interaction (Continued)

The results of Section 1a have a natural extension to the case when S is a compact subset of Euclidean space. To avoid technical difficulties we replace the transition matrix P in (1.1) by an absolutely continuous transition operator $P(x, dy) = p(x, y)\, dy$, with iterates $P^n(x, dy) = p^{(n)}(x, y)\, dy$, where dy is the Lebesgue measure, such that

$$p(x, y) \geqslant 0, \qquad \int_S p(x, y)\, dy = \int_S p(x, y)\, dx = 1,$$
$$p^{(n_1)}(x, y)\, p^{(n_2)}(y, x) > 0 \qquad \text{for} \quad x, y \in S, \tag{1.10}$$

where n_1 and n_2 may depend on x and y. Since S is *compact* we consider the motion, according to $\exp t(P - I)$, of a *finite* number n of particles on S, and define the interaction just as in (1.2) and (1.3), where of course the density Φ is normalized so that

$$\int_{S^n} \Phi(x)\, dx = 1, \qquad dx = dx_1 \ldots dx_n\, .$$

Then we still obtain a Markov process $x(t)$ on S^n, whose generator is defined by its action on continuous functions $f : S \to \mathbf{R}$; namely,

$$\Omega f(x) = \int_{S^n} \Omega(x, dy) f(y)$$
$$= \sum_{k=1}^{n} c_k(x) \int_S p(x_k\, , \eta)\{f[s_k(\eta)\, x] - f(x)\}\, d\eta. \tag{1.11}$$

The proof that Φ is the density of an invariant measure for the Markov process $x(t)$ is identical to the proof of (1.6), and by standard ergodic theorems (Ref. [5], Chapter VI) one can conclude that (1.4) holds, in the form

$$\lim_{t\to\infty} P_a[x(t) \in A] = \int_A \Phi(x)\, dx, \tag{1.12}$$

for all $a \in S^n$, and all Borel sets $A \subset S^n$. Conceptually the situation is even simpler than in Section 1a since with probability one no two particles will ever occupy the same point of S if the process is started with an absolutely continuous initial distribution, such as Φ.

Of course, it is possible to impose the speed change interaction (1.3) also on Markov processes with continuous paths. We shall take S to be the circle group T, and consider the two most interesting examples.

EXAMPLE 1.1. Consider n standard Brownian motions on T, interacting according to the speed change $c_k(x)$ given in (1.3). Thus the trajectory $x_k(t)$ of the k-th particle is described by the stochastic differential equations

$$dx_k(t) = \sqrt{c_k(x)}\, db_k(t), \qquad x = x(t) = (x_1(t), \ldots, x_n(t)),$$

$$b_k(t) = \textit{independent standard Brownian motions on } T, \qquad k = 1, 2, \ldots, n. \tag{1.13}$$

Equivalently, the generator of the Markov process $x(t)$ on T^n is given by

$$\Omega f(x) = \tfrac{1}{2} \sum_{k=1}^{n} c_k(x) \frac{\partial^2 f(x)}{\partial x_k^2}, \qquad x \in T^n, \quad f \in C^2(T^n), \tag{1.14}$$

since

$$\Omega f(x) = \lim_{t \downarrow 0} t^{-1} E_x\{f[x(t)] - f(x)\}$$

$$= \lim_{t \downarrow 0} t^{-1} E_x \tfrac{1}{2} \sum_{i=1}^{n} \sum_{j=1}^{n} \frac{\partial^2 f(x)}{\partial x_i \partial x_j} \sqrt{c_i(t)\, c_j(t)}\, b_i(t)\, b_j(t).$$

To prove that Φ is the density of an invariant measure observe that

$$\Omega^* \Phi(x) = \sum_{k=1}^{n} \frac{\partial^2}{\partial x_k^2} [c_k(x)\, \Phi(x)] = 0, \qquad x \in T^n, \tag{1.15}$$

since the function $c_k(x)\, \Phi(x)$ *is independent of* x_k. But an integration by parts shows that

$$\int_{T^n} f(x)\, \Omega^* \Phi(x)\, dx_1, \ldots, dx_n = \int_{T^n} \Phi(x)\, \Omega f(x)\, dx_1, \ldots, dx_n = 0$$

for all $f \in C^2(T^n)$. Since Ω is the generator of the diffusion $x(t)$ on T^n this implies that Φ is the density of the (unique) invariant measure and that (1.12) holds.

EXAMPLE 1.2. Consider *completely deterministic motion* on T^n, according to the differential equations

$$\frac{dx_k}{dt} = c_k(x), \qquad 1 \leqslant k \leqslant n. \tag{1.14}$$

Let

$$x(t) = (x_1(t), x_2(t), \dots, x_n(t)) = T_t x, \qquad x = (x_1, \dots, x_n) \in T^n$$

denote the semigroup of the motion, i.e., the solution of (1.14) determined by the initial conditions $x(0) = x$. The ergodic theory of such a (deterministic) flow is very complicated [1], but it is easy to prove that the measure defined on T^n by the density Φ is invariant under the flow. We define $T_t{}^*\Phi$ through

$$\int_{T^n} \Phi(x)\, g(T_t x) = \int_{T^n} g(x)\, T_t{}^*\Phi(x), \qquad g \in C(T^n). \tag{1.15}$$

For arbitrary $g \in C^2(T^n)$

$$\lim_{t\downarrow 0} \frac{1}{t} \int g(x)[T_t{}^k\Phi(x) - \Phi(x)] = \lim_{t\downarrow 0} \int \Phi(x) \frac{g(T_t x) - g(x)}{t}$$

$$= \int \Phi(x) \sum_{k=1}^{n} \frac{\Phi g}{\partial x_k} \frac{dx_k}{dt} = \sum_{k=1}^{n} \int \Phi(x)\, c_k(x) \frac{\partial g}{\partial x_k}. \tag{1.16}$$

Here all integrals are over T^n with respect to Lebesgue measure. Since $\Phi(x)\, c_k(x)$ is independent of x_k, it is convenient to integrate first on x_k. But since $g \in C^2(T^n)$

$$\int_T \frac{\partial g}{\partial x_k}\, dx_k = 0, \qquad 1 \leqslant k \leqslant n.$$

Therefore the limit in (1.16) is zero.

It follows from the semigroup property of T_t that

$$\int g T_t{}^*\Phi = \int g\Phi$$

for all $g \in C^2(T^n)$. Since $C^2(T^n)$ is dense in $C(T^n)$ we have $T_t{}^*\Phi \equiv \Phi$, or, for Borel subsets $A \subset T^n$,

$$\int_A \Phi(x) = \int_{[x:T_t x \in A]} \Phi(x). \tag{1.17}$$

However it is not known what other invariant probability measures there are.

In the special case when $U : T \to \mathbf{R}$

$$\Phi(x) = Z^{-1} \exp - \tfrac{1}{2} \sum_{1}^{n} \sum_{1}^{n} U(x_i - x_j),$$
$$c_k(x) = \exp \sum_{i=1}^{n} U(x_k - x_i), \tag{1.18}$$

the last two examples are of interest in statistical mechanics. It seems plausible that the Markovian interaction motion of ***infinitely many particles*** on R (either Brownian motion or deterministic) with speed change given by $c_k(x)$ in (1.18)[4] should have for invariant measure the thermodynamic limit of a sequence Φ_n of measures on T as, simultaneously, n and the radius of T tend to infinity. Since more information is available concerning the thermodynamic limit in the lattice case, we defer the detailed discussion to Section 4b.

2b. Zero Range Interaction (Continued)

We consider a denumerably infinite set S, and just as in Section 1 we take as given a doubly stochastic matrix $P(x, y)$, $x, y \in S$, which satisfies (1.1), and a function $\varphi : \mathbf{N}^+ \to \mathbf{R}^+$, and a speed $c_k(x)$, defined as in (2.2) in terms of φ. It is not entirely obvious that the interaction (2.2) imposed on the trajectories of $\exp t[P - I]$ leads to a well-defined Markovian motion of infinitely many particles on S. Therefore we return momentarily to the case of $|S| < \infty$, in order to suggest results for infinite S which will be interesting enough to justify considerable mathematical effort.

Returning to the case of $|S| < \infty$, let Σ_n denote the set of possible configurations of n particles on S, and consider the measures of (2.8), i.e.,

$$\mu_n(\sigma) = n! \prod_{\xi \in S} \frac{1}{p[\sigma(\xi)]\, \sigma(\xi)!}, \qquad \sigma \in \Sigma_n . \tag{2.12}$$

[4] Assuming that $U(x) \to 0$ sufficiently fast as $|x| \to \infty$.

Now we can define a probability measure Φ on $\Sigma = \bigcup_{n=0}^{\infty} \Sigma_n$ by

$$\Phi(\sigma) = \frac{(n!)^{-1} \rho^n \mu_n(\sigma)}{\sum_{k=0}^{\infty} \frac{\rho^k}{k!} \mu_k(\Sigma_k)}, \qquad \text{for} \quad \sigma \in \Sigma_n, \tag{2.13}$$

where ρ is a positive parameter (representing density) subject only to the condition that the denominator in (2.13) converges. It is clear that Φ is a probability measure on the countable set Σ; its restriction to each Σ_n is invariant under the interaction process σ_t; and Φ is a convex combination of these restrictions. Therefore Φ itself is invariant under the interaction. Observe that *Φ is nothing but product measure on $\Sigma = \mathbf{N}^S$*, where $\mathbf{N} = \{0, 1, 2,...\}$. For (2.13) may be written as

$$\Phi(\sigma) = \prod_{\xi \in S} f[\sigma(\xi)], \qquad \sigma \in \Sigma \tag{2.14}$$

where f is the probability density on $\mathbf{N}$ represented by[5]

$$f(k) = \alpha\rho^k \frac{1}{\varphi(1)\,\varphi(2) \cdots \varphi(k) \cdot k!} \qquad \text{for} \quad k \geqslant 1, \;\; f(0) = \alpha, \tag{2.15}$$

with α such that $\sum_0^\infty f(k) = 1$.

It is therefore reasonable to conjecture that this state of affairs persists when the set S is infinite. To obtain a more complete statement let $\Sigma = \mathbf{N}^S$, $\mathfrak{F}$ be the collection of all subsets of $\mathbf{N}$, f be the probability measure on $\mathbf{N}$ in (2.15), and $(\Sigma, \mathscr{B}, \mu)$ be the Cartesian product, over the index set S, of the probability spaces $(\mathbf{N}, \mathfrak{F}, f)$. Thus μ is the product measure defined by

$$\mu\{\sigma : \sigma(x_i) = k_i\,, i = 1,..., n\} = \prod_{i=1}^{n} f(k_i), \qquad x_i \in S, \quad n \geqslant 0. \tag{2.16}$$

Then, *given P satisfying* (1.1), $\varphi : \mathbf{N}^+ \to \mathbf{R}^+$, and $c_k(x)$ *defined* by (2.2) *there is a unique Markov process σ_t with state space Σ, and probability*

[5] Note that a slight restriction has been imposed on the function φ. It must not tend to zero too fast, in other words, there must exist some $\rho > 0$ such that the series

$$\sum_1^\infty \frac{\rho^k}{k!} \frac{1}{\varphi(1)\,\varphi(2),..., \varphi(k)} < \infty.$$

In general we will have a choice of all ρ in some interval

$$0 < \rho \leqslant R < \infty \qquad \text{or} \qquad 0 < \rho < R \leqslant \infty.$$

measure $P_\sigma[\cdot]$, determined by P, φ, and c_k, for which μ is an invariant probability measure, i.e.,

$$\int_\Sigma \mu(d\sigma)\, P_\sigma[\sigma_t \in B] = \mu(B) \qquad \text{for all} \quad B \in \mathscr{B}, \quad t \geqslant 0. \tag{2.17}$$

Suppose therefore that at time zero we put a random number of particles $\sigma_0(x)$ at the points x of S, independently and according to the measure f; suppose we let particles move according to the interaction process determined by $\exp t\{P - I\}$ and the speed change: "a particle at x together with $k - 1$ other particles has probability $\varphi(k)\, P(x, y)\, dt$ of making a transition in time dt to the point y." Then *the numbers $\sigma_t(x)$ of particles at $x \in S$ at a later time $t > 0$ are independent random variables distributed according to f.*

We do not even have a full proof of the existence of σ_t as a Markov process. A crucial fact in the necessary construction is that the expected number of particles arriving at a fixed point x in time dt equals the expected number of particles departing from x in time dt, and that *both are finite*. Assuming the initial measure μ they are

$$\sum_{y \neq x} \int \mu(d\sigma)\, \sigma(y)\, \varphi[\sigma(y)]\, P(y, x), \qquad \sum_{y \neq x} \int \mu(d\sigma)\, \sigma(x)\, \varphi[\sigma(x)]\, P(x, y), \tag{2.18}$$

respectively. They are both finite since the definition of the measure f in (2.15) implies that $\Sigma k\varphi(k) f(k) < \infty$. They are equal because P is doubly stochastic.

Turning to the proof of (2.17), which will contain an unfortunate gap, let $\mathscr{L}$ denote the space of bounded functions $h : \Sigma \to \mathbf{R}$, *which only depend on a finite number of coordinates.* Thus $h \in \mathscr{L}$ means that h is uniquely determined by its restriction to $\mathbf{N}^A$ for some finite subset A of S. Equation (2.17) may be expressed as

$$\frac{\partial}{\partial t} \int_\Sigma \mu(d\sigma)\, E_\sigma[h(\sigma_t)] = 0, \qquad t \geqslant 0, \quad h \in \mathscr{L}. \tag{2.19}$$

We shall only be able to prove, however, that (2.19) holds for the right-hand derivative at $t = 0$, or

$$\int_\Sigma \mu(d\sigma)\, \Omega h(\sigma) = 0, \tag{2.20}$$

where

$$\Omega\varphi(\sigma) = \lim_{t \downarrow 0} \frac{1}{t}\, E_\sigma[h(\sigma_t) - h(\sigma)] \tag{2.21}$$

exists for all $h \in \mathscr{L}$, and for *almost every* σ, (a.e. with respect to the measure μ).

To bridge the gap between (2.20) and (2.19) observe that the Markov property of σ_t gives

$$\frac{\partial}{\partial t}\int \mu(d\sigma)\, E_\sigma h(\sigma_t) = \frac{\partial}{\partial s}\int \mu(d\sigma)\, E_\sigma h(\sigma_{t+s})|_{s=0}$$

$$= \frac{\partial}{\partial s}\int \mu(d\sigma)\, E_\sigma E_{\sigma_s} h(\sigma_t)|_{s=0}$$

$$= \frac{\partial}{\partial s}\int \mu(d\sigma)\, E_\sigma g(\sigma_s)|_{s=0} = \int \mu(d\sigma)\, \Omega g(\sigma),$$

where $g(\sigma) = E_\sigma[h(\sigma_t)]$.

We would be finished, therefore, if (2.20) and (2.21) were valid for every g of the form $g(\sigma) = E_\sigma[h(\sigma_t)]$, $t \geqslant 0$, $h \in \mathscr{L}$. Such g are certainly not in $\mathscr{L}$, and therefore the proof requires more information about the Markov process σ_t than is available.

The action of the generator Ω on functions in $\mathscr{L}$ can be described by the formula

$$\Omega h(\sigma) = \sum_{x\in S}\sum_{y\in S}[h(\sigma - e_x + e_y) - h(\sigma)]\,\sigma(x)\,\varphi[\sigma(x)]\,P(x, y). \tag{2.22}$$

Here e_x denotes an element of Σ defined by $e_x(y) = \delta(x, y)$, and addition in Σ is defined coordinate-wise. For a fixed pair x, y in S the summand in (2.22) represents the contribution to the limit in (2.21) from a transition of a particle from x to y in the time interval $(0, t)$. This follows from the definition of the speed in (2.2). Next we check that the sum in (2.22) is absolutely convergent, for each $h \in \mathscr{L}$, for σ a.e. (with respect to μ). To see this let A be the finite subset of S, such that $h(\sigma)$ is determined by $\sigma(x)$ for $x \in A$. Then the sum in (2.22) can be decomposed into

$$\sum_{x\in S}\sum_{y\in S} = \sum_{x\in A}\sum_{y\in S-A} + \sum_{x\in S-A}\sum_{y\in A} + \sum_{x\in A}\sum_{y\in A}.$$

The last of these is a finite sum. If $|h| \leqslant M$, then the first sum (with the summands replaced by their absolute values) is dominated by

$$2M\sum_{x\in A}\sum_{y\in S}\sigma(x)\,\varphi[\sigma(x)]\,P(x, y) \leqslant 2M\sum_{x\in A}\sigma(x)\,\varphi[\sigma(x)],$$

which converges a.e. (μ) since it is nonnegative, and its integral with respect to μ is, in view of (2.18),

$$2M \mid A \mid \sum_{k=1}^{\infty} k\varphi(k) f(k) < \infty.$$

The proof for the second double sum is similar, but depends on the fact that P is doubly stochastic. To complete the proof of (2.22) it suffices to remark that all other contributions to (2.21) are negligible. They must represent the events of more than one transition in time $(0, t)$ either to or from the finite set A. For each of these possibilities one readily obtains the upper bound of the order t times the expected number of such transitions in time $(0, t)$, which gives a total contribution of order t^2.

To evaluate the integral in (2.20), let $\mu_{x,y}$ denote the product measure induced by f on $\Sigma_{x,y} = \mathbf{N}^{S-\{x \cup y\}}$. Then, in view of (2.22),

$$\int_{\Sigma} \mu(d\sigma)\, \Omega h(\sigma) = \sum_{x \in S} \sum_{y \in S} \int_{\Sigma_{x,y}} \mu_{x,y}(d\sigma) \sum_{i=1}^{\infty} \sum_{j=0}^{\infty} A(i, j, x, y, \sigma)\, P(x, y). \quad (2.23)$$

If σ_{xy} is the projection of σ on $\Sigma_{x,y}$, then

$$A(i, j, x, y, \sigma) = f(i) f(j) [h(\sigma_{xy} + (i-1)\, e_x + (j+1)\, e_y)$$
$$- h(\sigma_{xy} + ie_x + je_y)]\, i\varphi(i).$$

But $i\varphi(i) f(i) = \rho f(i-1)$, so that

$$\sum_{i=0}^{\infty} \sum_{j=0}^{\infty} A(i, j, x, y, \sigma) = \rho \sum_{m=0}^{\infty} \sum_{n=0}^{\infty} [h(\sigma_{xy} + me_x + (n+1)\, e_y)$$
$$- h(\sigma_{xy} + (m+1)\, e_x + ne_y)] f(m) f(n) = B(x, y, \sigma) - B(y, x, \sigma),$$

if

$$B(x, y, \sigma) = \rho \sum_{m=0}^{\infty} \sum_{n=0}^{\infty} h(\sigma_{xy} + me_x + (n+1)\, e_y) f(m) f(n).$$

Observe now that

$$\int_{\Sigma_{x,y}} \mu_{x,y}(d\sigma)\, B(x, y, \sigma) = B(y), \qquad y \in S$$

is independent of x, and that, as a function of y, it is constant outside the set A. Calling this constant β, and substituting into (2.23),

$$\int_\Sigma \mu(d\sigma)\,\Omega h(\sigma) = \sum_{x\in S}\sum_{y\in S} [B(y) - B(x)]\,P(x,y)\,1_{[x\in A]\cup[y\in B]}$$

$$= \sum_{x\in S}\sum_{y\in S} [B(y) - \beta]\,P(x,y) - \sum_{x\in S}\sum_{y\in S} [B(x) - \beta]\,P(x,y)$$

$$= \sum_{y\in S} [B(y) - \beta] - \sum_{x\in S} [B(x) - \beta] = 0.$$

Here the next to last step used the essential fact that P is doubly stochastic.

The point process defined on Σ by the probability measure μ thus *appears* to be invariant under our Markovian motion with zero-range interaction.

A complete construction of the interaction, and a proof of the invariance of μ, will be given by Holley in [9] under the assumptions that

$$S = \mathbf{Z}, \qquad P(x,y) = P(0, y-x),$$

$$\sum_{m=\infty}^{\infty} |m|\,P(0,m) < \infty,$$

and that the speed change satisfies

$$\lim_{m\to\infty} m\varphi(m) < \infty.$$

It has proved fruitful in other related studies [7, 15] to look at such an equilibrium system from the point of view of one distinguished *tagged particle*. One hopes then to be able to assert that *the system, as seen from the trajectory x_t of the tagged particle, will appear to be governed by the same probability law at all $t \geqslant 0$*. It is quite clear what this means when S is an Abelian group and P is group invariant ($P(x,y) = P(0, y-x)$). One would then hope for the following state of affairs. "There are two probability measures f and g on $\mathbf{N}$ with the following property. Put a random number of particles at the origin $0 \in S$ at time zero, according to the probability measure g, and tag one of them. At all other points $x \neq 0$ put, independently, a random number of particles according to the probability measure f. Let x_t, $t \geqslant 0$ denote the trajectory of the tagged particle, and let $\sigma_t(x)$ denote the number of particles at x, at time t. *Then at each* $t \geqslant 0$ *the random variables* $\sigma_t(x_t)$ *and* $\sigma_t(x_t + x)$ *with* $x \neq 0$ *are all mutually independent, the first with measure* g, *and the others with measure* f." It can be shown that this statement is correct if f is the measure in (2.15) with arbitrary ρ, but subject to the additional hypothesis that $\sum kf(k) < \infty$, and if g is defined as

$$g(k) = \frac{kf(x)}{\sum_1^\infty kf(k)}, \qquad k \geqslant 1. \tag{2.24}$$

Note that g is not just the restriction, properly renormalized, of f to $\mathbf{N}^+$.

Qualitatively (2.24) says that when we are looking at the tagged particle we are likely to be looking at a site with more than the "usual" number of particles. The following heuristic derivation of (2.24) can be made precise. Let $S = \mathbf{Z}$, let N_A denote the number of particles in the interval $[-A, A] \subset \mathbf{Z}$, and let $M_A(k)$ be the number of particles in $[-A, A]$ which sit on a site containing exactly k particles. Then we should have

$$\begin{aligned} g(k) &= \lim_{A\to\infty} \frac{M_A(k)}{N_A} \\ &= \frac{\lim_{A\to\infty}(2A+1)^{-1} M_A(k)}{\lim_{A\to\infty}(2A+1)^{-1} N_A} \\ &= \left[\sum kf(k)\right]^{-1} \lim_{A\to\infty}(2A+1)^{-1} M_A(k) \\ &= \left[\sum kf(k)\right]^{-1} \lim_{A\to\infty}(2A+1)^{-1} EM_A(k) \\ &= \left[\sum kf(k)\right]^{-1} EM_0(k) = \left[\sum kf(k)\right]^{-1} kf(k). \end{aligned}$$

EXAMPLE 2.4. Let $\varphi(k) = 1/k$, as in Example 2.2 of Section 2a. Let $S = \mathbf{Z}$, and take for P the transition function of a random walk on $\mathbf{Z}$, with

$$\sum xP(0, x) = 0, \qquad 0 < \sum x^2P(0, x) = \sigma^2 < \infty.$$

The special form of φ makes possible a simplified description: *Each site* $x \in \mathbf{Z}$ *has an exponential alarm clock with mean one, independent of all the others; when it rings at a site* x *and the site is not empty, then a particle is selected at random and moved to another site* y *with probability* $P(x, y)$. At time zero we put a random number of particles at each site $x \neq 0$

according to $f(k) = (1-\rho)\rho^k$, $k \geqslant 0$. At the origin we put $k \geqslant 1$ particles according to $g(k) = (1-\rho)^2 k\rho^{k-1}$, $k \geqslant 1$, and tag one of them. Let x_t denote the trajectory of the tagged particle. The system is then *in equilibrium, as seen from* x_t. Consequently x_t *is a process with stationary increments.* Further

$$E[x_t] = 0, \qquad E[x_t^2] = \sigma^2(1-\rho)\,t, \qquad t \geqslant 0, \tag{2.25}$$

i.e., *the interaction reduces the variance by a factor of* $(1-\rho)$ *from its value for independent particle motions.* The first part of (2.25) is clear. The second follows from

$$E[x_t^2] = E\left[\sum_{i=1}^{N_t} y_i\right]^2,$$

where N_t is the number of jumps of the tagged particle in time t and y_i are the jumps. By a form of Wald's lemma (Ref. [5], Chapter VII)

$$E[x_t^2] = E[y_i^2]\,E[N_t] = \sigma^2 E[N_t].$$

Since x_t has independent increments, $E[N_t]$ is a linear function of t. Therefore

$$\begin{aligned} E[x_t^2] &= \sigma^2 t E[N_1] = \sigma^2 t \lim_{h\downarrow 0} \frac{E[N_h]}{h} \\ &= \sigma^2 t \sum_{k=1}^{\infty} \varphi(k)\,g(k) = \sigma^2 t(1-\rho)^2 \sum_{1}^{\infty} \rho^{k-1} = \sigma^2 t(1-\rho). \end{aligned}$$

In fact (as pointed out by H. Kesten) one can even prove the central limit theorem

$$\lim_{t\to\infty} P[x_t < \alpha\sigma\sqrt{t(1-\rho)}] = \frac{1}{\sqrt{2\pi}} \int_{-\infty}^{\alpha} \exp\left(-\frac{u^2}{2}\right) du.$$

3b. Simple Exclusion Interaction (Continued)

Concerning the exclusion interaction (only Case I though), when the space S is *countably infinite*, we shall exhibit the same phenomenon as in Section 2b: *For infinite* S *the equilibrium measure is product measure.* For a complete statement recall that each particle moves on S according to a Markov process with transition function $\exp t[P - I]$, where P satisfies[6]

$$P(x, y) \geqslant 0, \qquad \sum_{y \in S} P(x, y) = \sum_{x \in S} P(x, y) = 1, \qquad x, y \in S.$$

Each particle moves independently of the others, *except that transitions to occupied sites of S are suppressed.* For the configuration space we therefore take $\Sigma' = \{0, 1\}^S \subset \Sigma$, the space of all subsets (finite or infinite) $A \subset S$ with the obvious Borel field $\mathscr{B}$ generated by the cylinder sets $B = \{A : A \subset S, x \in A\}$. It can be shown that we obtain a *unique Markov process* A_t, $t \geqslant 0$, with state space Σ', representing the set of sites occupied at time t, for an *arbitrary initial state* $A_0 = A \in \Sigma'$. The corresponding probability measure will be denoted $P_A[\cdot]$. (The construction is easier than the corresponding one in Section 2b, but it is omitted.) Let μ denote the product probability measure on $(\Sigma', \mathscr{B})$ with density ρ, $0 < \rho < 1$, defined by

$$\mu[A : B \subset A \subset S] = \rho^{|B|} \qquad \text{for all} \quad B \subset S, \quad \text{with} \quad |B| < \infty. \tag{3.5}$$

(It is not hard to see that (3.5) holds *if and only if* μ is the Cartesian product of the measures ν on $\{0, 1\}$ with $\nu(1) = \rho$, $\nu(0) = 1 - \rho$.) We conjecture that for each ρ, $0 < \rho < 1$, *the measure* $\mu = \mu_\rho$ *defined by* (3.5) *is an invariant measure for the Markov processes* $\{A_t\}$, $t > 0$. In other words, *the interaction leaves invariant the point processes defined by the recipe: Each site of S, independently of all the others, is occupied by one particle with probability ρ, or empty with probability* $1 - \rho$.

This conjecture will be proved in Ref. [9] for the case when $S = \mathbf{Z}$ and $P(x, y) = P(0, y - x)$ with

$$\sum_{n=-\infty}^{\infty} |n| P(0, n) < \infty.$$

To make the conjecture plausible for arbitrary countable S we proceed as follows. Take a fixed $\sigma \in \Sigma'$ as the initial state $A_0 = C$ of the process A_t, $t \geqslant 0$. Thus σ is a subset C of S which may be finite or infinite. Let

$$f_t(A) = f_t(A; \sigma) = P_\sigma[A_t \supset A], \qquad \text{for} \quad A \subset S, \quad |A| < \infty. \tag{3.6}$$

[6] The irreducibility of P is only needed if one hopes to prove convergence to an invariant equilibrium measure.

We shall show that f_t satisfies the equation

$$\frac{\partial f_t(A)}{\partial t} = \sum_{x\in A}\sum_{y\in A^c} P(y,x)[f_t(A\backslash x \cup y) - f_t(A)] + \sum_{x\in A}\sum_{y\in A^c} [P(x,y) - P(y,x)]\, f_t(A \cup y),$$

$$\text{for all} \quad A \subset S, \quad |A| < \infty. \tag{3.7}$$

Now define (for fixed ρ, $0 < \rho < 1$)

$$g_t(A) = \int \mu(d\sigma)\, f_t(A;\sigma), \qquad A \subset S, \quad |A| < \infty. \tag{3.8}$$

Since g_t is a convex combination of the f_t it follows that g_t also satisfies (3.7). Moreover for g_t we have the initial condition $g_0(A) = \rho^{|A|}$, for every finite subset A of S. The invariance of the measure μ is equivalent to the assertion that $g_t(A) \equiv \rho^{|A|}$ for all $t \geqslant 0$ and all finite $A \subset S$. Indeed this function is a solution of (3.7) since

$$\sum_{x\in A}\sum_{y\in A^c} P(y,x)[\rho^{|A\backslash x\cup y|} - \rho^{|A|}] + \sum_{x\in A}\sum_{y\in A^c} [P(x,y) - P(y,x)]\, \rho^{|A|+1} = \frac{\partial}{\partial t}\rho^{|A|} = 0.$$

(The first sum is zero because $|A\backslash x \cup y| = |A|$ for all $x \in A$, $y \in A^c$, and the second because P is doubly stochastic.) Therefore the proof of the invariance of μ would be complete if we knew that (3.7) has a unique solution under the initial condition $f_0(A) = \rho^{|A|}$. This is precisely the difficulty encountered in Section 2b. However, in the important special case when P is symmetric ($P(x,y) = P(y,x)$ for all $x, y \in S$) we shall see below, in (3.13) and (3.14), that (3.7) takes the form $\partial/\partial t f_t(A) = \Omega f_t(A)$ with Ω a *bounded* operator. In that case (3.7) has a unique solution and the proof of the invariance of μ is complete.

For the proof of (3.7) let $\epsilon_x(t) = 1$ if x is occupied at time t and 0 otherwise. Similarly let $A_{xy} = A_{xy}(t, h)$ be the indicator of the event that a particle makes a transition from x to y in time $(t, t+h)$. Let $\Delta_x(t,h) = \epsilon_x(t+h) - \epsilon_x(t)$, and note that

$$\Delta_x(t,h) = \sum_{y\in S} (A_{yx} - A_{xy}) + R_x \tag{3.9}$$

where the error term R_x is a random variable such that $E_t \mid R_x \mid = O(h^2)$, and E_t is the conditional expectation with respect to the motion up to time t. Note also that

$$
\begin{aligned}
E_t[A_{x_1y_1}A_{x_2y_2}] &= O(h^2) \qquad \text{unless} \quad x_1 = x_2 \quad \text{and} \quad y_1 = y_2 \\
&= \epsilon_x(t)[1 - \epsilon_y(t)]\, P(x, y)\, h \qquad \text{if} \quad x_1 = x_2 = x, \quad y_1 = y_2 = y.
\end{aligned} \tag{3.10}
$$

To simplify the notation we write $\epsilon_x(t) = \epsilon_x$, $\Delta_x(t, h) = \Delta_x$, $x \in S$. By (3.6)

$$
f_t(A) = E_C \prod_{x\in A} \epsilon_x(t) = E_C \prod_{x\in A} \epsilon_x ,
$$

$$
\frac{\partial f_t(A)}{\partial t} = \lim_{h\downarrow 0} h^{-1}E_C \left\{ \prod_{x\in A} [\epsilon_x + \Delta_x] - \prod_{x\in A} \epsilon_x \right\}
$$

$$
= \lim_{h\downarrow 0} h^{-1}E_C \sum_{n=1}^{|A|} \sum_{\substack{B\subset A\\ |B|=n}} \prod_{x\in A-B} \epsilon_x \prod_{y\in B} \Delta_y .
$$

Writing $E_C = E_C E_t$,

$$
\frac{\partial f_t(A)}{\partial t} = E_C h^{-1} \sum_{n=1}^{|A|} \sum_{\substack{B\subset A\\ |A|=n}} \prod_{x\in A-B} \epsilon_x\, E_t \prod_{y\in B} \sum_z (A_{zy} - A_{yz}). \tag{3.11}
$$

Using (3.10) we get, when $\mid B \mid = 1$ and $B = \{a\}$,

$$
E_t \prod_{y\in B} \sum_z (A_{zy} - A_{yz}) = \sum_z [\epsilon_z(1 - \epsilon_a)\, P(z, a) - \epsilon_a(1 - \epsilon_z)\, P(a, z)] + O(h^2),
$$

and when $\mid B \mid = 2$ and $B = \{a, b\}$,

$$
\begin{aligned}
E_t \prod_{y\in B} \sum_z (A_{zy} - A_{yz}) &= \sum_{z_1} \sum_{z_2} E_t[A_{z_1a} - A_{az_1}][A_{z_2b} - A_{bz_2}] \\
&= -\epsilon_a(1 - \epsilon_b)\, P(a, b) - \epsilon_b(1 - \epsilon_a)\, P_{ba} + O(h^2).
\end{aligned}
$$

When $\mid B \mid \geqslant 3$ the contribution is negligible, being of order h^2. Substituting back into (3.11)

$$
\begin{aligned}
\frac{\partial f_t(A)}{\partial t} &= E_C \sum_{a\in A} \prod_{x\in A\setminus a} e_x \sum_y [\epsilon_y(1 - \epsilon_a)\, P(y, a) - \epsilon_a(1 - \epsilon_y)\, P(a, y)] \\
&\quad - E_C \sum_{\substack{\{a,b\}\subset A\\ a\neq b}} \prod_{x\in A\setminus\{a,b\}} \epsilon_x[\epsilon_a(1 - \epsilon_b)\, P(a, b) + \epsilon_b(1 - \epsilon_a)\, P(b, a)].
\end{aligned}
$$

At this point we interchange summation and expectation, obtaining

$$\frac{\partial f_t(A)}{\partial t} = \sum_{a\in A}\sum_{y}[f_t(A\backslash a \cup y) - f_t(A\cup y)]\,P(y,a)$$

$$- \sum_{a\in A}\sum_{y}[f_t(A) - f_t(A\cup y)]\,P(a,y)$$

$$- \sum_{\substack{\{a,b\}\subset A\\ a\neq b}} \{[f_t(A\backslash b) - f_t(A)]\,P(a,b) + [f_t(A\backslash a) - f_t(A)]\,P(b,a)\}.$$

Let us write the above sum formally as $\partial f_t(A)/\partial t = \mathrm{I} - \mathrm{II} - \mathrm{III}$. Now we split I as follows:

$$\mathrm{I} = \sum_{a\in A}\sum_{y\in S} = \sum_{a\in A}\sum_{y\in A^c} + \sum_{a\in A}\sum_{y\in A} = \mathrm{IV} + \mathrm{V}.$$

It is easily seen that $\mathrm{V} - \mathrm{III} = 0$. Also the sum II may be simplified to a sum over $a \in A$ and $y \in A^c$ (its other terms are zero). The net result is not yet Eq. (3.7), but

$$\frac{\partial f_t(A)}{\partial t} = \sum_{x\in A}\sum_{y\in A^c} P(y,x)[f_t(A/x\cup y) - f_t(A\cup y)]$$

$$+ \sum_{x\in A}\sum_{y\in A^c} P(x,y)[f_t(A\cup y) - f_t(A)].$$

However (3.7) is immediately obtained from it by using the doubly stochastic nature of P, which allows us to replace the sum $\Sigma\Sigma P(x,y)\,f_t(A)$ by $\Sigma\Sigma P(y,x)\,f_t(A)$.

Exercise. Give a shorter proof of (3.7) by proving, for an arbitrary $f: \Sigma' \to \mathbf{R}$ and depending only on a finite number of the coordinates of the argument $\sigma \in \Sigma'$, that

$$\lim_{t\downarrow 0} t^{-1}E_\sigma[f(\sigma_t) - f(\sigma)] = \sum_{x:\sigma(x)=1}\;\sum_{y:\sigma(y)=0} [f(\sigma - e_x + e_y) - f(\sigma)]\,P(x,y).$$

Here $\sigma_t = A_t$, $\sigma = A_0$, e_x is the state described by a single particle at x, and addition in Σ' is coordinate-wise. Equation (3.7) should result by specializing to $f(\sigma) = \prod_{x\in A}\sigma(x)$, for an arbitrary finite set A.

Important new phenomena appear under the *additional hypothesis that the transition operator P is symmetric*, i.e.,

$$P(x,y) = P(y,x), \qquad x, y \in S. \tag{3.12}$$

Now (3.7) reduces to

$$\begin{aligned}\frac{\partial f_t(A)}{\partial t} &= \sum_{x\in A}\sum_{y\in A^c} P(x,y)[f_t(A\backslash x\cup y) - f_t(A)] \\ &= \sum_{B\subset S} \Omega(A,B) f_t(B) = \Omega f_t(A), \qquad A\subset S, \quad |A| < \infty.\end{aligned} \tag{3.13}$$

Note that the operator Ω is precisely the infinitesimal generator of the interaction on a finite set S, determined in (3.2). Observe that its restriction to the Banach space $\mathscr{L}_n$ of bounded real functions on the subsets $A\subset S$ of cardinality n is a bounded linear operator.[7] Therefore we may write the unique solution of (3.13) as

$$f_t(A) = \sum_{B} e^{t\Omega}(A,B) f_0(B), \qquad t\geqslant 0, \qquad A\subset S, \qquad |A| < \infty \tag{3.14}$$

where the sum extends over all $B\subset S$ such that $|B| = |A|$ (by its definition in (3.13) and also (3.2), $\Omega(A,B) = 0$ when $|A| \neq |B|$.)

Recall that $f_t(A) = P_C[A_t \supset A]$ for a fixed but arbitrary initial set C, which may be finite or infinite. Therefore

$$f_0(B) = P_C[A_0 \supset B] = P_C[C\supset B] = \begin{cases} 1 & \text{if } C\supset B \\ 0 & \text{if } C\not\supset B. \end{cases}$$

It follows from (3.14) that

$$P_C[A_t\supset A] = \sum_{\substack{B:B\subset C\\ |B|<\infty}} e^{t\Omega}(A,B), \qquad t\geqslant 0, \qquad A\subset S, \qquad |A| < \infty. \tag{3.15}$$

Consider on the other hand, the Markovian motion A_t of a finite set of particles on S, such that the set occupied at time zero is $A_0 = A$, $|A| < \infty$. Then it follows from the elementary theory in Section 3a (valid even when $|S| = \infty$ as long as the number of particles is finite) that

$$P_A[A_t = B] = e^{t\Omega}(A,B). \tag{3.16}$$

(The motion A_t takes place on the countable state space of subsets of S which have all the same cardinality, viz., $|A|$, and the generator of this

[7] Its norm is at most $2|A| = 2n$.

Markov process is Ω, as shown in (3.2).) Equation (3.16) implies that

$$P_A[A_t \subset C] = \sum_{\substack{B:B\subset C \\ |B|<\infty}} e^{t\Omega}(A, B), \tag{3.17}$$

for all finite A and for arbitrary C. Comparison of (3.15) and (3.17) yields the law that

$$P_A[A_t \subset C] = P_C[A_t \supset A] \quad \textit{for} \quad |A| < \infty, \quad |C| \leqslant \infty. \tag{3.18}$$

The probability that all particles lie in C at time t, having started in configuration A (finite) is the same as that all points of the finite set A are occupied at time t, if at time zero the configuration was C (possibly infinite).

When $|A| > |C|$ the result is trivial in content, since both sides in (3.18) vanish. When $|A| = |C| < \infty$, (3.18) becomes $P_A[A_t = C] = P_C[A_t = A]$, which expresses a type of *time reversibility* of the motion. When $|A| < |C| = \infty$, then (3.18) is extremely useful, as shown in

EXAMPLE 3.1. Suppose that the initial configuration is distributed according to A_0 = set of even integers (we assume $S = \mathbf{Z}$). Then the probability $\pi(t)$ that the origin is occupied by a particle at time t is the probability that a single particle, starting at the origin, will be at an even integer at time t, viz.,

$$\pi(t) = \sum_{x=\text{even}} \exp t[P - I](0, x),$$

which will tend to $\frac{1}{2}$ under mild assumptions on P.

Thus (3.18) reduces difficult interaction problems to easier ones with less or no interaction (as in the above example). It should be an essential tool in proving an *ergodic theorem* to the effect that *for each* ρ, $0 < \rho < 1$,

$$\lim_{t\to\infty} P_C[A_t \supset B] = \rho^{|B|}, \tag{3.19}$$

for all initial configurations C which have "approximately" the density ρ, and for a nice class of transition operators $P(x, y)$ on S.

It can be shown (either from Theorem 6.5 of Ref. [7], or by a long calculation which generalizes (3.7) that the simple exclusion interaction is *strongly stationary* when S is an Abelian group and $P(x, y) = P(0, y - x)$. This means that the system is stationary from the point of view of the tagged particle: *Put particles at all points* $x \neq 0$, *independently, with probability* ρ, *and put a tagged particle at* 0. *Let* $x(t)$ *denote the* *trajectory of the tagged particle. For every finite set A, not containing 0, the probability that $A + x(t)$ is occupied at time t is then $\rho^{|A|}$ for all $t \geqslant 0$.*

As indicated in Section 2b this implies that the non-Markovian process $x(t)$ has stationary increments. This, in turn, implies that $Ex(t)$ is a linear function of t and that with probability one

$$\lim_{t\to\infty} \frac{x(t)}{t} = \lim_{h\downarrow 0} \frac{E[x(h)]}{h} = (1-\rho)\sum_y yP(0,y).$$

Here we have assumed that $S = \mathbf{Z}_n$. However it is not necessary that $P(x, y) = P(y, x)$.

EXAMPLE 3.2. Let $S = \mathbf{Z}$ and $P(x, y) = 1$ if $y = x + 1$, and 0 otherwise. Let $x(t)$ denote the trajectory of a tagged particle starting at $x(0) = 0$, while at time zero all other points are occupied with probability ρ. Then, as was discovered by H. Kesten, the process $x(t)$, $t \geqslant 0$, is Markovian, in fact, *it is the simple Poisson process* with speed $1 - \rho$, i.e., $Ex(t) = (1 - \rho)t$.

PROBLEM. Consider the same situation as in the example above, but let $P(x, y) = \frac{1}{2}$ when $|x - y| = 1$ (simple random walk). Then $Ex(t) \equiv 0$, while it is not known how to estimate the variance $\sigma^2(t) = Ex^2(t)$. There is reason to believe, by analogy with the results in Ref. [6], that the interaction has the effect of reducing the variance from a linear function of t (as is the case without interaction) to $\sigma^2(t) \sim c\sqrt{t}$, as $t \to \infty$. If so, for what c? T.M. Liggett has disproved this conjecture, and shown that $\sigma^2(t) \sim ct$ for large t (oral communication).

4b. SPEED CHANGE WITH EXCLUSION (CONTINUED)

We select a special interaction model from Section 4a for extension to infinite S, because of its evident interest in statistical mechanics. Let $S = \mathbf{Z}$, and let P denote the transition function of a symmetric *random walk* on $\mathbf{Z}$, such that

$$P(x, y) = P(y, x) = P(0, y - x), \qquad 0 < \sum |x|^2 P(0, x) < \infty, \qquad x, y \in \mathbf{Z}. \tag{4.7}$$

Let $V : \mathbf{N} \to R$ such that

$$\sum_{k=1}^{\infty} k\,|V(k)| < \infty, \tag{4.8}$$

and let V determine the pair interaction

$$U(x, y) = V(|x - y|), \qquad x, y \in Z. \tag{4.9}$$

On the Markovian motion of particles according to $\exp t[P - I]$ we superimpose the interaction which consists of

(i) *Suppression of transition to occupied sites*;

(ii) *A particle at x at time t moves with speed*

$$c(x, A_t) = \exp \sum_{y \in A_t} U(x, y), \tag{4.10}$$

just as in Section 4a, *where A_t is the set of all sites which are occupied at time t.*

We shall not attempt to prove the intuitively plausible fact that this interaction defines a unique *Markov process A_t, $t \geqslant 0$*, with state space $\Sigma' = \{0, 1\}^{\mathbf{Z}}$, the subsets of $\mathbf{Z}$. We proceed instead with an (unfortunately incomplete) demonstration that *the process A_t has a certain family of invariant measures μ_ρ, $0 < \rho < \infty$, on $(\Sigma', \mathscr{B})$*. (Here $\mathscr{B}$ is the same Borel field as in Section 3b.) For a complete proof, including the existence of the Markov process A_t, see Holley [9].

We consider finite intervals $S_N = [-N, N] \in \mathbf{Z}$, and let $\Sigma_N' = \{0, 1\}^{S_N}$ and $\mathscr{B}_N$ be the corresponding finite Borel field. On $(\Sigma_N', \mathscr{B}_N)$ we define, for each $\rho > 0$, the probability measure $\nu_\rho^{(N)}$ by

$$\nu_\rho^{(N)}(A) = \Theta_N^{-1} \exp -\frac{1}{2}\Big[\sum_{x \in A} \sum_{y \in A} U(x, y) + \rho\,|A|\Big], \qquad A \in \Sigma_N' \tag{4.11}$$

where

$$\Theta_N = \sum_{k=0}^{2N+1} e^{-\rho k/2} \sum_{B \in \Sigma_N', |B| = k} \exp -\frac{1}{2} \sum_{x \in B} \sum_{y \in B} U(x, y). \tag{4.12}$$

Finally we have to define projection and convergence of measures. The spaces Σ' and Σ_N' are given the compact product topology. If ν is a measure on $(\Sigma_N', \mathscr{B}_N)$ and $M \geqslant N$ then the projection $\Pi_{N,M}\nu$ is a measure on $(\Sigma_M', \mathscr{B}_M)$ defined by

$$\int_{\Sigma_N'} f(\sigma)\, \Pi_{N,M}\nu(d\sigma) = \int_{\Sigma_M'} f(\sigma)\, \nu(d\sigma)$$

for all $f \in C(\Sigma_M')$ such that $f(\sigma)$ depends only on the coordinates of σ in $[-N, N]$. In the same way we define the projection Π_N of a measure ν on $(\Sigma_N', \mathscr{B}_N)$ on the measure $\Pi_N\nu$ on $(\Sigma', \mathscr{B})$. Weak convergence of measures on $(\Sigma_M, \mathscr{B}_M)$ (denoted by $\Rightarrow$) is defined as convergence of functionals on $C(\Sigma_M')$. In this setting we have the theorem of D. Ruelle [12]:

"*For each* $\rho > 0$ *there exists a probability measure* μ_ρ *on* $(\Sigma', \mathscr{B})$, *and for every* $\rho > 0$ *and* $M \geqslant 1$ *there exists a probability measure* $\omega_\rho^{(M)}$ *on* $(\Sigma_M', \mathscr{B}_M)$ *such that*

(i) $\Pi_{M,N}\nu_\rho^{(N)} \Rightarrow \omega_\rho^{(M)}$ *as* $N \to \infty$,

(ii) $\omega_\rho^{(M)} = \Pi_M\mu_\rho$ *for all* $M \geqslant 1$."

To show why the measures μ_ρ are invariant for $\{A_t\}_{t\geqslant 0}$ we go back to the interaction of particles on S_N. Examination of (4.3) and (4.4) shows that ν_ρ^N is an invariant measure for the interaction on S_N, because it is a convex combination of invariant measures (the explicit formula is in (4.11) and (4.12)). There is one slight difficulty however: We cannot speak of interactions with a common P matrix for each N, but have to modify $P(x, y)$ to make it symmetric and stochastic on $[-N, N]$. In the case of simple random walk this needs to be done only at the end points of $[-N, N]$. Assume now that this has been done. Let us denote by $A_t^{(N)}$ and A_t the configuration Markov processes on Σ_N' and Σ', respectively, and let

$$Q_t^{(N)}(\sigma, B) = P_\sigma[A_t^N \in B], \qquad \sigma \in \Sigma_N', \quad B \in \mathscr{B}_N,$$
$$Q_t(\sigma, B) = P_\sigma[A_t \in B], \qquad \sigma \in \Sigma', \quad B \in \mathscr{B}.$$

Now we know that, for each $\rho > 0$

$$\nu_\rho^{(N)}(B) = \int_{\Sigma_N'} \nu_\rho^{(N)}(d\sigma)\, Q_t^{(N)}(\sigma, B), \qquad B \in \mathscr{B}_N, \tag{4.13}$$

and wish to use this to prove

$$\mu_\rho(B) = \int_{\Sigma'} \mu_\rho(d\sigma)\, Q_t(\sigma, B), \qquad B \in \mathscr{B}_N. \tag{4.14}$$

Ruelle's theorem would give the desired conclusion if we could assert that for each fixed $t > 0$ and $B \in \mathscr{B}$,

$$\lim_{N\to\infty} Q_t^{(N)}(\sigma^{(N)}, B) = Q_t(\sigma, B), \tag{4.15}$$

uniformly over $\sigma \in \Sigma'$. Here $\sigma^{(N)}$ denotes the projection of σ on Σ_N'. It would suffice to do all this for finite cylinder sets B, and then (4.15) simply says that whether a finite set is occupied at time t does not depend much on what the situation is very far away from this set at time zero. Incidentally this difficulty is very similar to those encountered in proving that $\{A_t\}$ exists as a Markov process in the first place.

The same problem in the plane ($\mathbf{Z}_2$) or space ($\mathbf{Z}_3$), instead of on the integers, has an entirely different character. There it is known [11] that the thermodynamic limit of the measures $\nu_\rho^{(N)}$ may fail to exist, or fail to depend continuously on the density parameter ρ. This is true for example for the nearest neighbor potential $U(x, y) = -1$ if $|x - y| = 1$, 0 otherwise. The physical interpretation is that the mathematical model of a gas represented by the Gibbs measures $\nu_\rho^{(N)}$ at different densities ρ exhibits the phenomenon of phase change. In terms of our Markov process A_t this means that for certain ρ *there exists no extremal*[8] *invariant probability measure* on Σ'. Intuitively this means (since $U \leqslant 0$) that particles slow down when they approach other particles. When there is a certain critical particle density this slowing down apparently has the effect of forming large clusters or regions of high particle density surrounded by large spaces of low density. In other words, separation occurs into a thick phase (liquid) and a thin phase (gas) in the case when no extremal invariant probability measure exists (two phases may coexist).

5b. Nearest Neighbor Interaction (Continued)

The theorem of Section 5a has a natural generalization from a finite cyclic group S to the infinite groups T, $\mathbf{Z}$, $\mathbf{R}$. Just as in the transition from Section 2a to Section 2b it is easy to guess the correct generalization. The invariant measure Φ in (5.3) is given for a fixed number (n) of particles. If one takes suitable convex combinations of the measures $\Phi = \Phi_n$ in (5.3) with respect to n one is led to *product measures*. One finds, in this way, that interaction leaves invariant certain point processes which have the property that the *spacings between adjacent particles are independent random variables*. The probability measure of the individual

[8] Nonextremal invariant measures aboud but do not represent a pure thermodynamic phase (Ref. [11], Chapter 6). All conceivable motions leave invariant the measure of density ρ defined by: "With probability ρ all sites are occupied, with probability $1 - \rho$ all are empty."

spacings is a constant multiple of $\rho^k \exp[-U(k)]$, where U is the interaction potential and ρ is an arbitrary positive density parameter. The thermodynamic passage to the limit from these measures to the corresponding product measures on $\mathbf{Z}$ and $\mathbf{R}$ is of interest in statistical mechanics. The limit represents the Gibbs distribution for an infinite one-dimensional gas with nearest neighbor interaction, for which explicit thermodynamic calculations are possible [8]. But it was not known that, as will be described now,[9] this Gibbs distribution is invariant under certain Markovian motions with interaction.

The main ingredients are: A bounded measurable function $U : \mathbf{R}^+ \to \mathbf{R}$, the *interaction potential*, and for notational convenience, the function

$$\varphi(x) = \exp U(x), \qquad x \in \mathbb{R}^+; \tag{5.7}$$

secondly, a symmetric probability measure ν on $\mathbf{R}$. This is used to define the transition probabilities of a random walk on $\mathbf{R}$ by

$$P(x, dy) = P(0, dy - x) = P(0, -dy + x) = \nu(dy - x), \qquad x, y \in \mathbf{R}. \tag{5.8}$$

Now P is used to define the compound Poisson transition function

$$\exp t(P - I)(x, dy) = e^{-t} \sum_0^\infty \frac{t^n}{n!} P^n(x, dy) \tag{5.9}$$

of the motion of individual particles, upon which the interaction is to be imposed. The interaction will be such as to *preserve order* of particles on $\mathbf{R}$. It is therefore possible to tag the individual particles by denoting their trajectories,

$$x(t) = \{x_k(t)\}_{k=-\infty}^{\infty}, \qquad t \geqslant 0, \ldots, < x_{-1}(t) < x_0(t) < x_1(t) < \cdots. \tag{5.10}$$

The *spacings between adjacent particles* at time t will be denoted by

$$\delta(t) = \{\delta_k(t)\}_{k=-\infty}^{\infty}, \qquad t \geqslant 0, \qquad \delta_k(t) = x_{k+1}(t) - x_k(t) > 0. \tag{5.11}$$

Since the spacings will be of primary interest, there will be no obstacle to assuming, for the sake of fixing initial coordinates, that $x_0(0) = 0$.

The random (non-Markovian) motion of the trajectories $x_k(t)$ takes

[9] We give only the theory for $\mathbf{R}$, which is somewhat richer than on $\mathbf{Z}$ (due to the remark following (5.15)). It should be obvious how to reformulate the conjectures for $\mathbf{Z}$, and complete proofs for this case are given in the sequel to this paper by Holley [9].

place according to the compound Poisson process in (5.9) with the following interaction. The *speed* of $x_k(t)$ at a fixed time t_0 is

$$c_k(t_0) = \varphi[\delta_{k-1}(t_0)]\, \varphi[\delta_k(t_0)], \tag{5.12}$$

i.e., $x_k(t)$ moves according to $\exp c_k(t_0)\, t[P - I]$ for t in the (positive) time interval $[t_0, t')$ until the first change occurs in the value of $c_k(t)$. Further, all *transitions are suppressed* which would upset the order of the particles on $\mathbf{R}$. Thus the only transitions permitted are those

$$\textit{from} \quad x_k \quad \textit{to} \quad x_k + y_k \quad \textit{with} \quad -\delta_{k-1} < y_k < \delta_k . \tag{5.13}$$

It is claimed (without proof) that the above motion defines a *Markov process* $\delta(t)$, $t \geqslant 0$, *with state space* $D = (\mathbf{R}^+)^{\mathbf{Z}}$. Now let $\mathscr{B}$ be the field of Borel sets of $\mathbf{R}^+$, and let $\mathscr{D}$ be the corresponding product Borel field for D. Let ρ be a positive *density parameter*, and

$$\begin{aligned} f_\rho(x) &= f(x) = \theta^{-1} \frac{e^{-\rho x}}{\varphi(x)} = \theta^{-1} \exp[-\rho x - U(x)], \qquad x \in \mathbf{R}^+ \\ \theta &= \int_0^\infty \exp[-\rho x - U(x)]\, dx. \end{aligned} \tag{5.14}$$

Then f_ρ defines a probability measure on $\mathbf{R}$. Let μ_ρ denote the corresponding product probability measure on $(D, \mathscr{D})$. Then for each $\rho > 0$, μ_ρ *is an invariant measure for the Markov process* $\delta(t)$, $t \geqslant 0$, i.e.,

$$\int_D \mu_\rho(d\delta)\, P_\delta[\delta(t) \in A] = \mu_\rho(A), \qquad A \in \mathscr{D}, \tag{5.15}$$

if $P_\delta[\cdot]$ denotes the probability measure of the Markov process $\delta(t)$ starting at $\delta(0) = \delta \in D$.

Remark. Extensions of this result are possible in two directions. The potential U need not necessarily be bounded, but then the proof that $\delta(t)$ exists and is Markovian becomes even more difficult. More important perhaps, the compound Poisson motion in (5.9) may be replaced by any symmetric process with stationary independent increments, such as Brownian motion. But in that case one has to make careful provision for describing the motion of two particles which collide (arrive simultaneously at the same point). Fortunately it has been shown by Harris [6], in a simpler case, how one can define the reflection (interchange of trajectories) of two particles which is necessary to preserve linear order. In fact the interaction model of Harris is precisely the present model for Brownian motion with the potential $U \equiv 0$.

The proof of (5.15) breaks down at the same point as in Sections 2b and 3b. It may be somewhat easier to proceed by the method of the thermodynamic limit, as outlined in (4.13), (4.14), and (4.15), the necessary analog of Ruelle's convergence theorem being quite obvious in this case. But for the sake of completeness we sketch the analog of the computation in Section 2b. Let $\mathscr{L}$ denote the class of bounded functions $h : D \to \mathbf{R}$ which depend only on a finite number of coordinates of their argument $\delta \in D$. Then we wish to show

$$\frac{\partial}{\partial t} \int_D \mu(d\delta)\, E_\delta h[\delta(t)] = 0, \qquad t \geqslant 0, \quad h \in \mathscr{L}, \tag{5.16}$$

but can only do this at the origin $t = 0$. First we can write down, for arbitrary $h \in \mathscr{L}$,

$$\lim_{t \downarrow 0} t^{-1} E_\delta\{h[\delta(t)] - h(\delta)\}$$
$$= \sum_{k=-\infty}^{\infty} \varphi(\delta_{k-1})\, \varphi(\delta_k) \int_{-\infty}^{\infty} \nu(dy)\, 1_{[-\delta_{k-1}<y<\delta_k]}[h(\delta + e_{k-1}y - e_k y) - h(\delta)]. \tag{5.17}$$

Here e_k denotes the element of D whose k-th coordinate is one, and all others zero. There is no question of convergence since only a finite number of terms in the sum of (5.17) are nonzero. The k-th term in this sum represents the contribution of a transition of the k-th particle from x_k to $x_k + y$, and it is easy to see that the possibility of more than one transition in the time interval $(0, t)$ does not contribute to the limit.

Finally, we show that (5.16) holds at $t = 0$, or

$$\sum_{k=-\infty}^{\infty} \int_D \mu(d\delta)\, \varphi(\delta_{k-1})\, \varphi(\delta_k)$$
$$\times \int_{-\infty}^{\infty} \nu(dy)\, 1_{[\ \delta_{k-1}<y<\delta_k]}[h(\delta + e_{k-1}y - e_k y) - h(\delta)] = 0. \tag{5.18}$$

In fact each term of the above sum will vanish. Therefore we fix k, let μ' denote the projection of the product measure μ on $D' = \mathbf{R}^{\mathbf{Z}-\{k-1,k\}}$, and write $\delta' = \delta - e_k\, \delta_k - e_{k-1}\, \delta_{k-1}$. Then the k-th term in (5.18) can be written

$$\int_{D'} \mu'(d\delta') \int_0^\infty \int_0^\infty f(s) f(t)\, \varphi(s)\, \varphi(t)$$
$$\times \int_{-\infty}^{\infty} \nu(dy)\, 1_{[-s<y<t]}\{h[\delta' + e_{k-1}(s + y) - e_k(t - y)]$$
$$- h[\delta' + e_{k-1}s + e_k t]\}\, ds\, dt.$$

Substitution of the formula for f in (5.14) yields

$$\theta^{-2} \int_{D'} \mu'(d\delta') \int_0^\infty \int_0^\infty e^{-\rho(s+t)}$$

$$\times \int_{-\infty}^\infty \nu(dy)\, 1_{[-s<y<t]} \{F(s+y, t-y) - F(s,t)\}\, ds\, dt$$

where (suppressing the dependence on δ')

$$F(s,t) = h[\delta' + e_{k-1}s + e_k t].$$

It will now suffice to show that for each δ'

$$\int_0^\infty \int_0^\infty \exp[-\rho(s+t)]$$

$$\times \int_{-\infty}^\infty \nu(dy)\, 1_{[-s<y<t]} \{F(s+y, t-y) - F(s,t)\}\, ds\, dt = 0.$$

Interchange of the order of integration yields an integral of the form $\int_{-\infty}^\infty \nu(dy)\, H(y)$. In view of the fact that the probability measure ν is symmetric about zero it suffices to check that H is an odd function. One finds

$$\begin{aligned} H(y) &= \int_{-y\vee 0} ds \int_{y\vee 0} dt\, e^{-\rho(s+t)}[F(s+y, t-y) - F(s,t)] \\ &= \int_{y\vee 0} du \int_{-y\vee 0} dv\, e^{-\rho(u+v)}[F(u,v) - F(u-y, v+y)] \\ &= -\int_{y\vee 0} ds \int_{-y\vee 0} dt\, e^{-\rho(s+t)}[F(s-y, t+y) - F(s,t)] \\ &= -H(-y). \end{aligned}$$

Thus we have made plausible that *the spacings of tagged particles on* **R** *remain independently distributed according to the density* f_ρ, *if this is the case at time zero.* It remains, however, to exhibit a point process on **R** which is left invariant under the motion of the particle system. There is, according to a theorem of Ryll-Nardzewski [13], a unique point process corresponding to each spacing density f. (The fact that f has finite mean is essential here.) This point process has the property that, conditional to there being a particle at a fixed point x, all the spacings between particles, to the right and to the left of x, are independent random variables with common probability density f. To avoid new notation, we describe this point process picturesquely. *Place the particle nearest to* 0 *on the right at* $+R$, *and the nearest one on the left at* $-L$, *with* R *and* L *chosen according to the probability density*

$$\operatorname{Prob}[R \in dx, L \in dy] = \frac{f(x+y)\,dx\,dy}{\int_0^\infty tf(t)\,dt}, \qquad x > 0, \qquad y > 0. \tag{5.19}$$

Then proceed to place particles to the right of R *with independent spacings according to* f, *and similarly to the left of* $-L$.

It follows from Theorem 6.5 of Ref. [7] that *this point process is invariant under the motion.*

EXAMPLE 5.1. Let $U(x) \equiv 0$, so that the interaction between particles consists solely of suppression of transitions which would change the order of particles. Then f_ρ is the exponential density, with mean ρ^{-1}, and the corresponding point process is the Poisson point process with density ρ. Thus the P.p.p. *is invariant under this motion in spite of the interaction.*

REFERENCES

1. V. I. ARNOLD AND A. AVEZ, "Ergodic Problems of Classical Mechanics," Benjamin, New York, 1968.
2. C. DERMAN, Some contributions to the theory of denumerable Markov chains, *Trans. Amer. Math. Soc.* **79** (1955), 541–555.
3. R. L. DOBRUSHIN, Description of a random field by means of conditional probabilities and its regularity conditions, *Probability Theory and its Applications* **13** (1968), 197–224.
4. R. L. DOBRUSHIN, I. I. PYATETSKI-SHAPIRO, AND N. B. VASILEV, "Markov Processes in an Infinite Product of Discrete Spaces," Soviet-Japanese Symposium in Probability Theory, Khavarovsk, 1969.
5. J. L. DOOB, "Stochastic Processes," Wiley, New York, 1953.
6. T. E. HARRIS, Diffusion with collisions between particles, *J. Appl. Probability* **2** (1965), 323–338.
7. T. E. HARRIS, Random measures and motions of point processes, to appear.
8. M. FISHER AND B. WIDOM, Decay of correlations in linear systems, *J. Chem. Phys.* **9** (1969), 3756–3772.
9. R. HOLLEY, A class of interactions in an infinite particle system, *Advances in Math.* **5** (1970), 291–309.
10. A. I. KHINCHIN, "Statistical Mechanics," Dover, New York, 1949.
11. D. RUELLE, "Statistical Mechanics," Benjamin, New York, 1969.
12. D. RUELLE, Statistical mechanics of a one-dimensional lattice gas, *Comm. Math. Phys.* **9** (1969), 267–278.

13. C. Ryll-Nardzewski, "Remarks on Processes of Calls," Proc. Fourth Berkeley Symp., Vol. 2, pp. 455–465, Univ. of California Press, 1961.
14. F. Spitzer, "Random Processes Defined Through the Interaction of an Infinite Particle System," "Lecture Notes in Mathematics," Vol. 89, pp. 201–223, Springer-Verlag, New York/Berlin, 1969.
15. F. Spitzer, Uniform motion with elastic collision of an infinite particle system, *J. Math. Mech.* **18** (1969), 973–990.
16. C. Stone, On a theorem of Dobrushin, *Ann. Math. Statist.* **39** (1968), 1391–1401.
17 T. Thedéen, Convergence and invariance questions for point systems in R_1 under random motion, *Ark. Mat.* **7** (1967), 211–239.

PRINTED IN BELGIUM BY THE ST. CATHERINE PRESS, TEMPELHOF, 37, BRUGES, LTD.

PAPERS DEDICATED TO FRANK SPITZER

A Useful Renormalization Argument

MAURY BRAMSON AND LAWRENCE GRAY[1]

We define a collection of 'generic' population models for which we prove a survival criterion by using a renormalization argument. These models can be compared with other more familiar models, leading to simple proofs of various survival results. In particular, we prove a generalization of Toom's Theorem concerning survival in multidimensional probabilistic cellular automata. Our technique should also be applicable to a variety of other discrete and continuous time models.

0. Introduction. Consider the discrete time version of the basic asymmetric one-dimensional contact process. In this model, one imagines that a population of particles is located at the sites (points) of the integer lattice $\mathbb{Z}$, with at most one particle per site. A site containing a particle is called **occupied**, and a site with no particles is called **vacant**. During each discrete time step, the population changes according to a two step procedure: first, a new particle is placed at each vacant site x for which the neighboring site $x+1$ is occupied; second, occupied sites (including newly occupied sites) are vacated with probability ε, independently of one another. Thus, the probability that x is occupied at time t, given the past, is $1-\varepsilon$ if x or $x+1$ is occupied at time $t-1$, and 0 otherwise. The parameter ε is called the **death rate**. This describes a simple model for a population of individuals with geometrically distributed lifetimes and with births occurring only at vacant sites immediately to the left of (in 'contact' with) an occupied site. (The name 'contact process' was first applied to a continuous time version of this model, with the population under consideration being a collection of diseased cells. For simplicity, we will restrict this introductory discussion mostly to discrete time models. The basic asymmetric one-dimensional discrete time contact process is equivalent to oriented site percolation on $\mathbb{Z}^2$.)

If the death rate ε is close to 1, it is easy to show that even if all the sites are occupied at time 0, the population will die out, in the sense that the probability any given site is occupied will converge to 0 as $t \to \infty$. On the other hand, it was shown by Stavskaya and Pyatetskii-Shapiro [12] that if $\varepsilon > 0$ is sufficiently small, the population does not die out if it starts with all sites occupied. Instead, there is a non-trivial equilibrium for the process

[1]Both authors partially supported by the National Science Foundation.

in which the probability that a given site is occupied is strictly positive. We describe this situation by saying that the contact process **survives** for all sufficiently small death rates.

A large proportion of the work done in the field of interacting particle systems has been concerned with the question of whether a given parametrized family of processes survives at certain parameter values. It is interesting that between 1968 and 1988, with the notable exception of the work of Holley and Liggett [10] concerning the one-dimensional contact process in continuous time, virtually all work concerned with showing survival relied on a common technique, the so-called contour method. This method was first introduced in a statistical mechanics setting by Peierls [11], much later developed to a very high level for discrete time systems by Toom [13], adapted to the continuous time setting for the contact process by Harris [9], and then developed for more general use in continuous time by Gray and Griffeath [8], Bramson and Gray [4], and Durrett and Gray [5].

In the work of Durrett and Griffeath [6], another idea borrowed from statistical mechanics (and percolation theory) made its way into the field of interacting particle systems, namely the block rescaling method. Durrett and Griffeath used this technique to get more detailed information about the survival probabilities of the contact process.

Block rescaling is a powerful tool for examining the survival of a large class of processes. Many processes can be compared with the contact process by using this method, thus allowing a relationship to be established between the survival of the contact process and the survival of these other processes. This technique was first applied in Bramson [1], Bramson and Durrett [2], and Bramson, Durrett, and Swindle [3]. In particular, this procedure allows one to prove survival for interesting models that were too complicated to yield to contour methods.

In the rescaling procedure used in [1], [2], and [3], the discrete time contact process (or at least a close relative) plays a special role. The point is that the contact process is a relatively simple process whose behavior is fairly well understood, and it is possible to find transformations for many other processes that allow them to be compared with the contact process. However, not every process that survives can be compared in this fashion with the contact process. An example is the discrete time process known as **Toom's model**. We describe one version of this model here. It is similar to the discrete time contact process, with two important differences. The first difference is that it is a two-dimensional model, that is, the sites are the points of the two-dimensional integer lattice $\mathbb{Z}^2$. Again, there is a two stage procedure for changing the state of the system from one time to the next. The second stage of this procedure is the same as before (vacate occupied sites with probability ε), but during the first stage, a new particle appears at a vacant site (x, y) if and only if *both* of the two neighboring sites $(x+1, y)$ and $(x, y+1)$ are occupied by particles. (As before, a given particle disappears with probability ε during any time step.) Thus it takes at least two particles, properly positioned, to produce a new particle at a vacant site, which is why the continuous time version of this model is sometimes said to have sexual reproduction, to distinguish it from the contact process in which reproduction is asexual. There is no known way to prove that Toom's model survives by comparing it to a basic contact process. Note that if we change the rule to 'a vacant site is occupied if and only if *at least one* of the two neighboring sites $(x+1, y)$ or $(x, y+1)$ is occupied', then it is easy to make a direct comparison to the asymmetric contact process in one dimension to prove survival for small death rates. This 'at least one' model is a two-dimensional contact process, with asexual reproduction. Like the one-dimensional contact process, it can be used as a comparison to many models, but not to Toom's model. Toom [13] used contour methods to prove survival for Toom's model, and Durrett and Gray [5] used contour methods to prove this and other interesting facts about the continuous time version. There are several variations on Toom's model in which more symmetry exists, but they all behave similarly, and none of them seems to be comparable in any useful way to basic contact processes.

One can imagine that Toom's model could be used as a comparison process in the same way as the basic contact process. That is, one should be able to find other processes whose survival can be proved by transforming them into Toom's model, using some version of the block rescaling technique. There are still other processes which survive, but which can neither be transformed into Toom's model, nor into the basic contact process. In fact many (but not all!) of the higher dimensional processes in the general class studied by Toom appear to be not comparable with the one- or two-dimensional models.

In this paper, we introduce a class of **generic population models** which can be used as comparison processes via the block rescaling technique. These models have two desirable properties. The first is that they form a sufficiently general class so that many interesting attractive systems that survive can be compared to them by an appropriate application of the block rescaling technique. In particular, we will be able to use them to give a much simpler proof of Toom's Theorem [13] concerning the survival of multidimensional discrete time systems. The second property of these generic models is that they lend themselves to *renormalization*, which is to say that when the block rescaling argument is applied to them in a natural way, they are transformed into themselves. Once this is done, an induction argument then leads to a necessary and sufficient criterion for the survival of generic models. Thus, the block rescaling technique accomplishes two purposes: it transforms models of interest into generic processes, and it then tells us which generic processes survive. In previous work, this second step has usually required a (sometimes quite difficult) contour argument.

Here is an overview of the rest of the paper. In Section 1 we construct the generic population models and indicate briefly through some examples why they work well as comparison processes. It will be seen that they are temporally *and spatially* continuous analogues of Toom's general class of probabilistic cellular automata. In Section 2, we carry out the renormalization procedure and establish necessary and sufficient conditions for the survival of generic processes (Theorem 1). In Section 3, we prove that a general class of processes, which includes the class studied by Toom, can be compared to generic processes (Theorem 2). As a consequence, we immediately get a generalization of Toom's Theorem (the Corollary to Theorem 2). This generalization does not seem to be amenable to contour methods. We believe that the Corollary to Theorem 2 is only one of many potential applications of generic processes. For the reader who wants to first look at the application discussed in Section 3, we remark that it is not necessary to read the proof of Theorem 1 in Section 2 before reading Section 3.

1. Construction of generic population models. We will denote the state at time t of a generic process by $\mathcal{A}_t$. In general, $\mathcal{A}_t$ will be a collection of compact convex $(d+1)$-sided polyhedral regions in $\mathbb{R}^d$. Each such region will correspond to some point $(\mathbf{x}, s)$ in space-time $\mathbb{R}^d \times [0,\infty)$, and for $t \geq s$, we will denote the state of this region by $D(\mathbf{x}, s; t)$. The space-time points $(\mathbf{x}, s)$ will be contained in some random subset of $\mathbb{R}^d \times [0,\infty)$. We will be interested in the union

$$A_t = \bigcup_{(\mathbf{x},s):s\leq t} D(\mathbf{x}, s; t).$$

When we make comparisons to population models like the contact process, the set A_t will correspond to the set of vacant sites, so we call A_t the **vacant region**. We will always take

$$\mathcal{A}_0 = \emptyset$$

as the initial state of our generic process. This corresponds to the state in which everything is occupied.

Let us fix unit vectors $\mathbf{n}_1, \ldots, \mathbf{n}_{d+1} \in \mathbb{R}^d$. We call these vectors **orientation vectors**. In general, each of the regions $D(\mathbf{x}, s; t)$ will be a convex polyhedron whose faces are perpendicular to the vectors $\mathbf{n}_i$. Let us introduce some notation that will be helpful in describing such sets. Let $\mathbf{p}$ be a point and $\mathbf{n}$ a unit vector, both in $\mathbb{R}^d$, and define

$$H(\mathbf{n}, \mathbf{p}) = \{\mathbf{x} \in \mathbb{R}^d : \langle \mathbf{x} - \mathbf{p}, \mathbf{n} \rangle \geq 0\},$$

where $\langle\ ,\ \rangle$ denotes the usual Euclidean inner product in $\mathbb{R}^d$. In more descriptive terms, $H(\mathbf{n}, \mathbf{p})$ is one of the two closed half-spaces whose boundary contains the point $\mathbf{p}$ and is perpendicular to the vector $\mathbf{n}$. The vector $\mathbf{n}$ points into the set $H(\mathbf{n}, \mathbf{p})$. In a generic process, all of the polyhedral regions will be of the form $\mathbf{p}+R(r)$, where $\mathbf{p} \in \mathbb{R}^d$, $r \geq 0$, and

$$R(r) = \bigcap_{i=1}^{d+1} H(\mathbf{n}_i, -r\mathbf{n}_i)$$

for some fixed unit vectors $\mathbf{n}_1, \ldots, \mathbf{n}_{d+1}$. Note that according to this definition, $\mathbf{p}+R(r)$ contains the closed ball of radius r centered at $\mathbf{p}$, and that r is the distance from $\mathbf{p}$ to each of the faces of $\mathbf{p}+R(r)$. We call $\mathbf{p}$ the **base** and r the **size** of $\mathbf{p}+R(r)$. The following proposition collects some elementary facts about regions of the form $\mathbf{p}+R(r)$.

Proposition 1. *If the set*

$$R = \bigcap_{i=1}^{d+1} H(\mathbf{n}_i, \mathbf{p}_i)$$

is nonempty, then it is bounded if and only if $\mathbf{n}_1, \ldots, \mathbf{n}_{d+1}$ span $\mathbb{R}^d$ and there exist strictly positive scalars $a_1, \ldots, a_{d+1}$ such that

$$a_1\mathbf{n}_1 + \cdots + a_{d+1}\mathbf{n}_{d+1} = \mathbf{0},$$

where $\mathbf{0}$ is the origin in $\mathbb{R}^d$. In this case, there exists a unique real number $r \geq 0$ and a unique vector $\mathbf{p}$ such that $R = \mathbf{p}+R(r)$, where $R(r)$ is defined above. The quantity r and the vector $\mathbf{p}$ are linear functions of the vector

$$\mathbf{y} = (\langle \mathbf{p}_1, \mathbf{n}_1\rangle, \ldots, \langle \mathbf{p}_{d+1}, \mathbf{n}_{d+1}\rangle).$$

Proof. Without loss of generality, assume that R contains the origin. Under this assumption, the convexity of R implies that R is unbounded if and only if there exists a nonzero vector $\mathbf{x}$ such that for all $a \geq 0$,

$$\langle a\mathbf{x} - \mathbf{p}_i, \mathbf{n}_i\rangle \geq 0 \quad \text{for} \quad i = 1, \ldots, d+1.$$

Since R contains the origin, the quantities $\langle -\mathbf{p}_i, \mathbf{n}_i\rangle$ are all nonnegative, so R is unbounded if and only if there exists a nonzero vector $\mathbf{x}$ such that for all $a \geq 0$,

$$\langle a\mathbf{x}, \mathbf{n}_i\rangle \geq 0 \quad \text{for} \quad i = 1, \ldots, d+1.$$

It is now an easy matter to finish the proof of the first part of the proposition.

For the second part of the proposition, consider the following system of d equations:

$$\langle \mathbf{p} - \mathbf{p}_{d+1}, \mathbf{n}_{d+1}\rangle = \langle \mathbf{p} - \mathbf{p}_i, \mathbf{n}_i\rangle \quad i = 1, \ldots, d.$$

The conditions on the vectors $\mathbf{n}_i$ imply that any d of them are linearly independent. Hence this system has a unique solution $\mathbf{p}$. This solution lies equidistant from the hyperplanes that bound the half-spaces that determine R. Let r be the common value of the quantities $\langle \mathbf{p} - \mathbf{p}_i, \mathbf{n}_i \rangle$. Note that both $\mathbf{p}$ and r are linear functions of the vector $\mathbf{y}$. If R is nonempty, it is straightforward to check that $R = \mathbf{p} + R(r)$ as claimed. (If R is empty, it turns out that $r < 0$.) ■

From now on, we assume that the vectors $\mathbf{n}_i$ satisfy the conditions of Proposition 1. Note that the intersection of any collection of n regions of the form $\mathbf{p} + R(r)$ can be written as an intersection of half-spaces that are oriented by the vectors $\mathbf{n}_i$. By grouping the half-spaces according to orientation, such an intersection may be rewritten as the intersection of $d+1$ half-spaces, each with a different orientation $\mathbf{n}_i$. Thus Proposition 1 has the following consequence.

Corollary. *Let $r_1, \dots, r_n$ be nonnegative numbers and let $\mathbf{p}_1, \dots, \mathbf{p}_n$ be vectors in $\mathbb{R}^d$. If the set*

$$R = \bigcap_{j=1}^{n} [\mathbf{p}_j + R(r_j)]$$

is nonempty, then there is exists a unique number $r \geq 0$ and a unique vector $\mathbf{p}$ such that $R = \mathbf{p} + R(r)$.

We wish to describe the dynamics of a generic process. We will give three different ways in which new polyhedral regions enter the collection $\mathcal{A}_t$, namely by death, overlap, and collision, and we will say how each type of polyhedral region evolves. Afterwards we will explain why there exists a process that corresponds to our description.

Deaths. In order to describe the first way in which new polyhedral regions enter the collection $\mathcal{A}_t$, we need to define certain point processes. Fix a parameter $\varepsilon > 0$, called the **death rate**, and a probability density μ on the positive integers, called the **size distribution.** For $k \geq 1$, let $\mathcal{P}_k$ be independent Poisson point processes in $\mathbb{R}^d \times [0, \infty)$ with intensity measure equal to $\varepsilon\mu(k)\lambda$, where λ is Lebesgue measure on $\mathbb{R}^d \times [0, \infty)$. In other words, the $\mathcal{P}_k$'s are random subsets of $\mathbb{R}^d \times [0, \infty)$ such that the random variables

$$\sharp(A \cap \mathcal{P}_k), \quad A \text{ Borel},$$

satisfy two properties: (i) $\sharp(A_i \cap \mathcal{P}_k)$, $i = 1, \dots, n, k = 1, 2, \dots$, are independent if the sets A_i are pairwise disjoint; and (ii) $\sharp(A \cap \mathcal{P}_k)$ is a Poisson random variable with mean $\varepsilon\mu(k)\lambda(A)$. (Here and throughout the paper, we use $\sharp S$ to denote the cardinality of a set S.)

We combine the sets $\mathcal{P}_k$ into a point process

$$\mathcal{P} = \bigcup_{k=1}^{\infty} \mathcal{P}_k.$$

Since the $\mathcal{P}_k$'s are independent, $\mathcal{P}$ is itself a Poisson point process with intensity measure $\varepsilon\lambda$. We will label a typical point in $\mathcal{P}$ by $(\mathbf{x}, s)$, where $\mathbf{x} \in \mathbb{R}^d$ and $s \geq 0$. For $(\mathbf{x}, s) \in \mathcal{P}_k$, let

$$\Delta(\mathbf{x}, s) = k.$$

The quantities $\Delta(\mathbf{x}, s)$, $(\mathbf{x}, s) \in \mathcal{P}$, are independent and identically distributed random variables, with probability density μ. For the main part of our work, we will often assume that the size distribution μ satisfies the following condition:

$$\exists \theta > 0 : \sup_k [\exp(\theta k)\, \mu(k)] < \infty. \tag{1-1}$$

It is possible to get by with less, but the extra work required does not seem to be worth it.

For each point $(\mathbf{x}, s) \in \mathcal{P}$, we say that a **death** occurs at $(\mathbf{x}, s)$. If a death occurs at $(\mathbf{x}, s)$, a new region is added to the collection $\mathcal{A}_s$. It is defined by

$$D(\mathbf{x}, s; s) = \mathbf{x} + R(\Delta(\mathbf{x}, s)).$$

We call any region produced in this way a **death region**, and say that it arises due to the death at $(\mathbf{x}, s)$. The evolution of a death region after time s is determined by fixed real numbers $\alpha_1, \ldots, \alpha_{d+1}$, called **occupation rates**. For each $(\mathbf{x}, s) \in \mathcal{P}$ and $t \geq s$, the corresponding death region $D(\mathbf{x}, s; t)$ at time t is given by

$$D(\mathbf{x}, s; t) = \bigcap_{i=1}^{d+1} H(\mathbf{n}_i, \mathbf{x} + ((t - s)\alpha_i - \Delta(\mathbf{x}, s))\mathbf{n}_i).$$

Thus, $D(\mathbf{x}, s; t)$ starts at time s as the set $\mathbf{x} + R(\Delta(\mathbf{x}, s))$, and as time progresses, its ith face moves inward at rate α_i (if α_i is negative, the movement is actually outward). To picture this, the reader might want to think about the 2-dimensional case, in which each of the regions is a triangle with moving edges. Also, it is helpful to think about the case in which all of the α_i's are all equal to some positive number α, for which it is easy to see that the region $D(\mathbf{x}, s; t)$ vanishes after time $t = s + (\Delta(\mathbf{x}, s)/\alpha)$.

We now define a process

$$N_t = \bigcup_{(\mathbf{x}, s) \in \mathcal{P} : s \leq t} D(\mathbf{x}, s; t).$$

We call N_t the **noninteractive region** at time t. It is merely the union of all the death regions that exist at time t and does not involve any of the other two types of regions yet to be defined. When we complete our construction, it will be apparent that N_t is a subset of the vacant region A_t.

Before going on to describe the other two types of regions, we prove a result concerning the behavior of N_t. This result implies the 'easy half' of our main theorem, Theorem 1 in Section 2. Consider the following condition on the orientation vectors $\mathbf{n}_i$ and occupation rates α_i:

Generic eroder condition. *There exists a time* $t > 0$ *such that*

$$\bigcap_{i=1}^{d+1} H(\mathbf{n}_i, (\alpha_i t - 1)\mathbf{n}_i) = \emptyset.$$

Note that if all of the α_i's are positive, the generic eroder condition is satisfied, whereas it fails if they are all less than or equal to 0. Other cases depend on the orientation vectors. The following result says that if the generic eroder condition fails, then the noninteractive vacant region 'takes over'.

Proposition 2. *If the generic eroder condition does not hold, then for all* $\varepsilon > 0$,

$$P(\mathbf{y} \in N_t) \to 1 \text{ as } t \to \infty,$$

uniformly in $\mathbf{y} \in \mathbb{R}^d$.

Proof. The proof is based on the following claim:

CLAIM. If the eroder condition does not hold, then there exists a nonzero vector $\mathbf{v} \in \mathbb{R}^d$ such that for all $(\mathbf{x}, s)$ in $\mathcal{P}$ and $t \geq s$,

$$\mathbf{x} + B_1((t-s)\mathbf{v}) \subseteq D(\mathbf{x}, s; t),$$

where $D(\mathbf{x}, s; t)$ is the death region corresponding to $(\mathbf{x}, s)$, and where for any $\mathbf{z} \in \mathbb{R}^d$ and $\delta > 0$, $B_\delta(\mathbf{z})$ is the closed ball of radius δ centered at $\mathbf{y}$.

Let us show how to finish the proof of the proposition by applying the claim. Fix $\mathbf{x} \in \mathbb{R}^d$ and $t \geq 0$. Note that the condition $\mathbf{y} \in \mathbf{x} + B_1((t-s))\mathbf{v})$ is equivalent to the condition $(\mathbf{x}, s) \in B_1(\mathbf{y} - (t-s)\mathbf{v}) \times \{s\}$. It follows from the claim and our construction of the set N_t that $\mathbf{y} \in N_t$ if

$$\mathcal{P} \cap \left[\bigcup_{s \leq t} (B_1(\mathbf{y} - (t-s)\mathbf{v}) \times \{s\})\right] \neq \emptyset.$$

The cardinality of the set on the left side of the above expression is a Poisson random variable with mean $\varepsilon t \lambda(B_1(\mathbf{0}))$. Since this quantity goes to ∞ as $t \to \infty$, the proposition now follows.

It remains to prove the claim. Fix $(\mathbf{x}, s) \in \mathcal{P}$. By Proposition 1, for all $t \geq 0$ such that $D(\mathbf{x}, s; t)$ is nonempty, there exists a point $\mathbf{p}(t)$ and a nonnegative real number $r(t)$ such that

$$D(\mathbf{x}, s; t) = \mathbf{p}(t) + R(r(t)).$$

The hyperplanes determining $D(\mathbf{x}, s; t)$ move at constant linear rates which do not depend on $(\mathbf{x}, s)$ or $\Delta(\mathbf{x}, s)$, so it also follows from the last part of Proposition 1 that $\mathbf{p}(t)$ and $r(t)$ are also changing linearly: $\mathbf{p}(t) = \mathbf{x} + (t - s)\mathbf{v}$ and $r(t) = -\alpha(t-s) + \Delta(\mathbf{x}, s)$, for some vector $\mathbf{v}$ and real constant α, neither of which depends on $(\mathbf{x}, s)$ or $\Delta(\mathbf{x}, s)$. That is

$$D(\mathbf{x}, s; t) = \mathbf{x} + (t-s)\mathbf{v} + R(-\alpha(t-s) + \Delta(\mathbf{x}, s))$$

for all $t \geq 0$ such that $D(\mathbf{x}, s; t)$ is nonempty. In particular, this statement holds for sufficiently small $t > 0$. We may apply this same argument to the set

$$\bigcap_{i=1}^{d+1} H(\mathbf{n}_i, (\alpha_i t - 1)\mathbf{n}_i)$$

to conclude that it equals $t\mathbf{v} + R(-\alpha t + 1)$ for sufficiently small $t > 0$. From this we see that α is nonpositive, since the eroder condition fails. It follows that $D(\mathbf{x}, s; t)$ is nonempty for all $t \geq 0$, and that

$$\mathbf{x} + (t-s)\mathbf{v} + R(\Delta(\mathbf{x}, s)) \subseteq D(\mathbf{x}, s; t).$$

Since $\Delta(\mathbf{x}, s) \geq 1$, $B_1(\mathbf{0}) \subseteq R(\Delta(\mathbf{x}, s))$ and the claim follows. ∎

The generic eroder condition is the continuum analogue of the eroder condition of Toom [13], and the proof just given is essentially the same as Toom's proof of the corresponding fact for discrete time lattice models (except for the part concerning the claim, which is a technical feature that becomes necessary in our continuous setting). In Section 3 we present a description of Toom's results. We now return to the construction of generic processes.

Overlap interactions. Whenever a new polyhedral region appears as the result of a death, it may be that it overlaps one or more already existing polyhedral regions. These already existing polyhedral regions may be any of the three types of regions that we are in the process of defining. When such an overlap occurs, further new regions are produced in a manner that we now describe. Suppose that there is a death at a space-time point $(\mathbf{y}_1, u_1)$, and let $D(\mathbf{y}_1, u_1; u_1)$ be the corresponding new region. Write $s = u_1$, and let $D(\mathbf{y}_2, u_2; s), \ldots, D(\mathbf{y}_n, u_n; s)$ be any collection of regions in $\mathcal{A}_s$ such that $u_k < s$ for $k = 2, \ldots, n$, and

$$\bigcap_{k=1}^{n} D(\mathbf{y}_k, u_k; s) \neq \emptyset.$$

By the Corollary to Proposition 1, this set may be written as $\mathbf{x} + R(r)$ for some $r \geq 0$ and $\mathbf{x} \in \mathbb{R}^d$. We assume that the collection of regions $D(\mathbf{y}_k, u_k; s)$ is the maximal collection satisfying the above description with intersection equal to $\mathbf{x}+R(r)$ (there will be only one such maximal collection for each different set $\mathbf{x}+R(r)$ formed in this way). We write $D(\mathbf{x},s;s) = \mathbf{x}+R(r)$, and add this new region to our collection $\mathcal{A}_s$. We say that it appears due to an **overlap interaction**. Any region that appears due to an overlap interaction is called an **overlap region**.

Note that if the death region $D(\mathbf{y}_1, u_1; u_1)$ overlaps m already existing regions when it appears, then as many as $2^m - 1$ new overlap regions may be produced as a result. We have avoided one possible ambiguity that may appear in the labelling of these regions (when two different sub-collections of these m regions have the same intersection with $D(\mathbf{y}_1, u_1; u_1)$) by requiring that the collections of regions $D(\mathbf{y}_k, u_k; s)$ be maximal, as described in the preceding paragraph. There is another possible ambiguity that can arise when two different overlap regions are produced with the same base point. This second possibility has probability zero (see the discussion of existence given later in this section). The possibility that $m = \infty$ is not excluded. It can be shown that m is finite almost surely if (1-1) is satisfied, but as we don't really need this fact, we omit the proof.

The evolution of an overlap region is determined by the occupation rates α_i and a fixed quantity $\beta < 0$ called the **interaction rate** (as we will see, specifying that β be negative means that the corresponding movement of faces will be outward). We assume that

$$\beta \le \min_i \alpha_i.$$

For $(\mathbf{x}, s)$ and $(\mathbf{y}_k, u_k)$ as in the preceding paragraph, and $i = 1, \dots, d+1$, define

$$H_i(\mathbf{y}_k, u_k; t) = \text{the half-space corresponding to the } i^{\text{th}} \text{ face of } D(\mathbf{y}_k, u_k; t).$$

Let

$$\tau_i = \inf\left\{ t \ge s : H(\mathbf{n}_i, \mathbf{x} + (\beta - r)\mathbf{n}_i) \supseteq \bigcup_{k=1}^{n} H_i(\mathbf{y}_k, u_k; t) \right\},$$

and define a function $\gamma_i(\mathbf{x}, s; t)$ which is continuous and piecewise linear in $t \ge s$ and determined by the conditions

$$\text{(1-2)} \qquad \gamma_i(\mathbf{x}, s; s) = 0 \quad \text{and} \quad \frac{d\gamma_i}{dt}(\mathbf{x}, s; t) = \begin{cases} \beta & t < \tau_i \\ \alpha_i & t > \tau_i. \end{cases}$$

We allow the possibility that $\tau_i = \infty$, which can happen when $\alpha_i = \beta$. The overlap region $D(\mathbf{x}, s; t)$ is now defined as

$$D(\mathbf{x}, s; t) = \bigcap_{i=1}^{d+1} H(\mathbf{n}_i, \mathbf{x} + (\gamma_i(\mathbf{x}, s; t) - r)\mathbf{n}_i)$$

for all $t \geq s$. In descriptive terms, this region starts at time s as $\mathbf{x} + R(r)$, after which its ith face moves outward at speed $-\beta$ until time τ_i, at which time it has caught all of the corresponding faces of the regions $D(\mathbf{y}_k, u_k; t)$. The time τ_i is defined in such a way that at time τ_i, the ith face of each of the regions $D(\mathbf{y}_k, u_k; \tau_i)$ is moving at rate α_i. From time τ_i on, the ith face of $D(\mathbf{x}, s; t)$ also moves at rate α_i.

Collision interactions. The third way in which a new region appears is similar to the second. Fix s and suppose that $D(\mathbf{y}_1, u_1; s), \ldots, D(\mathbf{y}_n, u_n; s)$ are regions in the collection $\mathcal{A}_s$ such that $u_k < s$ for $k = 1, \ldots, n$ and

$$\bigcap_{k=1}^{n} D(\mathbf{y}_k, u_k; t) = \emptyset \quad \text{for} \quad t < s.$$

(In this formula and elsewhere, we interpret $D(\mathbf{y}_k, u_k; t)$ as the empty set for $t < u_k$.) Thus we have n regions that existed and had empty intersection prior to time s. Suppose that their intersection is not empty at time s. By the Corollary to Proposition 1, this intersection is of the form $\mathbf{x} + R(r)$ for some $\mathbf{x} \in \mathbb{R}^d$ and $r \geq 0$. All regions change continuously in our time evolution, so the intersection at time s cannot contain interior points of any of the regions. It follows that $r = 0$, so the intersection equals $\mathbf{x} + R(0) = \{\mathbf{x}\}$. We say that a **collision** occurs at $(\mathbf{x}, s)$. We assume that the collection of regions $D(\mathbf{y}_k, u_k; s)$, $k = 1, \ldots, n$, is maximal in the sense that it contains all regions that existed prior to s and have $\mathbf{x}$ as a boundary point. As with overlap interactions, we allow the possibility that $n = \infty$. Under condition (1-1), this occurs with probability 0.

We start a new region $D(\mathbf{x}, s; s) = \{\mathbf{x}\}$ at time s, and say that this region arises due to a **collision interaction**. For $t \geq s$, this new region is of the form

$$D(\mathbf{x}, s; t) = \bigcap_{i=1}^{d+1} H(\mathbf{n}_i, \mathbf{x} + \gamma_i(\mathbf{x}, s; t)\mathbf{n}_i),$$

where $\gamma_i(\mathbf{x}, s; t)$ is defined as in (1-2).

Summary of the description of generic processes. The state of a generic process $\mathcal{A}_t$ is a collection of regions $D(\mathbf{x}, s; t)$ of three types. For all three types of regions, we have

$$D(\mathbf{x}, s; t) = \bigcap_{i=1}^{d+1} H_i(\mathbf{x}, s; t),$$

where

$$\text{(1-3)} \qquad H_i(\mathbf{x}, s; t) = H(\mathbf{n}_i, \mathbf{x} + (\gamma_i(\mathbf{x}, s; t) - r(\mathbf{x}, s))\mathbf{n}_i)$$

for certain continuous piecewise linear functions $\gamma_i(\mathbf{x}, s; \cdot)$ and nonnegative numbers $r(\mathbf{x}, s)$. If the region arises due to a death, $r(\mathbf{x}, s) = \Delta(\mathbf{x}, s)$. In the case of a collision, $r(\mathbf{x}, s) = 0$. For overlap interactions, $r(\mathbf{x}, s)$ is the 'size' of the overlap. For regions arising due to a death, $\gamma_i(\mathbf{x}, s; t) = \alpha_i(t-s)$. In the two other cases, condition (1-2) says that $\gamma_i(\mathbf{x}, s; \cdot)$ starts at 0 at time $t = s$, then has slope β until some time τ_i, after which the slope is α_i. The vacant region A_t is the union of all of the regions in $\mathcal{A}_t$. As such, the vacant region involves regions of all three types, as opposed to the noninteractive region N_t which only involves death regions. It is clear that

$$N_t \subseteq A_t.$$

Our construction has been designed to ensure the following:

Proposition 3. *Suppose that $\mathcal{A}_t$ is a generic process defined in terms of orientation vectors $\mathbf{n}_i$, occupation rates α_i, interaction rate β, death rate ε, and size distribution μ satisfying (1-1). Let $D(\mathbf{y}_1, u_1; t), \ldots, D(\mathbf{y}_n, u_n; t)$ be a collection of regions in $\mathcal{A}_t$ with nonempty intersection, and define*

$$s = \inf\left\{u : \bigcap_{k=1}^{n} D(\mathbf{y}_k, u_k; u) \neq \emptyset\right\}.$$

Then there exists a region $D(\mathbf{x}, s; u), u \geq s$, with $(\mathbf{x}, s)$ possibly being one of the points $(\mathbf{y}_k, u_k)$, such that

$$\bigcap_{k=1}^{n} D(\mathbf{y}_k, u_k; u) \subseteq D(\mathbf{x}, s; u) \quad \text{for all} \quad u \in [s, t], \tag{1-4}$$

and such that for all $i = 1, \ldots, d+1$, either

$$\frac{d\gamma_i}{du}(\mathbf{x}, s; u) = \beta \quad \text{for all} \quad u \in [s, t), \tag{1-5}$$

or

$$\bigcup_{k=1}^{n} H_i(\mathbf{y}_k, u_k; t) \subseteq H_i(\mathbf{x}, s; t). \tag{1-6}$$

This proposition says that an overlap or collision region grows at speed $-\beta$ in the direction $-\mathbf{n}_i$ until the half-space that determines the region in that direction contains all of the corresponding half-spaces of the regions that initiated the overlap or collision. The proposition is a straightforward consequence of our description of the appearance and evolution in time of the three types of regions. Note particularly the role played by the relationships given in (1-2) and (1-3). In fact, these relationships were carefully created with Proposition 3 in mind. More simplistic approaches which might also lead to a result like Proposition 3, such as setting $\beta = -\infty$

so that (1-6) is satisfied already at time $t = s$, run into trouble later on when we want to prove survival under the generic eroder condition.

The collection of regions $D(\mathbf{y}_k, u_k; u)$ in the statement of Proposition 3 need not be the same as the maximal collection that was used in the definition of overlap and collision regions. However, if $D(\mathbf{y}, u; u)$ is an overlap or collision region, and if the collection of regions $D(\mathbf{y}_k, u_k; u)$ is the maximal collection that was used in the definition of $D(\mathbf{y}, u; u)$, then (1-2) ensures that

$$H_i(\mathbf{x}, s; t) \subseteq \bigcup_{k=1}^{n} H_i(\mathbf{y}, u_k; t) \tag{1-7}$$

for all $i = 1, \ldots, d+1$ and $t \geq s$. Thus, for this special case, we may replace '$\subseteq$' by '$=$' in (1-6).

Existence. It may not be immediately obvious that there exists a process $\mathcal{A}_t$ corresponding to the prescription given above. The main problem is that it is conceivable that inconsistencies arise due to the fact that we are working in an unbounded space. We also have the minor problem that our notation does not distinguish between two different regions that arise at the same time s and have the same initial base $\mathbf{x}$. One way to overcome these problems is by defining the process in terms of a finite volume limit. More precisely, for each j, we construct a process in which we only allow deaths to occur at points $(\mathbf{x}, s) \in \mathcal{P}$ such that $||\mathbf{x}|| \leq j$, according to the prescription given above. Since Poisson point processes are almost surely finite on bounded sets, we do not need to worry about infinitely many regions arising before any finite time. Thus, a straightforward inductive procedure can be used to define the process up to time t for any $t \in [0, \infty)$ and any $j \geq 0$. Our only concern is that more than one region might arise at a point $(\mathbf{x}, s)$ in one of these processes. This could happen if two deaths occur at the same point in space-time, or if a death occurs at the same point $(\mathbf{x}, s)$ in space-time where a collision also occurs, or if two different overlap regions arise at time s with the same base $\mathbf{x}$. The first possibility has probability zero because the different Poisson point processes $\mathcal{P}_k$ are independent. The second possibility has probability zero because there are only finitely many collision points up to any finite time. A little thought reveals that for each fixed s, the set of points $\mathbf{y}_1$ such that a death at $(\mathbf{y}_1, s)$ can produce two different regions with the same base $\mathbf{x}$ is a finite set. It follows that the third possibility also has probability zero. The proof that these possibilities all have probability zero relies on the fact that the probability that a Borel set B contains two or more points of the random set $\mathcal{P}$ is $O(\lambda(B)^2)$.

We now take a limit as $j \to \infty$. The set of points $(\mathbf{x}, s)$ at which deaths occur converges to $\mathcal{P}$. Unfortunately, the set of points $(\mathbf{x}, s)$ at which overlap and collision regions arise does not necessarily increase as $j \to \infty$. Due to the maximality conditions imposed on the collections of regions $D(\mathbf{y}_k, u_k; s)$ used in the definitions of overlap and collision regions, a given point $(\mathbf{x}, s)$ may correspond to a region for one value of j but not for some larger value of j. However, it can be shown by an elementary but tedious proof (using the fact about Poisson point processes mentioned in the last sentence of the preceding paragraph) that the probability is zero that there exists a point $(\mathbf{x}, s)$ and integers $j_1 < j_2 < j_3$ such that there is an overlap region corresponding to $(\mathbf{x}, s)$ for $j = j_1$ and $j = j_3$ but not for $j = j_2$. A similar fact can be proved for collision regions. Thus, the set of points $(\mathbf{x}, s)$ at which overlap regions arise has a limit as $j \to \infty$, as does the set of points at which collision regions arise. It is easy to see that for those points $(\mathbf{x}, s)$ that lie in the limit sets, the corresponding quantities $\gamma_{i,j}(\mathbf{x}, s; t)$ form a decreasing sequence for each fixed i and t, so the corresponding regions $D_j(\mathbf{x}, s, \zeta; t)$ form an increasing sequence for fixed t. Therefore, the limits $\gamma_i(\mathbf{x}, s; t)$ and $D(\mathbf{x}, s; t)$ exist for all $t \geq s$. We leave it to the reader to check that for these limits, conditions (1-2) and (1-3) are satisfied. Existence follows.

It is natural to also ask about uniqueness. We are content that at least one process exists that agrees with our prescription, since uniqueness is not needed in our applications. A graphical proof of uniqueness similar to the uniqueness proof in Gray [7] can be given under condition (1-1).

Note that in the construction, the randomness comes only from the random sets $\mathcal{P}_k$. Once these point processes are fixed, the remaining objects $(D(\mathbf{x}, s; t), \mathcal{A}_t, A_t$, etc.) are determined. This observation allows us to make certain comparisons. We state one such comparison here. It can be proved by first considering the finite volume processes and then taking limits.

Proposition 4. *Suppose that $\{\mathcal{P}_k\}$ and $\{\mathcal{P}'_k\}$ are two collections of Poisson point processes defined as above. Assume that they are defined jointly on the same probability space, and that for each k, $\mathcal{P}_k \subseteq \mathcal{P}'_k$. Also assume that we have two sets of occupation rates and interaction rates such that $\alpha_i \geq \alpha'_i$, and $\beta \geq \beta'$. Let $\mathcal{A}_t$ and $\mathcal{A}'_t$ be the corresponding generic processes, with vacant regions A_t and A'_t. Then $A_t \subseteq A'_t$.*

We conclude this section with two examples.

Example 1. We consider the general one-dimensional model ($d = 1$). In this case, polyhedral regions are intervals, so $\mathcal{A}_t$ is a collection of closed intervals, and A_t is their union. Note that in this one-dimensional setting, overlap and collision regions are always contained in the union of the corresponding death regions. Thus, as far as the set A_t is concerned, when a collision or overlap occurs between two regions, we treat them as being part of a single interval, ignoring the motions of the overlapping endpoints and also ignoring the collision or overlap regions that are produced. The remaining endpoints are the endpoints of the disjoint intervals whose union is A_t. The left endpoints of these intervals move inward at some rate α_ℓ and their right endpoints move inward at rate α_r. Note that except when a death causes an interval of A_t to jump in size, the intervals of A_t expand, shrink, or stay the same size, depending on whether $\alpha_\ell + \alpha_r$ is negative, positive, or zero. The generic eroder condition is that $\alpha_\ell + \alpha_r > 0$. The generic analogue to the asymmetric contact process has rates $\alpha_\ell = 0$ and $\alpha_r = 1$.

Example 2. We give the parameters for the generic analogue of Toom's model. The orientation vectors are $\mathbf{n}_1 = (1,0), \mathbf{n}_2 = (0,1)$, and $\mathbf{n}_3 = -(\sqrt{2}, \sqrt{2})/2$, with corresponding occupation rates $\alpha_1 = \alpha_2 = 0$ and $\alpha_3 = \sqrt{2}/2$. The interactive rate is $\beta = -1$. A typical death region is a right triangle whose vertical and horizontal edges do not move, and whose hypotenuse moves inward. The generic eroder condition is satisfied, since such triangles vanish after one time unit. If one were to simulate Toom's model and its generic analogue on a computer, using the distance between pixels as the unit of measurement, the two would look very similar. The main difference would be that vacant area increases somewhat faster in the generic model when the regions that arise from interactions are in their 'β-growth stage'.

2. The renormalization argument and the main theorem. Let A_t be a generic process, defined as in the preceding section in terms of occupation rates α_i, interaction rate β, orientation vectors $\mathbf{n}_i$, size distribution μ, and death rate $\varepsilon \geq 0$. We think of each of α_i, β, $\mathbf{n}_i$, and μ as being fixed, and ε as being a variable parameter. Our main result is

Theorem 1. *If the generic eroder condition does not hold, then for all* $\varepsilon > 0$

$$\lim_{t \to \infty} P(\mathbf{y} \in A_t) = 1$$

uniformly in $\mathbf{y} \in \mathbb{R}^d$. *If the generic eroder condition holds, and if in addition the size distribution* μ *satisfies (1-1), then*

$$\lim_{\varepsilon \searrow 0} \limsup_{t \to \infty} P(\mathbf{y} \in A_t) = 0$$

uniformly in $\mathbf{y}$.

Proof. Since $N_t \subseteq A_t$, the first half of the theorem follows from Proposition 2. In order to prove the second half of the theorem, we first simplify matters somewhat. Let α and $\mathbf{v}$ be as in the proof of Proposition 2. We change coordinates linearly according to the following transformation:

$$(\mathbf{x}, t) \to (\mathbf{x} - t\mathbf{v}, t).$$

Note that this transformation does not change the time variable, nor does it change distances in the spatial variables. Once this change of coordinates is made, the occupation rates α_i are all changed into α. Of course, the interaction rate β is also changed, resulting in $d+1$ possibly different interaction rates. By Proposition 4, we may replace these different interaction rates by their minimum, which will be some negative number. By changing our time scale if necessary, we may assume that this new interaction rate is less than or equal to -1. Since the eroder condition holds, $\alpha > 0$ (before and after the change in time scale). By applying Proposition 4 again, we may assume that $\alpha \leq 1$. For the remainder of the proof, we will assume that these modifications have been made, namely, that $\alpha_i = \alpha \in (0,1]$ and $\beta \leq -1$. We also fix $\theta \in (0,1]$ such that (1-1) holds, and let

$$C = \sup_k [\mu(k)\exp(\theta k)].$$

Here is an outline of the proof. We will first partition the points in $\mathcal{P}$ into clusters defined in terms of the proximity of these points to one another. We will prove that the clusters are finite almost surely. To each cluster of cardinality 2 or greater, we will associated a point in space-time, and the resulting set of points will be called $\mathcal{P}^{(1)}$. We will use the random set $\mathcal{P}^{(1)}$ to define a new generic process $\mathcal{A}_t^{(1)}$. This process is a renormalization of $\mathcal{A}_t$. We will need to take care of some technical details concerning the random set $\mathcal{P}^{(1)}$ that underlies the construction of $\mathcal{A}_t^{(1)}$ in order to know that $\mathcal{A}_t^{(1)}$ behaves in every essential way like a generic process. We then continue inductively to define successive renormalizations $\mathcal{A}_t^{(2)}, \mathcal{A}_t^{(3)}, \ldots$. It is then shown that the vacant region A_t of our original generic process is contained in $N_t \cup A_t^{(1)}$, the union of the noninteractive region of the original generic process and the vacant region of the renormalized generic process. An inductive argument then allows us to obtain an upper bound for the probability that a point $\mathbf{y}$ lies in A_t in terms of upper bounds for the probabilities that $\mathbf{y}$ lies in the union of the noninteractive regions of the sequence of renormalized generic processes. This upper bound is shown to converge to 0 as $\varepsilon \searrow 0$, uniformly in $\mathbf{y}$ and t.

Definition of clusters. First we need some notation. Recall that $B_\delta(\mathbf{x})$ is the closed ball in $\mathbb{R}^d$ of radius δ centered at $\mathbf{x}$. Choose an integer $M \geq 1$ such that $R(1) \subseteq B_M(0)$. (Note that $R(r) \subseteq B_{Mr}(0)$ for all $r \geq 0$.) When measuring distance between points in $\mathbb{R}^d \times [0,\infty)$, we will use the max-norm:

$$\|(\mathbf{x},s) - (\mathbf{y},u)\|_{\max} = \max\{|s-u|, |x_1 - y_1|, \ldots, |x_d - y_d|\},$$

where $\mathbf{x} = (x_1, \ldots, x_d)$ and $\mathbf{y} = (y_1, \ldots, y_d)$. For $(\mathbf{x},s) \in \mathbb{R}^d \times [0,\infty)$, we write

$$Q(\delta,\mathbf{x},s) = [x_1 - \delta, x_1 + \delta) \times \cdots \times [x_d - \delta, x_d + \delta) \times ([s-\delta, s+\delta) \cap [0,\infty)),$$

which is a cube with sides of length 2δ, centered at $(\mathbf{x},s)$, and truncated at $t = 0$ if necessary. For $(\mathbf{x},s) \in \mathcal{P}$, define

$$S(\mathbf{x},s) = \text{ the interior of } Q\left(\frac{5M^2|\beta|\Delta(\mathbf{x},s)}{\alpha}, \mathbf{x}, s\right).$$

We think of $S(\mathbf{x}, s)$ as the cube of influence of the point $(\mathbf{x}, s)$. The quantity $5M^2|\beta|\Delta(\mathbf{x}, s)/\alpha$ is chosen for the proof of the relationship $A_t \subseteq N_t \cup \mathcal{A}_t^{(1)}$. For now, the reader should note that since the size of a death region $D(\mathbf{x}, s; t)$ shrinks at rate α and hence reaches 0 at time $s + \Delta(\mathbf{x}, s)/\alpha$, the quantity has been chosen large enough so that $D(\mathbf{x}, s; t) \subseteq S(\mathbf{x}, s)$ for all $t \geq s$. Let

$$\mathfrak{S}(\mathbf{x}, s) = \text{ the component of } \bigcup_{(\mathbf{y},u)\in\mathcal{P}} S(\mathbf{y}, u) \quad \text{containing } S(\mathbf{x}, s)$$

and

$$\mathcal{C}(\mathbf{x}, s) = \mathfrak{S}(\mathbf{x}, s) \cap \mathcal{P}.$$

We call $\mathcal{C}(\mathbf{x}, s)$ the **cluster** containing $(\mathbf{x}, s)$.

Finiteness of clusters. In order to show that the clusters are finite in cardinality almost surely, we will mainly rely on two basic properties of Poisson point processes which we now describe. The first is quite simple:

$$P(\sharp(B \cap \mathcal{P}) \geq 2) = O(\lambda(B)^2) \tag{2-1}$$

for Borel sets B. To state the second property, let $Q_1, \ldots, Q_m$ be disjoint cubes in $\mathbb{R}^d \times [0, \infty)$. Then for all positive integers $k_1, \ldots, k_m$,

$$P\left(\bigcap_{j=1}^{m}\{Q_j \cap \mathcal{P}_{k_j} \neq \emptyset\}\right) \leq \prod_{j=1}^{m}[\varepsilon\mu(k_j)\lambda(Q_j)]. \tag{2-2}$$

The reason we mention these properties now is that later in the proof, we will define clusters in terms of point processes that are not Poisson point processes, but which do satisfy versions of (2-1) and (2-2). The reader should take care to observe that these are the only special properties of Poisson point processes used in the following argument.

Fix a cube $Q \subseteq \mathbb{R}^d \times [0, \infty)$. We wish to obtain an upper bound for the probability that Q has nonempty intersection with some cluster with infinitely many points. For each positive integer k_0, define

$$G_0(k_0) = Q \cap \mathcal{P}_{k_0} \text{ and } G_0 = F_0 = \bigcup_{k_0=1}^{\infty} G_0(k_0).$$

Now proceed inductively. For positive integers $k_0, \ldots, k_n$, let

$$G_n(k_0, \ldots, k_n) = \{(\mathbf{x}, s) \in \mathcal{P}_{k_n} \setminus F_{n-1} : S(\mathbf{x}, s) \cap S(\mathbf{y}, u) \neq \emptyset \text{ for some } (\mathbf{y}, u) \in G_{n-1}(k_0, \ldots, k_{n-1})\}$$

$$G_n = \bigcup_{k_0,\ldots,k_n\geq 1} G_n(k_0, \ldots, k_n)$$

$$F_n = F_{n-1} \cup G_n.$$

We may think of G_n as the nth generation of a family of particles with multiple types. The particles are points $(\mathbf{x}, s)$ in $\mathcal{P}$. The type of a particle $(\mathbf{x}, s)$ is $\Delta(\mathbf{x}, s)$. A sequence of points $(\mathbf{x}_j, s_j), j = 0, \ldots, n$, such that

$$(\mathbf{x}_0, s_0) \in G(k_0),\ (\mathbf{x}_1, s_1) \in G(k_0, k_1),\ \ldots,\ (\mathbf{x}_n, s_n) \in G(k_0, \ldots, k_n),$$

is called an **ancestry of type** $(k_0, \ldots, k_n)$ if for each $j = 0, \ldots, n-1$,

$$S(\mathbf{x}_j, s_j) \cap S(\mathbf{x}_{j+1}, s_{j+1}) \neq \emptyset.$$

Every point in $G_n(k_0, \ldots, k_n)$ is the last point of one or more ancestries of type $(k_0, \ldots, k_n)$.

We will use (2-1) and (2-2) to obtain an upper bound for the probability that $G_n(k_0, \ldots, k_n)$ is nonempty, which will in turn give us an upper bound for the probability that G_n is nonempty and thus help us to analyze the size of the clusters that intersect Q. To this end, fix $\delta > 0$ and let

$$Q(\delta, \mathbf{y}_i, u_i),\ i = 1, 2, \ldots,$$

be a partition of $\mathbb{R}^d \times [0, \infty)$ into disjoint cubes with sides of length 2δ. For positive integers $k_0, \ldots, k_n$, let $I(k_0, \ldots, k_n)$ be the collection of all sequences $(i_0, \ldots, i_n)$ of distinct positive integers such that

$$Q(\delta, \mathbf{y}_{i_0}, u_{i_0}) \cap Q \neq \emptyset$$

and

$$\|(\mathbf{y}_{i_j}, u_{i_j}) - (\mathbf{y}_{i_{j+1}}, u_{i_{j+1}})\|_{\max} \leq 2\delta + 5M^2|\beta|(k_j + k_{j+1})/\alpha$$

for $j = 0, \ldots, n-1$. To understand this last expression, note that for any two consecutive points $(\mathbf{x}_j, s_j)$ and $(\mathbf{x}_{j+1}, s_{j+1})$ of an ancestry of type $(k_0, \ldots, k_n)$, the max-norm distance $\|(\mathbf{x}_j, s_j) - (\mathbf{x}_{j+1}, s_{j+1})\|_{\max}$ is at most $5M^2|\beta|(k_j + k_{j+1})/\alpha$. Thus, for any ancestry $(\mathbf{x}_0, s_0), \ldots, (\mathbf{x}_n, s_n)$ of type $(k_0, \ldots, k_n)$, as long as the points in the ancestry lie in different cubes of the partition, there exists a sequence $(i_0, \ldots, i_n) \in I(k_0, \ldots, k_n)$ such that

$$Q(\delta, \mathbf{y}_{i_j}, u_{i_j}) \cap \mathcal{P}_{k_j} \neq \emptyset,\ j = 0, \ldots, n. \tag{2-3}$$

The points in an ancestry of type $(k_0, \ldots, k_n)$ necessarily lie inside the cube

$$Q((5M^2|\beta|(k_0 + \cdots + k_n)/\alpha) + r, \mathbf{p}, t)$$

if $Q = Q(r, \mathbf{p}, t)$, so the probability that any two of the points in the ancestry lie in the same partitioning cube goes to 0 as $\delta \to 0$ by a straightforward application of (2-1). By (2-2), the probability of the event in (2-3) is bounded above by

$$\prod_{j=0}^{n} [\varepsilon(2\delta)^{d+1} \mu(k_j)]. \tag{2-4}$$

It follows that

$$\begin{aligned}
&P(G_n(k_0,\ldots,k_n)\neq\emptyset)\\
&\le \lim_{\delta\to 0}\sum_{(i_0,\ldots,i_n)\in I(k_0,\ldots,k_n)} P\left(\bigcap_{j=1}^{n}\{Q(\delta,\mathbf{y}_{i_j},u_{i_j})\cap\mathcal{P}_{k_j}\neq\emptyset\}\right)\\
&\le \varepsilon\mu(k_0)\lambda(Q)\prod_{i=1}^{n}[\varepsilon(5M^2|\beta|(k_{i-1}+k_i)/\alpha)^{d+1}\mu(k_i)].
\end{aligned}$$

This last expression is (crudely) bounded above by

$$\lambda(Q)(\varepsilon(10M^2|\beta|/\alpha)^{d+1})^{n+1}\prod_{i=0}^{n}[k_i^{2d+2}\mu(k_i)].$$

Since by elementary calculus

$$k^{2d+2}\exp(-k\theta/2)\le\left(\frac{2d+2}{\theta}\right)^{2d+2},$$

we have by (1-1)

$$\frac{P(G_n(k_0,\ldots,k_n)\neq\emptyset)}{\lambda(Q)}\le \left(\frac{K_1C\varepsilon|\beta|^{d+1}}{\alpha^{d+1}\theta^{2d+2}}\right)^{n+1}\exp(-\frac{\theta}{2}(k_0+\cdots+k_n)), \tag{2-5}$$

where

$$K_1=2(10M^2)^{d+1}(2d+2)^{2d+2}.$$

Summing (2-5) on $k_0,\ldots,k_n$, we obtain

$$\frac{P(G_n\neq\emptyset)}{\lambda(Q)}\le\left(\frac{K_1C\varepsilon|\beta|^{d+1}}{\alpha^{d+1}\theta^{2d+3}}\right)^{n+1}. \tag{2-6}$$

In order for a cluster to be infinite and also contain at least one point in Q, either $G_n\neq\emptyset$ for all n, or $G_n(k_0,\ldots,k_n)$ has infinite cardinality for some n and some $k_0,\ldots,k_n$. Since the points in $G_n(k_0,\ldots,k_n)$ are confined by the construction to the intersection of $\mathcal{P}$ and a certain ball centered at the middle of Q, the sets $G_n(k_0,\ldots,k_n)$ are finite almost surely (this follows from (2-1)). There are only countably many such sets. The probability in (2-6) converges to 0 as $n\to\infty$ for small $\varepsilon>0$, certainly for

$$\varepsilon\le\frac{\alpha^{d+1}\theta^{2d+3}}{8K_1C|\beta|^{d+1}}. \tag{2-7}$$

(We have left some extra room in this condition because of what comes later.) Assuming (2-7), it follows that the probability that $G_n \neq \emptyset$ for all n is 0, so any cluster that intersects Q is finite almost surely. It is elementary to conclude that all clusters are finite almost surely for ε satisfying (2-7). We assume from now on in the proof that (2-7) is satisfied.

The first renormalization. In this part of the proof and the next, we will carry out a procedure that will be the first step in an induction. Our eventual goal will be to define point processes $\mathcal{P}_k^{(\ell)}$ and corresponding generic processes $\mathcal{A}_t^{(\ell)}$, such that analogues of (2-1) and (2-2) are satisfied.

To begin the first stage of the induction, we define

$$\mathcal{P}^{(1)} = \{(\mathbf{x}, s) \in \mathcal{P} : \sharp(\mathcal{C}(\mathbf{x}, s)) \geq 2 \text{ and } u \geq s \text{ for all } (\mathbf{y}, u) \in \mathcal{C}(\mathbf{x}, s)\}.$$

A point is in $\mathcal{P}^{(1)}$ if it is the 'oldest' member of a cluster containing at least 2 points. The fact that we only take points from clusters of size two or more will imply that for small $\varepsilon > 0$, the points in $\mathcal{P}^{(1)}$ are considerably sparser than the ones in $\mathcal{P}$. Since the clusters are finite almost surely, every cluster with at least 2 points contains a unique oldest member almost surely; there is therefore a one-to-one correspondence between clusters of cardinality greater than or equal to 2 and points in $\mathcal{P}^{(1)}$. For $(\mathbf{x}, s) \in \mathcal{P}^{(1)}$, let

$$\Delta^{(1)}(\mathbf{x}, s) = \lfloor 2 \inf\{r \geq 0 : \mathfrak{S}(\mathbf{x}, s) \subseteq B_r(\mathbf{x}) \times [s - r, s + r]\} \rfloor,$$

where $\lfloor \cdot \rfloor$ denotes 'greatest integer less than or equal to', and let

$$\mathcal{P}_k^{(1)} = \{(\mathbf{x}, s) \in \mathcal{P}^{(1)} : \Delta^{(1)}(\mathbf{x}, s) = k\}.$$

Also let

$$\alpha^{(1)} = \alpha/2 \text{ and } \beta^{(1)} = 2\beta.$$

We now have all the ingredients for defining a new generic process $\mathcal{A}_t^{(1)}$, using the sets $\mathcal{P}_k^{(1)}$ in place of the Poisson point processes $\mathcal{P}_k$, and the quantities $\Delta^{(1)}(\mathbf{x}, s)$ in the place of the quantities $\Delta(\mathbf{x}, s)$. We set all the occupation rates equal to $\alpha^{(1)}$, and the interactive rates equal to $\beta^{(1)}$. We use the same orientation vectors $\mathbf{n}_i$ as before. Since $\mathcal{P}^{(1)} \subseteq \mathcal{P}$, the finite volume construction outlined in Section 1 can be carried out, even though the sets $\mathcal{P}_k^{(1)}$ are not Poisson point processes. Thus, we can construct a collection $\mathcal{A}_t^{(1)}$ of polyhedral regions $D^{(1)}(\mathbf{x}, s; t)$, determined by the half spaces $H_i^{(1)}(\mathbf{x}, s; t)$ that move according to functions $\gamma_i^{(1)}(\mathbf{x}, s; t)$. We also have the vacant regions $A_t^{(1)}$ and noninteractive vacant regions $N_t^{(1)}$, all in analogy with what was done in the construction of the original process $\mathcal{A}_t$. The process $\mathcal{A}_t^{(1)}$ is our renormalization of $\mathcal{A}_t$. We remark that $\Delta^{(1)}(\mathbf{x}, s)$ has been chosen large enough so that for any point $(\mathbf{y}, u) \in \mathcal{C}(\mathbf{x}, s)$, $D(\mathbf{y}, u; u) \subseteq D^{(1)}(\mathbf{x}, s; u)$. This follows from the fact that the size of $D^{(1)}(\mathbf{x}, s; t)$ shrinks at rate $\alpha^{(1)} \leq 1/2$, so that the space-time set $\mathfrak{S}(\mathbf{x}, s)$ is contained in the union of the space-time sets $D^{(1)}(\mathbf{x}, s; t) \times \{t\}$, $t \geq s$.

Showing that $\mathcal{P}^{(1)}$ satisfies analogues of (2-1) and (2-2). We wish to continue the renormalization procedure inductively. We cannot do this immediately because the point processes $\mathcal{P}_k^{(1)}$ are not Poisson point processes. We must first prove the appropriate versions of (2-1) and (2-2). Since $\mathcal{P}^{(1)} \subseteq \mathcal{P}$, the following analogue of (2-1) is immediate:

$$\text{(2-1a)} \qquad P(\sharp(B \cap \mathcal{P}^{(1)}) \geq 2) = O(\lambda(B)^2) \quad \text{for Borel sets } B.$$

We must now find $\varepsilon^{(1)} > 0$, $\theta^{(1)} > 0$, $C^{(1)} < \infty$, and quantities $\mu^{(1)}(k)$ such that

$$\text{(1-1a)} \qquad \sup_k \left[\exp(\theta^{(1)} k)\, \mu^{(1)}(k)\right] = C^{(1)},$$

and such that for all collections of disjoint cubes $Q_1, \ldots, Q_m$ and positive integers $k_1, \ldots, k_m$,

$$\text{(2-2a)} \qquad P\left(\bigcap_{j=1}^{m} \{Q_j \cap \mathcal{P}_{k_j}^{(1)} \neq \emptyset\}\right) \leq \prod_{j=1}^{m} [\varepsilon^{(1)} \mu^{(1)}(k_j) \lambda(Q_j)].$$

We choose

$$\varepsilon^{(1)} = \varepsilon^2 \quad \theta^{(1)} = \frac{\theta\alpha}{80 M^2 |\beta|}$$

$$C^{(1)} = \frac{32 K_1^2 C^2 \beta^{2d+2}}{\alpha^{2d+2} \theta^{4d+6}} \quad \mu^{(1)}(k) = C^{(1)} \exp(-\theta^{(1)} k).$$

(Note that $\mu^{(1)}$ automatically satisfies (1-1a). It is not necessary for the numbers $\mu^{(1)}(k)$ to sum to 1.)

We first prove (2-2a) for the case $m = 1$. Let Q_1 be a cube in $\mathbb{R}^d \times [0, \infty)$. We partition Q_1 into small disjoint cubes with sides of length 2η. Let Q be one such cube, and let the sets $G_n(k_0, \ldots, k_n)$ be defined in terms of Q as in the preceding part of this proof (where we showed that the clusters are finite). Now look back at our definition of $\mathcal{P}_k^{(1)}$. If $(\mathbf{x}, s)$ is a point in $\mathcal{P}_k^{(1)}$, then some other point $(\mathbf{y}, u)$ in $\mathcal{C}(\mathbf{x}, s)$ lies outside the cube $Q(K, \mathbf{x}, s)$, where

$$K = \frac{k}{2} - (\Delta(\mathbf{x}, s) + \Delta(\mathbf{y}, u)) 5 M^2 |\beta| / \alpha.$$

If $(\mathbf{x}, s) \in Q$ and Q contains no other points of $\mathcal{P}$, then there is an ancestry from $(\mathbf{x}, s)$ to $(\mathbf{y}, u)$ of type $(k_0, \ldots, k_n)$ for some $n \geq 1$ and integers $k_0, \ldots, k_n$ (the case $n = 0$ can only occur if Q contains at least 2 points of $\mathcal{P}$). By the definition of ancestries, the max-norm distance between the ith point and the $(i+1)^{\text{st}}$ point in such an ancestry is no more than $(k_i + k_{i+1})5M^2|\beta|/\alpha$. Therefore, by the triangle inequality,

$$\|(\mathbf{x}, s) - (\mathbf{y}, u)\|_{\max} \leq (k_0 + 2k_1 + 2k_2 + \cdots + 2k_{n-1} + k_n)5M^2|\beta|/\alpha.$$

It follows that if Q contains at most one point of $\mathcal{P}$ and $Q \cap \mathcal{P}_k^{(1)} \neq \emptyset$, then there exists an integer $n \geq 0$ and positive integers $k_0, \ldots, k_n$ such that $G_n(k_0, \ldots, k_n)$ is nonempty and

$$(k_0 + 2k_1 + \cdots + 2k_{n-1} + k_n)5M^2|\beta|/\alpha \geq K,$$

where K is defined above. Since in this case $\Delta(\mathbf{x}, s) = k_0$ and $\Delta(\mathbf{y}, u) = k_n$, we have

$$k_0 + \cdots + k_n \geq Ak, \quad \text{where} \quad A = \frac{\alpha}{20M^2|\beta|}.$$

Since the probability that Q contains two or more points of $\mathcal{P}$ is $O(\lambda(Q)^2)$ by (2-1), we have by (2-5)

$$\begin{aligned}
&\frac{P(Q \cap \mathcal{P}_k^{(1)} \neq \emptyset)}{\lambda(Q)} - O(\lambda(Q)) \\
&\quad \leq \sum_{n=1}^{\infty} \sum_{k_0 + \cdots + k_n \geq Ak} \frac{P(G_n(k_0, \ldots, k_n) \neq \emptyset)}{\lambda(Q)} \\
&\quad \leq \sum_{n=1}^{\infty} \sum_{k_0 + \cdots + k_n \geq Ak} \left(\frac{K_1 C\varepsilon|\beta|^{d+1}}{\alpha^{d+1}\theta^{2d+2}}\right)^{n+1} \exp\left(-\frac{\theta}{2}(k_0 + \cdots + k_n)\right) \\
&\quad \leq \exp(-A\theta k/4) \sum_{n=1}^{\infty} \left(\frac{4K_1 C\varepsilon|\beta|^{d+1}}{\alpha^{d+1}\theta^{2d+3}}\right)^{n+1}.
\end{aligned}$$

Now sum the geometric series. By (2-7) and our choice of $\varepsilon^{(1)}$, $\theta^{(1)}$, $C^{(1)}$, and $\mu^{(1)}$, we obtain

$$P(Q \cap \mathcal{P}_k^{(1)} \neq \emptyset) \leq \varepsilon^{(1)}\mu^{(1)}(k)\lambda(Q) + O(\lambda(Q)^2). \tag{2-8}$$

The inequality in (2-8) holds for each of the cubes Q that partition the big cube Q_1. If we sum over all of the small cubes Q contained in Q_1 and let $\eta \searrow 0$, we obtain (2-2a) for $m = 1$.

Now consider general $m \geq 1$ in (2-2a). By subadditivity, we may assume that the cubes Q_j all have sides of length less than 1. Decompose the events $\{Q_j \cap \mathcal{P}_{k_j}^{(1)} \neq \emptyset\}$ into events

$$G_{n_j j}(k_{0j}, \ldots, k_{n_j j}) \neq \emptyset$$

with

$$k_{0j} + \cdots + k_{n_j j} \geq Ak_j$$

as above. Recall from the definition of $\mathcal{P}^{(1)}$ that only one point from a cluster can be a member of $\mathcal{P}^{(1)}$. Also recall from that definition that two points in $\mathcal{P}$ that lie in the same cube with sides of length less than 1 necessarily lie in the same cluster (since according to our assumptions, the quantities $M, |\beta|$, and $1/\alpha$ are all greater than or equal to 1). Therefore, if each Q_j contains a point in $\mathcal{P}^{(1)}$, then the sets

$$\bigcup_{n_j} \bigcup_{k_{0j},\ldots,k_{n_jj}} G_{n_jj}(k_{0j},\ldots,k_{n_jj})$$

are disjoint.

Now we proceed as in the case $m = 1$. We have

$$P\left(\bigcap_{j=1}^{m}\{Q_j \cap \mathcal{P}_{k_j}^{(1)} \neq \emptyset\}\right) \leq$$

$$\left(\sum_{n_1=0}^{\infty} \sum_{k_{01}+\cdots+k_{n_11}\geq Ak_{n_1}} \cdots \sum_{n_m=0}^{\infty} \sum_{k_{0m}+\cdots+k_{n_mm}\geq Ak_{n_m}}\right)' P\left(\bigcap_{j=1}^{m}\{G_{n_jj}(k_{0j},\ldots,k_{n_jj}) \neq \emptyset\}\right),$$

where the notation $(\Sigma \ldots \Sigma)'$ means that we only sum over terms for which the corresponding sets $G_{n_jj}(\cdots)$ are disjoint. To complete the proof of (2-2a), it is enough to show that under this disjointness assumption,

$$\text{(2-9)} \quad \frac{P\left(\bigcap_{j=1}^{m}\{G_{n_jj}(k_{0j},\ldots,k_{n_jj}) \neq \emptyset\}\right)}{\lambda(Q_1)\cdots\lambda(Q_m)} \leq \prod_{j=1}^{m}\left[\left(\frac{K_1 C\varepsilon|\beta|^{d+1}}{\alpha^{d+1}\theta^{2d+2}}\right)^{n_j+1} \exp\left(\frac{\theta}{2}(k_{0j}+\cdots+k_{n_jj})\right)\right],$$

since the rest of the proof is then the same as in the case $m = 1$. Decompose each of the events $\{G_{n_jj}(\cdots) \neq \emptyset\}$ in (2-9) into events involving the small partitioning cubes that were used in the derivation of (2-5). This allows us to write the intersection in (2-9) as a disjoint union of events, each of which is an intersection of events of the form $\{Q(\delta, \mathbf{y}_i, u_i) \cap \mathcal{P}_k \neq \emptyset\}$ for various values of i and k. Because of the disjointness mentioned just before (2-9), the integers i involved in any one of these intersections are distinct. This fact allows us to use (2-2) to bound the probabilities of these intersections by products analogous to the product that appears in (2-4). Now summing over the decomposition and letting $\delta \to 0$ leads to (2-9), just as in the derivation of (2-5).

Successive renormalizations. Let us summarize what we have just accomplished. Using only (2-1), (2-2), and (2-7), we constructed point processes $\mathcal{P}_k^{(1)}$ which satisfy (2-1a) and (2-2a) for certain parameters $\varepsilon^{(1)}$, $\theta^{(1)}$, $C^{(1)}$, and $\mu^{(1)}(k)$. The generic process $\mathcal{A}_t^{(1)}$ corresponding to the point processes $\mathcal{P}_k^{(1)}$ and with rates $\alpha_i^{(1)}$ and $\beta^{(1)}$ is our first renormalization of $\mathcal{A}_t$. As long as the appropriate analogue of (2-7) is satisfied at each stage, we may continue to renormalize, thus inductively obtaining a sequence of collections of point processes $\{\mathcal{P}_k^{(\ell)}; k \geq 1\}$ and a sequence of corresponding generic processes $\mathcal{A}_t^{(\ell)}$, $\ell \geq 1$. The corresponding parameters are

$$\alpha^{(\ell)} = \alpha^{(\ell-1)}/2 = \alpha/2^\ell \qquad \beta^{(\ell)} = 2\beta^{(\ell-1)} = 2^\ell \beta \qquad \varepsilon^{(\ell)} = (\varepsilon^{(\ell-1)})^2 = \varepsilon^{2^\ell}$$

(2-10)

$$\theta^{(\ell)} = \frac{\theta^{(\ell-1)}\alpha^{(\ell-1)}}{80M^2|\beta^{(\ell-1)}|} \geq \frac{\theta\alpha^\ell}{(80M^2|\beta|)^\ell 4^{1+2+\cdots+(\ell-1)}} \geq \frac{\theta\alpha^\ell}{(M^2|\beta|)^\ell 2^{\ell^2+7\ell}}$$

$$C^{(\ell)} = \frac{32(K_1C^{(\ell-1)})^2(\beta^{(\ell-1)})^{2d+2}}{(\alpha^{(\ell-1)})^{2d+2}(\theta^{(\ell-1)})^{4d+6}} \qquad \mu^{(\ell)}(k) = C^{(\ell)}\exp(-\theta^{(\ell)}k).$$

(We write $\mathcal{A}_t^{(0)} = \mathcal{A}_t$, $\varepsilon^{(0)} = \varepsilon$, etc.) We remark that as before, clusters $\mathcal{C}^{(\ell-1)}(\mathbf{x}, s)$ are used to construct the point process $\mathcal{P}^{(\ell)}$. In this construction, the appropriate cubes of influence are of the form

$$\mathcal{S}^{(\ell-1)}(\mathbf{x}, s) = \text{ the interior of } Q\left(\frac{5M^2|\beta^{(\ell-1)}|\Delta^{(\ell-1)}(\mathbf{x}, s)}{\alpha^{(\ell-1)}}, \mathbf{x}, s\right).$$

The renormalizations satisfy

(2-1b) $$P(\sharp(B \cap \mathcal{P}^{(\ell)}) \geq 2) = O(\lambda(B)^2) \quad \text{for Borel sets } B$$

and

(2-2b) $$P\left(\bigcap_{j=1}^{m}\{Q_j \cap \mathcal{P}_{k_j}^{(\ell)} \neq \emptyset\}\right) \leq \prod_{j=1}^{m}[\varepsilon^{(\ell)}\mu^{(\ell)}(k_j)\lambda(Q_j)]$$

for disjoint cubes $Q_1, \ldots, Q_m$.

In order to be able to continue the induction at each stage, we only need the following analogue of (2-7):

(2-7b) $$\varepsilon^{(\ell)} \leq \frac{(\alpha^{(\ell)})^{d+1}(\theta^{(\ell)})^{2d+3}}{8K_1C^{(\ell)}|\beta^{(\ell)}|^{d+1}}.$$

We will be able to obtain (2-7b) because of the simple but important fact that $\varepsilon^{(\ell)}$ goes to 0 very fast as $\ell \to \infty$. The quantities $\alpha^{(\ell)}, \theta^{(\ell)}$ go to 0 much more slowly, just as the quantity $|\beta^{(\ell)}|$ goes to ∞ much more slowly. The quantity $C^{(\ell)}$ is the only parameter that grows at a rate that is comparable to the rate at which $\varepsilon^{(\ell)}$ decreases, but for ε small enough, $\varepsilon^{(\ell)}$ will dominate $C^{(\ell)}$.

We use (2-10) to substitute for $\alpha^{(\ell)}$ in terms of α, for $\beta^{(\ell)}$ in terms of β, and the given lower bound for $\theta^{(\ell)}$ in terms of α, β, and θ. After these substitutions it is apparent that (2-7b) is implied by

$$\varepsilon^{(\ell)} \leq \frac{(\alpha/|\beta|)^{(2d+3)\ell+d+1}(\theta/M^{2\ell})^{2d+3}}{K_1 C^{(\ell)} 2^{(2d+3)(\ell^2+7\ell)+3}}.$$

It is clear that this last inequality is in turn implied by

$$\varepsilon^{(\ell)} \leq \frac{1}{K_2^{(\ell+1)^2+4} C^{(\ell)}} \tag{2-11}$$

for some constant K_2 which depends on α, β, θ, M, and d. After we put further assumptions on the constant K_2, we will prove inductively that (2-11) is satisfied for all ℓ if it is satisfied for $\ell = 0$. (Our choice of the exponent in (2-11) is dictated by the way in which the induction argument works.)

By using (2-10) as in the preceding paragraph, it is elementary to check that there exists some constant K_3 depending only on α, β, θ, M, and d such that

$$C^{(\ell)} \leq K_3^{\ell^2+1}(C^{(\ell-1)})^2.$$

Let us assume that the constant K_2 from the preceding paragraph is chosen large enough so that $K_2 \geq (K_3)^2$. Assume inductively that (2-11) is satisfied for some $\ell \geq 0$. Then

$$\begin{aligned}
\varepsilon^{(\ell+1)} &= (\varepsilon^{(\ell)})^2 \\
&\leq \frac{1}{(C^{(\ell)})^2 K_2^{2(\ell+1)^2+8}} \\
&\leq \frac{1}{C^{(\ell+1)}(K_2^{2(\ell+1)^2+8}/K_3^{(\ell+1)^2+1})} \\
&\leq \frac{1}{C^{(\ell+1)}(K_2)^{(3(\ell+1)^2+15)/2}}.
\end{aligned}$$

Since the exponent on the constant K_2 in the last line is bounded below by $(\ell+2)^2+4$, the inductive step in the proof of (2-11) is completed. In other words, if (2-11) is satisfied for $\ell = 0$, it is satisfied for all $\ell \geq 0$. We conclude that for all $\ell \geq 0$, the point processes $\mathcal{P}_k^{(\ell)}$ and corresponding generic processes $\mathcal{A}_t^{(\ell)}$ can be constructed so that (2-1b), (2-2b), and (2-7b) are satisfied, provided the original death rate ε satisfies

$$\varepsilon < \frac{1}{(K_2)^5 C}. \tag{2-12}$$

For the rest of the proof, we will assume that (2-12) holds.

Comparing the different renormalizations. Let $N_t^{(\ell)}$ and $A_t^{(\ell)}$ be respectively the noninteractive and vacant regions corresponding to the generic process $\mathcal{A}_t^{(\ell)}$ constructed in the preceding part of the proof. We will prove that for all $\ell \geq 0$,

$$A_t^{(\ell)} \subseteq N_t^{(\ell)} \cup A_t^{(\ell+1)} \quad \text{a.s. for all } t \geq 0. \tag{2-13}$$

This relationship between the different renormalized processes is the key to the entire proof. Our definitions of clusters and of the sizes $\Delta^{(1)}(\mathbf{x}, s)$ were designed with (2-13) in mind. As we will see in the final part of the proof, (2-13) will enable us to analyze the vacant region A_t in terms of the collection of noninteractive regions $N_t^{(\ell)}$ for $\ell = 0, 1, 2, \ldots$. It is relatively easy to work with the noninteractive regions because, for each ℓ, their faces all move inward at rate $\alpha^{(\ell)}$. Because of the rapidity with which $\varepsilon^{(\ell)}$ goes to 0 as ℓ increases, the noninteractive regions are extremely sparse for large ℓ and small ε. These simple facts motivate the proof of the theorem.

Because of the inductive way in which the renormalization is carried out for each ℓ, it is sufficient to prove (2-13) for the case $\ell = 0$. We write N_t for $N_t^{(0)}$, A_t for $A_t^{(0)}$, etc. It will be seen that certain parts of the proof of (2-13) are most properly done in terms of the finite volume processes that were used in Section 1 to show the existence of generic processes. It is certainly sufficient to prove (2-13) with A_t replaced by the finite volume process $A_{t,j}$, since A_t is the limit of $A_{t,j}$ as $j \to \infty$. The main advantage of finite volume processes is that there are only finitely many regions present in such a process at any finite time t. However, almost all of the notation acquires an extra subscript when working with the finite volume processes, so we will be slightly informal here and continue to speak in terms of the limit A_t. We will indicate those places in the argument where complete rigor would require dealing first with the finite volume processes and then taking limits.

For each point $(\mathbf{x}, s)$ corresponding to a region $D(\mathbf{x}, s; t)$ in the generic process $\mathcal{A}_t$, we wish to define a point $(\xi(\mathbf{x}, s), \sigma(\mathbf{x}, s)) \in \mathcal{P}$ which marks 'the most recent death involved' in the appearance of the region $D(\mathbf{x}, s; t)$. For death regions $D(\mathbf{x}, s; t)$ we simply let $(\xi(\mathbf{x}, s), \sigma(\mathbf{x}, s)) = (\mathbf{x}, s)$. For overlap and collision regions, we recall the maximal collection of regions $D(\mathbf{y}_k, u_k; s)$ used in the definitions of overlap and collision regions in Section 1. Assuming that $\xi(\mathbf{y}_k, u_k)$ and $\sigma(\mathbf{y}_k, u_k)$ are defined for each k, choose i such that $\sigma(\mathbf{y}_i, u_i) = \max_k \sigma(\mathbf{y}_k, u_k)$, and then define

$$(\xi(\mathbf{x}, s), \sigma(\mathbf{x}, s)) = (\xi(\mathbf{y}_i, u_i), \sigma(\mathbf{y}_i, u_i)).$$

In the case of an overlap region $D(\mathbf{x}, s; s)$, we have $(\xi(\mathbf{x}, s), \sigma(\mathbf{x}, s)) = (\mathbf{y}_1, s)$, since in the definition of an overlap region, $D(\mathbf{y}_1, u_1; s)$ is a death region with $u_1 = s$. For collision regions $D(\mathbf{x}, s; s)$, complete rigor would require that we work with the finite volume process, both so that the maximum in the definition of $\sigma(\mathbf{x}, s)$ is actually achieved and so that the induction implied in the phrase "assuming that $\xi(\mathbf{y}_k, u_k)$ and $\sigma(\mathbf{y}_k, u_k)$ are defined" can be successfully carried out.

We claim the following:

$$\text{(2-14)} \quad \text{if} \quad \mathbf{p} \in D(\mathbf{x}, s; t), \quad \text{then} \quad \|\mathbf{p} - \xi(\mathbf{x}, s)\| \le M[\Delta(\xi(\mathbf{x}, s), \sigma(\mathbf{x}, s)) + |\beta|(t - \sigma(\mathbf{x}, s))].$$

This statement is obvious for death regions $D(\mathbf{x}, s; t)$, since M has been chosen so that $D(\mathbf{x}, s; s) \subseteq B_{\Delta(\mathbf{x},s)M}(\mathbf{x})$, and since death regions shrink in size. For overlap and collision regions $D(\mathbf{x}, s; t)$, let $(\mathbf{y}_i, u_i)$ be as in the preceding paragraph. Then (2-14) is true, provided it is true with $(\mathbf{x}, s)$ replaced by $(\mathbf{y}_i, u_i)$, because $D(\mathbf{x}, s; s) \subseteq D(\mathbf{y}_i, u_i; s)$, and after time s, the radius of the smallest ball that contains $D(\mathbf{x}, s; t)$ increases no faster than rate $M|\beta|$. The induction argument implied by the phrase "provided it is true with $(\mathbf{x}, s)$ replaced by $(\mathbf{y}_i, u_i)$" is properly carried out in the finite volume setting.

We complete the proof of (2-13) (for $\ell = 0$) by showing that for any overlap or collision region $D(\mathbf{y}, u; t)$, there exists a region $D^{(1)}(\mathbf{x}, s; t) \in \mathcal{A}_t^{(1)}$ such that

$$\text{(2-15)} \quad H_i(\mathbf{y}, u; v) - (\alpha(v - \sigma(\mathbf{y}, u))/2)\mathbf{n}_i \subseteq H_i^{(1)}(\mathbf{x}, s; v) \quad \text{for all } v \in [u, t] \text{ and } i = 1, \ldots, d+1.$$

This last expression is just a way of saying that the ith face of the region $D(\mathbf{y}, u; v)$ is at least $\alpha(v - \sigma(\mathbf{y}, u))/2$ units inside of the corresponding face of the region $D^{(1)}(\mathbf{x}, s; v)$. The significance of of the factor $\alpha/2$ will be seen in the proof of (2-17) below. It is clear that (2-13) (for $\ell = 0$) holds if (2-15) is true for all overlap and collision regions $D(\mathbf{y}, u; t)$, since any death region $D(\mathbf{y}, u; t)$ is contained in N_t.

Let $D(\mathbf{y}, u; t)$ be a collision or overlap region, and let $D(\mathbf{y}_k, u_k; u)$, $k = 1, \ldots, n$, be the maximal collection of regions used in the definition of $D(\mathbf{y}, u; t)$ in Section 1 (we know that n is finite in the finite volume setting). We now make the assumption that (2-15) is satisfied for all overlap and collision regions that arise prior to time u. More precisely, we assume that for any overlap or collision region $D(\mathbf{y}', u'; t')$ such that $u' < u$ and $t' \le t$, there exists a region $D^{(1)}(\mathbf{x}', s'; t')$ such that (2-15) is satisfied with $(\mathbf{y}, u)$ replaced by $(\mathbf{y}', u')$, $(\mathbf{x}, s)$ replaced by $(\mathbf{x}', s')$, and t replaced by t'. As above, we are setting up an induction that can be rigorously carried out in the finite volume setting. For this reason, we refer to the assumption just made as the 'inductive hypothesis'. Note that in particular, this inductive hypothesis applies with $D(\mathbf{y}', u'; t') = D(\mathbf{y}_k, u_k; t)$ for all k such that $D(\mathbf{y}_k, u_k; t)$ is not a death region, since for each such k, it follows from our construction of generic processes that $u_k < u$. For these values of k, let $(\mathbf{x}_k, s_k)$ be the point $(\mathbf{x}', s')$ provided by the inductive hypothesis.

We now break the argument into three cases: (i) none of the regions $D(\mathbf{y}_k, u_k; u)$ is a death region; (ii) only $D(\mathbf{y}_1, u_1; u)$ is a death region; (iii) $D(\mathbf{y}_1, u_1; u), \dots, D(\mathbf{y}_m, u_m; u)$ are death regions for some m with $2 \le m \le n$, but none of the remaining regions is a death region. It will be seen that cases (i) and (iii) are relatively easy. Case (ii) is the critical case. The constants used in the definition of clusters were chosen with this case in mind.

First consider case (i). By Proposition 3 and (1-7), there is a region $D^{(1)}(\mathbf{x}, s; t)$ such that

$$\bigcap_{k=1}^{n} D^{(1)}(\mathbf{x}_k, s_k; v) \subseteq D^{(1)}(\mathbf{x}, s; v) \quad \text{for} \quad v \in [u, t], \tag{1-4a}$$

and either

$$\frac{d\gamma_i^{(1)}}{dt}(\mathbf{x}, s; v) = \beta^{(1)} \quad \text{for} \quad v \in [u, t), \tag{1-5a}$$

or

$$\bigcup_{k=1}^{n} H_i^{(1)}(\mathbf{x}_k, s_k; t) = H_i^{(1)}(\mathbf{x}, s; t). \tag{1-6a}$$

We claim that this region $D^{(1)}(\mathbf{x}, s; t)$ satisfies (2-15). If (1-6a) holds, (2-15) follows from the inductive hypothesis (with $(\mathbf{y}', u') = (\mathbf{y}_k, u_k)$ and $(\mathbf{x}', s') = (\mathbf{x}_k, s_k)$) and the fact that $\sigma(\mathbf{y}, u) \ge \sigma(\mathbf{y}_k, u_k)$ for all k. Suppose that (1-5a) holds instead. By (1-4a),

$$\begin{aligned} H_i(\mathbf{y}, u; u) - &\frac{\alpha}{2}(u - \sigma(\mathbf{y}, u))\mathbf{n}_i \\ &= \bigcap_{k=1}^{n} H_i(\mathbf{y}_k, u_k; u) - \frac{\alpha}{2}(u - \sigma(\mathbf{y}, u))\mathbf{n}_i \\ &\subseteq \bigcap_{k=1}^{n} H_i^{(1)}(\mathbf{x}_k, s; u) \quad \subseteq H_i^{(1)}(\mathbf{x}, s; u). \end{aligned} \tag{2-16}$$

Since the derivative of $\gamma_i(\mathbf{y}, u; v)$ with respect to v is at least $\beta = \beta^{(1)}/2$, it follows from (1-5a) that the distance between the ith face of $D(\mathbf{y}, u; v)$ and the corresponding face of $D^{(1)}(\mathbf{x}, s; v)$ increases at least at rate $|\beta^{(1)} - \beta| = |\beta| \ge 1 > \alpha/2$. This fact combined with (2-16) gives us (2-15).

Next consider case (ii). The argument for this case will be the same as the previous case, once we find a region $D^{(1)}(\mathbf{x}_1, s_1; t)$ such that for all $i = 1, \dots, d+1$ and $v \in [u, t]$,

$$H_i(\mathbf{y}_1, u_1; v) - (\alpha(v - \sigma(\mathbf{y}_1, u_1))/2)\mathbf{y}_i \subseteq H_i^{(1)}(\mathbf{x}_1, s_1; v). \tag{2-15a}$$

(Since $D(\mathbf{y}_1, u_1; u)$ is a death region, our inductive hypothesis does not help us directly in this case. It would do us no good to try to strengthen the inductive hypothesis, since (2-15a) is not true in general for death regions.) Recall that the faces of the death region $D(\mathbf{y}_1, u_1; t)$ all move inward at rate α, and that the faces of regions in the process $\mathcal{A}_t^{(1)}$ all either move outward, or they move inward at rate $\alpha^{(1)} = \alpha/2$. Thus, if a region $D^{(1)}(\mathbf{x}_1, s_1; v)$ contains a death region $D(\mathbf{y}_1, s_1; v)$ at some time v, then for each i the distance between the ith faces of these two regions increases at least at rate $\alpha/2$ after time v. Hence, to prove (2-15a), it is sufficient to prove that there exists a region $D^{(1)}(\mathbf{x}_1, s_1; t)$ such that for some $v \in [u_1, u]$,

$$D(\mathbf{y}_1, u_1; v) \subseteq D^{(1)}(\mathbf{x}_1, s_1; v). \tag{2-17}$$

To prove (2-17), we first need to find some region $D(\mathbf{z}, w; v)$ with $w < u$ and $v \in [u_1, u]$, such that

$$D(\mathbf{z}, w; v) \cap D(\mathbf{y}_1, u_1; v) \neq \emptyset \quad \text{and} \quad (\xi(\mathbf{z}, w), \sigma(\mathbf{z}, w)) \neq (\mathbf{y}_1, u_1).$$

(It will become apparent later how such a region will help us.) If there is at least one value of $k \geq 2$ such that $(\xi(\mathbf{y}_k, u_k), \sigma(\mathbf{y}_k, u_k)) \neq (\mathbf{y}_1, u_1)$, then we can let $(\mathbf{z}, w) = (\mathbf{y}_k, u_k)$ and $v = u$, since in this case $u_k < u$ and $D(\mathbf{y}_1, u_1; u)$ has nonempty intersection with all of the regions $D(\mathbf{y}_k, u_k; u)$. If on the other hand, $(\xi(\mathbf{y}_k, u_k), \sigma(\mathbf{y}_k, u_k)) = (\mathbf{y}_1, u_1)$ for all $k \geq 2$, then it follows from the definitions that all of the regions $D(\mathbf{y}_k, u_k; u)$ for $k \geq 2$ necessarily arose due to an overlap interaction caused by the appearance of the death region $D(\mathbf{y}_1, u_1; u_1)$. In this case, the overlap was necessarily between $D(\mathbf{y}_1, u_1; u_1)$ and some region $D(\mathbf{z}, w; u_1)$ with $w < u_1 \leq u$, so we have our desired region $D(\mathbf{z}, w; v)$ with $v = u_1$. We have now chosen the value of v for which (2-17) will be satisfied. It remains to use $D(\mathbf{z}, w; v)$ to find the appropriate region $D^{(1)}(\mathbf{x}_1, s_1; v)$.

Let us abbreviate by writing $(\xi, \sigma) = (\xi(\mathbf{z}, w), \sigma(\mathbf{z}, w))$. Note that both of the points (ξ, σ) and $(\mathbf{y}_1, u_1)$ are members of $\mathcal{P}$. If they lie in the same cluster, then there exists a point $(\mathbf{x}_1, s_1) \in \mathcal{P}^{(1)}$ such that $\mathcal{C}(\mathbf{y}_1, u_1) = \mathcal{C}(\xi, \sigma) = \mathcal{C}(\mathbf{x}_1, s_1)$. Now recall how the quantity $\Delta^{(1)}(\mathbf{x}_1, s_1)$ is defined. The remark at the end of the paragraph following the definition of $\Delta^{(1)}(\cdot)$ implies that $\Delta^{(1)}(\mathbf{x}_1, s_1)$ has been chosen large enough so that (2-17) is satisfied.

To finish case (ii), we must prove (2-17) under the assumption that $(\mathbf{y}_1, u_1)$ and (ξ, σ) do not lie in the same cluster. It follows from the definition of clusters that the death regions $D(\mathbf{y}_1, u_1; v)$ and $D(\xi, \sigma; v)$ are disjoint (see the comment made at the end of the section of the proof in which clusters were defined). But $D(\mathbf{y}_1, u_1; v)$ and $D(\mathbf{z}, w; v)$ are not disjoint, so $(\mathbf{z}, w) \neq (\xi, \sigma)$, implying that $D(\mathbf{z}, w; v)$ is not a death region. Since $w < u_1 \leq u$, the inductive hypothesis applies to $D(\mathbf{z}, w; v)$, giving us a region $D^{(1)}(\mathbf{x}_1, s_1; v)$ such that for all $i = 1, \ldots, d+1$,

$$H_i(\mathbf{z}, w; v) - (\alpha(v - \sigma))/2)\mathbf{n}_i \subseteq H_i^{(1)}(\mathbf{x}_1, s_1, v). \tag{2-15b}$$

We will show that the region $D^{(1)}(\mathbf{x}_1, s_1; v)$ found in the preceding paragraph satisfies (2-17). Since we have assumed that $(\mathbf{y}_1, u_1)$ and (ξ, σ) are not in the same cluster, it follows from the definition of clusters that either

$$\|\mathbf{y}_1 - \xi\| \geq 5M^2|\beta|(\Delta(\mathbf{y}_1, u_1) + \Delta(\xi, \sigma))/\alpha, \tag{2-18}$$

or

$$|u_1 - \sigma| \geq 5M^2|\beta|(\Delta(\mathbf{y}_1, u_1) + \Delta(\xi, \sigma))/\alpha. \tag{2-19}$$

We wish to show that both statements imply

$$v - \sigma \geq 4M(\Delta(\mathbf{y}_1, u_1) + \Delta(\xi, \sigma))/\alpha. \tag{2-20}$$

To see that (2-18) implies (2-20), we use (2-14), with $(\mathbf{x}, s)$ replaced by $(\mathbf{z}, w)$, t replaced by v, and with $\mathbf{p}$ equaling any point in $D(\mathbf{y}_1, u_1; v) \cap D(\mathbf{z}, w; v)$. Since the death region $D(\mathbf{y}_1, u_1; v)$ is contained in a ball of radius $M\Delta(\mathbf{y}_1, u_1)$, it follows from (2-14) that

$$\|\mathbf{y}_1 - \xi\| \leq M(\Delta(\mathbf{y}_1, u_1) + \Delta(\xi, \sigma) + |\beta|(v - \sigma)).$$

Combine this inequality with (2-18) to obtain (2-20). To see that (2-19) implies (2-20), it is sufficient to show that $u_1 > \sigma$ if (2-19) holds, since $v \geq u_1$ (and, of course, since $5M^2|\beta| > 4M$). We know that $u \geq \sigma$ and that the death region $D(\mathbf{y}_1, u_1; v)$ vanishes after time $v = u_1 + \Delta(\mathbf{y}_1, u_1)/\alpha$. If both $\sigma > u_1$ and (2-19) held, then the death region $D(\mathbf{y}_1, u_1; v)$ would be empty, contradicting the fact that it has nonempty intersection with $D(\mathbf{y}_k, u_k; v)$. This contradiction establishes the desired conclusion.

We have just proved that if $(\mathbf{y}_1, u_1)$ and (ξ, σ) do not lie in the same cluster, then (2-20) holds. Therefore, by (2-15b),

$$H_i(\mathbf{z}, w; v) - 2M(\Delta(\mathbf{y}_1, u_1) + \Delta(\xi, \sigma))\mathbf{n}_i \subseteq H_i^{(1)}(\mathbf{x}_1, s_1; v)$$

for $i = 1, \ldots, d+1$. Now use the fact that $D(\mathbf{y}_1, u_1; v)$ is contained in a ball of radius $M\Delta(\mathbf{y}_1, u_1)$ and has nonempty intersection with $D(\mathbf{z}, w; v)$ to conclude that (2-17) is satisfied.

Finally, we consider case (iii). As in case (ii), it is sufficient to prove the analogue of (2-17) for the regions $D(\mathbf{y}_k, u_k; u)$, $k = 1, \ldots, m$; namely, that for each of the death regions $D(\mathbf{y}_k, u_k; u)$, there exists a time $v \in [u_k, u]$ and a region $D^{(1)}(\mathbf{x}_k, s_k; v)$ such that $D(\mathbf{y}_k, u_k; v) \subseteq D^{(1)}(\mathbf{x}_k, s_k; v)$. Since the death regions $D(\mathbf{y}_k, u_k; t)$ have nonempty intersection for $k = 1, \ldots, m$, their corresponding cubes of influence $S(\mathbf{y}_k, u_k)$ have nonempty intersection (this follows from the comments made at the end of the section of the proof where clusters are defined). Therefore, the points $(\mathbf{y}_k, u_k)$, $k = 1, \ldots, m$, lie in the same cluster. Since $m \geq 2$, there is some point $(\mathbf{x}_1, s_1) \in \mathcal{P}^{(1)}$ that is also a member of that cluster. Let $(\mathbf{x}_k, s_k) = (\mathbf{x}_1, s_1)$ for $k = 1, \ldots, m$. As explained in case (ii), $\Delta^{(1)}(\mathbf{x}_1, s_1)$ has been chosen large enough so that the analogue of (2-17) for the regions $D(\mathbf{y}_k, u_k; u)$ is satisfied for $k = 1, \ldots, m$. This completes case (iii), so the proof of (2-15) and hence of (2-13) is finished.

The final upper bounds. It follows from (2-13) that

$$P(\mathbf{y} \in A_t) \le \limsup_{\ell\to\infty} P(\mathbf{y} \in A_t^{(\ell)}) + \sum_{\ell=0}^{\infty} P(\mathbf{y} \in N_t^{(\ell)}). \tag{2-21}$$

We will complete the proof of the theorem by obtaining appropriate bounds for the terms on the right side of (2-21). More precisely, we will show that the constant K_2 can be chosen sufficiently large so that for $\varepsilon > 0$ satisfying (2-12),

$$\limsup_{\ell\to\infty} P(\mathbf{y} \in A_t^{(\ell)}) = 0 \quad \text{for fixed } t, \text{ uniformly in } \mathbf{y}, \tag{2-22}$$

and that

$$\lim_{\varepsilon \searrow 0} \sum_{\ell=0}^{\infty} P(\mathbf{y} \in N_t^{(\ell)}) = 0 \quad \text{uniformly in } t \text{ and } \mathbf{y}. \tag{2-23}$$

We prove (2-23) first. Note that if $\mathbf{y} \in N_t^{(\ell)}$, then there exists a point $(\mathbf{x}, s) \in \mathcal{P}_k^{(\ell)}$ for some positive integer k such that

$$(\mathbf{x}, s) \in Q((Mk/\alpha^{(\ell)}), \mathbf{y}, t).$$

The reason for this is that the death region $D^{(\ell)}(\mathbf{x}, s; t)$ is always contained in the ball of radius Mk centered at $\mathbf{x}$, and furthermore, this death region vanishes after time $k/\alpha^{(\ell)}$. (We are also relying on the obvious inequality $Mk/\alpha^{(\ell)} \ge (k/\alpha^{(\ell)}) \vee (Mk)$.) Therefore, by (2-2b) with $m = 1$,

$$\begin{aligned} P(\mathbf{y} \in N_t^{(\ell)}) &\le \sum_{k=1}^{\infty} P(Q((Mk/\alpha^{(\ell)}), \mathbf{y}, t) \cap \mathcal{P}_k^{(\ell)} \ne \emptyset) \\ &\le \sum_{k=1}^{\infty} \varepsilon^{(\ell)} (2Mk/\alpha^{(\ell)})^{d+1} \mu^{(\ell)}(k) \\ &= \sum_{k=1}^{\infty} \varepsilon^{(\ell)} (2Mk/\alpha^{(\ell)})^{d+1} C^{(\ell)} \exp(-\theta^{(\ell)} k) \\ &\le \frac{\varepsilon^{(\ell)} 2^{2d+3} M^{d+1} C^{(\ell)} (d+1)!}{(\alpha^{(\ell)} \theta^{(\ell)})^{d+1}}. \end{aligned} \tag{2-24}$$

We have used the fact that

$$\sum_{k=1}^{\infty} k^{d+1} \exp(-ak) \le (2/a)^{d+2}\,(d+1)!$$

for $0 < a < 1$.

Now use (2-10) to substitute for $\alpha^{(\ell)}, \beta^{(\ell)}$, and $\theta^{(\ell)}$ in terms of α, β, and θ. It follows, as in the argument preceding (2-12), that we can find a constant K_4 depending on α, β, θ, M, and d such that the expression on the right side of (2-24) is bounded above by

$$\varepsilon^{(\ell)} C^{(\ell)} K_4^{\ell^2+1}.$$

After substituting in (2-24), we obtain

$$P(\mathbf{y} \in N_t^{(\ell)}) \le \frac{K_4^{\ell^2+1}}{K_2^{(\ell+1)^2+4}}. \tag{2-25}$$

Note that the right side of (2-25) does not depend on t or $\mathbf{y}$. As $\varepsilon \searrow 0$, we can let $K_2 \to \infty$ in accordance with (2-12). Thus (2-23) follows.

We now prove (2-22). Define an auxiliary process $\hat{N}_t^{(\ell)}$ by

$$\hat{N}_t^{(\ell)} = \bigcup_{(\mathbf{x},s)\in\mathcal{P}^{(\ell)}:s\le t} \; \bigcap_{i=1}^{d+1} H(\mathbf{n}_i, \mathbf{x} + (2^\ell\beta(t-s) - \Delta^{(\ell)}(\mathbf{x},s))\mathbf{n}_i).$$

This process is just like $N_t^{(\ell)}$, except that all of the occupation rates $\alpha_i^{(\ell)}$ have been replaced by the value $\beta^{(\ell)} = 2^\ell\beta$. The points at which deaths occur are the same for the processes $N_t^{(\ell)}$ and $\hat{N}_t^{(\ell)}$, so it is clear that $N_t^{(\ell)} \subseteq \hat{N}_t^{(\ell)}$. If we define the generic process $\hat{A}_t^{(\ell)}$ corresponding to $\hat{N}_t^{(\ell)}$, Proposition 4 implies that $A_t^{(\ell)} \subseteq \hat{A}_t^{(\ell)}$. A little thought based on the details of the construction given in Section 1 reveals that $\hat{A}_t^{(\ell)} = \hat{N}_t^{(\ell)}$. Thus

$$P(\mathbf{y} \in A_t^{(\ell)}) \le P(\mathbf{y} \in \hat{N}_t^{(\ell)}).$$

Now note that if $\mathbf{y} \in \hat{N}_t^{(\ell)}$, then $\mathbf{y}$ lies in some death region $\hat{D}^{(\ell)}(\mathbf{x}, s; t)$, where $(\mathbf{x}, s) \in \mathcal{P}_k^{(\ell)}$ for some $k > 0$. The faces of such a region move outward at speed $2^\ell|\beta|$, so $(\mathbf{y}, t) \in Q(M(2^\ell|\beta|(t-s)+k), \mathbf{x}, s)$, or equivalently,

$$(\mathbf{x}, s) \in Q(M(2^\ell|\beta|(t-s)+k), \mathbf{y}, t).$$

Noting that $M(2^\ell|\beta|(t-s)+k) \le 2^{\ell+1}M|\beta|(t\vee 1)k$, it follows from (2-2b), as in (2-24), that

$$\begin{aligned} P(\mathbf{y} \in A_t^{(\ell)}) &\le P(\mathbf{y} \in \hat{N}_t^{(\ell)}) \\ &\le \sum_{k=1}^{\infty} \varepsilon^{(\ell)} (2^{\ell+2} M k|\beta|(t\vee 1))^{d+1} C^{(\ell)} \exp(-\theta^{(\ell)} k) \\ &\le \varepsilon^{(\ell)} 2^{2d+3} (2^{\ell+2}(t\vee 1)M|\beta|/\theta^{(\ell)})^{d+1} C^{(\ell)} (d+1)!\,. \end{aligned}$$

As in the proof of (2-23), we may choose a constant K_5 depending only on α, β, θ, M, and d such that the term on the right side of this last expression is bounded above by

$$\frac{K_5^{\ell^2+1}(t \vee 1)^{d+1}}{K_2^{(\ell+1)^2+3}}. \tag{2-26}$$

By choosing $\varepsilon > 0$ sufficiently small, we can choose $K_2 > K_5$ in accordance with (2-12). For such a choice of ε and K_2, the expression in (2-26) converges to 0 as $\ell \to \infty$. Since $\mathbf{y}$ does not appear in this expression, (2-22) follows. ∎

3. Toom's Theorem and other applications. Now that we know precisely which families of generic processes survive for small ε, we can use them as comparison processes for other models of interest. In this section, we will show how to carry out this comparison for the general class of discrete time processes studied by Toom [13]. As a result, we obtain a relatively easy proof of a generalization of the main theorem in [13].

TOOM PROCESSES. The models most directly related to our generic processes are certain probabilistic cellular automata that we call Toom processes. This is the class of processes to which Toom's Theorem applies. The two discrete time processes mentioned in the introduction, namely the basic discrete time contact process and Toom's model, are both in this class.

To fix notation, we will briefly describe the construction of Toom processes. A **(d-dimensional) Toom process with range r** is defined in terms of a **Toom rule**, which is a function

$$\varphi : \{\text{subsets of } \mathbb{Z}^d\} \to \{\text{subsets of } \mathbb{Z}^d\}$$

satisfying four conditions:

(1) (monotonicity) $S_1 \subseteq S_2 \Rightarrow \varphi(S_1) \subseteq \varphi(S_2)$;
(2) (translation invariance) $\varphi(S) + \mathbf{x} = \varphi(S + \mathbf{x})$;
(3) (range r) $\mathbf{x} \in \varphi(S) \Leftrightarrow \mathbf{x} \in \varphi(S \cap B_r(\mathbf{x}))$;
(4) (nontriviality) $\varphi(\emptyset) = \emptyset$ and $\varphi(\mathbb{Z}^d) = \mathbb{Z}^d$.

Fix a parameter value $\tilde{\varepsilon} \in [0,1]$, the **death rate**. For $(\mathbf{x}, t) \in \mathbb{Z}^d \times \{1, 2, 3, \ldots\}$, let $M(\mathbf{x}, t)$ be independent identically distributed Bernoulli random variables with $P(M(\mathbf{x}, t) = 0) = \tilde{\varepsilon}$, and let

$$\tilde{\mathcal{P}} = \{(\mathbf{x}, t) : M(\mathbf{x}, t) = 0\}.$$

We now define the Toom process with rule φ, death rate $\tilde{\varepsilon}$ and initial state $\tilde{A}_0$, where $\tilde{A}_0$ is is an arbitrary subset of $\mathbb{Z}^d$. For $t = 0, 1, 2, \ldots$, define inductively

$$\tilde{A}_{t+1} = \varphi(\tilde{A}_t) \cup \{\mathbf{x} : (\mathbf{x}, t+1) \in \tilde{\mathcal{P}}\}.$$

It is also natural to consider a somewhat more general class of processes as follows. For $(\mathbf{x}, t) \in \mathbb{Z}^d \times \{1, 2, 3, \dots\}$ let $\tilde{\Delta}(\mathbf{x}, t)$ be independent identically distributed positive integer-valued random variables, with probability density $\tilde{\mu}$. Let φ, $\tilde{\mathcal{P}}$, and $\tilde{A}_0$ be as above. The **generalized Toom process** with rule φ, death rate $\tilde{\varepsilon}$, initial state $\tilde{A}_0$ and size density $\tilde{\mu}$ is defined inductively by

$$\tilde{A}_{t+1} = \varphi(\tilde{A}_t) \cup \left(\bigcup_{(\mathbf{x}, t+1) \in \mathcal{P}} (B_{\tilde{\Delta}(\mathbf{x}, t+1)}(\mathbf{x}) \cap \mathbb{Z}^d) \right).$$

We may interpret a generalized Toom process as follows. The system follows the rule φ except near the random points $(\mathbf{x}, t) \in \tilde{\mathcal{P}}$. At such a point $(\mathbf{x}, t)$ in space-time, a 'catastrophe' occurs at time t which vacates all the sites within a certain random distance of $\mathbf{x}$.

We have used notation that is analogous to that used for generic processes. It is more standard to think of the state of one of these processes as a configuration of 0's and 1's located at the sites of the integer lattice $\mathbb{Z}^d$. In our notation, $\tilde{A}_t$ is the set of sites at which there is a 0 at time t, or in terms of population models, the vacant region.

Toom [13] gave an elegant criterion for survival at small death rates in families of Toom processes parametrized by the death rate. Fix a Toom rule φ with range r. For $t = 1, 2, 3, \dots$, let

$$\varphi^t = \varphi \circ \cdots \circ \varphi \quad (t \text{ times}).$$

Toom's eroder condition. *For all finite sets $A \subseteq \mathbb{Z}^d$,*

$$\varphi^t(A) = \emptyset$$

for some value of $t \in \{1, 2, 3, \dots\}$ which may depend on A.

Toom's Theorem is as follows: Let $\tilde{A}_t$ be the Toom process with rule φ, death rate $\tilde{\varepsilon}$, and initial state $\tilde{A}_0 = \emptyset$. If Toom's eroder condition holds, then

$$\lim_{\tilde{\varepsilon} \searrow 0} \limsup_{t \to \infty} P(0 \in \tilde{A}_t) = 0,$$

whereas if Toom's eroder condition does not hold, the limit is 1. We will use generic processes and our Theorem 1 to prove the hard half of Toom's Theorem, namely the first part (the proof of the easy half is essentially the same as the proof of Proposition 2). In fact, as we will see, the same proof works for generalized Toom processes, provided the size distribution $\tilde{\mu}$ satisfies (1-1). Toom has communicated to one of us (L. Gray) that he was not able to prove the more general result, since contour methods do not seem to work well with generalized Toom processes.

Toom processes have a simple property that makes it easy for us to compare them to generic processes. For any point $\mathbf{p}$ and unit vector $\mathbf{n}$ in $\mathbb{R}^d$, let

$$\tilde{H}(\mathbf{n},\mathbf{p}) = H(\mathbf{n},\mathbf{p}) \cap \mathbb{Z}^d.$$

We say that a unit vector $\mathbf{n} \in \mathbb{R}^d$ is **rationalizable** if there exists a vector $\mathbf{x} \in \mathbb{Z}^d$ such that $\mathbf{n} = \mathbf{x}/||\mathbf{x}||$.

Proposition 5. *Let* $\mathbf{n}$ *be a rationalizable unit vector, and let* φ *be a Toom rule with range* r. *Let*

$$\alpha = \sup\{a : \varphi(\tilde{H}(\mathbf{n},\mathbf{0})) \subseteq \tilde{H}(\mathbf{n},a\mathbf{n})\}.$$

Then $\alpha \in [-r,r]$ *and*

$$\varphi^t(\tilde{H}(\mathbf{n},\mathbf{0})) = \tilde{H}(\mathbf{n},\alpha t\mathbf{n})$$

for all $t = 0,1,2,\ldots$.

This simple fact, which is also used by Toom [13], follows easily from the four properties in the definition of a Toom rule. We will not give the proof here.

For any rationalizable unit vector $\mathbf{n}$ and rule φ, we let $\alpha(\mathbf{n},\varphi)$ be the quantity α given in Proposition 5. We now state an equivalent form of Toom's eroder condition:

Proposition 6. *Fix a* d*-dimensional Toom rule* φ *with range* r. *Then Toom's eroder condition is satisfied if and only if there exist* $d+1$ *rationalizable unit vectors* $\mathbf{n}_i \in \mathbb{R}^d$ *satisfying the conditions of Proposition 1, and a real number* $t \geq 0$, *such that*

$$\bigcap_{i=1}^{d+1} H(\mathbf{n}_i,(\alpha(\mathbf{n}_i,\varphi)t-1)\mathbf{n}) = \emptyset. \tag{3-1}$$

This alternate form of the eroder condition is in fact the one that Toom used to prove his theorem. Note that if we set $\alpha_i = \alpha(\mathbf{n}_i,\varphi)$, (3-1) is the same as our generic eroder condition. The proof of Proposition 6 is easy in one direction, since (3-1) obviously implies Toom's eroder condition. To prove the opposite direction, one first shows that Toom's condition implies that there is a finite collection of rationalizable vectors $\mathbf{n}_1,\ldots,\mathbf{n}_k$ such that (3-1) is satisfied with $d+1$ replaced by k. Then, assuming that $\alpha(\mathbf{n}_1,\varphi) = \max_i \alpha(\mathbf{n}_i,\varphi)$, one projects onto the $d-1$-dimensional hyperplane that forms the boundary of $H(\mathbf{n}_1,(\alpha(\mathbf{n}_1,\varphi)t-1)\mathbf{n}_1)$. An inductive argument completes the proof. The details are left to the reader.

We now define, for each Toom rule that satisfies (3-1), a corresponding set of parameters for a generic process that will be used as a comparison process. Suppose that φ is a rule with range r that satisfies condition (3-1) for a given set of unit vectors $\mathbf{n}_i$. Let $\tilde{\mu}$ be a size density that satisfies (1-1), and define another size density μ by

$$\mu(\Delta + k) = \tilde{\mu}(k),$$

where Δ is the smallest integer greater than or equal to $2r + \sqrt{d}/2$. Choose $\tilde{\varepsilon} \geq 0$ and then let ε be the solution to the equation $\tilde{\varepsilon} = 1 - \exp(-\varepsilon)$. The reason for these choices of μ and ε will become clear later. Let $\tilde{A}_t$ be the generalized Toom process with rule φ, death rate $\tilde{\varepsilon}$, size density $\tilde{\mu}$ and initial state $\tilde{A}_0 = \emptyset$. Also let A_t be the vacant region of the generic process with orientations $\mathbf{n}_i$, occupation rates $\alpha_i = \alpha(\mathbf{n}_i, \varphi)$, interaction rate $\beta = -r$, death rate ε, and size density μ. We say that A_t is a **canonical comparison process** for $\tilde{A}_t$.

Theorem 2. *Let $\tilde{A}_t$ be the generalized Toom process with rule φ, death rate $\tilde{\varepsilon}$, size density $\tilde{\mu}$ satisfying (1-1), and initial state $\emptyset$. Suppose that φ satisfies (3-1) for a given set of unit vectors $\mathbf{n}_i$, and let A_t be the corresponding canonical comparison process. Then the processes A_t and $\tilde{A}_t$ may be defined jointly ('coupled') so that*

$$\tilde{A}_t \subseteq A_t$$

for all $t \geq 0$.

Proof. Let the process A_t be constructed as in Section 1. For $\mathbf{x} = (x_1, \dots, x_d) \in \mathbb{Z}^d$ and $t = 1, 2, 3, \dots$, define

$$M(\mathbf{x}, t) = \begin{cases} 0 & \text{if } \mathcal{P} \cap Q(\frac{1}{2}, \mathbf{x}, t - \frac{1}{2}) \neq \emptyset \\ 1 & \text{otherwise,} \end{cases}$$

where $\mathcal{P}$ is the Poisson point process with intensity parameter ε used in the construction of A_t. Note that $Q(\frac{1}{2}, \mathbf{x}, t - \frac{1}{2})$ is a unit cube. The random variables $M(\mathbf{x}, t)$ are independent identically distributed Bernoulli, with

$$P(M(\mathbf{x}, t) = 0) = \tilde{\varepsilon} = 1 - \exp(-\varepsilon).$$

For $(\mathbf{x}, t)$ such that $M(\mathbf{x}, t) = 0$, let

$$\hat{\Delta}(\mathbf{x}, t) = -\Delta + \sup\{k : \mathcal{P}_k \cap Q(\frac{1}{2}, \mathbf{x}, t - \frac{1}{2}) \neq \emptyset\}.$$

Note that these random variables are independent and identically distributed. We use them to construct a generalized Toom process $\hat{A}_t$ with rule φ and initial state $\emptyset$. Unfortunately, the common distribution of the random variables $\hat{\Delta}(\mathbf{x}, t)$ is not the same as $\tilde{\mu}$. However, it dominates $\tilde{\mu}$ in the appropriate sense, so that it is straightforward to construct a generalized Toom process $\tilde{A}_t$ with rule φ, death rate $\tilde{\varepsilon}$, size density $\tilde{\mu}$ and initial state $\emptyset$ so that $\tilde{A}_t \subseteq \hat{A}_t$. We leave the details to the reader. Note that Δ has been chosen large enough so that

$$\text{(3-2)}\quad M(\mathbf{x},t)=0 \quad \text{and} \quad \hat{\Delta}(\mathbf{x},t)=k \quad \Rightarrow$$
$$\exists\,(\mathbf{y},u)\in\mathcal{P}_{k+\Delta}\cap Q(\tfrac{1}{2},\,\mathbf{x},\,t-\tfrac{1}{2}) \text{ such that } B_{k+r}(\mathbf{x})\subseteq D(\mathbf{y},u;t).$$

To see that this is the case, note that for $(\mathbf{y},u)$ as in (3-2), (i) $\|\mathbf{x}-\mathbf{y}\|\le\sqrt{d}/2$, and (ii) the size of $D(\mathbf{y},u;t)$ is at least $k+r+\sqrt{d}/2$ since $t-u\le 1$.

The comparison. We will prove by induction on $t=0,1,2,\ldots,$ that

$$\text{(3-3)}\qquad \mathbf{p}\in\hat{A}_t \Rightarrow B_r(\mathbf{p})\subseteq A_t,$$

from which it follows that $\hat{A}_t\subseteq A_t$, as desired. The case $t=0$ is obvious since both processes have initial state $\emptyset$. Now let t be a positive integer. By (3-2), if $\mathbf{p}$ is a site that is vacated at time t due to a 'catastrophe' at some site $\mathbf{x}$ such that $\hat{\Delta}(\mathbf{x},t)=k$ and $\mathbf{p}\in B_k(\mathbf{x})$, then $B_r(\mathbf{p})\subseteq A_t$. By the construction of Toom processes, all other points $\mathbf{p}\in\hat{A}_t$ are in $\varphi(\hat{A}_{t-1})$, so to complete the proof of (3-3), it is sufficient to prove

$$\text{(3-4)}\qquad \mathbf{p}\in\varphi(\hat{A}_{t-1})\Rightarrow B_r(\mathbf{p})\subseteq A_t.$$

Let $\mathbf{p}_1,\ldots,\mathbf{p}_n$ be those points in $\hat{A}_{t-1}$ that are within distance r of $\mathbf{p}$:

$$\hat{A}_{t-1}\cap B_r(\mathbf{p})=\{\mathbf{p}_1,\ldots,\mathbf{p}_n\}.$$

By the inductive hypothesis, there exist regions $D(\mathbf{y}_k,u_k;t-1)$ such that

$$\text{(3-5)}\qquad B_r(\mathbf{p}_k)\subseteq D(\mathbf{y}_k,u_k;t-1)$$

for $k=1,\ldots,n$. Since the distance between each $\mathbf{p}_j$ and $\mathbf{p}$ is less than or equal to r, we also have

$$\text{(3-6)}\qquad \mathbf{p}\in\bigcap_{k=1}^{n}D(\mathbf{y}_k,u_k;t-1).$$

In order to prove (3-4), we first need the following statement:

$$\text{(3-7)}\qquad B_r(\mathbf{p})\subseteq\bigcup_{k=1}^{n}H_i(\mathbf{y}_k,u_k;t)\quad\text{for all}\quad i=1,\ldots,d+1.$$

(Refer back to (1-3) for the definition of the half-spaces $H_i(\mathbf{y}_k,u_k;t)$.) To prove (3-7), we choose real numbers q_{ik} such that

$$H_i(\mathbf{y}_k,u_k;t-1)=H(\mathbf{n}_i,q_{ik}\mathbf{n}_i)$$

for $i = 1, \dots, d+1$ and $k = 1, \dots, n$. Let

$$q_i = \min\{q_{ik} : k = 1, \dots, n\}.$$

Since all of the functions γ_i in (1-3) have slopes which are bounded above by α_i,

$$H(\mathbf{n}_i, (q_i + \alpha_i)\mathbf{n}_i) \subseteq \bigcup_{k=1}^{n} H_i(\mathbf{y}_k, u_k; t) \quad \text{for } i = 1, \dots, d+1. \tag{3-8}$$

Thus it is enough to show that $B_r(\mathbf{p})$ is contained in the set on the left of (3-8) for each i. Fix i. By (3-5), $\{\mathbf{p}_1, \dots, \mathbf{p}_n\} \subseteq H(\mathbf{n}_i, (q_i + r)\mathbf{n}_i)$. Assuming that the hypothesis of (3-4) holds, it follows from the finite range and monotonicity properties that

$$\mathbf{p} \in \varphi(\{\mathbf{p}_1, \dots, \mathbf{p}_n\}) \subseteq \varphi(H(\mathbf{n}_i, (q_i + r)\mathbf{n}_i) \cap \mathbb{Z}^d).$$

By Proposition 5,

$$\mathbf{p} \in H(\mathbf{n}_i, (q_i + r + \alpha_i)\mathbf{n}_i),$$

from which it follows that

$$B_r(\mathbf{p}) \subseteq H(\mathbf{n}_i, (q_i + \alpha_i)\mathbf{n}_i),$$

completing the proof of (3-7).

To complete the proof of the theorem, we apply Proposition 3. Let $D(\mathbf{x}, s; u), u \geq s$, be a region satisfying (1-4) and either (1-5) or (1-6), with the regions $D(\mathbf{y}_k, u_k; u)$ being as above. The existence of such a region $D(\mathbf{x}, s; u)$ is guaranteed by Proposition 3. For all i such that (1-5) is satisfied, (3-6) and (1-4) imply that $B_r(\mathbf{p}) \subseteq H_i(\mathbf{x}, s; t)$ (since $\beta = -r$). For all i such that (1-6) is satisfied, (3-7) implies the same result. Proposition 3 says that every i must satisfy either (1-5) or (1-6), so

$$B_r(\mathbf{p}) \subseteq \bigcap_{i=1}^{d+1} H_i(\mathbf{x}, s; t) = D(\mathbf{x}, s; t) \subseteq A_t,$$

and the proof is complete. ■

Note that the size density μ of the canonical comparison process satisfies (1-1) if the size density $\tilde{\mu}$ of the corresponding generalized Toom process satisfies (1-1), and that the generic eroder condition is satisfied for the comparison process if Toom's eroder condition is satisfied for the rule of the generalized Toom process. Therefore, we have the following generalization of Toom's Theorem:

Corollary. *Let φ be a Toom rule which satisfies (3-1), and let $\tilde{\mu}$ be a discrete probability density on $\{1, 2, 3, \dots\}$ satisfying (1-1). Let $\tilde{A}_t$ be the* *corresponding generalized Toom process with death rate $\tilde{\varepsilon}$ and initial state $\tilde{A}_0 = \emptyset$. Then*

$$\lim_{\tilde{\varepsilon} \searrow 0} \limsup_{t \to \infty} P(\mathbf{0} \in \tilde{A}_t) = 0.$$

APPLICATIONS TO OTHER MODELS. The class of Toom processes does not include all of the finite range translation invariant attractive (monotone) models that are of interest. In a Toom process, the conditional probability that there is a change at a site $\mathbf{x}$ at time t, given the state of the process at time $t-1$, is either 1, $1-\tilde{\varepsilon}$, $\tilde{\varepsilon}$, or 0. When $\tilde{\varepsilon} = 0$, the process is deterministic, so we may view Toom processes as $\tilde{\varepsilon}$ perturbations of deterministic cellular automata. The Corollary to Theorem 2 says that if a Toom rule φ satisfies Toom's eroder condition, then the deterministic process is stable under perturbation by $\tilde{\varepsilon}$ when the initial state is 'all occupied'.

In general Markovian discrete time models A_t, the conditional probability that there is a change at time t, given $A_{t-1} = A$, equals some quantity $c(\mathbf{x}, A)$ which may not be close to either 0 or 1. This is more in line with what happens in continuous time, where the rate at which there is a change at a site $\mathbf{x}$ at time t, conditioned on $A_{t-} = A$, is given by a flip rate $c(\mathbf{x}, A)$.

One can ask about the stability of these more general models under various types of perturbations of the rates $c(\mathbf{x}, A)$. We suggest that generic processes are quite useful for determining stability properties. In a future paper, we intend to prove stability for various classes of systems that don't fit into the context of the present paper. We have already found certain eroder conditions that lead to some new stability results. These eroder conditions also apply to the sexual contact process, and relatively simple proofs of both of the main stability results in Durrett and Gray [5] can be given. We are however a long way from finding necessary and sufficient conditions for general systems. In general, one needs some probabilistic analogue of Proposition 5.

REFERENCES

[1] M. Bramson, *Survival of nearest particle systems with low birth rate*, Ann. Probab. **17** (1989), 433-443

[2] M. Bramson and R. Durrett, *A simple proof of the stability criterion of Gray and Griffeath*, Probab. Th. Rel. Fields **80** (1988), 293-298

[3] M. Bramson, R. Durrett, and G. Swindle, *The statistical mechanics of crabgrass*, Ann. Probab. **17** (1989), 444-481

[4] M. Bramson and L. Gray, *The survival of branching annihilating random walk*, Z. Wahrsch. Verw. Gebiete **68** (1985), 447-460

[5] R. Durrett and L. Gray, *Some peculiar properties of a particle system with sexual reproduction*, preprint. Preliminary version in P. Tautu, *Stochastic spatial processes*, Lecture Notes in Mathematics No. 1212, Springer-Verlag, New York, 1986

[6] R. Durrett and D. Griffeath, *Supercritical contact processes on* $\mathbb{Z}$, Ann. Probab. **11** (1983), 1-15

[7] L. Gray, *Duality for general attractive spin systems, with applications in one dimension*, Ann. Probab. **14** (1986), 371-396

[8] L. Gray and D. Griffeath, *A stability criterion for attractive nearest-neighbor spin systems on* $\mathbb{Z}$, Ann. Probab. **10** (1982), 67-85

[9] T. Harris, *Contact interactions on a lattice*, Ann. Probab. **2** (1974), 969-988

[10] R. Holley and T. Liggett, *The survival of contact processes*, Ann. Prob. **6** (1978), 198-206

[11] R. Peierls, *On Ising's model of ferromagnetism*, Proc. Cambridge Phil. Soc. **36** (1936), 477-481

[12] O. Stavskaya and I. Pyatetskii-Shapiro, *On homogeneous nets of spontaneously active elements*, Systems Theory Res. **20** (1971), 75-88

[13] A. Toom, *Nonergodic multidimensional systems of automata*, Problems Inform. Transmission **10** (1974), 239-246

M. Bramson
Dept. of Mathematics
University of Wisconsin
Madison, WI 53706

L. Gray
School of Mathematics
University of Minnesota
Minneapolis, MN 55455

Capture Problems For Coupled Random Walks

MAURY BRAMSON[1] AND DAVID GRIFFEATH [2]

1. Introduction. N predator random walks $X_k(t)$ stalk a prey random walk $Y(t)$. By conspiring among themselves and observing their prey, the predators try to minimize the *capture time* T:

$$T = \min\{t : X_k(t) = Y(t) \text{ for some } k\}.$$

The $X_k(t)$ start from respective positions x_k, and $Y(t)$ starts from y, where $x_k \neq y$ for all k. By a *random walk*, we will mean a continuous-time simple random walk on the d-dimensional integers that jumps after rate 1 exponential holding times. Thus we seek joint probability laws, or *couplings*, $\mathcal{P}$ for $(X_1(t), \ldots, X_N(t), Y(t))$ such that each marginal process is a random walk on $\mathbf{Z}^d$, and such that T is a finite random variable. Capture constitutes one generalization from coupling a pair of processes (cf. [7]) to coupling several. When dealing with two marginals it is customary to require that they remain together ("on the diagonal") after the coupling time T. Our analogous requirement is:

$$\mathcal{P}(\text{ For some } k,\ X_k(t) = Y(t) \text{ for all } t \geq T\ ) = 1\,. \tag{1.1}$$

Thus the sample paths of $\mathcal{P}$ can be thought of as movies in which all the predators and the prey, observed individually, wander around randomly, but such that some predator eventually captures the prey and remains with it thereafter.

In this context, we classify the coupling $\mathcal{P}$ as

(1.2)

- *unsuccessful* if $\mathcal{P}(T = \infty) > 0$,
- *successful* if $\mathcal{P}(T < \infty) = 1$,
- *efficient* if $\mathbf{E}^{\mathcal{P}}[T] < \infty$,
- *quick* if $\mathcal{P}(T > t) \leq \exp\{-Ct\}$ for sufficiently large t (some $C > 0$),
- *complete* if $\mathcal{P}(T \leq t) = 1$ for some $t < \infty$.

---

[1] Partially supported by the National Science Foundation.

[2] Partially supported by the National Science Foundation.

We view the dimension d, the number of predators N, and the initial locations x_k, y, as parameters. For given parameter values, we seek a $\mathcal{P}$ that is as far down the above classification as possible. Of course, more effective capture strategies are to be expected as we allow for greater dependency between the coordinate processes, so we divide our study into several parts.

As a warm-up, Section 2 treats a two-parameter family of capture problems in one dimension. This family provides simple examples of successful, efficient and quick couplings, and constitutes a rare instance in which the expected capture times are explicitly computable.

Section 3 analyzes in some detail couplings such that all the individual random walks are *independent*. Depending on the dimension, and the number of predators and their locations, we are able to determine the classification of each coupling, *except* in the one-dimensional case when N predators "chase" the prey. In that case we prove that 3 predators cannot capture in finite expected time, but appeal to simulation for compelling evidence that 4 predators can.

Next, Section 4 discusses *Markov capture*, where $\mathcal{P}$ is a Markov chain on $(\mathbf{Z}^d)^{N+1}$ with well-behaved marginals. Here our treatment is less thorough. The main objective is to illustrate by example the greater possibility for efficient and quick capture once the predators are allowed to observe their prey and conspire.

Finally, in Section 5 we turn to the general case, where the predators can exploit information such as the past or future trajectory of the prey. Rather surprisingly, in any dimension d, for suitably large N and suitable geometry of the initial locations, we are able to construct a capture $\mathcal{P}$ that is complete. This last construction has the important property that the number of predators remains bounded as the distance between the predators and prey increases. Consequently, complete capture can be used to gain information about certain systems of *interacting* random walks. In fact, such an application to two-particle annihilating random walks [2] was the original motivation for our study.

2. Simple Examples. We begin our discussion of capture problems by considering the case of two predators that surround the prey in one dimension. That is, we assume $d = 1$, $n = 2$, and $x_1 < y < x_2$. For simplicity, we let the prey $Y(t)$ move independently of the predator process $(X_1(t), X_2(t))$ until T, and suppose that the latter process, until T, has independent increments on $\mathbf{Z}^2$ chosen from the following two-parameter family $(0 \le \alpha, \beta \le 1,\ \alpha + \beta \le 1)$:

$(\Delta X_1, \Delta X_2)$	*at rate*
$(\rightarrow, \rightarrow)$	$\frac{1}{2}\alpha$
$(\leftarrow, \leftarrow)$	$\frac{1}{2}\alpha$
$(\rightarrow, \leftarrow)$	$\frac{1}{2}\beta$
$(\leftarrow, \rightarrow)$	$\frac{1}{2}\beta$
$(\rightarrow,\ \ )$	$\frac{1}{2}(1-\alpha-\beta)$
$(\leftarrow,\ \ )$	$\frac{1}{2}(1-\alpha-\beta)$
$(\ \ , \rightarrow)$	$\frac{1}{2}(1-\alpha-\beta)$
$(\ \ , \leftarrow)$	$\frac{1}{2}(1-\alpha-\beta)$

Thus, until capture occurs, the predators simultaneously displace in parallel at total rate α, simultaneously displace in opposite directions at total rate β, and otherwise make independent transitions. Once a predator meets the prey, the two walks are "pasted together" (i.e., make identical transitions) to ensure (1.1). Note that the three marginal processes of such a coupling $\mathcal{P}$ execute rate 1 random walks, as required. Our first result classifies the couplings in this two-parameter family according to the scheme (1.2), and explicitly evaluates the expected capture times.

Theorem 1. *Let $x_1 < y < x_2$, and let T be the capture time for the one-dimensional Markovian coupling with parameters α, β described above. Then $\mathcal{P}$ is always successful, efficient unless $\beta = 1$, and quick if $\alpha = 1$. Moreover, the expected capture time can be computed exactly:*

$$\text{(2.1)} \qquad \mathbf{E}[T] = \frac{(y - x_1)(x_2 - y)}{1 + \alpha - \beta} \qquad (= \infty \text{ if } \beta = 1).$$

Proof. The coupling $\mathcal{P}$ is always successful since the distance between the prey and (either) one of the predators undergoes a recurrent (rate 2) random walk until they meet. If $\alpha = 1$, then the distance between the predators remains fixed for all time, so the prey is effectively executing a rate 2 random walk on an interval of length $x_2 - x_1$ until capture occurs. In this case the tail of T is exponential by standard Markov chain theory, and (2.1) agrees with the well-known gambler's ruin solution. If $\beta = 1$, then $\mathbf{E}[T] = \infty$; we omit the proof of this fact since a similar but more interesting result will be derived in Section 3.

Our main task, then, is to prove (2.1) for $\beta < 1$. To begin, set $L(t) = Y(t) - X_1(t)$ and $R(t) = X_2(t) - Y(t)$. Observe that $(L(t), R(t))$ has independent increments on $\mathbf{Z}^2$ until T, and that

$$T = \min\{t : L(t) = 0 \text{ or } R(t) = 0\}.$$

Let G denote the generator (or so-called "Q-matrix") of $(L(t), R(t))$. Note that G has jump rates

$$q_{(l,r),(l',r')} = q_{\,l'-l,r'-r} \qquad (l > 0, r > 0)$$

of the form

$$q_{1,-1} = q_{-1,1} = \frac{1+\alpha}{2},$$

$$q_{1,1} = q_{-1,-1} = \frac{\beta}{2},$$

$$q_{1,0} = q_{-1,0} = q_{0,1} = q_{0,-1} = \frac{1-\alpha-\beta}{2}.$$

On the infinite domain $D = \{(l,r) : l > 0,\ r > 0\}$,

$$m(l,r) = \mathbf{E}[T \mid L(0) = l, R(0) = r]$$

is a solution of the Poisson equation (cf. [6])

$$Gm = -1 \tag{2.2}$$

with boundary conditions

$$m \equiv 0 \text{ on } \partial D = \{(l,r) : l = 0 \text{ or } r = 0\}. \tag{2.3}$$

Setting $\overline{m}(l,r) = (1+\alpha-\beta)^{-1} \cdot l\,r$, some algebra yields

$$\begin{aligned} G\overline{m} = (1+\alpha-\beta)^{-1} \cdot \Bigg(& \frac{1+\alpha}{2} \cdot ((l+1)(r-1) + (l-1)(r+1) - 2lr) \\ & + \frac{\beta}{2} \cdot ((l+1)(r+1) + (l-1)(r-1) - 2lr) \\ & + \frac{1-\alpha-\beta}{2} \cdot ((l+1)r + (l-1)r + l(r+1) + l(r-1) - 4lr) \Bigg) \\ = -1, & \end{aligned}$$

in agreement with formula (2.1). But D is unbounded, so an approximation argument is needed to ensure that we have exhibited the desired (i.e., minimal) solution of (2.2)-(2.3).

To this end, for $n \geq 1$, introduce the truncated domains

$$D_n = \{(l,r) : l > 0, r > 0, \text{ and } l + r < n\}$$

with boundary

$$\partial D_n = \{\, (l,r) \, : l = 0 \text{ or } r = 0 \text{ or } l + r = n \,\},$$

and corresponding exit times

$$T_n = \min \{ t \, : (\, L(t) \, , \, R(t) \,) \in \partial D_n \, \} .$$

Approximate $m(l,r)$ by

$$m_n(l,r) = \mathbf{E}[T_n \mid L(0) = l, R(0) = r] .$$

By monotone convergence, $m_n \uparrow m$. Moreover, since m_n satisfies (2.2) on D_n with 0 boundary conditions on ∂D_n, $\overline{m} - m_n$ is G-harmonic on D_n with non-negative boundary conditions on ∂D_n. Consequently,

$$\overline{m}(l,r) - m(l,r) = \lim_{n \to \infty} \Big(\overline{m}(l,r) - m_n(l,r) \Big) \geq 0 .$$

For the opposite inequality we need a suitable comparison function. A judicious choice is

$$\phi(l,r) = l \, r \, (l + r)^{\epsilon} .$$

We claim that for each G corresponding to a coupling $\mathcal{P}$ with $\beta < 1$, for all sufficiently small $\epsilon > 0$,

$$G\phi \leq 0 . \tag{2.4}$$

To check this, we compute:

$$\begin{aligned}
G\phi = {} & \\
& \frac{1+\alpha}{2} \cdot [(l+1)(r-1) + (l-1)(r+1) - 2lr] \, (l+r)^{\epsilon} \\
& + \frac{\beta}{2} \cdot [(l+1)(r+1)(l+r+2)^{\epsilon} + (l-1)(r-1)(l+r-2)^{\epsilon} - 2lr(l+r)^{\epsilon}] \\
& + \frac{1-\alpha-\beta}{2} \cdot \Bigg[\Big((l+1)r + l(r+1) \Big) (l+r+1)^{\epsilon} + \\
& \qquad \Big((l-1)r + l(r-1) \Big) (l+r-1)^{\epsilon} - 4lr(l+r)^{\epsilon} \Big) \Bigg] .
\end{aligned}$$

Factor out $(l+r)^\epsilon$ on the right, and linearize at $\epsilon = 0$ to get

$$G\phi =$$

$$(l+r)^\epsilon \cdot \Bigg(\frac{1+\alpha}{2} \cdot [(l+1)(r-1) + (l-1)(r+1) - 2lr]$$

$$+ \frac{\beta}{2} \cdot \left[(l+1)(r+1)\left(1 + \frac{2\epsilon}{l+r}\right) + (l-1)(r-1)\left(1 - \frac{2\epsilon}{l+r}\right) - 2lr\right]$$

$$+ \frac{1-\alpha-\beta}{2} \cdot \Bigg[\Big((l+1)r + l(r+1)\Big)\left(1 + \frac{\epsilon}{l+r}\right) +$$

$$\Big((l-1)r + l(r-1)\Big)\left(1 - \frac{\epsilon}{l+r}\right) - 4lr\Big)\Bigg] + o(\epsilon)\Bigg),$$

where the error term tends to 0 uniformly in l and r as $\epsilon \to \infty$. Simplifying, we find that

$$G\phi(l,r) = [-1 + \epsilon + (1+\epsilon)(\beta - \alpha) + o(\epsilon)]\,(l+r)^\epsilon .$$

Thus $G\phi \le 0$ provided that $\beta < 1$ and $\epsilon > 0$ is sufficiently small. So (2.4) holds. Next, let

$$\overline{m}_n(l,r) = \left(\frac{lr}{(1+\alpha-\beta)}\right) \left(\frac{n^\epsilon - (l+r)^\epsilon}{n^\epsilon}\right)$$

$$= \overline{m} - \frac{\phi}{(1+\alpha-\beta)\, n^\epsilon},$$

noting that, $\overline{m}_n \uparrow \overline{m}$ as $n \to \infty$. Observe also that $\overline{m}_n - m \le 0$ on ∂D_n, and that $G(\overline{m}_n - m) \ge 0$ on D_n by (2.4). It follows that for each n, $\overline{m}_n \le m$ on D_n. Finally, letting $n \to \infty$, we get

$$\overline{m}(l,r) - m(l,r) = \lim_{n\to\infty} \Big(\overline{m}_n(l,r) - m(l,r)\Big) \le 0,$$

completing the derivation of (2.1). ∎

A Little History. The simple capture examples of Theorem 1, and similar one-dimensional independent captures to be discussed in the next section, are intimately connected with exit problems for Brownian motion from certain infinite domains in several dimensions. We would like to explain some of the history of this connection.

As graduate students at Cornell, we learned from Frank Spitzer that the exit time τ of two-dimensional Brownian motion from the angular region $A = \{ x_2 \le \gamma |x_1| \}$, for $0 < \gamma < 1$, has expectation

$$\mathbf{E}_x[\tau] = \frac{\gamma^2 x_1^2 - x_2^2}{1 - \gamma^2} \tag{2.5}$$

starting from $x = (x_1, x_2)$. In particular, the smallest two-dimensional wedge with infinite expectation is a quadrant [10]. This beautiful result is very much in the spirit of Dynkin and Yushkevich [6], where one finds formulas for exit distributions and expected exit times from several simple regions. In particular, they derive the probability that Brownian motion leaves angle A through one of its two rays, using only symmetry and continuity, and derive the Cauchy exit density from the half-plane as a consequence. But (2.5), not mentioned there, remains unfamiliar to many probabilists.

Formula (2.5) extends to *cones* in higher dimensions. The exit time τ of d-dimensional Brownian motion from the Euclidean cone $C = \{||x||^2 \le (1 + \gamma^2)|x_1|^2 \}$, for $0 < \gamma < \sqrt{d-1}$, has expectation

$$\mathbf{E}_x[\tau] = \frac{(1 + \gamma^2)|x_1|^2 - ||x||^2}{(d-1) - \gamma^2}. \tag{2.6}$$

Frank Spitzer knew this generalization as well. A thorough and elegant treatment in the literature is by Burkholder [4]; his paper strongly suggests that in two dimensions, as $t \to \infty$,

$$P_x(\tau > t) \sim C(x)\, t^{-\theta(\gamma)}, \tag{2.7}$$

where

$$\theta = \frac{\pi}{4} \cdot \frac{1}{\arctan \gamma}, \tag{2.8}$$

and that τ should have a power law tail in higher dimensions as well. The proof of (2.7)-(2.8) was completed by De Blassie [5], who also established power law tails for the exit distributions of Brownian motion for a class of "generalized cones" in 3 or more dimensions. In his general framework, the constant θ is determined by the first eigenvalue of the Laplace-Beltrami operator on a subset of the unit ball determined by the generalized cone. This eigenvalue approach seems of limited utility for explicit evaluation of θ.

The connection between our capture examples and such results for exit times from cones can be summarized as follows. First consider the Brownian coupling $\tilde{\mathcal{P}}$ on $\mathbf{R}$, analogous to $\mathcal{P}$ on $\mathbf{Z}$, in which the predators and prey execute Brownian motions. Then the corresponding process $(\tilde{L}(t), \tilde{R}(t))$ of differences is a certain diffusion on the positive quadrant of $\mathbf{R}^2$. An appropriate linear transformation maps this diffusion to Brownian motion on a corresponding wedge A, with $0 < \gamma \leq 1$ since $\tilde{L}(t)$ and $\tilde{R}(t)$ are *negatively* correlated. Thus the Brownian captures can be classified by appeal to (2.5) and (2.7)-(2.8). More ingenuity is needed to make the corresponding comparisons for random walks directly, without appeal to Brownian motion. But results of McConnell [8] can be used to show that the capture time T in Theorem 1 is quick *only* if $\alpha = 1$. Moreover, with some work we believe that the methods of [9] and results of [5] could be used to obtain power law tails of the form

$$\mathcal{P}_{(l,r)}(T > t) \sim C(l,r)\, t^{-\theta(\alpha,\beta)}, \tag{2.9}$$

as $t \to \infty$, where

$$\theta = \frac{\pi}{4} \cdot \frac{1}{\arccos \frac{1}{2}\sqrt{3+\alpha-\beta}}, \tag{2.10}$$

whenever $\alpha \neq 1$.

About ten years ago we studied the behavior of coalescing independent random walks in one dimension [3]. A statistic of key importance in that work was the first hitting time T among three particles started from different locations. In essence, we wanted to show that the intermediate particle ("the prey") was captured by one of the two particles surrounding it ("the predators") at least as quickly (in distribution) as if the distances from the predators to the prey evolved independently. More precisely, we needed an estimate on the tail of T of the form

$$P(T > t) \leq \frac{Clr}{t} \qquad (\text{some } C > 0). \tag{2.11}$$

Although (2.11) is not hard to verify using correlation inequalities and the reflection principle (cf. [1]), another approach is to establish efficient capture by exploiting the connection with the Brownian motion problem of the previous paragraphs. This second approach is carried out for independent motions when $\alpha = \beta = 0$ in Theorem 1. According to (2.10), the correct power law for the tail should be $-\frac{3}{2}$ in this case. J. Gravner (private communication) has pointed out a simpler derivation of (2.1) in this independent situation. With $\alpha = \beta = 0$, and $\epsilon = 1$ in our proof, the linearization is exact and ϕ is harmonic. It follows that the Poisson equation on D_n is explicitly solvable:

$$\mathbf{E}[T_n] = \frac{lr\,[n-(l+r)]}{n}.$$

3. Independent Capture. In this section we let P denote the law of the product process $(X_1(t), \ldots, X_N(t), Y(t))$ consisting of independent random walks. Recall that the predators $X_k(t)$ start from respective positions

x_k, and the prey $Y(t)$ starts from y, where $x_k \neq y$ for all k. Let $\mathcal{P}$ be the corresponding (Markov) coupling in which each predator makes the same transitions as the prey once they meet, so that (1.1) holds. Clearly the laws of P and $\mathcal{P}$ agree up to the capture time T. We present here a fairly complete picture of the capture classification (1.2) for such "independent" couplings. There is one delicate and intriguing case in which, for lack of rigorous argument, we base our conclusions on computer simulation.

Theorem 2.

(*a*) $d \geq 3$. *For any* $N \geq 1$, *and all initial locations,* $\mathcal{P}$ *is unsuccessful.*

(*b*) $d = 2$. *For any* $N \geq 1$, *and all initial locations,* $\mathcal{P}$ *is successful but not efficient.*

(*c*) $d = 1$. *If* $N \geq 2$, *and* $x_j < y < x_k$ *for some* j, k, *then* $\mathcal{P}$ *is efficient.*

Proof. (*a*) Let P_x denote the law of a random walk $X(t)$ started at x. By transience in three or more dimensions, we can choose a distant x such that

$$P_x(X(t) = y \text{ for some } t \geq 0) < \frac{1}{N}. \tag{3.1}$$

Let G denote the event

$$G = \left(\bigcap_k \{X_k(1) = x \text{ and } X_k(t) \neq y \ \forall t \leq 1\}\right) \cap \{Y(t) = y \ \forall t \leq 1\}.$$

Clearly $P(G) > 0$, and by the Markov property,

$$\begin{aligned} P(T = \infty) &\geq P(G \cap \{T = \infty\}) \\ &\geq P(G) \cdot P(T = \infty \mid \bigcap_k \{X_k(0) = x\} \cap \{Y(0) = y\}). \end{aligned} \tag{3.2}$$

Now by symmetry,

$$\begin{aligned} &P(T < \infty \mid \bigcap_k \{X_k(0) = x\} \cap \{Y(0) = y\}) \\ &\qquad \leq N P(X_1(t) = Y(t) \text{ for some } t \mid X_1(0) = x, Y(0) = y). \end{aligned}$$

This last probability equals the left side of (3.1), since $X_1(t) - Y(t)$ is a rate 2 random walk. So N times this is less than 1. It follows that the left side of (3.2) is strictly positive. In other words $\mathcal{P}$ is unsuccessful.

(*b*) Successful capture follows immediately from the recurrence of random walk in two dimensions. To see that the capture cannot be efficient, define G as in (*a*), but with x any site different than y. Again, $P(G) > 0$, and by the Markov property, for $t > 1$,

$$P(T > t) \geq P(G) \cdot P(T > t - 1 \mid \bigcap_k \{X_k(0) = x\} \cap \{Y(0) = y\}).$$

Let Q be the law of the prey Y, and let $\omega(\cdot)$ denote a generic path up to time $t-1$. Then by independence and symmetry, the last probability equals

$$\int [P(X_1(s) \neq \omega(s) \text{ for all } s \leq t-1 \mid X_1(0) = x)]^N \, dQ(\omega \mid \omega(0) = y).$$

Applying Jensen's inequality, the integral is

$$\geq \left[\int P(X_1(s) \neq \omega(s) \text{ for all } s \leq t-1 \mid X_1(0) = x) dQ(\omega \mid \omega(0) = y) \right]^N$$

$$= [P(X_1(s) \neq Y(s) \text{ for all } s \leq t-1 \mid X_1(0) = x, Y(0) = y)]^N .$$

Recall that $X_1(s) - Y(s)$ is a rate 2 random walk. In two dimensions, random walk asymptotics for the hitting time τ of the origin are (cf. [11]):

$$P_z(\tau > t) \sim C_z \frac{1}{\log t} \quad \text{as } t \to \infty,$$

where C_z is some finite positive constant . We conclude that for $t > 1$,

$$P(T > t) \geq \left[\frac{c_{y-x}}{\log(t-1)} \right]^N$$

for some constant $c_{y-x} > 0$. Integrate over t to get $\mathbf{E}[T] = \infty$, as claimed.

(c) Clearly the capture time with N predators is stochastically smaller than the capture time with a pair of predators starting at x_j and x_k, $x_j < y < x_k$. But this latter capture is efficient according to the $\alpha = \beta = 0$ case of Theorem 1. ∎

The situations *not* covered by Theorem 2 are one-dimensional, with all of the predators initially situated to one side or the other of the prey: $d = 1$, $N \geq 1$, and (without loss of generality) $x_k < y$ for all k. $\mathcal{P}$ is successful whenever $N \geq 1$ for the same reason as in two dimensions. But the question of efficient capture is evidently more delicate when the predators, instead of surrounding the prey, must "chase" it. Suppose first that $N = 2$ and, as in the proof of Theorem 1, write $L(t) = Y(t) - X_1(t)$ and $R(t) = X_2(t) - Y(t)$. Now $L(t)$ and $R(t)$ are *positively* correlated. (If $x_1 = x_2$ this follows from Jensen's inequality, as in the proof of Theorem 2b.) So, in contrast to inequality (2.11), standard random walk asymptotics based on the reflection principle imply

$$P(T > t) = P(L(t) > 0 \textit{ and } R(t) > 0) \geq \frac{clr}{t} \quad (\text{some } c > 0),$$

where $l = y - x_1$ and $r = y - x_2$. Integrating this inequality we conclude that $\mathcal{P}$ is not efficient. Next, consider the case of 3 chasing predators. A naive heuristic, discounting the role of the positive correlations, argues that T might have a tail of order about $t^{-3/2}$, suggesting efficient capture. As our next result indicates, this heuristic is *wrong*.

Theorem 3. *If* $N = 3$, *and* $x_k < y$ *for all* k *(or* $x_k > y$ *for all* k*), then the independent capture* $\mathcal{P}$ *is not efficient.*

Proof. Our framework generalizes the $\alpha = \beta = 0$ case of Theorem 1 to three predators. Assume $x_k < y$ for $1 \le k \le 3$. Denote $\overline{m}(t) = Y(t) - X_1(t)$, $\overline{m}(t) = Y(t) - X_2(t)$ and $\overline{Z}(t) = Y(t) - X_3(t)$. Then $(\overline{X}(t), \overline{Y}(t), \overline{Z}(t))$ has independent increments on $\mathbf{Z}^3$ until T, and

$$T = \min\{t : \overline{X}(t) = 0 \ \mathit{or}\ \overline{Y}(t) = 0 \ \mathit{or}\ \overline{Z}(t) = 0\}.$$

The generator G of $(\overline{X}(t), \overline{Y}(t), \overline{Z}(t))$ has jump rates

$$q_{(x,y,z),(x',y',z')} = q_{x'-x,y'-y,z'-z} \qquad (x > 0,\ y > 0,\ z > 0)$$

of the form

$$q_{1,1,1} = q_{-1,-1,-1} = q_{1,0,0} = q_{-1,0,0} = q_{0,1,0}$$
$$= q_{0,-1,0} = q_{0,0,1} = q_{0,0,-1} = \tfrac{1}{2}\,.$$

Suppose that

$$m(x, y, z) = \mathbf{E}[T \mid \overline{X}(0) = x, \overline{Y}(0) = y, \overline{Z}(0) = z]$$

is finite at some point of the domain $D = \{\,(x, y, z)\ : x > 0, y > 0, z > 0\,\}$. Then by standard Markov chain theory, m is finite everywhere on D, and satisfies the Poisson equation (2.2) with boundary conditions

$$m \equiv 0 \ \text{ on } \ \partial D = \{(x, y, z) : x = 0 \ \ \mathit{or} \ \ y = 0 \ \ \mathit{or} \ \ z = 0\}\,,$$

For large $n < \infty$ and small $\epsilon > 0$, consider the truncated domains

$$D_n = \{(x, y, z) : x > 0, y > 0, z > 0 \ \ \mathit{and} \ \ x + y + z < n\,\}\,,$$

and, for $\delta, \epsilon > 0$, the corresponding symmetric comparison functions $\psi_{n,\delta,\epsilon}$:

$$\psi_{n,\delta,\epsilon}(x, y, z) = \frac{1}{12\delta}\Big[\big(\frac{1}{2} - \delta\big)(x + y + z)^2 - x^2 - y^2 - z^2\Big]$$
$$\cdot \left[1 - \left(\frac{x + y + z}{n}\right)^{\epsilon}\right].$$

It is easy to check that $\psi_{n,\delta,\epsilon} \le 0$ on ∂D_n. To evaluate $G\,\psi_{n,\delta,\epsilon}$, first compute the entries in the following table, linearizing at $\epsilon = 0$, as in the proof of Theorem 1 :

$\underline{u}$	$\underline{Gu}$
$(x+y+z)^2$	12
x^2	2
$(x+y+z)^{2+\epsilon}$	$[12+24\epsilon+o(\epsilon)]\,(x+y+z)^\epsilon$
$(x^2+y^2+z^2)(x+y+z)^\epsilon$	$[6+8\epsilon+o(\epsilon)]\,(x+y+z)^\epsilon$

For instance, if $u(x,y,z)=(x+y+z)^2$, then Gu is given by

$$\tfrac{1}{2}\Big[(x+y+z+3)^2+(x+y+z-3)^2+3(x+y+z+1)^2 \\ +3(x+y+z-1)^2-8(x+y+z)^2\Big]=12.$$

The rest of the computations are left to the reader. By linearity, after some algebra, we get

$$G\psi_{n,\delta,\epsilon}(x,y,z) \\ =\frac{1}{12\delta}\left[-12\delta+\frac{1}{n^\epsilon}(x+y+z)^\epsilon\cdot\Big(12\delta-\epsilon(4-24\delta)+o(\epsilon)\Big)\right] \quad \text{on } D_n\,.$$

Given $\delta>0$, we can choose $\epsilon=\epsilon(\delta)>0$ small enough that the term in large parentheses is ≥ 0, and hence $G\psi_{n,\delta,\epsilon}\geq -1$ on D_n. For this choice, $G(m-\psi_{n,\delta,\epsilon})\leq 0$ on D_n, and $m-\psi_{n,\delta,\epsilon}$ has non-negative boundary values. Hence $m\geq\psi_{n,\delta,\epsilon}$ *on* D_n. But

$$\lim_{\delta\to 0}\lim_{n\to\infty}\psi_{n,\delta,\epsilon(\delta)}(1,1,1)=\infty,$$

forcing $m(1,1,1)=\infty$, a contradiction. We conclude that $m\equiv\infty$, i.e., the capture $\mathcal{P}$ is *not* efficient. ∎

Theorem 3 begs the question:

In one dimension, how many chasing predators are needed for efficient capture?

Unfortunately, we have not been able to answer this question rigorously. In fact, we are unable to show that N predators suffice for *any* finite N. Nevertheless, based on extensive computer simulation, we conjecture that independent capture is efficient whenever $N\geq 4$. Let us first describe the computer experiments that prompt this conjecture, and then mention some of the obstacles to rigorous analysis.

Our simulations, using a program furnished by J.T. Cox, were carried out as follows. The N predators start at 0; the prey starts at 1. All execute simulated, synchronous, independent discrete-time random walks on $\mathbf{Z}$ that jump -1, 0 or $+1$ with probability $\frac{1}{3}$ each. This process runs until the first time $\hat{T}$ that some predator occupies the same site as the prey or jumps over the prey. The value of $\hat{T}$ is recorded for each such *trial* in which capture occurs by time 4000; otherwise the trial is terminated and $\{\hat{T} > 4000\}$ is true. An *experiment* consists of 100 million such trials. At the conclusion of the experiment, the empirical tail $Q(t)$ is tabulated:

$$Q(t) = \frac{\#\ \text{trials} : \hat{T} > t}{\text{total}\ \#\ \text{trials}}.$$

We assume that the true capture-time tail for the discrete-time model shares a common power-law tail with the original continuous-time model : as $t \to \infty$,

$$\mathcal{P}_{(x_1,\ldots,x_N,y)}(T > t) \sim C(x_1,\ldots,x_N,y)\, t^{-\theta_N},$$

and that Q approximates these asymptotics well as the number of trials gets large. Thus the graph of $\{(\log t, \lambda(t)); 1 \le t \le 4{,}000\}$ where

$$\lambda(t) = \frac{-\log Q(t)}{\log t},$$

should consist of points close to a line with slope θ_N. To estimate the exponent of the tail of T, we compute the slope of the least-squares line corresponding to λ.

In the case of 4 predators, one such experiment produced the graph and the estimate below:

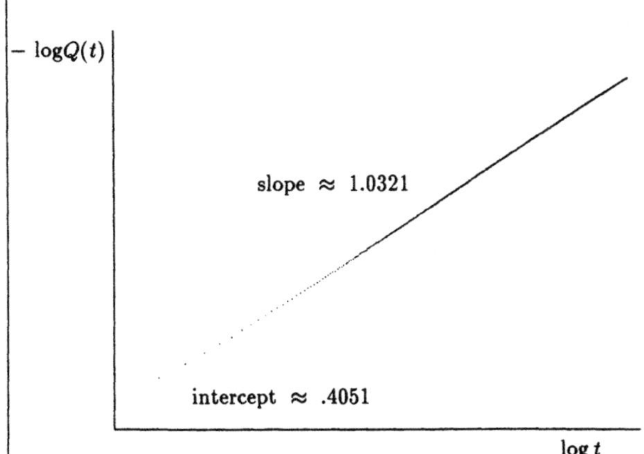

Four Predators

100 million trials of 4000 time units each

Evidently the data of this particular experiment suggest an exponent of about -1.03. To bolster our faith in the numbers, we performed 20 such experiments (100 million trials each). In summary, our results were as follows.

Four Predators

Aggregate data from 20 experiments of 100 million trials each

Mean slope: 1.032.
Range of slopes: [1.0261, 1.041]

Mean intercept: .40.
Range of intercepts: [.3445, .4365]

We feel that the data provide adequate evidence for the estimate:

$$\theta_4 \approx 1.032 \pm .01 .$$

In particular, an exponent in this range would imply (barely) efficient capture with four particles. Our conjecture is further supported by two considerations. First, the most substantial deviation from the least squares line occurs over the first few data points, which consistently lie above the line. As a consequence, our data should underestimate θ_4 if anything. Second, we are able to evaluate θ_1 and θ_2 exactly. $\theta_1 = \frac{1}{2}$, of course, by the reflection principle. The connection with Brownian motion and formula (2.7)-(2.8), which indicates the previously mentioned exponent $-\frac{3}{2}$ for independent 2-predator capture when the prey is surrounded, suggests that $\theta_2 = \frac{3}{4}$. Our simulation program reproduces these theoretical values to two decimal place accuracy when corresponding experiments are performed.

We have not analyzed θ_N for other values of N as thoroughly, but preliminary data suggest, for instance, $\theta_3 \approx .91$, $\theta_{10} \approx 1.4$. As $N \to \infty$, the exponents presumably tend *slowly* toward $-\infty$. However, we have been unable to prove efficient capture for *any* N. A partial result due to J. Gravner (private communication) comes tantalizingly close. Namely, suppose we slow down the prey so that it performs a rate μ random walk, $\mu < 1$. Then N predators capture efficiently provided that

$$\mu < \frac{N-2}{N-1} \qquad (N > 2).$$

We suspect that the Poisson equation approach offers the best prospects for establishing efficient capture when $N = 4$, or at least for some $N < \infty$. The three-dimensional generator G from the proof of Theorem 3 generalizes to N dimensions in a straightforward manner; one needs a good approximation to the solution of (2.2) for this G on $D = \{x : x > 0\} \subset \mathbf{Z}^N$.

As suggested in Section 2, the domains for which one knows simple solutions and comparison functions are *cones*, whereas D is a "wedge." In two dimensions these two types of domain coincide, but as the dimension N increases, the disparity becomes more and more pronounced. The idea behind our proof of Theorem 3 illustrates this point. By the connection with Brownian motion described at the end of Section 2, when $N = 3$ the diffusion analog $\tilde{G}$ of G maps to $\frac{1}{2}\Delta$ on a certain 3-dimensional wedge. As it turns out, the "critical cone" C (with $\gamma = \sqrt{2}$) exactly inscribes this wedge. Since the expected exit time from C is infinite by formula (2.6), the same holds for the wedge. Our proof of Theorem 3 is a translation of this comparison to discrete time and space; the computations work with so little room to spare because the comparison barely succeeds. (A simpler cone comparison in two dimensions shows that $\mathbf{E}[T] = \infty$, $\beta = 1$ in Theorem 1.) Unfortunately, when $N > 3$ the comparison between cones and wedges is so bad as to be worthless for our purposes.

4. Markov Capture. We now turn our attention to capture schemes where $\mathcal{P}$ is the law of a continuous time Markov chain on $(\mathbf{Z}^d)^{N+1}$. Thus the transitions of $(X_1(t), \ldots, X_N(t); Y(t))$ are determined by rates

$$q_{(x_1,\ldots,x_N;y)\,(x'_1,\ldots,x'_N;y')}\,.$$

In general, the predators and prey need not be random walks – they might just as well be Markov chains with some common rate function $p(z, z')$, $z, z' \in \mathbf{Z}^d$. We will assume that q satisfies the following conditions for all $x_1, \ldots, x_N, y$:

$$\sum_{x'_1,\ldots,x'_N} q_{(x_1,\ldots,x_N;y)\,(x'_1,\ldots,x'_N;y')} = p(y, y') \tag{4.1}$$

and for each $i : 1 \leq i \leq N$,

$$\sum_{x'_j : j \neq i,\, y'} q_{(x_1,\ldots,x_N;y)(x'_1,\ldots,x'_N;y')} = p(x_i, x'_i)\,. \tag{4.2}$$

Let us call a Markov chain $\mathcal{P}$ on $(\mathbf{Z}^d)^{N+1}$ that satisfies (1.1) and (4.1)-(4.2) a *Markov capture*. In words, conditions (4.1)-(4.2) mean that each marginal transition function does not depend on the current locations of the other marginals. Subject to these constraints, however, the predators can base their next transitions on their present positions, the present position of the prey, *and* the prey's next transition. For instance, all the predators are allowed to move "in parallel" with the prey, using identical displacements, if they desire. In this section we will construct various Markov captures that, for given d, N, x_k and y, are lower in the classification (1.2) than their independent counterparts.

A natural special class of Markov captures arises if the prey is *independent* of the predators until capture, as in the examples of Section 2. Interesting classification problems arise in that situation, where the predators are not allowed to "peek" at the prey's one-step transitions, but we will not discuss them here.

The first result of this section deals with the case of one predator. Henceforth we will assume that the predators and prey are random walks.

Proposition 1. *If $N = 1$, there is always a successful Markov capture, but no Markov capture is efficient.*

Proof. To obtain a successful capture, begin by letting the first coordinates of the predator and prey evolve independently while translating the remaining coordinates in parallel. Recurrence of one-dimensional random walk implies that, with probability one, the first coordinates agree at some finite time T_1. Next, let the second coordinates evolve independently while translating all others in parallel. At time $T_2 < \infty$ the first two coordinates agree. Proceeding coordinate-wise, the predator captures the prey at time $T_d < \infty$ a.s. A little thought reveals that this algorithm is described by a Markov chain with jump rates that depend on which of the corresponding coordinates agree.

To see that one-predator Markov capture cannot be efficient, consider the first coordinate processes $X^1(t)$, $Y^1(t)$ of the predator and prey respectively. We may for simplicity assume that $X^1(0) = 0$ and $Y^1(0) = 1$. (If $X^1(0) = Y^1(0)$, then consider one of the other coordinates.) Set

$$M(t) = Y^1(t) - X^1(t),$$

and

$$T^1 = \min\{t : X^1(t) = Y^1(t)\}.$$

Note that M is a martingale and T^1 is a stopping time with respect to the σ-algebra generated by $(X^1(t), Y^1(t))$. Hence, for any $t \geq 0$,

$$\mathbf{E}[M(T^1 \wedge t)] = 1.$$

Also,

$$\begin{aligned}\mathbf{E}[M(T^1 \wedge t)] &= \mathbf{E}[Y^1(T^1 \wedge t)] - \mathbf{E}[X^1(T^1 \wedge t)] \\ &\leq \mathbf{E}[|Y^1(T^1 \wedge t)|, T^1 > t] + \mathbf{E}[|X^1(T^1 \wedge t)|, T^1 > t] \\ &\leq 2\mathbf{E}[|Y^1(t)|, T^1 > t].\end{aligned}$$

Consequently,

$$\mathbf{E}[|Y^1(t)|, T^1 > t] \geq \tfrac{1}{2}. \tag{4.3}$$

Standard large deviation techniques involving moment generating functions and Chebyshev's inequality show that

$$P(|Y^1(t)| \geq z\sqrt{t}) \leq e^{-cz}$$

for $t \geq 1$, $z \geq 1$, and an appropriate $c > 0$. It follows that for all sufficiently large t,

$$\mathbf{E}[|Y^1(t)|, T^1 > t] \leq \sqrt{t} \int_{-\frac{1}{c}\log P(T^1>t)}^{\infty} ze^{-cz}\,dz$$

$$= \frac{1}{c}\sqrt{t}P(T^1 > t)\left(1 - \log P(T^1 > t)\right).$$

For t satisfying $P(T^1 > t) \leq \frac{1}{4}$, this is

$$\leq -\frac{2}{c}\sqrt{t}P(T^1 > t)\log P(T^1 > t).$$

Combining this last estimate with (4.3), we see that for large t,

$$\frac{c}{4\sqrt{t}} \leq -P(T^1 > t)\log P(T^1 > t).$$

Consequently, for a suitable $C < \infty$ and t sufficiently large,

$$P(T^1 > t) \geq \frac{C}{\sqrt{t}\log t}.$$

So $\mathbf{E}[T^1] = \infty$, i.e., the capture is not efficient. ■

With sufficiently many predators an efficient Markov capture is *always* possible, according to the next result.

Theorem 4. *In d dimensions, there is an efficient Markov capture for all initial locations provided that N exceeds $C(d)$.*

Proof. First, in one dimension, we exhibit an efficient Markov capture with three predators. Namely, consider the coupling $\mathcal{P}$ which, until T, has independent increments specified by the following table:

$$
\begin{array}{cc}
\underline{(\Delta X_1, \Delta X_2, \Delta X_3\,; \Delta Y)} & \underline{\textit{at rate}} \\
(\quad ,\rightarrow,\rightarrow;\rightarrow) & \frac{1}{6} \\
(\quad ,\leftarrow,\leftarrow;\leftarrow) & \frac{1}{6} \\
(\rightarrow, \quad ,\rightarrow;\rightarrow) & \frac{1}{6} \\
(\leftarrow, \quad ,\leftarrow;\leftarrow) & \frac{1}{6} \\
(\rightarrow,\rightarrow, \quad ;\rightarrow) & \frac{1}{6} \\
(\leftarrow,\leftarrow, \quad ;\leftarrow) & \frac{1}{6} \\
(\rightarrow, \quad , \quad ; \quad) & \frac{1}{6} \\
(\leftarrow, \quad , \quad ; \quad) & \frac{1}{6} \\
(\quad ,\rightarrow, \quad ; \quad) & \frac{1}{6} \\
(\quad ,\leftarrow, \quad ; \quad) & \frac{1}{6} \\
(\quad , \quad ,\rightarrow; \quad) & \frac{1}{6} \\
(\quad , \quad ,\leftarrow; \quad) & \frac{1}{6}
\end{array}
\tag{4.4}
$$

That is, a randomly chosen predator moves independently of the prey, while the other two predators displace in parallel with the prey. Note that $\mathcal{P}$ (suitably extended beyond T) is a Markov capture, the marginals are indeed random walks, and in this case (as opposed to independent capture) the process of differences $(\overline{X}(t), \overline{Y}(t), \overline{Z}(t)$, defined as in the proof of Theorem 3, is simply a three-dimensional random walk. Since the three coordinates are independent,

$$P(T > t) = O\left(t^{-\frac{3}{2}}\right),$$

so that $\mathbf{E}[T] < \infty$ and $\mathcal{P}$ is efficient. Note that this coupling contrasts with that having independent capture, which is not efficient.

Now let us describe an efficient two-dimensional capture, with $N = 9$ for simplicity. We divide the predators into 3 groups of 3. Within each group the first coordinates (denoted below by a superscript 1) evolve according to the one-dimensional coupling (4.4) until time

$$T^1 = \min\{t : X_i^1(s_i) = Y^1(s_i), s_i \le t,$$
$$\text{for some } i \text{ in each group of 3 predators}\}.$$

Initially the second coordinates of all 9 predators displace in parallel with the prey. As before, it is easy to see that

$$P(T^1 > t) \leq 3P(T > t) = O\left(t^{-\frac{3}{2}}\right),$$

so that $\mathbf{E}[T^1] < \infty$. After time T^1 displace all first coordinates in parallel and apply the one-dimensional coupling (4.4) to the second coordinates of the 3 predators whose first coordinates agree with those of the prey. Again by the one-dimensional result, the additional time T^2 required until some predator agrees with the prey in *both* coordinates has $\mathbf{E}[T^2] < \infty$. Thus the capture is efficient; here we use the fact that the second coordinates of the predators do not change relative to the prey until T^2.

This construction can be continued inductively to higher dimensions, with $C(d) = 3^d$. ∎

The $\alpha = 1$ example of Section 2 illustrates quick Markov capture in one dimension when the prey is between two predators. Can we devise similar Markov captures in higher dimensions provided that the prey is suitably surrounded? It is instructive to consider first the analogous problem for coupled Brownian motions.

Let $\mathcal{H}$ be the (Euclidean) closed convex hull of N locations $x_1, \ldots, x_N$ in $\mathbf{R}^d$. Assume for simplicity that $\mathcal{H}$ is d-dimensional. We claim that a Brownian prey starting from any position $y \in \mathcal{H}$ can be captured quickly by Brownian predators starting from the x_k. The construction is quite simple. Without loss of generality, assume that y belongs to the interior $\mathcal{H}^\circ$ of $\mathcal{H}$; otherwise the construction can be carried out in a suitable hyperplane. First, we translate all of the predators in parallel, driven by a common Brownian trajectory, while letting the prey evolve according to an independent Brownian motion. Let T_1 be the first time the prey intersects a $(d-1)$-dimensional hyperplane $\mathcal{H}_1$ formed by the positions of some of the predators. As in the one-dimensional case, T_1 has the same distribution as the exit time of a (speed 2) Brownian motion from $\mathcal{H}^\circ$; in particular, T_1 has an exponential tail since $\mathcal{H}$ is compact. After time T_1, for all the predators *and* the prey, displace in parallel the components of motion that are orthogonal to $\mathcal{H}_1$, move in parallel the components in the direction of $\mathcal{H}_1$ for all the predators, but let the component of the prey in the direction of $\mathcal{H}_1$ evolve independently. Let T_2 be the additional time until the prey intersects a $(d-2)$-dimensional hyperplane $\mathcal{H}_2$ formed by the positions of some of the predators. Again, T_2 has exponential tail. And so on. Finally, at time $T = T_1 + \ldots + T_d$, some predator captures the prey. The exponential rates for the tails of $T_1, \ldots, T_d$ are uniformly positive, and hence the capture is quick. Our strategy is reminiscent of Lindvall and Rogers [8].

In the context of coupled random walks, the above capture strategy works easily if 2^d predators start at the "corners" $(\pm n, \ldots, \pm n)$ *of* a "lattice cube," and the prey starts at some point y inside this cube. A bit more thought shows that the strategy also works with only $d+1$ predators surrounding the prey, provided that the predators start from an appropriate "lattice simplex." We choose to omit further details of the proof of the following result.

Theorem 5. *In d dimensions, with $N = d+1$ and for suitable initial locations of the predators and prey, there is a quick Markov capture. Specifically, for $1 \leq k \leq d$ and fixed integer $n \geq 1$, let e_k be the k'th unit vector, set $x_k = n\, e_k$, and consider the lattice simplex $\mathcal{S}_n \subset \mathbb{Z}^d$:*

$$\mathcal{S}_n = \{z = (z^1, \ldots, z^d) : z^i \geq 0 \;\; \forall i, \;\; \sum_i z^i \leq n\}.$$

If $y \in \mathcal{S}_n$, then there is a quick Markov capture starting from $(0, x_1, \ldots, x_d; y)$.

For coupled random walks with general initial locations $(x_1, \ldots, x_N; y)$, y surrounded by the x's, the question of whether a quick Markov capture exists appears surprisingly delicate. In contrast to the Brownian case, parallel displacement of predators and prey is not possible in every direction, but only in a small number of lattice-compatible directions (e.g., parallel to an axis or at a 45° angle to an axis when $d = 2$). Rather than delve into the intricacies of the situation, we restrict ourselves to some comments about the two-dimensional three-predator case. Even in this simple setting various types of behavior occur.

Consider the rule that translates all the predators in parallel while the prey moves independently. The lattice triangle formed by the predators moves but does not change shape. For which initial positions of the prey is this coupling quick? The answer, it turns out, depends on certain features of the triangle. By ***highways***, let us mean (lattice) line segments within the triangle that emanate from a vertex and are either parallel to an axis or at 45° to an axis. Any triangle will have 4 highways; the key feature to consider is the distribution of these highways among the vertices. Depending on this distribution, the "quick" positions for the prey will contain a region, determined by the intersections of the highways, that consists of two overlapping triangles, a single triangle, or perhaps has no "interior." In addition, the portions of the highways running from the triangular regions to the vertices are quick. Not quick are (up to 3) obtuse-triangular interior regions that share a side with the original triangle, and of course the exterior region. After some cataloging, one can obtain a complete classification without too much trouble. We invite the reader to analyze some examples, for instance, $x_1 = (6, 0)$, $x_2 = (0, 9)$, $x_3 = (-4, -2)$.

In cases where the capture rule just described fails to be quick, is quick capture possible by means of another scheme? We conclude this section with two examples, still in the two-dimensional three-predator setting, to show that the answer is sometimes yes and sometimes no.

Examples.

(*a*) $x_1 = (4, 0)$, $x_2 = (4,4)$, $x_3 = (-4,-2)$.
Quick Markov capture is possible if the prey starts at one of 13 lattice points distinct from the x_k. Three are obvious: (4,1), (4,2), and (4,3). Can the reader find the other 10? We expect that quick Markov capture is impossible if the prey starts from any point other than these 13 and the x_k. These include the remaining 5 points within the triangle. (It is not difficult to show that points outside the triangle are not quick.) Note that the scenario is different than when the predators move in parallel: in that case, only 3 points other than the x_k are quick.

(*b*) $x_1 = (3,-6)$, $x_2 = (0,9)$, $x_3 = (-1,-1)$.
We expect that quick Markov capture is impossible if the prey starts from any lattice point other than one of the x_k. For the coupling in which the predators move in parallel, this reflects the fact that no highways pass through x_1. Quick Markov capture should also be impossible when the predators start at nx_k, $n \geq 1$.

5. Complete Capture. Here we construct complete captures $\mathcal{P}$ that are not Markov in the sense of the previous section. In particular, predators can make use of the prey's past and future in determining their motion. However, for simplicity, our constructions assume that the trajectories of different predators are independent, i.e., the predators are not permitted to collaborate in their search.

Complete capture is often possible in rather general settings. This contrasts with the more intuitive Markov case, where capture will typically be no better than quick. We illustrate the distinction for a "toy" example – a continuous-time 2-state Markov chain – in Proposition 2 below.

In the more complicated context of random walks the situation is less clear. The state space is infinite and distributions of processes starting from different sites have tails that are not directly comparable. Proposition 3 gives general conditions under which the distributions of the predators overlap sufficiently with that of the prey to permit complete capture. The problem is thereby reduced to verifying said conditions for the processes at hand. Lemmas 1-3 and Proposition 4 do most of the dirty work for systems of random walks by comparing their tail distributions. Finally, Theorem 6 shows that if the prey is surrounded by sufficiently many predators initially, then one can employ Proposition 3 to construct a complete coupling.

As motivation, consider the continuous-time Markov chain on $\{1,2\}$ and Q-matrix $\begin{pmatrix} -1 & 1 \\ 1 & -1 \end{pmatrix}$. Assume that two predators and the prey move according to this chain, the predators starting from state 1, the prey from state 2. In this context, we now show that a capture can be complete only if it is not Markov, although it is easy to construct a complete capture.

Proposition 2. *Let $\mathcal{P}$ be a coupling of two-state Markov chains, as described above, with $N = 2$, $x_1 = x_2 = 1$, and $y = 2$.*

(a) If $\mathcal{P}$ is a Markov capture, then for all $t \geq 1$,

$$\mathcal{P}(T > t) \geq e^{-11t}.$$

Consequently $\mathcal{P}$ is not complete.

(b) There exists a complete capture $\mathcal{P}$ such that

$$\mathcal{P}(T > 1) = 0\,.$$

Proof. (*a*) If $\mathcal{P}$ is Markov, then by (4.1)-(4.2), for $t \leq .1$,

$$\mathcal{P}_{(1,1;2)}(X_1(s) \equiv 1, X_2(s) \equiv 1, Y(s) \equiv 2 \text{ for } 0 \leq s \leq .1)$$

$$\geq 1 - 3(1 - e^{-.1}) \geq e^{-1}.$$

Hence, by the Markov property, for $t \geq .1$,

$$\mathcal{P}(T > t) \geq \mathcal{P}_{(1,1;2)}(X_1(s) \equiv 1, X_2(s) \equiv 1, Y(s) \equiv 2 \text{ for } 0 \leq s \leq t)$$

$$\geq \left(e^{-1}\right)^{10t+1} \geq e^{-11t}\,.$$

(*b*) Assuming that the predators move independently, the explicit formula for the transition semigroup of a two-state Markov chain yields, for $s \geq 0$,

$$P_{(1,1;2)}(Y(s) = 1) = \frac{1}{2} - \frac{1}{2}e^{-2s}\,, \tag{5.1}$$

and

$$P_{(1,1;2)}(X_1(s) = X_2(s) = 1) = \left(\frac{1}{2} + \frac{1}{2}e^{-2s}\right)^2, \tag{5.2}$$

where P denotes any probability law for a valid coupling. It is easy to check that for $s \geq 1$, the probability in (5.1) is greater than 3/8 and the probability in (5.2) is less than 3/8. Setting $s = 1$, a little thought shows that $(X_1(1), X_2(1))$ and $Y(1)$ can be coupled so that

$$\{X_1(1) = X_2(1) = 1\} \subset \{Y(1) = 1\}$$

and

$$\{X_1(1) = X_2(1) = 2\} \subset \{Y(1) = 2\}\,.$$

As a consequence,

$$(5.3)\qquad \begin{aligned}&\{Y(1)=1\}\subset\{X_1(1)=1 \text{ or } X_2(1)=1\}, \text{ and}\\ &\{Y(1)=2\}\subset\{X_1(1)=2 \text{ or } X_2(1)=2\}.\end{aligned}$$

Denote this coupling at time 1 by $\mathcal{P}$. Extend $\mathcal{P}$ to times $t < 1$ in a simple but somewhat arbitrary way: let the predators and prey all evolve independently up to time 1 *conditioned* on their values at time 1. Extend $\mathcal{P}$ to times $t > 1$ by pasting the prey to whichever predator occupies the same state at time 1 (or pasting all three processes together if they all agree at time 1). According to (5.3), $T \le 1$ $\mathcal{P}$-a.s. Finally, in order to satisfy (1.1), redefine $\mathcal{P}$ so that the prey is pasted to one of the predators after time T even on $\{T < 1\}$. This completes the construction of the desired complete capture. ■

We note in passing that if a capture is not required to be Markov, then it is typically easy to construct one which is quick, even if there is only one predator $X(t)$. For example, let ξ_i, $i \ge 1$, be i.i.d. random variables equal to 0 or 1 with probability $\frac{1}{2}$ each. Define

$$\begin{aligned}\xi_i' &= \xi_{i+1} && \text{if } i \text{ is odd,}\\ &= \xi_{i-1} && \text{if } i \text{ is even.}\end{aligned}$$

Let $N(t)$ be an independent rate 1 Poisson process, and define

$$X(t)=\sum_1^{N(t)}\xi_i \quad\text{and}\quad Y(t)=\sum_1^{N(t)}\xi_i'+1.$$

Then X and Y are coupled rate $\frac{1}{2}$ Poisson processes starting from $x = 0$ and $y = 1$ respectively, and

$$Y(t)-X(t)=1 \quad\text{whenever}\quad N(t) \text{ is even.}$$

We easily conclude that the tail of the capture time T decreases exponentially fast in $N(t)$, and hence in t. So this capture is quick. Needless to say, one cannot usually construct a complete capture with only one predator; if this is possible then the distributions of the predator and prey must be identical after some time $t < \infty$.

Our next result, Proposition 3, provides general conditions that ensure enough overlap between the distributions of predators and prey at a fixed time to allow capture. This time will typically be sufficiently large that the distributions have all "spread out." The setup for Proposition 3 is as follows. We are given integers $m, n \ge 1$, and independent $\mathbf{Z}^d$-valued random variables X_k, $1 \le k \le m+n$, with respective densities $p_k(x)$. In addition, we let $\Psi : (\mathbb{Z}^d)^n \to \mathbb{Z}^d$ denote a map which selects one of the n coordinates of its argument, and set

$$\psi(x) = P\Big(\Psi(X_{m+1}, \dots, X_{m+n}) = x\Big).$$

In the main result of this section, Theorem 6, Y will be the position of the prey at that time t, with $Y(0) = 0$, and $X_1, \dots, X_m$ will be the positions at a suitable time t of random walks that started "on all sides" of the prey (but typically far away). Also, in that application $n = 2$, $\Psi(x_{m+1}, x_{m+2})$ will be the point x_{m+1} or x_{m+2} with smaller norm, and the value δ of the following criterion will be $\frac{2}{5}$.

Proposition 3. *In the context of the preceding paragraph, let $q(x)$ be a probability density on $\mathbb{Z}^d$, and $\delta \in [0,1]$. Assume that*

$$\delta\psi(x) \le q(x) \le p_1(x) + \delta \sum_{k=2}^{m} p_k(x) + \delta\psi(x) \quad \text{for all } x. \tag{5.4}$$

Then there is a random variable Y with density q, and a coupling $\mathcal{P}$ of $(X_1, \dots, X_{m+n}, Y)$ such that

$$\mathcal{P}(Y = X_k \ \text{ for some } \ k) = 1. \tag{5.5}$$

Before proceeding to the proof, we note that something like the first two terms in the upper bound of (5.4) is required: one needs sufficient mass among the $p_k(x)$ to "cover" $q(x)$ for each x. Once we have used $p_1, \dots, p_m$, there will be a mass of size δ yet to be assigned in our procedure. The lower bound of (5.4) ensures that the distribution ψ can handle the remaining probability.

Proof. We first adjoin independent events $A_k(x)$, $0 \le k \le m$, with corresponding, appropriately chosen probabilities $r_k(x)$ to the underlying probability space. Using the X_k and this auxiliary structure, we will define a partition of the whole space into disjoint events $C_k(x)$, with respective probabilities $c_k(x)$. We will define the desired random variable Y so that:

$$\begin{aligned} &Y = X_k = x \ \text{ on } \ C_k(x), \quad 1 \le k \le m, \\ &Y = \Psi(X_{m+1}, \dots, X_{m+n}) = x \ \text{ on } \ C_0(x). \end{aligned} \tag{5.6}$$

Our goal will be to show that

$$\mathcal{P}(Y = x) = \sum_{k=0}^{m} c_k(x) = q(x). \tag{5.7}$$

The proposition will then be a consequence of (5.6) and (5.7).

The $r_k(x)$, $C_k(x)$, and $c_k(x)$ are determined by the following recursive scheme. For $k = 1$, take

$$r_1(x) = \left[\frac{q(x) - \delta\psi(x)}{p_1(x)}\right] \wedge 1 . \tag{5.8}$$

(If the denominator on the right side equals 0, set $r_1(x) = 0$.) We need $r_1(x) \geq 0$. Choosing $A_1(x)$ as above, set

$$C_1(x) = \{X_1 = x\} \cap A_1(x) . \tag{5.9}$$

By independence,

$$c_1(x) = p_1(x) r_1(x) . \tag{5.10}$$

Denote $C_k = \bigcup_x C_k(x)$, $c_k = P(C_k) = \sum_x c_k(x)$. Next, for $2 \leq k \leq m$, put

$$r_k(x) = \left[\frac{q(x) - \delta\psi(x) - \sum_{j=1}^{k-1} c_j(x)}{\left(1 - \sum_{j=1}^{k-1} c_j\right) p_k(x)}\right] \wedge 1 . \tag{5.11}$$

(Again, if the denominator on the right side is 0, set $r_k(x) = 0$.) We will need to show that $r_k(x) \geq 0$, so $A_k(x)$, chosen as above, makes sense. Next, set

$$C_k(x) = \left(\bigcup_{j=1}^{k-1} C_j\right)^c \cap \{X_k = x\} \cap A_k(x) . \tag{5.12}$$

By independence,

$$c_k(x) = \left(1 - \sum_{j=1}^{k-1} c_j\right) p_k(x) r_k(x) . \tag{5.13}$$

We also need to account for the mass remaining after $c_1(x), \ldots, c_m(x)$ have been assigned. To this end, for $k = 0$, we let

$$r_0(x) \equiv 1 \tag{5.14}$$

and

$$C_0(x) = \left(\bigcup_{j=1}^{m} C_j\right)^c \cap \{\Psi(X_{m+1}, \ldots, X_{m+n}) = x\} . \tag{5.15}$$

Once again by independence,

$$c_0(x) = \left(1 - \sum_{j=1}^{m} c_j\right)\psi(x). \tag{5.16}$$

Let us offer a few words of explanation at this point. The $r_k(x)$ are designed to regulate the proportion of the k'th component at x that is chosen for coupling. In essence, if the numerator of (5.11) is strictly less than the denominator, then $p_k(x)$ is large enough to "fill Y up at x" by assigning enough mass from $p_k(x)$ to obtain $q(x)$. In this case $p_{k+1}(x), \ldots, p_m(x)$ are not needed, so

$$r_{k+1}(x) = \ldots = r_m(x) = 0\,.$$

When $r_k(x) = 1$, on the other hand, we assign the "maximum amount of mass available" at $\{X_k = x\}$, namely $c_k(x)$. The succeeding terms $r_{k+1}(x), \ldots, r_m(x)$ then deal with the residual mass at x after $c_k(x)$ is subtracted. The right side of (5.4) ensures that the $p_k(x)$, $1 \leq k \leq m$, are large enough so that, by setting $r_k(x) = 1$, the total amount of mass assigned to x is at least $q(x)$. The left side of (5.4) ensures that the mass at x allotted by $\psi(x)$ does not already exceed $q(x)$.

Proceeding with the proof, we need to make sure that $r_k(x) \geq 0$ for all k, x, so that the events $A_k(x)$ are well-defined. If so, then the C_k evidently form a partition with probabilities given by (5.10), (5.13) and (5.16). Defining Y as in (5.6), the proof will then be complete once we show (5.7).

To see that the $r_k(x)$ are non-negative, it suffices to check that, for all k, $1 \leq k \leq m$,

$$\sum_{j=1}^{k-1} c_j(x) \leq q(x) - \delta\psi(x)\,; \tag{5.17}$$

for (5.21) below, we will also need the case $k = m + 1$. Note that if (5.17) holds for some k, then summation over x yields

$$\sum_{j=1}^{k-1} c_j \leq 1 - \delta. \tag{5.18}$$

The argument for (5.17) is an easy induction. When $k = 1$, (5.17) is just the left inequality of (5.4). Now assume that (5.17) holds for $k \leq m$. From the definitions of $c_{k+1}(x)$ and $r_{k+1}(x)$,

$$c_{k+1}(x) \leq q(x) - \delta\psi(x) - \sum_{j=1}^{k} c_j(x)\,,$$

so (5.17) holds for $k + 1$.

Let us turn to (5.7). We proceed with the argument outlined earlier. Suppose first that the numerator in (5.8) or (5.11) is strictly less than the denominator for some k, $1 \le k \le m$. Then

$$c_k(x) = q(x) - \delta\psi(x) - \sum_{j=1}^{k-1} c_j(x)$$

and

$$\sum_{j=1}^{m} c_j(x) \ge \sum_{j=1}^{k} c_j(x) = q(x) - \delta\psi(x) . \tag{5.19}$$

Otherwise, $r_k(x) = 1$ for $1 \le k \le m$, and so

$$\sum_{k=1}^{m} c_k(x) = p_1(x) + \sum_{k=2}^{m} \left(1 - \sum_{j=1}^{k-1} c_j\right) p_k(x) .$$

Making use of (5.18) and then the right inequality of (5.4), we see that

$$\sum_{k=1}^{m} c_k(x) \ge p_1(x) + \delta \sum_{k=2}^{m} p_k(x) \ge q(x) - \delta\psi(x) , \tag{5.20}$$

which is the same inequality as (5.19). Combining (5.17), (5.19) and (5.20), we conclude that

$$\sum_{k=1}^{m} c_k(x) = q(x) - \delta\psi(x), \tag{5.21}$$

and, summing over x,

$$\sum_{k=1}^{m} c_k = 1 - \delta . \tag{5.22}$$

Appealing to (5.16), (5.21) and (5.22), we have

$$\sum_{k=0}^{m} c_k(x) = \sum_{k=1}^{m} c_k(x) + \left(1 - \sum_{k=1}^{m} c_k\right)\psi(x)$$

$$= (q(x) - \delta\psi(x)) + \delta\psi(x) = q(x) .$$

We have verified (5.7) and the proposition is proved. ∎

The next three lemmas, preparatory to Proposition 4 below, estimate transition probabilities of a continuous-time rate 1 simple random walk. They will be used to verify the lower bound of (5.4) for predator-prey systems of random walks (in any dimension). The first lemma gives an upper bound on the tail of a one-dimensional random walk $Y(t)$. Below, we let P and $\mathbf{E}$ denote the law and expectation, respectively, of a walk starting from 0.

Lemma 1. *For all $t > 0$ and $x \geq 0$,*

$$P(Y(t) \geq x) \leq e^{-\frac{x^2}{2(t+x)}}.$$

Proof. Since

$$\mathbf{E}\left[e^{\theta Y(t)}\right] = \exp\left\{t\left(\frac{e^{\theta} + e^{-\theta}}{2} - 1\right)\right\},$$

Chebyshev's inequality implies that for $\theta \geq 0$,

$$P(Y(t) \geq x) \leq \exp\left\{t\left(\frac{e^{\theta} + e^{-\theta}}{2} - 1\right) - \theta x\right\}.$$

Set $\theta = \log(1 + \frac{x}{t})$. Then the right side becomes

$$\exp\left\{\frac{x}{2}\left(1 - \frac{t}{t+x} + 2\log\frac{t}{t+x}\right)\right\} \leq \exp\left\{-\frac{x^2}{2(t+x)}\right\}. \qquad \blacksquare$$

Our second lemma, which will be applied in Lemma 3, deals with the positive and negative parts of the random walks $Y(t)$, denoted $Y^+(t)$ and $Y^-(t)$. These are obtained by counting the positive and negative jumps, respectively.

Lemma 2.
(*a*) *For every y, $P(Y^-(t) \leq y \mid Y(t) = x)$ is increasing in x.*
(*b*) *Given any constants C_1, C_2,*

$$P(Y^-(t) \leq y \mid Y(t) = x) \geq C_3,$$

for appropriate $C_3 > 0$, provided $y \geq \frac{t}{2} - C_1\sqrt{t}$, $x \geq -C_2\sqrt{t}$, and t is sufficiently large.

Proof. (*a*) Abbreviate

$$A^+(z) = P(Y^+(t) = z),$$
$$A^-(z) = P(Y^-(t) = z).$$

We can write

$$\Big(P(Y^-(t) \le y \mid Y(t) = x)\Big)^{-1} = \frac{\sum_{j=0}^{\infty} A^-(j)A^+(x+j)}{\sum_{j=0}^{y} A^-(j)A^+(x+j)}$$

$$= 1 + \frac{\sum_{j=y+1}^{\infty} A^-(j)A^+(x+j)}{\sum_{j=0}^{y} A^-(j)A^+(x+j)}. \tag{5.23}$$

Since Y^+ is a rate $\frac{1}{2}$ Poisson process,

$$\frac{A^+(z+1)}{A^+(z)} = \frac{t}{2(z+1)},$$

which is decreasing in z. Hence, expression (5.23) is

$$= 1 + \frac{\dfrac{A^+(x+y+1)}{A^+(x+y)} \sum_{j=y+1}^{\infty} A^-(j)A^+(x+j)}{\dfrac{A^+(x+y+1)}{A^+(x+y)} \sum_{j=0}^{y} A^-(j)A^+(x+j)}$$

$$\ge 1 + \frac{\sum_{j=y+1}^{\infty} A^-(j)A^+(x+j+1)}{\sum_{j=0}^{y} A^-(j)A^+(x+j+1)}$$

$$= \Big(P(Y^-(t) \le y \mid Y(t) = x+1)\Big)^{-1}.$$

Thus, (*a*) holds.

(b) Using part (*a*), it is easy to check that

$$P(Y^-(t) \le y \mid Y(t) = x) \ge P(Y^-(t) \le y \mid Y(t) \le x).$$

By the hypotheses on x and y, the right side is

$$\geq P\left(\frac{t}{2} - (C_1 + 1)\sqrt{t} \leq Y^-(t) \leq \frac{t}{2} - C_1\sqrt{t}\right) \cdot$$

$$P\left(Y^+(t) \leq \frac{t}{2} - (C_1 + C_2 + 1)\sqrt{t}\right) \Big/ P\left(Y(t) \leq x\right).$$

Apply the central limit theorem to obtain a positive lower bound for each probability in the numerator provided t is sufficiently large. Using the trivial bound 1 for the denominator, claim (b) is proved. ■

Our final lemma shows that the tail of a random walk does not decrease too quickly. Here $X(t)$ and $Y(t)$ are one-dimensional walks starting from x_0 and 0, respectively.

Lemma 3. *Assume that* $0 < x_0 \leq C_2\sqrt{t}$. *Then for all sufficiently large* t,

$$\frac{P(Y(t) = x)}{P(X(t) = x)} \geq \left(C_4 \exp\left\{-C_5\frac{x}{\sqrt{t}}\right\}\right) \wedge 1 \tag{5.24}$$

for all $x \in \mathbb{Z}$ *and appropriate constants* $C_4, C_5 > 0$.

Proof. Inequality (5.24) is equivalent to

$$\frac{P(Y(t) = x + x_0)}{P(Y(t) = x)} \geq \left(C_4' \exp\left\{-C_5\frac{x}{\sqrt{t}}\right\}\right) \wedge 1 \tag{5.25}$$

for some $C_4' > 0$. Since the left side of (5.25) has 1 as a trivial bound when $x \leq -x_0$, we can assume that $x > -x_0$. First, set $y = \lfloor \frac{t}{2} - C_1\sqrt{t} \rfloor$ in Lemma 2(b) to obtain

$$P(Y^-(t) \leq \frac{t}{2} - C_1\sqrt{t} \mid Y(t) = x) \geq C_3 > 0.$$

Consequently,

$$\frac{P(Y(t) = x + x_0)}{P(Y(t) = x)} \geq C_3 \frac{P(Y(t) = x + x_0, Y^-(t) \leq \frac{t}{2} - C_1\sqrt{t})}{P(Y(t) = x, Y^-(t) \leq \frac{t}{2} - C_1\sqrt{t})}.$$

We can rewrite the right side as

$$C_3 \frac{\sum_{j=0}^{l(t)} q_j P(Y^+(t) = x + j + x_0)}{\sum_{j=0}^{l(t)} q_j P(Y^+(t) = x + j)},$$

where $q_j = A^-(j) = P(Y^-(t) = j)$, and $l(t) = \lfloor \frac{t}{2} - C_1\sqrt{t} \rfloor$. Since Y^+ is rate $\frac{1}{2}$ Poisson, the last expression is

$$C_3 \frac{\displaystyle\sum_{j=0}^{l(t)} q_j \frac{(t/2)^{j+x_0}}{(x+j+x_0)!}}{\displaystyle\sum_{j=0}^{l(t)} q_j \frac{(t/2)^j}{(x+j)!}} \geq \min_{0\leq j\leq l(t)} \frac{\dfrac{(t/2)^{j+x_0}}{(x+j+x_0)!}}{\dfrac{(t/2)^j}{(x+j)!}}$$

$$\geq \frac{(t/2)^{x_0}}{(x+l(t)+x_0)^{x_0}} \geq \left(1+\frac{2x}{t}\right)^{-x_0} \geq \exp\left\{-2C_2\frac{x}{\sqrt{t}}\right\},$$

the final inequality holding because $x_0 \leq C_2\sqrt{t}$. This implies (5.25), or equivalently (5.24), as desired. ∎

We now apply Lemmas 1 and 3 to prove Proposition 4, which bounds the tail of d-dimensional random walk $Y(t)$ from below in terms of the tail of the "minimum" of two other independent walks $X_1(t)$ and $X_2(t)$. Given $x = (x^1, \ldots, x^d) \in Z^d$, introduce $\|x\| = \sup_i |x^i|$, and set

$$X_{\min}(t) = X_1(t) \text{ if } \|X_1(t)\| \leq \|X_2(t)\|, \tag{5.26}$$
$$= X_2(t) \text{ if } \|X_1(t)\| > \|X_2(t)\|.$$

Proposition 4. *Assume that* $Y(0) = 0$, *and that* $\|X_k(0)\| \leq C_6\sqrt{t}$ $(k = 1, 2)$ *for some* C_6. *Then for all sufficiently large* t *and a suitable* C_7,

$$P(Y(t) = x) \geq P(X_{\min}(t) = x) \tag{5.27}$$

whenever $\|x\| \geq C_7\sqrt{t}$.

Proof. We first note that

$$P(X_{\min}(t) = x) \leq P(X_1(t) = x)P(\|X_2(t)\| \geq \|x\|) + P(X_2(t) = x)P(\|X_1(t)\| \geq \|x\|).$$

Consequently,

$$\frac{P(X_{\min}(t) = x)}{P(Y(t) = x)} \leq \frac{P(X_1(t) = x)P(\|X_2(t)\| \geq \|x\|)}{P(Y(t) = x)} + \frac{P(X_2(t) = x)P(\|X_1(t)\| \geq \|x\|)}{P(Y(t) = x)}.$$

We will show that the first term on the right side is $\leq \frac{1}{2}$; the same argument applies, of course, to the second term. This will prove (5.27).

Let us rewrite

$$\frac{P(X_1(t)=x)}{P(Y(t)=x)}=\prod_{i=1}^{d}\frac{P(X_1^i(t)=x^i)}{P(Y^i(t)=x^i)}$$

where the superscripts denote coordinates. By Lemma 3, the product is at most

$$C_4^d\exp\left\{dC_5\frac{||x||}{\sqrt{t}}\right\} \tag{5.28}$$

for all x and large t. On the other hand, for $||x||\geq C_7\sqrt{t}$ with $C_7\geq 2C_6$,

$$\begin{aligned}P(||X_2(t)||\geq ||x||)&\leq P(||Y(t)||\geq \frac{1}{2}||x||)\\&\leq 2dP(||Y^1(t)||\geq \frac{1}{2}||x||).\end{aligned}$$

By Lemma 1, the last term is

$$\leq 2d\exp\left\{-\frac{||x||^2}{8(t+||x||)}\right\}. \tag{5.29}$$

Using bounds (5.28) and (5.29), we conclude that

$$\begin{aligned}&\frac{P(X_1(t)=x)P(||X_2(t)||\geq ||x||)}{P(Y(t)=x)}\\&\qquad\leq 2dC_4^d\exp\left\{dC_5\frac{||x||}{\sqrt{t}}-\frac{||x||^2}{8(t+||x||)}\right\}.\end{aligned}$$

It is easy to check that for $||x||\geq C_7\sqrt{t}$, and C_7 sufficiently large compared to C_5, the right side is less than $\frac{1}{2}$, as desired. ■

We come now to Theorem 6. In essence, the theorem states that if the prey Y is initially "surrounded" by appropriately many predators X_k, $1\leq k\leq m+2$, where $m=3\cdot 2^d$, and if all the predators start within distance $C_6\sqrt{t}$ of the prey, for a suitable C_6, then Y and $(X_1,\ldots,X_{m+2})$ can be coupled so that $Y(t)=X_k(t)$ for some k.

This result is used in Bramson and Lebowitz ([2], Lemma 5.2) to couple "matter" particles from a 2-type annihilating particle system and random walks from a Poisson system having several times the density. Using elementary large deviation estimates, it is shown that at large times, over suitable large cubes, the Poisson system nearly always has several times as many particles as the number of matter particles in the annihilating system. Applying the complete capture provided by Theorem 6, it follows that, after an appropriate additional amount of time, the Poisson system contains the matter in nearly all regions. With further work, this comparison implies that the opposite particle type ("anti-matter") does not decay faster than at a given rate. Even quick capture would not be sufficient for the needed estimates.

In order to prove Theorem 6, we will verify condition (5.4) of Proposition 3. Most of the groundwork has now been prepared in Lemmas 1-3 and Proposition 4. Setting $n = 2$ in Proposition 3, and

$$\Psi(x_1, x_2) = x_{\min} \tag{5.30}$$

as in (5.26), it will follow that the lower bound of (5.4) holds for the tails of the distributions. The upper bound of (5.4) will hold for the tails because there are always predators that are initially closer to distant points than the prey is. Both the lower and upper bounds away from the tails will follow from the local central limit theorem.

More formally, the framework for Theorem 6 is as follows. Let $I = \{u = (u^1, \ldots, u^d) : u^i = \pm 1\}$, and let $D^u(R)$, $u \in I$, denote the cube with sides of length R in which the i'th coordinate extends from 0 to $u^i R$. Labeling the first $3 \cdot 2^d$ predators as $X^{(u,l)}$, $u \in I$, $1 \leq l \leq 3$, we assume that

$$X^{(u,l)}(0) \in D^u(C_6\sqrt{t}) \quad \text{for all } u,\ l.$$

In words, three predators start in each of the 2^d "orthants" surrounding the origin, all initial positions with ℓ^∞ norm at most $C_6\sqrt{t}$. We assume that two additional predators, X_{m+1} and X_{m+2}, have starting locations that satisfy

$$\|X_{m+1}(0)\| \leq C_6\sqrt{t}, \quad \|X_{m+2}(0)\| \leq C_6\sqrt{t},$$

and that the prey starts at the origin, $Y(0) = 0$.

Theorem 6. *If $C_6 > 0$ is sufficiently small and t is large enough, then there is a coupling $\mathcal{P}$ of Y and $(X_1, \ldots, X_{m+2})$ such that for all $s \geq t$,*

$$Y(s) = X_k(s) \quad \textit{for some } k,\ 1 \leq k \leq m+2.$$

Proof. Choosing Ψ as in (5.30) and $\delta = \frac{2}{5}$, we will show that condition (5.4) of Proposition 3 holds at time t. Thus we can invoke that result to construct a coupling $\mathcal{P}$ at time t. The extension of $\mathcal{P}$ to other times is not difficult. The law at times $s < t$ can be prescribed using conditional independence, as in Proposition 2. Times $s > t$ can be handled by pasting the prey to the same predator that has captured it by time t.

Let us first consider the upper bound of (5.4). Note that for each site $x \in \mathbf{Z}^d$, there is a $u \in I$ so that if $x_0 \in D^u(C_6\sqrt{t})$, then

$$|x^i - x_0^i| \leq |x^i| \text{ or } |x^i| \leq C_6\sqrt{t} \quad \text{for each } i, 1 \leq i \leq d. \tag{5.31}$$

Let X_0 denote a random walk starting from x_0. It is well known that if the first inequality of (5.31) holds for a given coordinate i, then

$$P(Y^i(t) = x^i) \le P(X_0^i(t) = x^i). \tag{5.32}$$

(For instance, one can use the reflection principle.) If, on the other hand, the second inequality of (5.31) holds, then one can apply the local central limit theorem to deduce that for large t,

$$\begin{aligned} P(Y^i(t) = x^i) &\sim \frac{1}{\sqrt{2\pi t}} e^{-|x^i|^2/2t}, \\ P(X_0^i(t) = x^i) &\sim \frac{1}{\sqrt{2\pi t}} e^{-|x^i - x_0^i|^2/2t}. \end{aligned} \tag{5.33}$$

If $C_6 > 0$ is chosen small enough, this implies that

$$P(Y^i(t) = x^i) \le \left(\tfrac{6}{5}\right)^{\frac{1}{d}} P(X_0^i(t) = x^i) \tag{5.34}$$

for $|x^i| \le C_6 \sqrt{t}$. Combining (5.32) and (5.34) for $1 \le i \le d$, we conclude that

$$P(Y(t) = x) \le \tfrac{6}{5} P(X_0(t) = x) \tag{5.35}$$

for small C_6, large t, and u chosen so that x_0 satisfies (5.31). Sum the contributions of the predators $X^{(u,l)}$, $1 \le l \le 3$, to see that the right side of (5.4) is at least

$$\delta \sum_{l=1}^{3} P(X^{(u,l)}(t) = x) \ge \tfrac{2}{5} \cdot \tfrac{5}{6} \cdot 3P(Y(t) = x) = P(Y(t) = x).$$

Evidently the upper bound of (5.4) is satisfied for large t if $C_6 > 0$ is chosen small.

Most of the work for the lower bound of (5.4) has been done previously. By inequality (5.27) of Proposition 4, for a suitable C_7 and large t,

$$P(Y(t) = x) \ge \psi(x)$$

whenever $\|x\| \ge C_7 \sqrt{t}$. This gives the lower bound of (5.4) for large $\|x\|$. Since $\|X_{m+1}(0)\| \le C_6 \sqrt{t}$ and $\|X_{m+2}(0)\| \le C_6 \sqrt{t}$, we can use the local central limit theorem again for small $\|x\|$. Applying (5.33) for $\|x\| \le C_7 \sqrt{t}$, we get the following analog of (5.35) for small C_6 and large t:

$$P(Y(t) = x) \ge \tfrac{4}{5} P(X_{m+j}(t) = x), \quad j = 1, 2.$$

Consequently,

$$P(Y(t) = x) \geq \tfrac{2}{5}\left(P(X_{m+1}(t) = x) + P(X_{m+2}(t) = x)\right)$$
$$\geq \delta\psi(x)\,.$$

So the lower bound of (5.4) holds for all x. We have verified both inequalities of (5.4), so Proposition 3 provides the desired complete coupling. ■

Acknowledgments

We would like to thank Ted Cox, who provided the simulation program **predview** *that produced the* **Figure** *and generated the corresponding data. Thanks also to Greg Lawler and Tom Polaski for preliminary simulations of the same sort, and to Don Burkholder for the reference to De Blassie [5]. Finally, we gratefully acknowledge the assistance of Janko Gravner, especially in connection with the proof of Theorem 3 and the examples at the end of Section 4.*

REFERENCES

[1] R. Arratia, *Limiting point processes for rescaling of coalescing and annihilating random walks on* $\mathbb{Z}^d$, Ann. Probab. **9** (1981), 909-936.

[2] M. Bramson and J.L. Lebowitz, *Asymptotic behavior of densities for two-particle annihilating random walks*, to appear in J. Stat. Physics.

[3] M. Bramson and D. Griffeath, *Clustering and dispersion rates for some interacting particle systems on* $\mathbb{Z}^1$, Ann. Probab. **8** (1980), 183-213.

[4] D.L. Burkholder, *Exit times of Brownian motion, harmonic majorization, and Hardy spaces*, Adv. Math. **26** (1977), 182-205.

[5] R.D. De Blassie, *Exit times from cones in* $\mathbf{R}^n$ *of Brownian motion*, Z. Wahr. verw. Gebiete **74** (1987), 1-29.

[6] E.B. Dynkin and A.A. Yushkevich, *Markov Processes: Theorems and Problems*, Plenum Press, NY, 1969.

[7] D. Griffeath, *Coupling methods for Markov processes*, Studies in Probability and Ergodic Theory, ed. Rota. Academic Press, New York, NY, 1978, 1-43.

[8] T. Lindvall and L.C.G. Rogers, *Coupling of multidimensional diffusions by reflection*, Ann. Probab. **14** (1986), 860-872.

[9] T.R. McConnell, *Exit times of* N*-dimensional random walks*, Z. Wahr. verw. Gebiete **67** (1984), 213-233.

[10] F. Spitzer, *Some theorems concerning 2-dimensional Brownian motion*, Trans. Amer. Math. Soc. **271** (1958), 719-731.

[11] F. Spitzer, *Principles of Random Walk*, Graduate Texts in Mathematics **34**. Springer-Verlag, New York, 1976.

Maury Bramson & David Griffeath
Department of Mathematics
University of Wisconsin
Madison WI 53706
bramson@math.wisc.edu
griffeat@math.wisc.edu

Nonlinear Voter Models

J.T. COX[1] AND R. DURRETT[2]

We investigate generalizations of the voter model in which the flip rates are a nonlinear function of the number of neighbors with a different opinion. These new models exhibit a wide variety of limiting behavior from clustering to ergodicity.

1. Introduction. The inspiration (or excuse) for this paper comes from Spitzer's paper *Interaction of Markov processes.* In that work Spitzer introduced five models, computed their behavior on finite sets, and made a number of conjectures concerning the behavior of the infinite systems that inspired a lot of subsequent work. Attempting to follow in Spitzer's footsteps, we will here describe a number of variations of the voter model, report on a few preliminary results, and make a number of conjectures. One of the many differences from Spitzer's work, is that most of our conjectures are motivated by computer simulations.

In all the processes we consider the state at time t is $\xi_t : \mathbb{Z}^d \to \{0,1\}$. We think of the sites of $\mathbb{Z}^d$ as occupied by persons and $\xi_t(x)$ as the opinion of the voter at x at time t. Our voters are very simple minded and change their opinion at a rate a_k if k of their neighbors $y \in x+\mathcal{N}$ have the opposite opinion. Here and throughout the paper $\mathcal{N}$ is a finite set that is symmetric with respect to all reflections and irreducible (i.e., the group it generates is $\mathbb{Z}^d$).

Example 1. The basic voter model. In this case $a_k = ck$ where c is a normalizing constant whose value is not important and changes from paper to paper. The basic facts about this process were established by Holley and Liggett [9]. To state their results we need two definitions that we will use throughout the paper. We say that *clustering* occurs if for all $x, y \in \mathbb{Z}^d$ with $x \neq y$, we have $P(\xi_t(x) \neq \xi_t(y)) \to 0$ as $t \to \infty$. We say that *coexistence* occurs if there is a stationary distribution that concentrates on configurations with infinitely many 0's and 1's.

Theorem 1. *Consider the basic voter model. (i) In dimensions $d = 1,2$*

---

[1]Partially supported by the National Science Foundation.

[2]Partially supported by the National Science Foundation, the National Security Agency, and the Army Research Office through the Mathematical Sciences Institute at Cornell University.

clustering occurs starting from any initial configuration. (ii) In $d \geq 3$ coexistence occurs.

To prove (ii), Holley and Liggett showed that if we start from product measure with density θ, that is, $\{\xi_0^\theta(x) = 1\}$ are independent and have probability θ, then as $t \to \infty$, ξ_t^θ converges in distribution ($\Rightarrow$) to ν_θ, a stationary distribution. Let δ_i denote the pointmass on the configuration $\xi(x) \equiv i$. (i) implies that in $d = 1, 2$ the only stationary distributions are convex combinations of δ_0 and δ_1, while in $d \geq 3$ there is a one parameter family of stationary distributions. Holley and Liggett have shown in $d \geq 3$ all stationary distributions are convex combinations of the ν_θ.

The main theme of our investigations is "How does the behavior reported in Theorem 1 change if we change the a_k?" Our first modification is

Example 2. The threshold voter model. In this system $a_0 = 0$ and $a_k = 1$ if $k > 0$. To explain the name of the process and mislead the reader about its properties we recall that in the general class of models called *contact processes* 1's become 0's at rate 1 independent of the state of the neighbors. In the *basic contact process*, 0's become 1's at rate λ times the number of neighbors that are 1. In the *threshold contact process*, 0's become 1's at rate λ if at least one neighbor is 1. The qualitative properties of these two contact processes are almost identical.

Theorem 2. *Consider the basic or threshold contact process. There is a critical value $\lambda_c = \inf\{\lambda : P(\xi_t \not\equiv 0 \text{ for all } t) > 0\}$ so that (i) If $\lambda \leq \lambda_c$, then $\xi_t \Rightarrow \delta_0$. (ii) Let $\tau = \inf\{t : \xi_t \equiv 0\}$. If $\lambda > \lambda_c$, then*

$$\xi_t \Rightarrow P(\tau < \infty)\,\delta_0 + P(\tau = \infty)\,\xi_\infty^1$$

where ξ_∞^1 is the limit starting from $\xi_0^1 \equiv 1$.

Bezuidenhout and Grimmett [1] proved the last result for the basic contact process. It is easy to see that their argument extends to the threshold contact process and many other processes that are additive in the sense of Griffeath [7].

On the basis of the last result the reader might leap to the conclusion that Theorem 1 applies to the threshold voter model as well, but this is wrong.

Theorem 3. *Consider the threshold voter model in $d = 1$ with $\mathcal{N} = \{x : 0 < |x| \leq M\}$. If $M \geq 4$ then coexistence occurs.*

Proof: We begin by proving the result for $M \geq 7$ and then improve the argument to get the indicated result. Let $I_k = \{4k, 4k+1, 4k+2, 4k+3\}$. We say that I_k is *occupied* and set $\zeta_t(k) = 1$ if at least one site in I_k is $= 1$ at time t. Occupied sites become vacant (i.e., $\zeta_t(k) = 0$) at rate ≤ 1. If k is vacant and $k-1$ or $k+1$ is occupied then k becomes occupied at rate 4. The process ζ_t thus dominates the threshold contact process with $\mathcal{N} = \{-1,1\}$ and $\lambda = 4$, which in turn dominates the basic contact process with $\mathcal{N} = \{-1,1\}$ and $\lambda = 2$. Holley and Liggett [10] have shown that the last process survives (i.e., has $\xi_t \not\equiv 0$ for all t) with positive probability. To construct a stationary distribution, we start the threshold contact process from product measure with density 1/2, run the system to time T, take the Cesaro average of the distributions at times $0 \le t \le T$, and extract a convergent subsequence. The result, call it ν, is a stationary distribution. (See Liggett [12] Proposition 1.8 on p. 10). To see that it is nontrivial note that $\{\zeta_0(k) = 1\}$ are independent and have probability 15/16, so Theorem 2 implies that when the threshold contact process starts from this initial distribution it converges to ξ^1_∞. It is known that ξ^1_∞ concentrates on configurations with infinitely many 1's so ν does as well. To see that there must be infinitely many 0's as well notice each approximation is symmetric under interchange of 0 and 1 so the limit is.

Let λ^b_c and λ^t_c denote the critical values for the basic and threshold contact processes respectively. The last proof shows that if $[(M+1)/2] \ge \lambda^t_c$ then there is a nontrivial stationary distribution. Then we used $\lambda^t_c \le 2\lambda^b_c \le 4$. Liggett [13] has shown that $\lambda^t_c \le 2.18$ so there is a nontrivial stationary distribution for $M \ge 5$. By breaking the system into alternating blocks of size 2 and 3 and using another result of Liggett [13] one can prove this for $M = 4$. ■

Numerical results (see Dickman and Burschka [4]) indicate that $\lambda^t_c \approx 1.742$. If $\lambda^t_c < 2$ then by using blocks of size 2 we get the conclusion for $M = 3$. These comparisons will never prove the result for $M = 2$ but computer simulations make it "clear" that the result is true in this case. Finally, as we will prove at the end of Section 2, the result is false when $M = 1$.

Theorem 4. *Consider the threshold voter model in $d = 1$ with $\mathcal{N} = \{x-1, x+1\}$. Clustering occurs starting from any initial configuration.*

In $d = 2$ if we let $I_k = \{(k,-1),(k,0),(k,1)\}$, proceed as in the proof of Theorem 3, and use $\lambda^t_c \le 2.18$ we see coexistence occurs for $\mathcal{N} = \{x : 0 < \|x\|_\infty \le 2\}$ or $\mathcal{N} = \{x : 0 < \|x\|_1 \le 3\}$. The conclusion should hold when $\mathcal{N} = \{x : \|x\|_1 = 1\}$. To argue this we observe that the set of 1's in the threshold voter model dominates the set of 1's in the threshold contact process with $\lambda = 1$ so coexistence occurs if $\lambda^t_c < 1$. To bound λ^t_c in $d > 1$ when $\mathcal{N} = \{x : \|x\|_1 = 1\}$ we look at a one dimensional projection: $\pi_t(k) = 1$ if $\xi_t(x) = 1$ for some x with $x_1 + \cdots + x_d = k$. It is easy to see that 1's in π become 0's at rate ≤ 1 and 0's become 1's at rate $\ge d\lambda$ if at least one neighbor is 1. From this and Liggett's bound it follows immediately that

$$\lambda^t_c(\mathbb{Z}^d) \le 2.18/d.$$

The right–hand side is < 1 in $d \ge 3$ and just misses in $d = 2$. The last argument comes from Holley and Liggett [10] who used it to prove $\lambda^b_c(\mathbb{Z}^d) \le$

$2/d$. In $d = 2$ the last upper bound is 1 while the critical value is about 0.41, so it seems reasonable to believe that $\lambda_c^b(\mathbb{Z}^2) < 1$. We summarize our speculations in

Conjecture 1. *Coexistence occurs in the threshold voter model in all cases except the one dimensional nearest neighbor case.*

With coexistence established in a number of cases, it is natural to try to describe the set of stationary distributions. We are not able to answer this question for the threshold voter model so we consider a more general class of models

Example 3. Threshold voter models with spontaneous deaths. As the name suggests we modify the threshold voter model so that 1's become 0's at rate $a_k + \delta$ when k neighbors disagree. An easy modification of the proof of Theorem 3 gives

Theorem 5. *Consider the threshold voter model in $d = 1$ with $\mathcal{N} = \{x : 0 < |x| \le M\}$ and $M \ge 4$. There is a $\delta_M > 0$ so that if we add spontaneous deaths at rate $\delta < \delta_M$ then coexistence occurs.*

Similar conclusions hold in $d > 1$ with the obvious restrictions on M. The behavior in Theorem 5 contrasts sharply with

Theorem 6. *Consider the basic voter model plus spontaneous deaths at rate $\delta > 0$. In any dimension and for any initial configuration we have $P(\xi_t(x) = 1) \le \exp(-\delta t)$.*

The last conclusion follows immediately from the fact that the (additive) dual of the voter model plus death is coalescing random walk plus death. See Griffeath [7] Chapter I for the duality.

Theorems 5 and 6 suggest that the threshold voter model behaves more like the contact process than the voter model. The next result is more evidence for that opinion.

Theorem 7. *Consider the threshold voter model with spontaneous deaths at a positive rate. Suppose ξ_0 is translation invariant and has $P(\xi_0 \equiv 0) = 0$. Then as $t \to \infty$, $\xi_t \Rightarrow \xi_\infty^{1/2}$, the limit starting from product measure with density 1/2. In particular, all translation invariant stationary distributions are convex combinations of δ_0 and $\xi_\infty^{1/2}$.*

The last result holds in any dimension, for any neighborhood set, and for any value of δ, although in some cases $\xi_\infty^{1/2} = \delta_0$. Theorem 7 follows from a generalization of a result of Harris [8], which says roughly that if a process has a cancellative dual $\tilde{\xi}_t$ with the property that $P(0 < |\tilde{\xi}_t| < K) \to 0$ as $t \to \infty$ for any $K < \infty$ (and satisfies some mild technical conditions) then the conclusion of Theorem 7 holds. We will describe the concept of cancellative duality in Section 2 and sketch the proof of this result. The threshold voter model has a cancellative dual but we do not know how to show

Conjecture 2. *Consider the threshold voter model and suppose coexistence occurs. Suppose* ξ_0 *is translation invariant and has* $P(\xi_0 \equiv 0) = 0$ *and* $P(\xi_0 \equiv 1) = 0$, *then as* $t \to \infty$, $\xi_t \Rightarrow \zeta_\infty^{1/2}$, *the limit starting from product measure with density 1/2.*

If Conjecture 1 is true there is only exceptional case which is the subject of

Conjecture 3. *Consider the threshold voter model in* $d = 1$ *with* $\mathcal{N} = \{-1, 1\}$. *If we let* ξ_0^θ *have product measure with density* θ *then* $\xi_t^\theta \Rightarrow f(\theta)\,\delta_0 + (1 - f(\theta))\,\delta_1$.

A logical next step after the threshold-1 voter model is to look at

Example 4. The threshold-2 voter model. As the name may suggest, $a_0 = a_1 = 0$, and $a_k = 1$ for $k \geq 2$. A straightforward generalization of the proof of Theorem 3 gives

Theorem 8. *Consider the threshold-2 voter model in* $d = 1$ *with* $\mathcal{N} = \{x : 0 < |x| \leq M\}$. *If* $M \geq 47$, *then coexistence occurs.*

Proof: Let $I_k = \{kL, \ldots (k+1)L - 1\}$. We say that I_k is occupied at time n and set $\zeta_n(k) = 1$ if there are at least four 1's in I_k at time $n\delta$ and at least two of these 1's survive until time $(n+1)\delta$. By arguments in the proof of Theorem 3, it suffices to show that $\{(k, n) : \zeta_n(k) = 1\}$ dominates the set of wet sites in supercritical oriented site percolation. To make the comparison we first give a special construction of the threshold-2 voter model. For each $x \in \mathbb{Z}$ and $i = 0, 1$, let $\{T_n^{x,i}, n \geq 1\}$ be independent rate one Poisson processes. At time $T_n^{x,i}$ we flip from $1 - i$ to i if at least two neighbors are in state i. We introduce the explicit construction so that we can define the process starting from any state at any time and we can define an associated *independent flips process* in which the site x always flips from $1 - i$ to i at time $T_n^{x,i}$ We say that the site (n, k) is open if (a) the independent flips process starting from all sites 0 at time $(n-1)\delta$ has four 1's in I_k at time $n\delta$ and (b) at least two of these do not flip in the independent flips process before time $(n+1)\delta$. The last definition is motivated by the observation that if the threshold voter model has range $M = 2L - 1$ and $\zeta_{n-1}(k-1) = 1$ or $\zeta_{n-1}(k+1) = 1$ then the set of 1's in I_k dominates the independent flips process starting from all sites 0 at time $(n-1)\delta$ (a). Straightforward but somewhat tedious calculations show that if we pick $\delta = .4$ and $L = 24$ then sites are open with probability $> .819$, the upper bound on the critical value for oriented site percolation given in Durrett [6] and the result follows. ∎

From the other direction, it is clear that the model with $M = 1$ is stupid (only isolated 1's or 0's can flip). Computer simulations make it clear that clustering occurs when $M = 2, 3$ and there is coexistence for $M \geq 4$ but we do not know how to prove this. Turning now to d=2 we have

Conjecture 4. *Consider the threshold-2 voter model in d=2 with* $\mathcal{N} = \{x : \|x\|_1 = 1\}$. *If we let* ξ_0^θ *have product measure with density* $\theta < 1/2$ *then* $\xi_t^\theta \Rightarrow \delta_0$.

Of course the last conjecture implies that the limit is δ_1 when $\theta > 1/2$. Since the limit changes drastically as θ passes through 1/2 we say that the process ***bifurcates*** when the conclusion of Conjecture 4 is true. To see why Conjecture 4 should be true for small θ observe that if the set of 1's at time t is contained in a rectangle then this will be true for all time. The last property breaks down if we consider the larger "Moore neighborhood" and we believe the conclusion breaks down as well.

Conjecture 5. *Consider the threshold-2 voter model in d=2 with* $\mathcal{N} = \{x : \|x\|_\infty = 1\}$. *If we let* $\xi_0^{1/2}$ *have product measure with density 1/2 then* $\xi_t^{1/2}$ *converges to a nontrivial stationary distribution* $\xi_\infty^{1/2}$ *as* $t \to \infty$.

Again computer simulations support this conjecture and suggest that $\xi_t^\theta \Rightarrow \xi_\infty^{1/2}$ whenever $\theta \in (0, 1)$.

We close this lengthy "introduction" with the models that launched this investigation.

Example 5. Geometric voter models. In this case $a_k = (\theta^k - 1)/(\theta^n - 1)$ where $n = |\mathcal{N}|$ and $\theta > 1$. The denominator is chosen so that $a_n = 1$ and $a_k \to k/n$ as $\theta \to 1$, i.e., we recover the basic voter model in the limit. The last fact is true for many other formulas we could write down. The property that makes this one special is that if we add spontaneous flips at rate $1/(\theta^n - 1)$, i.e., consider flip rates $\hat{a}_k = \theta^k/(\theta^n - 1)$ then the result is a stochastic Ising model. See formula (2.4) on p.191 of Liggett [12] and take $J_R = \beta = (1/2)\log\theta$ when $R = \{x, y\}$ and y is a nearest neighbor of x, $J_R = 0$ otherwise. It is known that this stochastic Ising model is *nonergodic*, i.e., has more than one stationary distribution, when $\beta > (1/2)\log(1+\sqrt{2}) \approx .4406$. Since adding spontaneous flips should make the model more ergodic,

Conjecture 6. *Consider the geometric voter model with* $\theta > (1+\sqrt{2})$ *and spontaneous flips at rate* $\delta \geq 0$. *The process bifurcates when* $\delta = 0$ *and is nonergodic when* δ *is small.*

It seems likely that the last result is true for all $\theta > 1$. One of the difficulties in proving Conjecture 5 is that the nonergodicity conclusion is probably not true if we add spontaneous deaths instead. To explain why we believe this we look at

Example 6. The corners model. Suppose $d = 2$, $\mathcal{N} = \{x : \|x\|_1 = 1\}$, and our voters change their opinions at a rate = the number of diagonally adjacent pairs of neighbors (e.g. $x+(1,0)$ and $x+(0,-1)$) with the opposite opinion. In this case $a_0 = 0$, $a_1 = 0$, $a_3 = 2$, and $a_4 = 4$ but a_2 may be 0 or 1. The idea behind this rule is that if we have a finite cluster of 1's that is a solid blob (i.e., the set of 1's and the complimentary set of 0's are each connected) then the rate of change of the size of the cluster can be computed by looking at the contour that separates the 1's from the surrounding 0's. In particular, if the contour is oriented in a clockwise fashion, then the cluster decreases at a rate = the number of left turns and increases at a rate = the number of right turns. Since the number of left turns minus the number of right turns = 4 and a cluster of size k has a contour of length at least $4\sqrt{k}$ there is a small drift toward 0, which for large k will be overwhelmed by a one sided perturbation.

Remark. The geometric voter model also makes sense for $\theta < 1$. In Section 2 we will see that in this case the model has a cancellative dual. The reasoning used for the threshold voter model in the discussion after Theorem 7 applies to this case so we conjecture the conclusion of Theorem 7 holds as well.

2. Cancellative duality. In this section we will sketch the proof of Theorem 7 and discuss Conjectures 1 and 2. The key to these developments is the notion of "cancellative duality" which we treat following Chapter III of Griffeath [7]. Readers familiar with this material or who prefer the approach in Section III of Chapter 4 of Liggett [12] can skip to formula (3) now. We begin by defining a graphical representation, use it to define a process, and then reverse time to get a dual process. For $x \in \mathbb{Z}^d$, $S \subset \mathbb{Z}^d$, with $S \in \mathcal{S}$ let $\{T_n^{x,S}, n \geq 1\}$ be independent Poisson processes with rate $\beta(S)$. At time $T_n^{x,S}$ draw arrows from $x+y$ to x for each $y \in S - \{0\}$. If $0 \notin S$ we write a δ at x. We say that (y,t) can be reached from $(x,0)$ and write $(x,0) \to (y,t)$ if there is a path from $(x,0)$ to (y,t), that is, a pair of sequences $x_0 = x, \ldots, x_n = y$ and $s_0 = 0 < s_1 < \ldots s_n < s_{n+1} = t$ so that

(i) for $1 \leq m \leq n$ there is an arrow from x_{m-1} to x_m at time s_m
(ii) for $1 \leq m \leq n+1$ there are no δ's in (s_{m-1}, s_m)

In words, $(x,0) \to (y,t)$ if there is a path from $(x,0)$ to (y,t) that goes up but not through δ's and across arrows in the direction of their orientation. Since the intervals in (ii) are open we are allowed to touch δ's at the endpoints of vertical segments in the path.

To construct our process from its graphical representation we let

$$N_t^x(y) = \text{ the number of paths from } (x,0) \text{ to } (y,t)$$
$$N_t^A(y) = \sum_{x \in A} N_t^x(y)$$
$$\eta_t^A(y) = (-1)^{N_t^A(y)}$$
$$\zeta_t^A = \{y : \eta_t^A(y) = 1\}$$

In words, $\eta_t^A(y) = -1$ if there is an odd number of paths from $A \times \{0\}$ to (y,t). Since 0 is an even number we interpret -1 as occupied and 1 as vacant.

Remark. The open intervals and the strict inequality $s_n < s_{n+1}$ in the definition cause something slightly unexpected. The processes we just constructed is left continuous since a δ or arrow occuring exactly at time t will have no effect. To bring ourselves back into line with the theory of Markov processes we modify the process to make it right continuous.

To see what sort of process we have constructed suppose there is just one set S and $\beta(S) = 1$. In this case the flip rate at 0 when the configuration is η is

$$c_S(0,\eta) = \tfrac{1}{2}\left(1 - \prod_{y \in S\Delta\{0\}} \eta(y)\right)$$

where $S\Delta\{0\}$ is the symmetric difference of S and $\{0\}$, i.e., $S - \{0\}$ if $0 \in S$ and $S \cup 0$ if $0 \notin S$. To check this we consider several cases:

(a) $0 \in S$. The state changes if there are an odd number of -1's in $S - \{0\}$.
(b) $0 \notin S$, $\eta(0) = 1$. The δ at 0 has no effect. There will be a birth at 0 if and only if there are an odd number of particles in S which in this case is the same as having an odd number of particles in $S \cup 0$.
(c) $0 \notin S$, $\eta(0) = -1$. In this case the δ will kill the particle at 0 and it will stay dead if and only if there are an even number of -1's in $S - \{0\}$, i.e., an odd number of -1's in $S \cup 0 = S\Delta 0$.

When we superimpose independent Poisson processes, the rates at 0 add so the flip rate at 0 for the process we constructed are $\sum_S \beta(S)c_S(0,\eta)$ Our next step is to consider several examples.

Example 1. The basic voter model. Suppose for simplicity $a_k = k$. In this case we let $\beta(\{x\}) = 1$ for each $x \in \mathcal{N}$. Above we argued that a flip occurs if and only if there are an odd number of -1's in $S\Delta\{0\}$, which is $\{x, 0\}$ in this case, so flips occur at 0 if and only if $\eta(x) \neq \eta(0)$ and it is easy to see that the basic voter model results.

To get a model in which the birth and death rates just depend on the number of occupied neighbors it is natural to let $\beta(S)$ depend only on $|S|$ and $|S \cap \{0\}|$. It is easy to check that if $|\mathcal{N}| = n$ and we let $\beta(S) = 1$ for all $S \subset \mathcal{N} \cup \{0\}$ with $0 \in S$ and $|S| = m$ then we get a process with

$$b_j = d_j = \sum_{\text{odd } i \le j} \binom{j}{i}\binom{n-j}{m-1-i}$$

while if let $\beta(S) = 1$ for all $S \subset \mathcal{N} \cup \{0\}$ with $0 \notin S$ and $|S| = m$ then we get a process with

$$b_j = \sum_{\text{odd } i \le j} \binom{j}{i}\binom{n-j}{m-i} \qquad d_j = \sum_{\text{even } i \le j} \binom{j}{i}\binom{n-j}{m-i}$$

To get a voter model we want $a_j = b_j = d_{n-j}$, that is, we want $|S|$ to be odd in either case. Writing $k = |S - \{0\}|$, it follows that if

$$M(k,j) = \sum_{\text{odd } i \le j} \binom{j}{i}\binom{n-j}{k-i}$$

then any a_j of the form

$$a_j = \sum_{k=1}^{n} \alpha_k M(k,j)$$

with $\alpha_k \ge 0$ determines a nonlinear voter model with a cancellative construction.

Examples 2 and 5. Threshold and Geometric Voter Models. Let $a_j = 1 - \theta^j$ with $0 \le \theta < 1$. When $\theta = 0$ this is the threshold voter model. ($0^0 = 1$.) When $\theta > 0$ this is the version of the geometric voter model mentioned at the end of the introduction. Taking $\alpha_k = (1-\theta)^k(1+\theta)^{n-k}2^{1-n}$ shows that these examples have a cancellative construction.

$$\begin{aligned}\sum_{k=1}^{n} \alpha_k M(k,j) &= \frac{(1+\theta)^n}{2^{n-1}} \sum_{\text{odd } i \le j} \binom{j}{i}\left[\frac{1-\theta}{1+\theta}\right]^i \sum_{k=1}^{n} \binom{n-j}{k-i}\left[\frac{1-\theta}{1+\theta}\right]^{k-i} \\ &= \frac{(1+\theta)^n}{2^{n-1}} \sum_{\text{odd } i \le j} \binom{j}{i}\left[\frac{1-\theta}{1+\theta}\right]^i \left[1 + \frac{1-\theta}{1+\theta}\right]^{n-j} \\ &= 2^{1-j} \sum_{\text{odd } i \le j} \binom{j}{i}(1-\theta)^i(1+\theta)^{j-i} \\ &= 2^{-j}\left([(1+\theta)+(1-\theta)]^j - [(1+\theta)-(1-\theta)]^j\right) \\ &= 1 - \theta^j\end{aligned}$$

The authors would like to thank *Mathematica* for helping come up with this guess.

The interest in defining a process on a graphical representation is that it allows us to define a dual process by reversing time. We say there is a dual path from (y,t) to $(x,0)$ and write $(y,t) \xrightarrow{*} (x,0)$ if $(x,0)$ can be reached by a path that goes down but not through δ's and across arrows in a direction *opposite* their orientation. As before we are allowed to touch δ's at the endpoints of vertical segments of the path. To construct the dual processes let

$$\begin{aligned}\hat{N}_s^{(y,t)}(x) &= \text{ the number of dual paths from } (y,t) \text{ to } (x,t-s) \\ \hat{N}_s^{(B,t)}(x) &= \sum_{y \in B} \hat{N}_s^{(y,t)}(y) \\ \hat{\eta}_s^{(B,t)}(x) &= (-1)^{\hat{N}_s^{(B,t)}(x)} \\ \hat{\zeta}_s^{(B,t)} &= \{y : \eta_s^{(B,t)}(y) = -1\}\end{aligned}$$

It follows immediately from the definitions of path and dual path that

$$N_t^x(y) = \hat{N}_t^{(y,t)}(x)$$

and summing over $x \in A$, $y \in B$ gives

$$\sum_{y \in B} N_t^A(y) = \sum_{x \in A} \hat{N}_t^{(B,t)}(x)$$

If B is finite the right hand side of the last equation is, and reducing things mod 2 gives

$$\sum_{y \in B} (N_t^A(y) \bmod 2) = \sum_{x \in A} (\hat{N}_t^{(B,t)}(x) \bmod 2) \quad \bmod 2$$

that is

$$(1) \qquad \{\zeta_t^A \cap B \text{ is odd }\} = \{\hat{\zeta}_t^{(B,t)} \cap A \text{ is odd }\}$$

The dual process $\hat{\zeta}_s^{(B,t)}$ is only defined for $s \in [0,t]$. Because of this it is convenient to define a version of the dual process $\tilde{\zeta}_s^B$ that lives on its own graphical representation. For $x \in \mathbb{Z}^d$, $S \subset \mathbb{Z}^d$, with $S \in \mathcal{S}$ let $\{\tilde{T}_n^{x,S}, n \geq 1\}$ be independent Poisson processes with rate $\beta(S)$. At time $\tilde{T}_n^{x,S}$ draw arrows from x to $x+y$ for each $y \in S - \{0\}$. If $0 \notin S$ we write a δ at x. We define paths as we did before and let

$$\tilde{N}_t^x(y) = \text{ the number of paths from } (x,0) \text{ to } (y,t)$$

$$\tilde{N}_t^A(y) = \sum_{x \in A} \tilde{N}_t^x(y)$$

$$\tilde{\eta}_t^A(y) = (-1)^{\tilde{N}_t^A(y)}$$

$$\tilde{\zeta}_t^A = \{y : \tilde{\eta}_t^A(y) = -1\}$$

A little thought reveals that the dual is an *annihilating branching process.* That is, (i) each $x \in \tilde{\zeta}_t^A$ branches into $x + S$ at rate $\beta(S)$, and (ii) if an offspring is sent to an occupied site annihilation occurs and an empty site results. It should be clear from the definition that

$$\{\tilde{\zeta}_s^B, 0 \leq s \leq t\} \stackrel{d}{=} \{\hat{\zeta}_s^{(B,t)}, 0 \leq s \leq t\}$$

so

$$(2) \qquad P(\zeta_t^A \cap B \text{ is odd }) = P(\tilde{\zeta}_t^B \cap A \text{ is odd })$$

To see the usefulness of the last equation, let $A \stackrel{d}{=} \nu_{1/2}$, i.e., start with product measure with density 1/2. In this case, (2) says

$$(3) \qquad P(\zeta_t^{1/2} \cap B \text{ is odd }) = P(\zeta_0^{1/2} \cap \tilde{\zeta}_t^B \text{ is odd }) = \tfrac{1}{2} P(\tilde{\zeta}_t^B \neq \emptyset)$$

Since $\emptyset$ is an absorbing state for the dual, the right hand side decreases to a limit as $t \to \infty$. The finite dimensional distributions that appear in (3) are enough to determine the distribution of $\zeta_t^{1/2}$, (see Griffeath [7], p. 69) so (3) implies $\zeta_t^{1/2}$ converges weakly to a limit $\zeta_\infty^{1/2}$, which must be a stationary distribution.

Turning to the special case of nonlinear voter models with $a_0 = 0$ we see that if $A = \mathbb{Z}^d$ then $\zeta_t^A = \mathbb{Z}^d$ so

$$P(\tilde{\zeta}_t^B \text{ is odd }) = \begin{cases} 1 & \text{if } |B| \text{ is odd} \\ 0 & \text{if } |B| \text{ is odd} \end{cases} .$$

In words, all transitions preserve the parity of the dual. From this it follows immediately that $P(|\tilde{\zeta}_t^{\{0\}}| \text{ is odd }) = 1$ so (3) implies $P(0 \in \zeta_t^{1/2}) = 1/2$. The last fact does not keep $\zeta_\infty^{1/2}$ from being trivial. In $d \leq 2$ the voter model has $\zeta_\infty^{1/2} = (1/2)\delta_0 + (1/2)\delta_1$. To test to see if this occurs we look at $\tilde{\zeta}_t^{\{x,y\}}$. If $P(\tilde{\zeta}_t^{\{x,y\}} \neq \emptyset \text{ for all } t) > 0$ for some $x \neq y$ we say that *the dual survives.* It is clear from (3) that $\zeta_\infty^{1/2}$ is nontrivial if and only if the dual survives.

Intuitively, if the dual survives then it must grow, i.e.,

$$P(0 < |\tilde{\zeta}_t^{\{x,y\}}| \leq K) \to 0 \quad \text{as } t \to \infty \text{ for each } K < \infty,$$

but we have not been able to prove this even for the threshold voter model. The next proof should explain our interest in this conclusion.

Sketch of the proof of Theorem 7. The threshold voter model plus deaths at rate δ can be constructed by taking the recipe in Example 2 and adding Poisson processes $\{T_n^{x,\emptyset} : n \geq 1\}$ at rate δ. The dual process has spontaneous deaths at rate $\delta > 0$ so it is easy to see that for all $B \subseteq \mathbb{Z}^d$ and $K < \infty$

$$(4) \qquad P(0 < |\tilde{\zeta}_t^B| \leq K) \to 0 \quad \text{as } t \to \infty$$

The proof of Lemma (9.14) in Harris [8] generalizes easily to show that if ζ_0 is translation invariant and has no mass on $\emptyset$ and $\epsilon > 0$ then there is $K(\epsilon)$ so that if $|D| \geq K(\epsilon)$ then $P(\zeta_1 \cap D \neq \emptyset) \geq 1 - \epsilon$. From this it follows immediately that if $|D| \geq L \cdot K(\epsilon/L)$ then $P(|\zeta_1 \cap D| \geq L) \geq 1 - \epsilon$. Combining the last observation with (4) and using the fact that spontaneous deaths occur at a positive rate it is not hard to conclude that

$$(5) \qquad P(\zeta_2 \cap \tilde{\zeta}_t^B \text{ is odd }) \approx \tfrac{1}{2} P(\tilde{\zeta}_t^B \neq \emptyset)$$

(See Bramson, Ding, and Durrett [2] for details.) An easy generalization of (3) shows that the left hand side of (5) is $P(\zeta_{t+2} \cap B \text{ is odd})$. The right hand side is $P(\zeta_t^{1/2} \cap B \text{ is odd})$, so the desired result follows. ■

Our last topic is the threshold voter model in d=1 with $\mathcal{N} = \{-1, 1\}$.

Proof of Theorem 4. By calculations above, this process can be constructed from a graphical representation in which $S = \{1\}, \{-1\}$, and $\{-1, 0, 1\}$ each have rate 1/2. The system thus has a dual process that Sudbury[15] calls *double branching annihilating random walk*, or DBAP for short. In the dual, a particle at x jumps to $x+1$ or $x-1$ at rate 1/2 each, and at rate 1/2 gives birth to two new particles that are sent to $x+1$ and $x-1$. If either transition results in a particle landing on an occupied site, annihilation occurs and an empty site results. The competing forces of birth and annihilation might suggest that a nontrivial equilibrium is possible but this is false. To prove this we use Sudbury's observation that the DBAP is the boundary of the *swapping voter model.* Let $\eta : 1/2 + \mathbb{Z} \to \{0, 1\}$ be a system in which for each $z \in 1/2 + \mathbb{Z}$ the following transitions each occur at rate 1/2:

(i) set $\eta(z) = \eta(z+1)$
(i) set $\eta(z+1) = \eta(z)$
(iii) swap the values at z and $z+1$.

The first two transitions are those of a voter model, the third is that of *stirring* or *the simple exclusion process.* We claim that $\eta_t^* = \{x : \eta_t(x - 1/2) \neq \eta_t(x + 1/2)\}$ is a DBAP. To see this, check that if $z = x - 1/2$ then the induced transitions in η_t^* are

(i) jump from x to $x-1$
(ii) jump from x to $x+1$
(iii) branch to $x-1$ and $x+1$

Let ξ_t denote the threshold voter model. The duality equation implies that

$$(6) \quad P(\zeta_t \cap \{x, y\} \text{ is odd }) \leq P(\tilde{\zeta}_t^{\{x,y\}} \neq \emptyset) = P(\eta_t^{[x+1/2, y-1/2]} \not\equiv 0),$$

where $\eta_t^{[a,b]}$ is the swapping voter model starting with 1's on the interval [a,b] and 0's otherwise. To prove that the last quantity goes to zero we observe that in the swapping voter model, as in the basic voter model, the number of sites occupied by 1's is a time change of simple random walk and hence eventually becomes 0.

Discussion of Conjecture 2. Having proved that clustering occurs in the one dimensional nearest neighbor case, it is natural to ask if the system converges to a convex combination of δ_0 and δ_1 when we start from product measure with density θ. In view of Theorem 3, it suffices to show that the density of 1's has a limit. The duality equation (3) implies

$$P(x \in \zeta_t^\theta) = P(\tilde{\zeta}_t^{\{x\}} \cap \zeta_0^\theta \text{ is odd })$$

In the basic voter model $\tilde{\zeta}_t^{\{x\}}$ is always a singleton so the right-hand side is θ. In the threshold voter model the dual can grow so the density of 1's is not conserved. We believe that if configurations that differ by a translation are identified in the dual chain then it is positive recurrent but we do not know how to prove this.

Acknowledgement. The authors would like to thank Claudia Neuhauser for several useful conversations concerning the material presented here.

REFERENCES

[1] C. Bezuidenhout and G. Grimmett *The critical contact process dies out,* Ann. Prob., to appear

[2] M. Bramson, W.D. Ding, and R. Durrett *Annihilating branching processes*, Stoch. Processes Appl., to appear

[3] R.C. Brower, M.A. Furman, and M. Moshe *Critical exponents for Reggeon quantum spin model*, Physics Letters **76B** (1978), 213–219

[4] R. Dickman and M.A. Burschka *Nonequilibrium critical poisoning in a single-species model*, Physics Letters A, **127** (1988), 132–137

[5] R. Durrett *Lecture Notes on Particle Systems and Percolation* Wadsworth, Pacific Grove, CA, 1988

[6] R. Durrett *Stochastic growth models: bounds on critical values*, preprint

[7] D. Griffeath *Additive and Cancellative Interacting Particle Systems*, Springer Lecture Notes in Math, Vol. 724

[8] T.E. Harris *On a class of set-valued Markov processes*, Ann. Prob. **4** (1976), 175–194

[9] R. Holley and T.M. Liggett *Ergodic theorems for weakly interacting systems and the voter model*, Ann. Prob. **3** (1975), 643–663

[10] R. Holley and T.M. Liggett *The survival of contact processes*, Ann. Probab., **6** (1978), 198–206

[11] R. Holley and T.M. Liggett *Generalized potlatch and smoothing processes*, Z. Warsch. verw. Gebiete **55** (1981), 165–195

[12] T.M. Liggett *Interacting Particle Systems*, Springer, New York, 1985

[13] T.M. Liggett *The periodic threshold contact process*, this volume

[14] F. Spitzer *Interaction of Markov Processes*, Adv. in Math. **5** (1970), 246-290

[15] A. Sudbury *The branching annihilating process: an interacting particle system*, Ann. Prob, to appear

J. T. Cox
Department of Mathematics
Syracuse University
Syracuse, NY 13210

R. Durrett
Department of Mathematics
Cornell University
Ithaca, NY 14853

On the Long Term Behavior of Finite Particle Systems: A Critical Dimension Example

J.T. Cox[1] and Andreas Greven[2]

We show that the two-dimensional voter model provides an example of a suitably modified version of the large finite systems scheme of [6] and [9].

1. Introduction. The goal of this paper is to present a critical dimension example of the large finite systems scheme developed in [6] and extended in [9]. This scheme compares the long term behavior of large finite particle systems with corresponding infinite particle systems in the case where the finite and infinite systems have *different* ergodic behavior (this is often the case in high dimensions). However, for many models, it is the case that in low dimensions, finite and infinite systems apparently have the same ergodic behavior. Our purpose here is to show that even in such cases suitable modifications of the scheme can hold. We will illustrate this by considering the voter model in two dimensions. In this first section we will review the finite systems scheme, summarize facts about the voter model and its ergodic theory, review results on clustering of the two dimensional voter model, and then present our result (Theorem 1 below).

The finite systems scheme

Let η_t be a Markov process with state space $\Xi = \{0,1\}^{\mathbf{Z}^d}$. We think of η_t as a *particle system;* $\eta_t(x) = 1$ means that site x is occupied by a particle at time t, $\eta_t(x) = 0$ means that site x is vacant at time t. We refer to η_t as the *infinite system* since it has infinitely many components, or sites. Consider now *finite systems* η_t^N. Let $\Lambda(N) = [-N,N]^d \cap \mathbf{Z}^d$, and let η_t^N be a Markov process with state space $\Xi^N = \{0,1\}^{\Lambda(N)}$ obtained by restricting in some natural way the dynamics of η_t to $\Lambda(N)$. η_t and η_t^N are

---

[1] Work supported in part by the National Science Foundation, and by the SFB 123, Universität Heidelberg.
[2] Work supported in part by the SFB 123, Universität Heidelberg

also connected through their initial states by assuming that if the initial distribution of η_t is μ, then the initial distribution of η_t^N is the restriction of μ to Ξ^N for $N = 1, 2, \ldots$. Let $\mathcal{M}$ be the set of translation invariant measures on Ξ and for $\theta \in I$ let $\mathcal{M}_e(\theta)$ be the set of shift ergodic μ in $\mathcal{M}$ with density θ (i.e. $E_\mu(\eta(x)) \equiv \theta$). Let $\mathcal{L}$ denote probability law, let $\Rightarrow$ denote weak convergence, and let δ_θ be the unit point mass at θ. We will write $f(N) \ll g(N)$ if $f(N)/g(N) \to 0$ as $N \to \infty$.

The finite systems scheme requires the following four ingredients:

- A family $\{\nu_\theta, \theta \in I\}$ of extremal invariant measures for η_t.
- Functions $\hat{\theta}^N : \Xi^N \to I$, $N = 1, 2, \ldots$ (consistent estimators of θ).
- A time scale $\beta(N)$, $\beta(N) \to \infty$ as $N \to \infty$ such that if the initial distribution is $\mu \in \mathcal{M}_e(\theta)$, then as $N \to \infty$,

$$\mathcal{L}(\hat{\theta}^N(\eta^N_{t(N)})) \Rightarrow \delta_\theta \quad \text{if} \quad t(N) \ll \beta(N)$$

and

$$\mathcal{L}(\hat{\theta}^N(\eta^N_{t(N)})) \not\Rightarrow \delta_\theta \quad \text{if} \quad t(N) = \beta(N).$$

- A Markov process Z_s on I with probability transition semigroup Q_s.

The large finite systems scheme with these ingredients is realized if whenever

$$T(N) \to \infty \quad \text{and} \quad T(N)/\beta(N) \to s \in [0, \infty] \quad \text{as} \quad N \to \infty \tag{1.1}$$

then the following two statements hold:

$$\mathcal{L}(\hat{\theta}^N(\eta^N_{T(N)})) \Rightarrow \mathcal{L}(Z_s) \tag{1.2}$$

and

$$\mathcal{L}(\eta^N_{T(N)}) \Rightarrow \int_I Q_s(\theta, d\theta')\nu_{\theta'}. \tag{1.3}$$

Letting $\nu_\theta(s)$ denote the right-hand side of (1.3), the convergence in (1.3) means that for any finite $A \subset \mathbf{Z}^d$ and $\xi \in \Xi$,

$$\lim_{t \to \infty} P\left(\eta^N_{T(N)}(x) = \xi(x) \text{ for all } x \in A\right) = \nu_\theta(s)\big(\eta(x) = \xi(x) \text{ for all } x \in A\big).$$

Roughly speaking, the scheme presents the following picture. (1.2) says that $\beta(N)$ is the scale on which $\hat{\theta}^N(\eta_t^N)$ fluctuates smoothly. (1.3) is an integrated version of the statement that conditional on $\hat{\theta}^N(\eta^N_{T(N)}) = \theta'$, the distribution of $\eta^N_{T(N)}$ is locally given by $\nu_{\theta'}$. Thus, for large t and N, the distribution of η_t^N is the invariant measure $\nu_{\theta'}$ with density $\theta' = \hat{\theta}(\eta_t^N)$. That is, the evolution of the components is *slaved* to the evolution of the density, which only changes on the large time scale $\beta(N)$. (This type of behavior is known for systems of differential equations with fast and slow variables.) Taking $s = 0$, we see that for $t \ll \beta(N)$

$$\mathcal{L}(\eta_t^N) \approx \mathcal{L}(\eta_t) \approx \nu_\theta.$$

Thus $\beta(N)$ determines the scale on which we can see a difference in the local behavior of the infinite and finite systems. See [6] and [9] for more details, where several specific models are treated.

If the set $\{\nu_\theta, \theta \in I\}$ consists only of point masses on traps, then the scheme degenerates and gives little or no information. This is the case for the voter model in dimensions one and two. However, we will show that a modified version of the scheme still holds in dimension two (the critical case), and gives useful information concerning the differences in the finite and infinite systems. We begin by reviewing the voter model.

The voter model and its ergodic theory

The voter model η_t is the Markov process on Ξ determined by:

$$\eta_t(x) \to 1 - \eta_t(x) \quad \text{at rate} \quad (2d)^{-1}\#\{y : y \sim x,\ \eta_t(y) \neq \eta_t(x)\},$$

where $y \sim x$ means $|y - x| = 1$. [14] is an excellent source for a complete technical description of the voter model. To summarize the basic ergodic theory of the voter model let $\mathcal{I}$ denote the set of invariant measures for η_t, let $\mathcal{I}_e$ the set of extremal elements of $\mathcal{I}$, and let $I = [0, 1]$. Let $\mathbf{0}$ denote the element of Ξ which is identically 0, and let $\mathbf{1}$ denote the element of Ξ which is identically 1.

Theorem 0. *If $d \leq 2$ then $\mathcal{I}_e = \{\delta_\mathbf{0}, \delta_\mathbf{1}\}$. If $d \geq 3$ then $\mathcal{I}_e = \{\nu_\theta, \theta \in I\}$, where $\nu_\theta \in \mathcal{M}_e(\theta)$. Furthermore, if the initial distribution of η_t belongs to $\mathcal{M}_e(\theta)$,* then as $t \to \infty$,

$$\mathcal{L}(\eta_t) \Rightarrow \begin{matrix} (1-\theta)\delta_\mathbf{0} + \theta\delta_\mathbf{1}, & d \leq 2, \\ \nu_\theta, & d \geq 3. \end{matrix} \tag{1.4}$$

Thus, the voter model clusters in low dimensions and is stable in high dimensions.

In [4] and [6] it is shown that the $d \geq 3$ voter model fits into the finite systems scheme, with

$$\begin{aligned} \hat{\theta}^N(\eta) &= \frac{1}{\#\Lambda(N)} \sum_{x \in \Lambda(N)} \eta(x), \\ \beta(N) &= (2N+1)^d \\ Z_s &= \text{ the diffusion on } I \text{ with generator } \frac{1}{G}u(1-u)\frac{d^2}{du^2}. \end{aligned} \tag{1.5}$$

(G is the expected number of visits to the origin by a simple random walk starting at the origin.) We consider now the low dimensional setting, $d \leq 2$.

For the $d = 1$ voter model there is an invariance principle in [1] (see also [3] and Section V.3 of [14]) which provides a very precise picture of the clustering exhibited in (1.4). Roughly speaking, space should be scaled by $t^{1/2}$ at time t to obtain a nondegenerate limit. This invariance principle can be used to understand the behavior of large finite systems (see Section 7 of [4] for work along these lines); the situation is rather different from the general finite systems scheme outlined above. For the $d = 2$ voter model there is no invariance principle, since correlations and clusters exist on multiple scales. Thus $d = 2$ is a *critical dimension* for the voter model. We will review in more detail some features of the $d = 2$ model, and then show that the general framework of the finite systems scheme, appropriately modified, still holds.

Clustering in the critical dimension

Assume now that $d = 2$. Define a family of exchangeable measures $\{\nu_\theta^\alpha,\ \theta \in I, 0 < \alpha \leq 1\}$ on Ξ by the formula

$$\nu_\theta^\alpha\big(\eta(x) = 1 \text{ for all } x \in A\big) = \int_I P_{\hat\alpha}(\theta, d\rho)\, \rho^{\#A},$$

where $\hat\alpha = \log(1/\alpha)$ and P_s is the probability transition semigroup of the Wright–Fisher diffusion Y_s on I with generator

$$\frac{1}{2}u(1-u)\frac{d^2}{du^2} \tag{1.6}$$

We now spread the sites apart in order to avoid the clustering of (1.4) and to see clusters of smaller size than those of width comparable to N. For a fixed $\alpha \in (0, 1]$ define ζ_t, the $t^{\alpha/2}$ thinning of η_t, by

$$\zeta_t(x) = \eta_t\big([xt^{\alpha/2}]\big), x \in \mathbf{Z}^2.$$

Theorem 4 of [8] asserts that if the initial distribution of η_t is product measure with density θ, then

$$\mathcal{L}(\zeta_t) \Rightarrow \nu_\theta^\alpha \quad \text{as} \quad t \to \infty. \tag{1.7}$$

This shows that correlations exist on multiple scales. Furthermore, (1.7) is more like the $d \geq 3$ case of (1.4), and it is the finite systems version of this statement that we consider below.

Main result

First define the finite systems. Let η_t^N be the Markov process on $\Xi^N = \{0,1\}^{\Lambda(N)}$ which makes transitions

$$\eta_t^N(x) \to 1 - \eta_t^N(x) \text{ at rate } (2d)^{-1} \#\{y : y \overset{N}{\sim} x,\ \eta_t^N(y) \neq \eta_t^N(x)\},$$

where $y \overset{N}{\sim} x$ means that $|(y-x) \bmod (2N+1)| = 1$, $x, y \in \Lambda(N)$. It is the case that (for $d \leq 2$) if the initial distribution of η_t^N is the restriction to Ξ^N of any $\mu \in \mathcal{M}_e(\theta)$, then for any $T(N) \to \infty$,

$$\mathcal{L}(\eta_{T(N)}^N) \Rightarrow (1-\theta)\delta_0 + \theta\delta_1. \tag{1.8}$$

This seems to indicate that there is little difference between the evolution of the finite systems and the infinite systems. But we can obtain more precise information by considering a spread out version of η_t^N. Fix $\alpha \in (0,1]$ and define the thinned finite systems ζ_t^N by

$$\zeta_t^N(x) = \eta_t^N([xN^\alpha]).$$

Let

$$\beta(N) = (2N+1)^2 \log(2N+1),$$

let $T(N)$ satisfy (1.1), and let $\hat{\theta}^N$, Z_s and Q_s be as in (1.5), but with $G = 2/\pi$. Then the $d = 2$ case of Theorem 2 of [4] states that if the initial distribution of η_t^N is the restriction to Ξ^N of product measure with density θ, then

$$\mathcal{L}(\hat{\theta}^N(\eta_{T(N)}^N)) \Rightarrow \mathcal{L}(Z_s) \text{ as } N \to \infty,$$

where $Z_0 = \theta$. Here is our main result.

Theorem 1. *Under the above assumptions,*

$$\mathcal{L}(\zeta_{T(N)}^N) \Rightarrow \int_I Q_s(\theta, d\theta')\nu_{\theta'}^\alpha \text{ as } N \to \infty.$$

This shows that $\beta(N)$ determines a time scale on which we can see the difference in correlations exhibited by the finite and infinite systems. We expect that by using the techniques of Section 6 of [6] it should be possible to show that the conclusion of Theorem 1 holds whenever the initial distributions of the η_t^N are obtained by restriction to Ξ^N of any fixed $\mu \in \mathcal{M}_e(\theta)$.

In the next section we prove Theorem 1 using results from [4] and [8]. In Section 3 we state a hitting time result for random walk on $\Lambda(N)$ which follows from the work in Section 2. In Section 4 we make some remarks concerning a few other models.

2. Proof of Theorem 1. Our chief tool is duality. The voter model η_t is dual to a system of coalescing random walks, and the Wright–Fisher diffusion is dual to a death process. We begin by describing the duality between the voter model and coalescing random walk. Coalescing random walk ξ_t^A on $\mathbf{Z}^2$ is the process obtained by placing a particle at each point of A and allowing the particles to perform independent random walks, requiring that particles coalesce when they attempt to occupy the same site. Let $\xi_t^{N,A}$ denote the corresponding system on $\Lambda(N)$. (By random walk we always mean continuous time, rate one, simple symmetric random walk on $\mathbf{Z}^2$ or $\Lambda(N)$). If the initial distribution of η_t is product measure with density θ, and the initial distribution of η_t^N is the restriction to Ξ^N of product measure with density θ, then the duality equations are (see [10] or [14]):

$$P\big(\eta_t(x) = 1 \text{ for all } x \in A\big) = E\theta^{\#\xi_t^A} \tag{2.1}$$

and

$$P\big(\eta_t^N(x) = 1 \text{ for all } x \in A\big) = E\theta^{\#\xi_t^{N,A}} \tag{2.2}$$

The dual of the Wright–Fisher diffusion Y_t with generator given in (1.6) is the pure death process D_t on $\mathbf{N} = \{1, 2, \dots\}$ which makes transitions

$$n \to n-1 \quad \text{at rate} \quad \binom{n}{2}.$$

The processes Y_t and D_t are dual in the sense that:

$$E_\theta Y_t^n = E_n \theta^{D_t} \quad \text{for} \quad \theta \in I, \quad n \in \mathbf{N}. \tag{2.3}$$

The subscript θ denotes the initial point of Y_t, the subscript n denotes the initial point of D_t. This equation is discussed in [8] and [15]; it follows easily from a generator calculation.

Our first step is to rephrase the conclusion of Theorem 1 using duality. Recall that Q_s is the semigroup for the generator given in (1.5), with $G = 2/\pi$, and P_s is the semigroup for the generator given in (1.6), so $Q_s = P_{\pi s}$. For finite $A \subset \mathbf{Z}^2$, with $n = \#A$,

$$\begin{aligned}\int_I Q_s(\theta, d\theta')\nu_{\theta'}^{\alpha}\big(\eta(x) = 1 \text{ for all } x \in A\big) &= \int_I P_{\pi s}(\theta, d\theta') \int_I P_{\hat\alpha}(\theta', d\rho)\rho^n \\ &= \int_I P_{\pi s + \hat\alpha}(\theta, d\rho)\rho^n \\ &= E_\theta\big(Y_{\pi s + \hat\alpha}\big)^n,\end{aligned}$$

where the subscript θ denotes the initial point of Y_s. Thus, by the duality equations (2.1) and (2.2), in order to prove Theorem 1 it is sufficient to prove that for distinct $x_1, \dots, x_n$, with $A(N) = \{[x_1 N^\alpha], \dots, [x_n N^\alpha]\}$,

$$E\theta^{\#\xi_{T(N)}^{N,A(N)}} \to \sum_{k=1}^{n} \theta^k q_{\pi s + \hat\alpha}(n, k) \tag{2.4}$$

where $q_t(n,k) = P_n(D_t = k)$. Equivalently, our problem reduces to proving

$$(2.5) \qquad P\big(\#\xi^N_{T(N)}(A(N)) = k\big) \to q_{\pi s+\hat{\alpha}}(n,k).$$

We need two results from [4] and [8]. The first result is Theorem 3 of [8], which states:

(A) For fixed n, uniformly in $B = \{y_1, \dots, y_n\}$ such that $t^{\alpha/2}/\log t \le |y_i - y_j| \le t^{\alpha/2}\log t$ for $i \ne j$,

$$P\big(\#\xi_t(B) = k\big) \to q_{\hat{\alpha}}(n,k) \quad \text{as} \quad t \to \infty.$$

The second result we need is Theorem 5 of [4].

(B) If $N/\sqrt{\log N} \ll a_N \ll N$, and $T(N)/N^2 \log N \to s$, then uniformly in $B = \{y_1, \dots, y_n\} \subset \Lambda(N)$ such that $|y_i - y_j| \ge a_N$ for $i \ne j$,

$$P\big(\#\xi^{N,B}_{T(N)} = k\big) \to q_{\pi s}(n,k) \quad \text{as} \quad N \to \infty.$$

Proof of (2.5). Assume $0 < \alpha < 1$; the $\alpha = 1$ case can be handled separately. Choose a_N, b_N, and $t(N)$ such that

$$\frac{N^2}{\log N} \ll a_N^2 \ll t(N) \ll b_N^2 \ll N^2.$$

We will prove that for $k \le j \le n$

$$(2.6) \qquad P\big(\#\xi^{N,A(N)}_{t(N)} = j,\ \#\xi^{N,A(N)}_{T(N)} = k\big) \to q_{\hat{\alpha}}(n,j)q_{\pi s}(j,k)$$

as $N \to \infty$, which is enough to prove (2.5). The idea behind (2.6) is the following. Up until time $t(N)$ the finite and infinite systems ξ_t and ξ^N_t more or less agree, so (A) implies that by time $t(N)$ there are j particles left with probability $q_{\hat{\alpha}}(n,j)$. At time $t(N)$, with overwhelming probability, the remaining j particles are spaced at least distance a_N apart. Now (B) implies that the probability that these j particles coalesce to k particles in time $T(N) - t(N)$ is $q_{\pi s}(j,k)$. Here are the details.

We construct the systems ξ_t and ξ^N_t in terms of a common set of random walks. Let $\{X_t(x), t \ge 0, x \in \mathbf{Z}^2\}$ be a collection of independent random walks on $\mathbf{Z}^2$, with $X_0(x) = x$. For $x \in \Lambda(N)$ let $X^N_t(x) = X_t(x) \bmod (2N+1)$, $X^N_t(x) \in \Lambda(N)$, so that $X^N_t(x)$ is a random walk on $\Lambda(N)$. By choosing some collision ordering we can define $\xi^{A(N)}_t$ in terms of the $X_t([x_i N^\alpha])$, and hence $\xi^{N,A(N)}_t$ in terms of the $X^N_t([x_i N^\alpha])$. Now define

$$\tau_N = \inf\{t : |X_t([x_i N^\alpha])| \ge b_N \ \text{ for some } \ i,\ 1 \le i \le n\},$$
$$G_N = \{|X^N_{t(N)}([x_i N^\alpha]) - X^N_{t(N)}([x_{i'} N^\alpha])| \ge a_N \ \text{ for all } \ i \ne i'\}.$$

Note that $\xi_t^{N,A(N)} = \xi_t^{A(N)}$ for $t < \tau_N$. Furthermore, standard random walk estimates easily give

$$P(\tau_N > t(N)) \to 1 \tag{2.7}$$

and

$$P(\bar{G_N}) \to 1. \tag{2.8}$$

Let us write $o(1)$ for any quantity that tends to zero as $N \to \infty$. By (2.7) we have

$$\begin{aligned} P(\#\xi_{t(N)}^{N,A(N)} = j) &= P(\#\xi_{t(N)}^{N,A(N)} = j,\ \tau_N > t(N)) + o(1) \\ &= P(\#\xi_{t(N)}^{A(N)} = j,\ \tau_N > t(N)) + +o(1) \\ &= P(\#\xi_{t(N)}^{A(N)} = j) + o(1). \end{aligned}$$

Thus by (A),

$$P(\#\xi_{t(N)}^{N,A(N)} = j) \to q_{\hat{\alpha}}(n, j). \tag{2.9}$$

Now suppose that $B = \{y_1, \dots, y_j\} \subset \Lambda(N)$ satisfies $|y_i - y_{i'}| \geq a_N$ for $i \neq i'$. Then by (B), uniformly in such B,

$$P(\#\xi_{T(N)-t(N)}^{N,B} = k) \to q_{\pi s}(j, k). \tag{2.10}$$

Thus

$$\begin{aligned} P(\#\xi_{t(N)}^{N,A(N)} = j,\ \xi_{T(N)}^{N,A(N)} = k) &= \sum_{\#B=j} P(\xi_{t(N)}^{N,A(N)} = B, G_N) \cdot \\ &\qquad \cdot P(\xi_{T(N)-t(N)}^{N,B} = k) + o(1) \\ &= \sum_{\#B=j} P(\xi_{t(N)}^{N,A(N)} = B, G_N) q_{\pi s}(j, k) \\ &\qquad + o(1) \\ &= \sum_{\#B=j} P(\xi_{t(N)}^{N,A(N)} = B) q_{\pi s}(j, k) + o(1) \\ &= P(\xi_{t(N)}^{N,A(N)} = j) q_{\pi s}(j, k) + o(1) \\ &= q_{\hat{\alpha}}(n, j) q_{\pi s}(j, k) + o(1) \end{aligned}$$

where we have used first (2.10) and then (2.9). This proves (2.6).

3. A Hitting Time Result. Let X_t^N be a random walk on $\Lambda(N)$, and let H_N be the hitting time of the origin,

$$H_N = \inf\{t : X_t^N = 0\}.$$

The $n = 2$ case of (B) asserts if $N/\sqrt{\log N} \ll a_N \ll N$, then uniformly in $x \in \Lambda(N)$ such that $|x| \geq a_N$, $P_x(H_N > s(2N+1)^2 \log(2N+1)) \to \exp(-\frac{\pi s}{2})$ as $n \to \infty$. Rephrasing this, if $\mathcal{E}(\lambda)$ denotes the exponential distribution with parameter λ, then uniformly in $|x| \geq a_N$,

$$\mathcal{L}\left(\frac{H_N}{(2N+1)^2 \log(2N+1)}\right) \Rightarrow \mathcal{E}\left(\frac{\pi}{2}\right). \tag{3.1}$$

The proof of the $n = 2$ case of (2.5) "fills in" the region $|x| < a_N$. Namely, if $x_n = [xN^\alpha]$, then

$$P_{x_N}(H_N > s(2N+1)^2 \log(2N+1)) \to 1 - \alpha + \alpha\, e^{-\frac{\pi s}{2}}.$$

We can rewrite this as

$$\mathcal{L}_{x_N}\left(\frac{H_N}{(2N+1)^2 \log(2N+1)}\right) \Rightarrow (1-\alpha)\delta_0 + \alpha\mathcal{E}\left(\frac{\pi}{2}\right). \tag{3.2}$$

Note that α is the limiting probability that a random walk on $\mathbf{Z}^2$ starting at x_N does *not* visit the origin by time $(2N+1)^2 \log(2N+1)$. (3.2) should be compared with a result of Spitzer. His result (see Theorem 7 of [4]) is that for $d \geq 3$, if $x = (1, 0, \dots, 0) \in \mathbf{Z}^d$, and α is the probability starting at x that a random walk *never* visits the origin, then

$$\mathcal{L}_x\left(\frac{H_N}{(2N+1)^d}\right) \Rightarrow (1-\alpha)\delta_0 + \alpha\mathcal{E}\left(\frac{1}{G}\right). \tag{3.3}$$

4. A conjecture. It is possible to consider other models and make appropriately modified finite systems type statements for them. For instance, it was shown in [6] that for critical branching random walk, the high dimensional ($d \geq 3$) case fits the finite systems scheme, while the low dimensional case is more like (1.8). Dimension two can be considered the critical dimension, but in this case a result like our Theorem 1 seems unlikely. Instead, we expect the differences between the finite and infinite systems to be revealed on the time scale $\beta(N) = (2N+1)^2$, and that the differences are most clearly exhibited by the behavior of the *occupation time functionals* studied in [7] (which treated Brownian motion instead of random walk). We will not formulate a precise version of this, but will instead concentrate on what seems a more interesting example, the critical contact process.

The one-dimensional contact process η_t with parameter λ is the Markov process on $\Xi = \{0,1\}^{\mathbf{Z}}$ which makes transitions

$$\eta_t(x) \to 0 \text{ at rate one if } \eta_t(x) = 1,$$
$$\eta_t(x) \to 1 \text{ at rate } \lambda\big[\eta_t(x-1) + \eta_t(x+1)\big] \text{ if } \eta_t(x) = 0.$$

Let $|\eta| = \sum_{x\in Z}\eta(x)$ and let τ be the hitting time of the empty configuration, $\tau = \inf\{t > 0 : |\eta_t| = 0\}$. Let * denote the element of Ξ which has a single one at the origin and zeros elsewhere. There is a critical value $\lambda_c \in (0,\infty)$ with the property that, starting in configuration *, $P(\tau = \infty) = 0$ for $\lambda \le \lambda_c$ and $P(\tau = \infty) > 0$ for $\lambda > \lambda_c$. If $\lambda \le \lambda_c$ then δ_0, the point mass on the empty configuration, is the unique invariant measure for η_t. If $\lambda > \lambda_c$ there are exactly two extremal invariant measures, δ_0 and the *upper invariant measure* $\bar{\nu}$. See [14] for an extensive treatment of the contact process, and [2] for the proof that the critical contact process dies out.

The finite systems η_t^N are defined similarly on $\Lambda(N)$, except that we do not treat $\Lambda(N)$ as a torus, but instead permanently set $\eta_t^N(-N-1) = \eta_t^N(N+1) = 0$. In [6] it is shown that the finite systems scheme holds for $\lambda > \lambda_c$ for appropriate choices of the ingredients. The question here is what can be said about the critical case, since both the finite and infinite systems die out in this case. To attempt to answer this we introduce the contact process seen from the right edge.

Define

$$\begin{aligned} r(\eta) &= \sup\{x \in \mathbf{Z} : \eta(x) = 1\}, \\ X &= \{\eta \in \Xi : r(\eta) < \infty,\ |\eta| = \infty\}, \\ \tilde{X} &= \{\eta \in X : r(\eta) = 0\}, \end{aligned}$$

and the mapping S,

$$S\eta(x) = \eta(x + r(\eta)).$$

Then $S\eta_t$ is the contact process *seen from the right edge*. For each $\lambda \ge \lambda_c$ there is a unique invariant measure for $S\eta_t$ which concentrates on $\tilde{X}$ (see [12] for $\lambda > \lambda_c$ and [5] for $\lambda = \lambda_c$). Consider $S\eta_t^N$, the finite system seen from the right edge, where we set $S\eta_t^N \equiv 0$ if $\eta_t^N \equiv 0$.

Suppose now that $\lambda = \lambda_c$, that the initial state of η_t^N is

$$\eta_t^N(x) = \begin{matrix} 1, & \text{if } -N \le x \le 0, \\ 0, & \text{if } 0 < x \le N \end{matrix}$$

and that $\tilde{\mu}$ is the unique invariant measure for $S\eta_t$ which concentrates on $\tilde{X}$. We make the following

Conjecture. There is a sequence $\beta(N) \to \infty$ as $N \to \infty$ such that

$$\mathcal{L}(S\eta^N_{T(N)}) \Rightarrow \begin{matrix} \tilde{\mu}, & T(N) \ll \beta(N), \\ \delta_0, & T(N) \gg \beta(N). \end{matrix}$$

Based on the results in [11], it seems likely that if $\beta(N)$ exists, then $\beta(N) \sim N^\gamma$ for some $1 < \gamma < 4$.

REFERENCES

[1] R. Arratia, *Coalescing Brownian motion and the voter model on $\mathbb{Z}$*, Ph.D. Dissertation, University of Wisconsin, Madison, 1979.

[2] C. Bezuidenhout and G. Grimmett, *The critical contact process dies out*, Ann. Probab. **18** (1990), 1462–1482. 1990.

[3] M. Bramson and D. Griffeath, *Clustering and dispersion rates for some interacting particles systems on $\mathbb{Z}$*, Ann. Probab. **8** (1980), 183–213.

[4] J.T. Cox, *Coalescing random walks and voter model consensus times on the torus in $\mathbb{Z}^d$*, Ann. Probab. **17** (1989), 1333-1366.

[5] J.T. Cox, R. Durrett, and R. Schinazi, *The critical contact process seen from the right edge,* to appear, Prob. Th. Rel. Fields (1990).

[6] J.T. Cox and A. Greven, *On the long term behavior of some finite particle systems,* Prob. Th. Rel. Fields **85** (1990), 195-237.

[7] J.T. Cox and D. Griffeath, *Occupation times for critical branching Brownian motions,* Ann. Probab. **13** (1985), 1108-1132.

[8] J.T. Cox and D. Griffeath, *Diffusive clustering in the two dimensional voter model,* Ann. Probab. **14** (1987), 347-370.

[9] D. Dawson and A. Greven, *Multiple time scale analysis of interacting diffusions,* Preprint, SFB 123, Universität Heidelberg, 1990.

[10] R. Durrett, *Lecture Notes on Particle Systems and Percolation,* Wadsworth & Brooks/Cole, Pacific Grove, CA, 1988.

[11] R. Durrett, R. Schonmann, and N. Tanaka, *The contact process on a finite set, III: The critical case,* Ann. Probab. **17** (1990), 1303-1321.

[12] A. Galves and E. Presutti, *Edge fluctuations for the one dimensional supercritical contact process,* Ann. Probab. **15** (1987), 1131-1145.

[13] D. Griffeath, *Additive and Cancellative Interacting Particle Systems,* Springer Lecture Notes in Mathematics, Vol 724. Springer-Verlag, New York, 1979.

[14] T.L. Liggett, *Interacting Particle Systems,* Springer-Verlag, New York, 1985.

[15] S. Tavaré, *Line-of-descent and genealogical processes, and their applications in population genetics models,* Theoretical Population Biology **26** (1984), 119-164.

J. T. Cox
Department of Mathematics
Syracuse University
Syracuse, NY 13210
USA

A. Greven
Institut f. Mathematische Stochastik
Universität Göttingen
Lotzestrasse 13
3400 Göttingen, Germany

Large Deviation Lower Bounds For General Sequences Of Random Variables

BY A. DEACOSTA, P. NEY AND E. NUMMELIN

A large deviation lower bound is proved for general sequences of random variables, and is applied to prove such a bound for Markov-additive chains under minimal hypothesis.

1. An Approximation Scheme. We will say that a sequence of finite positive measures $\{\mu_n\}_0^\infty$ on $\mathbb{R}^d$ satisfies the large deviation lower bound (LDLB) with rate function $I : \mathbb{R}^d \to \mathbb{R}$, if for every *open* set $\Gamma \subset \mathbb{R}^d$

$$\underline{\lim} \frac{1}{n} \log \mu_n(n\Gamma) \geq -\inf\{I(x) : x \in \Gamma\}. \tag{1.1}$$

(This differs slightly from the usual terminology as in [6].)

Let $\hat{\mu}_n(\xi) = \int e^{\langle \xi, x \rangle} \mu_n(dx)$. The following useful result gives a general condition for (1.1) to hold. For terminology not explicitly defined here and other background see e.g. Ellis [2], chapter II.

Theorem 1. (Gärtner [3], Ellis [2]). *If*

$$\lim \frac{1}{n} \log \hat{\mu}_n(\xi) \stackrel{\text{def}}{=} \Lambda(\xi) < \infty \tag{1.2}$$

exists and is differentiable for all $\xi \in \mathbb{R}^d$, then $\{\mu_n\}$ satisfies the LDLB *with rate function $I(x) = \Lambda^*(x) = \sup_\xi[\langle \xi, x \rangle - \Lambda(\xi)]$, called the convex conjugate of Λ.*

The proof in [2] is for probability measures $\{\mu_n\}$, but the modification to the case of arbitrary finite positive measures is straightforward. The purpose of this note is to show that even when the hypotheses on Λ required in this theorem fail, it may be possible to approximate $\{\mu_n\}$ by measures that do satisfy such conditions, and thence conclude the LDLB for $\{\mu_n\}$. In this note we prove such a theorem, and apply it to LDLB's for Markov additive chains. The proof of Theorem 2 is a streamlining of some of the arguments in [1] and [5].

Theorem 2. *Assume that for each $k = 0, 1, \ldots$ there exists a sequence of measures $\{\mu_n^{(k)}; n = 0, 1, \ldots\}$ satisfying*

(i) $\mu_n^{(k)}(\cdot) \leq \mu_n^{(k+1)}(\cdot) \nearrow \mu_n(\cdot) \leq \infty$, *as* $k \to \infty$, *and*

(ii) $\lim_{n\to\infty} \frac{1}{n} \log \hat{\mu}_n^{(k)}(\xi) \stackrel{\text{def}}{=} \Lambda^{(k)}(\xi) < \infty$ *exists and is differentiable for* $\xi \in \mathbb{R}^d$ *(i.e. the hypothesis of Theorem 1 is satisfied for each k).*

Then $\lim_{k\to\infty} \Lambda^{(k)}(\xi) = \Lambda(\xi) \leq \infty$ *exists and* $\{\mu_n\}$ *satisfies the LDLB with rate* Λ^*.

Remark. By a truncation argument this theorem implies the lower bound in the classical Cramer theorem for i.i.d. random variables in $\mathbb{R}^d$ without any hypothesis.

Proof. Let $\mu_{n,\sigma}(\cdot)$ be the measure having generating function $\hat{\mu}_{n,\sigma} = \hat{\mu}_n \cdot \hat{\gamma}_{\sigma\sqrt{n}}$, where $\hat{\gamma}_\sigma(\xi) = \exp\{\frac{1}{2}\sigma^2\|\xi\|^2\}$ = the transform of a d-dimensional Gaussian. We call $\mu_{n,\sigma}$ the "smoothing" of μ_n. Similarly for $\mu_{n,\sigma}^{(k)}$. Clearly $\{\mu_{n,\sigma}^{(k)}\}$ also satisfies the hypotheses (i) and (ii) of the theorem, and

$$\Lambda_\sigma^{(k)}(\xi) \stackrel{\text{def}}{=} \lim_n \frac{1}{n} \log \hat{\mu}_{n,\sigma}^{(k)}(\xi) = \Lambda^{(k)}(\xi) + \frac{\sigma^2}{2}\|\xi\|^2 . \tag{1.3}$$

Since $\Lambda^{(k)}$ and $\Lambda_\sigma^{(k)}$ are monotone in k,

$$\Lambda_\sigma(\xi) \stackrel{\text{def}}{=} \lim_{k\to\infty} \Lambda_\sigma^{(k)}(\xi) = \Lambda(\xi) + \frac{\sigma^2}{2}\|\xi\|^2 \tag{1.4}$$

exists.

Now

$$\Lambda_\sigma^*(x) = \sup_\xi \left[\langle \xi, x\rangle - \Lambda(\xi) - \frac{\sigma^2}{2}\|\xi\|^2\right] \leq \Lambda^*(x) \quad \text{for } x \in \mathbb{R}^d, \tag{1.5}$$

and thus by the equality in (1.5), Λ_σ^* is increasing as $\sigma \searrow 0$. Hence

$$\begin{aligned} \lim_{\sigma\to 0} \Lambda_\sigma^*(x) &= \sup_{\sigma>0} \Lambda_\sigma^*(x) = \sup_{\sigma>0} \sup_\xi [\langle \xi, x\rangle - \Lambda(\xi) - \frac{\sigma^2}{2}\|\xi\|^2] \\ &= \sup_\xi \sup_{\sigma>0} [\langle \xi, x\rangle - \Lambda(\xi) - \frac{\sigma^2}{2}\|\xi\|^2] \\ &= \sup_\xi [\langle \xi, x\rangle - \Lambda(\xi)] = \Lambda^*(x), \qquad \text{for all } x \in \mathbb{R}^d. \end{aligned} \tag{1.6}$$

By a straightforward adaptation of the argument on page 12 of Varadhan [6], it follows from (1.6) that it is sufficient to prove the LDLB for the smoothed measures $\{\mu_{n,\sigma}\}$, with rate Λ_σ^*. We proceed to do so.

Note that the hypotheses of Theorem 2 imply that $\{\mu_{n,\sigma}^{(k)}; n = 0, 1, \ldots\}$ also satisfies the hypothesis of Theorem 1, and thus the LDLB (1.2) holds for $\{\mu_{n,\sigma}^{(k)}\}$ with rate $\Lambda_\sigma^{(k)*}$. Hence, letting $B(x,\epsilon) =$ the ϵ-ball centered at x, and observing that $\mu_{n,\sigma} \geq \mu_{n,\sigma}^{(k)}$ for all k, we have

$$(1.7) \qquad \varliminf \frac{1}{n} \log \mu_{n,\sigma}(nB(x,\epsilon)) \geq \varliminf \frac{1}{n} \log \mu_{n,\sigma}^{(k)}(nB(x,\epsilon)) \geq -\Lambda_\sigma^{(k)*}(x)\,.$$

If we can show that

$$(1.8) \qquad \Lambda_\sigma^{(k)*}(x) \searrow \Lambda_\sigma^*(x) \text{ as } k \to \infty \text{ for all } x \in \mathbb{R}^d\,,$$

then we will have (1.1) for $\{\mu_{n,\sigma}\}$, and hence the theorem.

First consider the case $x = 0$, namely show that

$$(1.9) \qquad \Lambda_\sigma^{(k)*}(0) \searrow \Lambda_\sigma^*(0) \text{ as } k \to \infty.$$

For $x \neq 0$ the result then follows by a translation argument (see e.g. [5]). By the definition of convex conjugate, (1.9) is equivalent to proving that

$$(1.10) \qquad \inf_{\xi \in \mathbb{R}^d} \Lambda_\sigma^{(k)}(\xi) \nearrow \inf_{\xi \in \mathbb{R}^d} \Lambda_\sigma(\xi) \text{ as } k \to \infty\,.$$

Define the level sets of Λ_σ

$$L_c(\Lambda_\sigma) = \{\xi \in \mathbb{R}^d : \Lambda_\sigma(\xi) \leq c\}\,,$$

and similarly define $L_c(\Lambda_\sigma^{(k)})$. Since $\Lambda_\sigma^{(k)}$ is continuous and $\Lambda_\sigma^{(k)} \nearrow \Lambda_\sigma$, Λ_σ is ℓ.s.c. Thus $L_c(\Lambda_\sigma)$ and $L_c(\Lambda_\sigma^{(k)})$ are closed. We claim they are also bounded and hence *compact.* (Note that we are here talking about the level sets of the Λ's, not the Λ^*'s.) Boundedness will follow from the fact that

$$(1.11) \qquad \Lambda_\sigma^{(k)}(\xi) \geq \alpha + \gamma\|\xi\|^2 \text{ for some } \alpha > -\infty, 0 < \gamma < \infty, \text{ and all } k\,.$$

To see (1.11) apply the Chebycheff inequality

$$(1.12) \qquad \hat{\mu}_{n,\sigma}^{(k)}(\xi) \geq e^{-nr\|\xi\|} \mu_{n,\sigma}^{(k)}(nB_r),$$

where B_r is a ball of radius r centered at 0. Now since $\{\mu_{n,\sigma}^{(k)}(\cdot)/\hat{\mu}_{n,\sigma}^{(k)}(0)\}$ are probability measures, we can conclude (e.g. by Theorem II.6.3 of Ellis [2]) that for r sufficiently large

$$(1.13) \qquad \frac{\mu_{n,\sigma}^{(k)}(nB_r)}{\hat{\mu}_{n,\sigma}^{(k)}(0)} \longrightarrow 1 \quad \text{as} \quad n \to \infty.$$

Now $\lim \frac{1}{n} \log \hat{\mu}_{n,\sigma}^{(k)}(0) = \Lambda_\sigma^{(k)}(0)$ and hence (1.12) and (1.13) imply that

$$\Lambda_\sigma^{(k)}(\xi) = \lim_{n\to\infty} \frac{1}{n} \log \hat{\mu}_{n,\sigma}^{(k)}(\xi) \geq -r\|\xi\| + \Lambda_\sigma^{(k)}(0).$$

This with (1.3) implies (1.11). The monotonicity of $\Lambda_\sigma^{(k)}$ (in k) then implies (1.11) uniformly in k.

Now since the $\Lambda_\sigma^{(k)}$ are continuous and increasing to Λ_σ as $k \to \infty$, it follows that for any compact set K

$$\inf_{\xi\in K} \Lambda_\sigma^{(k)}(\xi) \nearrow \inf_{\xi\in K} \Lambda_\sigma(\xi) \quad \text{as } k \to \infty. \tag{1.14}$$

Since the level sets of Λ_σ and $\Lambda_\sigma^{(k)}$ are compact we see that for sufficiently large, compact K

$$\inf_{\xi\in K} \Lambda_\sigma^{(k)}(\xi) = \inf_{\xi\in \mathbb{R}^d} \Lambda_\sigma^{(k)}(\xi) \quad \text{and} \quad \inf_{\xi\in K} \Lambda_\sigma(\xi) = \inf_{\xi\in \mathbb{R}^d} \Lambda_\sigma(\xi). \tag{1.15}$$

Together with (1.14), (1.15) implies that

$$-\Lambda_\sigma^{(k)^*}(0) = \inf_{\xi\in \mathbb{R}^d} \Lambda_\sigma^{(k)}(\xi) \nearrow \inf_{\xi\in \mathbb{R}^d} \Lambda_\sigma(\xi) = -\Lambda_\sigma^*(0),$$

which is (1.9).

This proves Theorem 2. ■

2. Application to Markov additive chains. Let $\mathbb{N} = \{0, 1, \ldots\}$ and let $\{(X_n, \sigma_n); n = 0, 1, \ldots\}$ be a Markov chain with state space $\mathbb{N} \times \mathbb{R}^d$ satisfying

$$\begin{aligned} &P\{X_{n+1} = j, \sigma_{n+1} \in \Gamma \mid X_n = i, \sigma_n = s\} \\ &\qquad = P\{X_1 = j, \sigma_1 \in \Gamma \mid X_0 = i\} \stackrel{\text{def}}{=} p_{ij}(\Gamma) \end{aligned} \tag{2.1}$$

for $i, j \in \mathbb{N}$ and Γ a Borel set $\subset \mathbb{R}^d$. The matrix of measures $P(\cdot) = \{p_{ij}(\cdot); i, j \in \mathbb{N}\}$ is called an MA kernel, and the process $\{(X_n, S_n)\}$ where $S_n = \sigma_0 + \cdots + \sigma_n$, is called an MA chain. A typical example is $S_n = \sum_{i=1}^n f(X_{i-1}, X_i)$ for $f : \mathbb{N}^2 \to \mathbb{R}^d$. We will say that $\{S_n\}$ satisfies the LD lower bound with rate function I if, for all $i, j \in \mathbb{N}$ and open sets $\Gamma \subset \mathbb{R}^d$,

$$\underline{\lim} \frac{1}{n} \log p_{ij}^{(n)}(n\Gamma) \geq -\inf_{v\in\Gamma} I(v). \tag{2.2}$$

Here $p_{ij}^{(n)}(\Gamma) = P\{X_n = j, S_n \in \Gamma \mid X_0 = i\}$.

Theorem 3. *If $P(\mathbb{R}^d)$ is an irreducible aperiodic matrix, then $\{S_n\}$ satisfies the LD lower bound with rate function $I(\cdot)$ to be specified below.*

Remarks.

(i) The rate function will be identified below as the convex conjugate of the limit of a sequence of *convergence parameters* (see [4]), which will play the role of the approximating sequences in Theorem 2.

(ii) The requirement that the underlying M.C. be irreducible is obviously necessary for the theorem. Note that there are *no other restrictions on* $\{X_n\}$ (such as recurrence), and *no restrictions whatever on the distributions of the* σ_n's, i.e. on the measures $p_{ij}(\cdot)$.

(iii) Note that in contrast to the general state space case in [5], no further minorization hypotheses is needed in the present case.

Proof. Fix some state $i_0 \in I$, and define the hit times of i_0 by

$$T_0 = \inf\{n : X_n = i_0\}, \quad = 0 \text{ if } X_0 = i_0, \text{ and}$$
$$T_{k+1} = \inf\{n > T_k : X_n = i_0\}, \quad k = 0, 1, \dots .$$

Also let

$$L_n = \max\{k \le n : \ X_k = i_0\},$$
$$V_n = n - L_n = \text{the elapsed time since the last visit to } i_0, \text{ at time } n.$$

Note that $V_{T_k} = 0$ for all k.

Define the generating function

$$\psi(\xi, \zeta) = E_{i_0}\{\exp[\langle \xi, S_{T_1}\rangle - \zeta T_1]; T_1 < \infty\}, \quad \xi \in \mathbb{R}^d, \quad \zeta \in \mathbb{R},$$

with convergence domain $\mathcal{W} = \{(\xi, \zeta) : \psi(\xi, \zeta) < \infty\}$.

Let

$$\Lambda(\xi) = \inf\{\zeta : \psi(\xi, \zeta) \le 1\}. \tag{2.3}$$

We will see that the rate function in Theorem 3 is $I(x) = \Lambda^*(x)$.

Let $\hat{p}_{ij}(\xi) = \int e^{\langle \xi, s\rangle} p_{ij}(ds)$.

Lemma. (From [3]). *If* $\{X_n\}$ *is irreducible and* $\mathcal{W}$ *is an open set, then the characteristic equation*

$$\psi(\xi, \zeta) = 1 \tag{2.4}$$

has a unique solution $\zeta \overset{\text{def}}{=} \Lambda(\xi)$, *which is analytic on* $\mathcal{D} = \{\xi : \Lambda(\xi) < \infty\}$, *and*

$$\lim_{n \to \infty} \frac{1}{n} \log \hat{p}^n_{ij}(\xi) = \Lambda(\xi). \tag{2.5}$$

In general, $\mathcal{W}$ need not be open. However if the additive components $\{\sigma_n\}$ and the times $\{V_n\}$ are bounded, then $\mathcal{W} = \mathbb{R}^{d+1}$ and the Lemma can be applied. We therefore proceed via an approximating sequence of truncations and Theorem 2.

Consider the Markov chain $\{\tilde{X}_n\} \stackrel{\text{def}}{=} \{(X_n, V_n)\}$ on the state space $\mathbb{E} \stackrel{\text{def}}{=} \mathbb{N} \times \mathbb{N}$. For $m \in \mathbb{N}$, let $\mathbb{N}_m = \{0, 1, \cdots, m\}$, and $\mathbb{E}_m = \mathbb{N} \times \mathbb{N}_m$. Let $\tilde{i} = (i, \ell) \in \mathbb{E}, \tilde{j} = (j, h) \in \mathbb{E}$, and define $\{_m\tilde{X}_n; n = 0, 1, \cdots\}$ to be the M.C. obtained from $\{\tilde{X}_n\}$ by "killing" the latter if and when it leaves $\mathbb{E}_m$; namely having transition function

$$(2.6) \qquad {}_m\tilde{p}_{\tilde{i},\tilde{j}} = \begin{cases} P\{X_{n+1} = j, V_{n+1} = h | X_n = i, V_n = \ell\} & \text{if } \tilde{i}, \tilde{j} \in \mathbb{E}_m \\ 0 & \text{otherwise.} \end{cases}$$

This is a substochastic matrix. Let $\tilde{\mathbb{E}}_m$ = the irreducibility class of the matrix (2.6) which contains the point $(i_0, 0)$. Note that

$$(2.7) \qquad \tilde{\mathbb{E}}_m \subset \tilde{\mathbb{E}}_{m+1}, \quad \text{and} \quad \cup_m \tilde{\mathbb{E}}_m = \mathbb{E}.$$

Finally, let $B_m = \{s \in \mathbf{R}^d : \|s\| \leq m\}$ and define the (sub-stochastic) MA-chain $\{(_m\tilde{X}_n, {}_mS_n); n = 0, 1, \cdots\}$ via the MA-kernel

$${}_m\tilde{p}_{\tilde{i},\tilde{j}}(\Gamma) = \begin{cases} P\{\tilde{X}_{n+1} = \tilde{j}, \sigma_{n+1} \in \Gamma \cap B_m | \tilde{X}_n = \tilde{i}\} & \text{for } \tilde{i}, \tilde{j} \in \tilde{\mathbb{E}}_m \\ 0 & \text{otherwise.} \end{cases}$$

Let $\tilde{\psi}_m(\xi, \zeta)$ be the generating function (2.3) for the $\{(_m\tilde{X}_n, {}_mS_n)\}$ process, with $\mathcal{W}_m$ and $\tilde{\Lambda}_m$ the corresponding convergence domains and parameters. Now $\mathcal{W}_m = \mathbf{R}^{d+1}$ and hence by the Lemma

$$(2.8) \qquad \lim_{n \to \infty} \frac{1}{n} \log {}_m\tilde{p}^n_{\tilde{i}\,\tilde{j}}(\xi) = \tilde{\Lambda}_m(\xi), \quad \xi \in \mathbb{R}^d.$$

Thus the hypotheses of Theorem 2 are clearly satisfied, and we conclude that $\lim_{m \to \infty} \tilde{\Lambda}_m(\xi) \stackrel{\text{def}}{=} \tilde{\Lambda}(\xi)$ exists and

$$(2.9) \qquad \underline{\lim} \frac{1}{n} \log \tilde{p}^n_{\tilde{i}\,\tilde{j}}(n\Gamma) \geq - \inf_{s \in \Gamma} \tilde{\Lambda}^*(s)$$

for open sets Γ.

It remains to show that (2.9) implies (2.2) with $I = \Lambda$ determined by (2.3). First write

$$\tilde{p}^n_{\tilde{i},j}(\cdot) = \sum_h \tilde{p}^n_{\tilde{i},(j,h)}(\cdot),$$

and note that $\tilde{p}^{(n)}_{(i,\ell),j}(\Gamma)$ is independent of ℓ. Hence

$$p^n_{ij}(\Gamma) = \tilde{p}^n_{\tilde{i},j}(\Gamma) \geq \tilde{p}^n_{\tilde{i},\tilde{j}}(\Gamma)$$

for $\tilde{j} = (j, h)$, and by (2.9)

$$\underline{\lim}\ \frac{1}{n} \log p_{ij}^{n}(n\Gamma) \geq - \inf_{s \in \Gamma} \tilde{\Lambda}^*(s).$$

Finally we need to argue that

$$\Lambda(\xi) = \tilde{\Lambda}(\xi) \quad \text{for} \quad \xi \in \mathbb{R}^d, \tag{2.10}$$

where $\tilde{\Lambda} = \lim \tilde{\Lambda}_m$ and $\tilde{\Lambda}_m$ is the unique solution of

$$\tilde{\psi}_m(\xi, \tilde{\Lambda}_m(\xi)) = 1,$$

and Λ is defined by (2.3). The proof of (2.10) is identical to that of Lemma 3.3 in [5].

This concludes the proof of Theorem 3. ■

REFERENCES

[1] A. de Acosta,, *Large deviations for vector valued functionals of a Markov chain: Lower Bounds*, Annals of Prob. **16** (1988), 925-960.

[2] R. Ellis, *Entropy, Large Deviations, and Statistical Mechanics*, Springer, New York, 1985.

[3] J. Gärtner, *On large deviations from the invariant measure.* Th. Prob. and Appl. **22** (1977), 24-39.

[4] P. Ney and E. Nummelin, *Markov additive processes I: Eigenvalue properties and limit theorems*, Ann. Prob. **15** (1984), 561-592.

[5] P. Ney and E. Nummelin, *Markov additive processes II: Large deviations*, Ann. Prob. **15** (1984), 593-609.

[6] S. R. S. Varadhan, *Large Deviations and Applications*, SIAM, Philadelphia, 1984.

A. de Acosta
Dept. of Mathematics
Case/Western Reserve University
Cleveland, OH

P. Ney
Dept. of Mathematics
University of Wisconsin
Madison, WI

E. Nummelin
Dept. of Mathematics
University of Helsinki
Helsinki, Finland

Asymptotic Laplace-Transforms

H. DINGES

Introduction. In 1954 Henry Daniels published his pioneering paper "*Saddlepoint approximations in statistics*". Daniels studies asymptotic expansions for the densities of the averages

$$\frac{1}{n}\,(Y_1 + \cdots + Y_n)$$

when the Y_j are i.i.d. with short tails. In the discrete case it is assumed that $\mathcal{L}(Y)$ is concentrated on $\mathbb{Z}$ and not on any proper subgroup of $\mathbb{Z}$. Here one is interested in the weights

$$\Pr(Y_1 + \cdots + Y_n = k)$$

for $\frac{k}{n}$ in some neighborhood of $\mathcal{E}Y$. In the continuous case it is assumed that $\mathcal{L}(Y)$, the law of Y, satisfies Cramér's condition. Using techniques similar to those in Daniels' paper Lugannani and Rice in 1980 obtained uniform asymptotic expansions for the distribution functions.

The title of Daniels' paper has become the name of a flourishing branch of the asymptotic theory of distributions although the original saddlepoint methods have lost their prominent rôle in the proofs.

The classical approach which concentrates on sums of independent random variables uses the affine structure very essentially; nonlinear transformations are foreign to the classical approach. It is clear that techniques which are based on characteristic functions are inappropriate when one has to deal with nonlinear transformations.

It turns out that the theory of uniform asymptotic expansions gains in transparency and applicability when one considers families of distributions which are more general than those of the averages of independent random variables. The idea of increasing sample size n is abandoned; the families which we are going to study are indexed by a parameter $\varepsilon \to 0$. The range of the parameter ε may be any set of positive numbers which has 0 as an accumulation point. (In the classical situations the parameter would be $\varepsilon = \frac{1}{n}$.) The objects of our theory have the form

$$\frac{1}{\sqrt{2\pi\varepsilon}} \cdot \exp\left(-\frac{1}{\varepsilon}\, K(x) - L_0(x) - \varepsilon L_1(x) - \ldots - \varepsilon^m L_m(x) + o(\varepsilon^m)\right)\, dx$$

uniformly in some neighborhood of a "center" x^*. The assumptions on the "entropy function" $K(x)$ and the functions $L_j(x)$ will be specified below. The families $\{f_\varepsilon(x)\, dx : \varepsilon \to 0\}$ are called **Wiener germs**. The exact definition will be given below. An explanation for the name can be given as follows:

The most elementary Wiener germ is the family of normal distributions $\{\mathcal{N}(z^*, \varepsilon\sigma^2) : \varepsilon \to 0\}$

$$h_\varepsilon(z)\, dz = \frac{1}{\sqrt{2\pi\varepsilon}} \exp\left(\frac{1}{2\varepsilon}\left(\frac{z - z^*}{\sigma}\right)^2\right) \frac{1}{\sigma}\, dz\ .$$

Let Z_ε be normal random variables with $\mathcal{E} Z_\varepsilon = z^*$, $\operatorname{var} Z_\varepsilon = \varepsilon\sigma^2$. Consider $X_\varepsilon = T(\varepsilon, Z_\varepsilon)$, where

$$T(\varepsilon, z) = T(z) + \varepsilon T_0(z) + \varepsilon^2 T_1(z) + \ldots + \varepsilon^{m+1} T_m(z) + o(\varepsilon^{m+1})$$

uniformly in some neighborhood of z^*. The distributions $\mathcal{L}(X_\varepsilon)$ restricted to some neighborhood of $x^* = T(z^*)$ then form a Wiener germ, provided that $T(\cdot)$ is a diffeomorphism and the $T_j(\cdot)$ satisfy certain smoothness assumptions. The smoothness assumptions will be specified below.

Roughly speaking: the Wiener germs are those families of distributions which we obtain when we transform the marginals of a Wiener process by the smooth transformations $T(\varepsilon, \cdot)$. (The full truth concerning morphisms is more complicated, unfortunately. Those Wiener germs which can be generated in the described way show slightly more smoothness than those in definition 1. The construction of transformations

$$A(\varepsilon, x) = A(x) + \varepsilon A_0(x) + \ldots + \varepsilon^{m+1} A_m(x) + o(\varepsilon^{m+1})$$

which map a given Wiener germ into the trivial Wiener germ $\{\mathcal{N}(0, \varepsilon) : \varepsilon \to 0\}$ is fairly delicate and will not be sketched here.)

Definition 1. Let U be an open subset of $\mathbb{R}^d$, $x^* \in U$, $m \in \mathbb{N}_0$. Let $\{f_\varepsilon(x)\, dx : \varepsilon \to 0\}$ be a family of distributions on U of the form

$$f_\varepsilon(x)\, dx = (2\pi\varepsilon)^{-d/2} \exp\left(-\frac{1}{\varepsilon}\, L(\varepsilon, x) + o(\varepsilon^m)\right)\, dx$$

$$= (2\pi\varepsilon)^{-d/2} \exp\left(-\frac{1}{\varepsilon}\, K(x) - L_0(x) - \ldots - \varepsilon^m L_m(x) - \varepsilon^m \widetilde{L}(\varepsilon, x)\right)\, dx\ .$$

a) We say that the family $\{f_\varepsilon(x)\,dx : \varepsilon \to 0\}$ is a **smooth Wiener germ** of order m on U with center x^* when

(i) $\widetilde{L}(\varepsilon, x) \to 0$ uniformly on every compact subset of U.

(ii) $K(x)$ is $(2m+2)$–smooth with $K''(x)$ positive definite.

(iii) $L_j(x)$ is $(2m-2j)$–smooth for $j = 0, 1, 2, \ldots, m$.

(iv) $\int\limits_{U^*} f_\varepsilon(x)\,dx = 1 + o(\varepsilon^m)$
for every compact $U^* \subset U$ which is a neighborhood of x^*.

(We say that a function is ℓ–smooth when it is ℓ–times continuously differentiable.)

b) $\frac{1}{\varepsilon}\, L(\varepsilon, x) = \frac{1}{\varepsilon}\, K(x) + L_0(x) + \ldots + \varepsilon^m L_m(x)$
is called the **differential exponent** for the Wiener germ.

Using the terminology of definition 1 a classical theorem due to H. Daniels can be stated very briefly. We first recall this result in traditional terminology. We restrict ourselves to the one–dimensional case. We mention the result for the discrete case; discrete Wiener germs will be treated elsewhere. Our ways of speaking emphasize that matters are different in the discrete case.

Theorem 1 (Daniels) *Let $Y_1, Y_2, \ldots$ be i.i.d. random variables whose distributions have short tails in the sense that*

$\psi(\vartheta) := \ln[\mathcal{E}\exp(\vartheta Y)]$ *is finite in some neighborhood V of the origin.*

The convex conjugate of the cumulant generating function $\psi(\vartheta)$, the so–called entropy function,

$$K(x) := \sup\{\vartheta x - \psi(\vartheta) : \vartheta \in V\}$$

is defined in the open set $U = \{x : x = \psi'(\vartheta)$ for some $\vartheta \in V\}$, the so–called accessible support of $\mathcal{L}(Y)$.

A) *When $\mathcal{L}(Y)$ satisfies Cramér's condition the densities $f^{(n)}(x)$ of*

$$\mathcal{L}\left(\frac{1}{n}\,(Y_1 + \ldots + Y_n)\right) \quad \textit{restricted to } U$$

have the asymptotic form

$$\sqrt{\frac{n}{2\pi}} \cdot \exp\left(- nK(x) - L_0(x) - n^{-1}L_1(x) - \ldots - n^{-m}L_m(x) + o(n^{-m})\right)$$

uniformly on every compact subset of U.

B) *When $\mathcal{L}(Y)$ is concentrated on $\mathbb{Z}$ and not on a proper subgroup of $\mathbb{Z}$ then uniformly for all $k \in \mathbb{Z}$ with $\frac{k}{n}$ in a compact subset of U*

$$\Pr(Y_1 + \ldots + Y_n = k)$$

$$= \frac{1}{\sqrt{2\pi n}} \exp\left(-n \cdot K\left(\frac{k}{n}\right) - K_0\left(\frac{k}{n}\right) - \ldots - \left(\frac{1}{n}\right)^m \cdot K_m\left(\frac{k}{n}\right) + o(n^{-m})\right) .$$

C) *The functions $L_j(x)$ and $K_j(x)$ can be expressed by the derivatives of order $\leq 2m - 2j$ of the entropy function $K(x)$.*

Daniels' theorem tells us that the averages

$$\left\{\mathcal{L}\left(\frac{1}{n}(Y_1 + \ldots + Y_n)\right) : n \to \infty\right\}$$

form a smooth Wiener germ when Cramér's condition is satisfied. Assertion B) suggests a definition of the concept of a discrete Wiener germ which will be studied in a forthcoming paper.

We now return to the d–dimensional case and introduce some more concepts.

Definition 2.

a) We say that a function $K(x)$ on an open convex set $U \subseteq \mathbb{R}^d$ is an **entropy function** of order m when $K''(x)$ is positive definite and $K(x)$ is $(2m+2)$–smooth on U.

b) On the open set $V = \{\vartheta : \vartheta = K'(x) \text{ for some } x \in U\}$ we define the **convex conjugate** of the entropy function $K(x)$

$$\psi(\vartheta) := \sup\{\vartheta x - K(x) : x \in U\} .$$

Well–known results from convex analysis tell us

$$\begin{aligned} U &= \{x : x = \psi'(\vartheta) \text{ for some } \vartheta \in V\} , \\ K(x) &= \sup\{\vartheta x - \psi(\vartheta) : \vartheta \in V\} \text{ for all } x \in U , \end{aligned}$$

$$\begin{aligned} K'(\psi'(\vartheta)) &= \vartheta \text{ for all } \vartheta \in V , \\ \psi'(K'(x)) &= x \text{ for all } x \in U . \end{aligned}$$

(We think of $\vartheta = K'(x)$ as a row vector and of $x = \psi'(\vartheta)$ as a column vector. The Hessian matrices $K''(x)$ and $\psi''(\vartheta)$ are considered as symmetric matrices.) Note that for every $\vartheta^* = K'(x^*) \in V$ the function $K^*(x) := K(x) - \vartheta^* x + \psi(\vartheta^*)$ vanishes of second order in the point x^*.

Definition 3.

a) When $K(x)$ is an entropy function of order m and K_j is $(2m-2j)$-smooth then

$$\frac{1}{\varepsilon} K(\varepsilon, x) = \frac{1}{\varepsilon} K(x) + K_0(x) + \varepsilon K_1(x) + \ldots + \varepsilon^m K_m(x)$$

is called an **admissible exponent** of order m on U.
(ε may be considered as an indeterminate in the sense of formal power series or as a parameter varying in $\mathbb{R}^+$.)

b) When there exists $x^* \in U$ with $K(x^*) = 0$ and $K'(x^*) = 0$ then this point x^* is called the **center** for the entropy function $K(\cdot)$ or also the center for the admissible exponent $\frac{1}{\varepsilon} K(\varepsilon, x)$.

c) We say that the admissible exponent is **normed in the sense of the continuous theory** when for every compact $U^* \subseteq U$ which is a neighborhood of the center we have

$$\int_{U^*} (2\pi\varepsilon)^{-d/2} \exp\left(-\frac{1}{\varepsilon} K(\varepsilon, x)\right) dx = 1 + o(\varepsilon^m) .$$

The admissible exponents $\frac{1}{\varepsilon} K(\varepsilon, x)$ which come up in Daniels' theorem are special in many ways. In particular they are normed in the sense of the continuous theory in the smooth situation A, but not generally in the situation B. The following theorem states that norming can be achieved by the addition of constants to the $K_j(\cdot)$. The significance of the norming constants is explained in theorem 3.

Theorem 2 *Let $K(x)$ be an entropy function of order m on U and let $\psi(\vartheta)$ on V be its convex conjugate. Let $\frac{1}{\varepsilon} K(\varepsilon, x)$ be an admissible exponent of order m. Then there exists an admissible exponent on V*

$$\frac{1}{\varepsilon} \psi(\varepsilon, \vartheta) = \frac{1}{\varepsilon} \psi(\vartheta) + \psi_0(\vartheta) + \varepsilon \psi_1(\vartheta) + \ldots + \varepsilon^m \psi_m(\vartheta)$$

such that for every $\vartheta^* \in V$ *the admissible exponent*

$$\frac{1}{\varepsilon}\, K(\varepsilon, \vartheta^*, x) = \frac{1}{\varepsilon}\, K(\varepsilon, x) - \frac{1}{\varepsilon}\, \vartheta^* x + \frac{1}{\varepsilon}\, \psi(\varepsilon, \vartheta^*)$$

is normed in the sense of the continuous theory.

Definition 4. When $\{f_\varepsilon(x)\, dx : \varepsilon \to 0\}$ is a Wiener germ of order m on U and $\frac{1}{\varepsilon}\, K(\varepsilon, x)$ is its differential exponent then the admissible exponent

$$\frac{1}{\varepsilon}\, \psi(\varepsilon, \vartheta) = \frac{1}{\varepsilon}\, \psi(\vartheta) + \psi_0(\vartheta) + \ldots + \varepsilon^m \psi_m(\vartheta) \quad \text{on } V$$

is called the **dual exponent** for the Wiener germ.

Note that the theorem tells us

$$\int \exp\left(\frac{1}{\varepsilon}\, \vartheta x\right) f_\varepsilon(x)\, dx = \exp\left(\frac{1}{\varepsilon}\, \psi(\varepsilon, \vartheta) + o(\varepsilon^m)\right) .$$

(The integral is to be extended over any compact subset of U which is a neighborhood of $x^* = \psi'(\vartheta)$).

Theorem 3 a) *Every admissible exponent* $\frac{1}{\varepsilon}\, K(\varepsilon, x)$ *on* U *induces an admissible exponent* $\frac{1}{\varepsilon}\, \psi(\varepsilon, \vartheta)$ *on the dual domain* V. *The order is the same.*

b) *For every admissible exponent* $\frac{1}{\varepsilon}\, \psi(\varepsilon, \vartheta)$ *on* V *there exists a unique admissible exponent* $\frac{1}{\varepsilon}\, K(\varepsilon, x)$ *such that* $\frac{1}{\varepsilon}\, \psi(\varepsilon, \vartheta)$ *is its dual in the sense of the continuous theory.*

c) *The derivatives of the* $\psi_j(\vartheta)$ *are determined by the derivatives of the* $K_j(x)$ *evaluated in* $x = \psi'(\vartheta)$, *and vice versa. (For* $\psi(\vartheta)$ *we also write* $\psi_{-1}(\vartheta)$ *and for* $K(x)$ *also* $K_{-1}(x)$.*)*

The proof of the theorem will be given elsewhere. It may be seen as an escalation of the well-known Method of Laplace.

Remarks.

1) When $\{f_\varepsilon(x)\, dx : \varepsilon \to 0\}$ is a Wiener germ then for every fixed ε the Laplace transform $\int \exp\left(\frac{1}{\varepsilon}\, \vartheta x\right) f_\varepsilon(x)\, dx$ is holomorphic in the variable ϑ. Note, however, that the $\psi_j(\vartheta)$ showing up in the dual exponent

have limited smoothness. We need all the smoothness in the differential exponent to get exactly the same smoothness for the "asymptotic Laplace transform" $\exp\left(\frac{1}{\varepsilon}\,\psi(\varepsilon,\vartheta)\right)$.

2) Let $\{d\mu_\varepsilon(x) : \varepsilon \to 0\}$ be a family of measures with

$$\int \exp\left(\frac{1}{\varepsilon}\,\vartheta x\right)\,d\mu_\varepsilon(x) = \exp\left(\frac{1}{\varepsilon}\,\psi(\varepsilon,\vartheta) + o(\varepsilon^m)\right)$$

where $\frac{1}{\varepsilon}\,\psi(\varepsilon,\vartheta)$ is an admissible exponent. One cannot conclude anything about the existence or smoothness of densities. The relation to the corresponding smooth Wiener germ

$$f_\varepsilon(x)\,dx = (2\pi\varepsilon)^{-d/2}\exp\left(-\frac{1}{\varepsilon}\,K(\varepsilon,x) + o(\varepsilon^m)\right)$$

is not very direct. This point will be taken up in a forthcoming paper on discrete Wiener germs on $\mathbb{R}^1$. (We know very little about d–dimensional discrete Wiener germs; the very special but nontrivial and presumably very important case of multinomial distributions is being studied elsewhere.)

3) Generally one should be aware that asymptotic Laplace transforms relate local behaviour in ϑ to local behaviour in $x = \psi'(\vartheta)$. Ordinary Fourier transforms reflect local behaviour (smoothness of the density etc.) in their behaviour at infinity and tail behaviour (existence of moments etc.) by their behaviour at the origin. This makes evident that the "asymptotic Laplace transform" does not fit into the classical frames.

Application. Let $\psi(\vartheta)$, $\vartheta \in (\vartheta_-, \vartheta_+)$, be the cumulant generating function of a probability distribution μ which does not necessarily satisfy Cramér's condition.
There exists a Wiener germ $\{f_\varepsilon(x)\,dx : \varepsilon \to 0\}$ of order m on the interval

$$U = (x_-, x_+) = \left(\lim_{\vartheta\to\vartheta_-}\psi'(\vartheta)\ ,\ \lim_{\vartheta\to\vartheta_+}\psi'(\vartheta)\right)$$

such that uniformly on every compact subset of $(\vartheta_-, \vartheta_+)$

$$\ln \int \exp\left(\frac{1}{\varepsilon}\,\vartheta x\right) f_\varepsilon(x)\,dx = \frac{1}{\varepsilon}\,\psi(\vartheta) + o(\varepsilon^m)\ .$$

On the other hand: When $Y_1, Y_2, \ldots$ are i.i.d. with distribution μ then for all $n \in \mathbb{N}$ we have

$$\ln \mathcal{E} \exp (\vartheta(Y_1 + \ldots + Y_n)) = n \cdot \psi(\vartheta) .$$

Thus $f_{1/n}(x)\,dx$ for large n may be considered as a smooth approximation to the distribution

$$\mathcal{L}\left(\frac{1}{n}(Y_1 + \ldots + Y_n)\right) .$$

This kind of interpretation is impossible when $\psi(\vartheta)$ is not holomorphic on $(\vartheta_-, \vartheta_+)$. It would be interesting to see what kind of exponents $\frac{1}{\varepsilon}\,\psi(\varepsilon, \vartheta) + o(\varepsilon^m)$ one could get as an approximate cumulant generating function

$$\ln \mathcal{E} \exp (\vartheta(Y_1 + \ldots + Y_n)) = n \cdot \psi(\vartheta) + \psi_0(\vartheta) + \ldots + n^{-m}\psi_m(\vartheta) + o(n^{-m})$$

when $Y_1, Y_2, \ldots$ is a stationary process.

REFERENCES

[1] N.G. de Bruijn, *Asymptotic Methods in Analysis*, North Holland, 1970

[2] H.E. Daniels, *Saddlepoint approximations in statistics*, Ann. Math. Stat. **25** (1954), 631–650

[3] H.E. Daniels, *Tail probability approximations*, Int. Stat. Rev. **55** (1987), 37–48

[4] H. Dinges, *Wiener Germs Applied to the Tails of M–Estimators*, Prob. Theory and Math. Stat., Vol. 1, pp. 411–418, Prohorov et al. (eds.), 1986 VNU Science Press

[5] H. Dinges, *Asymptotic Normality and Large Deviations*, Transactions of the Tenth Prague Conf. on Inf. Theory Statistical Decision Functions, Random Processes (1988), 13–21

[6] H. Dinges, *Special cases of second order Wiener germ approximations*, Probab. Th. Rel. Fields **83** (1989), 5-57

[7] R. Lugannani and S. Rice, *Saddlepoint approximation for the distribution of the sum of independent random variables*, Adv. Appl. Probab. **12** (1980), 475-490

Prof. Dr. Hermann Dinges
Fachbereich Mathematik
Universität Frankfurt
Postfach 11 19 32
D — 6000 Frankfurt

Higher Order Hydrodynamic Equations for a System of Independent Random Walks

R. DOBRUSHIN,[1] F. SOKOLOVSKII

We describe a construction that permits us to define hydrodynamic equations of higher order for a system of independent particles which perform homogeneous random walks. The main result states that, under some conditions on the increments of the random walk, the solution of the hydrodynamic equations of order n approximates the density of the original particle dynamics uniformly on time intervals $[0, \epsilon^{-(n-\gamma)}]$ for any $\gamma > 0$. Here ϵ is the small parameter used in the definition of the hydrodynamic limit.

1. Introduction. In this paper we consider a system of independent particles each of which performs a d-dimensional homogenous random walk with the same probability distribution for the increments. The problem of the derivation of the Euler hydrodynamic equation from the equations of classical Newtonian dynamics was first mathematically formulated in a pioneering paper of Morley [6]. He introduced an adequate notion of the hydrodynamic limit approach. But even now, more than 30 years after this paper, the real physical situation is too difficult for a complete mathematical analysis. A new direction connected with a study of the hydrodynamic limit approach for simpler models of infinite-particle dynamics arose in the last 10 years. It was initiated by the paper [10] devoted to the one-dimensional hard-rod system and by the paper [3] in which a simple case of independent one-dimensional random walks with Gaussian increments was considered. In the case when the mean value m of increments does not vanish a shift equation

$$\frac{\partial \rho}{\partial t} = -m \frac{\partial \rho}{\partial x} \tag{1.1}$$

for a density $\rho(x,t)$ of particles where $x \in \mathbb{R}^1$ is a space coordinate and $t \in [0, \infty)$ is a time parameter, arises in a hydrodynamic limit approach as a trivial analogue of the Euler equation. The case $m = 0$ requires

[1]The preparation of this paper was partly supported by U.S. Army Research Office through the Mathematical Science Institute of Cornell University, Contract DAAG29-85-C-0018.

another normalization in the hydrodynamic limit in which space and time are normalized in different scales. If the variance σ^2 of increments does not vanish a diffusion equation

$$\frac{\partial \rho}{\partial t} = \sigma^2 \frac{\partial^2 \rho}{\partial x^2} \tag{1.2}$$

arises after this new variant of the hydrodynamic limit approach.

A number of more complex cases of dynamics of dependent particles have been studied (see a review paper [1] and an excellent new book [9] for example). In all the cases considered the original variant of the hydrodynamic limit approach gives some differential (or integro-differential) equations of first order which are usually treated as an analogue of the Euler equation of the classical hydrodynamics. If the result of this limit approach (= "Euler term") vanishes then the second variant of the hydrodynamic limit approach is applicable and gives a hydrodynamic equation of diffusion type.

In the report [2] a new construction of hydrodynamic equations was discussed. It leads to hydrodynamic equations of a more general kind which include a sum of Euler-type and diffusion-type terms. For example in the case of Gaussian random walks discussed above this construction gives an equation

$$\frac{\partial \rho}{\partial t} = -m \frac{\partial \rho}{\partial x} + \varepsilon \sigma^2 \frac{\partial^2 \rho}{\partial x^2} \tag{1.3}$$

where ε is a small parameter, which characterises a typical ratio of micro- and macroscales and is included in the definition of the hydrodynamic limit approach. There is hope that the solutions of two-terms hydrodynamic equations of such type give a good approximation for the density of particles in the original particle dynamics for asymptotically larger intervals of time. It seems that it is possible to compare the hydrodynamic equations of such type with the Navier-Stokes equations of the classical hydrodynamics. As it was observed in [2] a similar approach can be applied to the construction of higher-order terms of hydrodynamic equations. But in [2] an erroneous conclusion was made that in general these higher-order terms vanish. This is not true and a purpose of this paper is to illustrate a real situation in the simplest case of a system of independent random walks. In this case the higher-order hydrodynamic equations turn out to be linear partial differential equations with coefficients which are the semi-invariants of the increments of the random walks. Under some restrictive conditions on signs of these semi-invariants we prove that the solution of the hydrodynamic equation of n-th order gives an asymptotically close approximation to the original density of multi-particle dynamics on the time interval $[0, K\varepsilon^{-(n-1)}]$, where $\varepsilon \to 0$ and K is any constant. If the conditions on signs of the semi-invariants are violated the solutions of the equations become divergent but the situation can be saved by a cut-off of the solutions at the high-frequency domain.

The real depth of the theory of random walks, which at first glance can seem a very simple subject, was brilliantly shown in a beautiful book by Spitzer [8]. We hope that this paper published in a volume devoted to Frank Spitzer can add something to it.

2. An independent system of random walks. Let F be a probability measure on the Euclidean space $\mathbb{R}^d$ describing the one-step distribution of a random walk. Let

$$F_n = F^{*n}, \qquad n \in \mathbb{N} \tag{2.1}$$

be its n-fold convolution describing distribution of the walk after n steps. We suppose that for all $h \in \mathbb{R}^d$ the *generating function*

$$G(h) = \int_{\mathbb{R}^d} \exp\{hx\} F(dx) < \infty . \tag{2.2}$$

The dynamics of a system of independent particles has to be in general described as a Markov process the one-time distribution of which are random point fields. However because in our case we have a unique conserved quantity — the number of particles — it is enough to follow the time-space evolution of the mean values of this number (see [2] for a detailed discussion). So we suppose that an initial density $\rho_0(x)$ is given which is a non-negative and smooth enough function of $x \in \mathbb{R}^d$. We let

$$L_k \overline{\rho}(x) = \int_{\mathbb{R}^d} \overline{\rho}(x-y) F_k(dy), \qquad x \in \mathbb{R}^d . \tag{2.3}$$

It is clear that a system of operators L_k defined for all non-negative smooth enough functions $\overline{\rho}(x)$, $x \in \mathbb{R}^d$ generates a semigroup

$$L_k L_m = L_{k+m}, \qquad k, m \in \mathbb{N} . \tag{2.4}$$

The function

$$\rho_k(x) = L_k \rho_0(x) \tag{2.5}$$

describes the density of particles of our system of independent particles at time k if the initial density is $\rho_0(x)$.

3. Hydrodynamic equations. Let $\{L_k, \; k \in \mathbb{N}\}$ be a semigroup of operators defined on a space of non-negative smooth enough functions of $x \in \mathbb{R}^d$. Let ε be a positive parameter going to zero (in all the following we will suppose that $0 < \varepsilon \leq 1$). Let

$$\overline{\rho}_0^{\varepsilon}(x) = \rho_0(\varepsilon x), \qquad x \in \mathbb{R}^d \tag{3.1}$$

be a system of initial densities obtained by a rescaling of a fixed initial density $\rho_0(x)$. Then we also make a rescaling of time and let

$$(3.2) \qquad \overline{\rho}^{\varepsilon}(x,t) = L_{t\varepsilon^{-1}}\overline{\rho}_0^{\varepsilon}(x), \qquad x \in \mathbb{R}^d,\ t \in \varepsilon\mathbb{N}.$$

At last making an inverse rescaling in the space we let

$$(3.3) \qquad \rho^{\varepsilon}(x,t) = \overline{\rho}^{\varepsilon}(\varepsilon^{-1}x,t), \qquad x \in \mathbb{R}^d,\ t \in \varepsilon\mathbb{N}.$$

The limit

$$(3.4) \qquad \lim_{\substack{\varepsilon\to 0\\ t'\to t}} \rho^{\varepsilon}(x,t') = \rho(x,t), \qquad x \in \mathbb{R}^d,\ t \in [0,\infty)$$

is called the *hydrodynamic evolution with the initial density* ρ_0 if of course this limit exists. The operator

$$(3.5) \qquad A_1\,\rho_0(x) = \lim_{t\to 0} t^{-1}\big(\rho(x,t) - \rho_0(x)\big), \qquad x \in \mathbb{R}^d$$

where $\rho(x,t)$ is defined by the relations (3.1)–(3.4) is called the *hydrodynamic operator* (if the limits (3.4) and (3.5) exist for a large enough class of non-negative smooth functions ρ_0). Finally the equations

$$(3.6) \qquad \frac{\partial \rho(x,t)}{\partial t} = A_1\,\rho(x,t), \qquad x \in \mathbb{R}^d,\ t \in [0,\infty)$$

is called the *hydrodynamic equation of the first order* (= *Euler type equation*). It is easy to understand that a system of transformations T_t : $\rho_0(\cdot) \to \rho(\cdot,t)$, $t \in [0,\infty)$ is in general a semigroup of transformations and so it is natural to expect that the solution of the equation (3.6) with the initial condition ρ_0 coincides with $\rho(x,t)$ defined by (3.4).

A definition of hydrodynamic operators A_n, $n = 2,3,\dots$ of higher orders will be given by a help of an induction. Suppose that operators $A_1,\dots,A_n$ have been already defined and that the equation

$$(3.7) \qquad \frac{\partial \rho_n^{\varepsilon}(x,t)}{\partial t} = \sum_{m=1}^{n} \varepsilon^{m-1} A_m \rho_n^{\varepsilon}(x,t), \qquad x \in \mathbb{R}^d,\ t \in [0,\infty)$$

have for any ε, $0 < \varepsilon \leq 1$ a unique solution with the initial condition

$$(3.8) \qquad \rho_n^{\varepsilon}(x,0) = \rho_0(x), \qquad x \in \mathbb{R}^d$$

in a corresponding class of solutions and for a corresponding large enough class of initial conditions. Then the operator

$$(3.9) \qquad A_{n+1}\,\rho_0(x) = \lim_{t\to 0}\Big(t^{-1} \lim_{\varepsilon\to 0} \varepsilon^{-n}\big(\rho^{\varepsilon}(x,t) - \rho_n^{\varepsilon}(x,t)\big)\Big), \qquad x \in \mathbb{R}^d$$

is called the *hydrodynamic operator of order* $n+1$ (if the limit (3.9) exists for a large enough class of functions ρ_0). The equation (3.7) is called the *hydrodynamic equation of order* n. It is possible to expect that hydrodynamic equations of higher orders give better approximations for the evolution of the density in the original particle dynamics in the sense that for any K, $0 < K < \infty$ and $C = C(K, n, \rho_0)$

$$|\rho^\varepsilon(x,t) - \rho_n^\varepsilon(x,t)| \leq Ct\varepsilon^n \quad \text{if} \quad t \leq K\varepsilon^{-n}, \ x \in \mathbb{R}^d . \tag{3.10}$$

(The constant C and as all the constants introduced in the following depend also on the semigroup $\{L_k, k \in \mathbb{N}\}$). It follows from (3.10) that for any $\gamma > 0$

$$\lim_{\varepsilon \to 0} \sup_{\substack{t : t \leq K\varepsilon^{-(n-\gamma)} \\ x \in \mathbb{R}^d}} |\rho^\varepsilon(x,t) - \rho_n^\varepsilon(x,t)| = 0 . \tag{3.11}$$

4. Formulation of the main result. We will use the following notations. By $\mathcal{R}$ we will denote the set of all *multi-indices* $R = (r_1, \dots, r_d)$, $r_i \in \mathbb{N}$, $i = 1, \dots, d$. Let

$$|R| = r_1 + \ldots + r_d , \qquad R \in \mathcal{R} \tag{4.1}$$

be the *order of a multi-index* R. Let

$$R! = r_1! \ldots r_d! , \qquad r \in \mathcal{R} . \tag{4.2}$$

(Here $0! = 1$). Let also

$$x^R = x_1^{r_1} \ldots x_d^{r_d} , \quad x = (x_1, \dots, x_d) \in \mathbb{R}^d, \qquad \mathbb{R} \in \mathcal{R} . \tag{4.3}$$

For any smooth enough function $\varphi(x)$, $x \in \mathbb{R}^d$ let

$$D^R\varphi = \frac{\partial^{|R|}\varphi}{\partial x_1^{r_1} \ldots \partial x_d^{r_d}} , \qquad R \in \mathcal{R} . \tag{4.4}$$

(Here $D^R\varphi = \varphi$ if $R = \phi$ is the multi-index of order $|\phi| = 0$). Finally fixing a probability measure F satisfying the condition (2.2) we denote by

$$\kappa_R = D^R \ln G(h) \Big|_{h=0} , \qquad R \in \mathcal{R} \tag{4.5}$$

the *semi-invariant of order* R of the measure F.

Main Theorem. *Fix a probability measure F defining a semigroup of operators (2.3) and an integer $n \geq 1$. Let the forms*

$$W_m(\lambda) = \sum_{R \in \mathcal{R} : |R| = m} (R!)^{-1} \kappa_R \lambda^r , \qquad \lambda \in \mathbb{R}^d , \ m = 1, 2, \ldots \tag{4.6}$$

and suppose that the functions W_m are non-positive for $m \equiv 0 \pmod 4$, $m \le n$ and that they are non-negative for $m \equiv 2 \pmod 4$, $m \le n$. Suppose also that the values $W_2(\lambda)$ are positive for all $\lambda \neq 0$ (i.e., the matrix of covariances for the probability distribution F is non-degenerate). Let

$$A_m = (-1)^m \sum_{R \in \mathcal{R} : |R| = m} (R!)^{-1} \kappa_R D^R , \quad m = 1, \dots . \tag{4.7}$$

Then for any initial condition ρ_0 which is square integrable in $\mathbb{R}^d$ the equation (3.7) has a unique solution which is square integrable with respect to $x \in \mathbb{R}^d$ for any fixed $t \in [0, \infty)$ and is n time continuously differentiable with respect to $t \in [0, \infty)$ for any fixed $x \in \mathbb{R}^d$.

Let

$$\psi^\gamma(x) = \exp\{-\gamma |x|^2\}, \qquad x \in \mathbb{R}^d, \ \gamma > 0 . \tag{4.8}$$

and

$$\rho_0^\gamma(x) = \rho_0(x) \exp\{-\gamma |x|^2\} . \tag{4.9}$$

(Of course $\rho_0^\gamma(x) \to \rho_0(x)$ if $\gamma \to 0$). Suppose that the function $\rho_0(x)$ is uniformly bounded in $x \in \mathbb{R}^d$. Let $\rho_n^{\varepsilon,\gamma}(x,t)$ be a solution of the eqution (3.7) with the initial condition ρ_0^γ. Then the limit

$$\lim_{\gamma \to 0} \rho_n^{\varepsilon,\gamma}(x,t) = \rho_n^\varepsilon(x,t) , \qquad x \in \mathbb{R}^d , \ t \in [0, \infty) \tag{4.10}$$

exists and is a solution of the equation (3.7) with the initial condition ρ_0.

Suppose now that the function ρ_0 is s time continuously differentiable and for some constant $H < \infty$ and any multi-index R of order $|R| \le s$ the derivative

$$\left| D^R \rho_0(x) \right| \le H, \qquad x \in \mathbb{R}^d. \tag{4.11}$$

Then there exists a value $s_0 = s_0(n,d)$ such that if the condition (4.11) is valid for $s = s_0$[3] then for any K, $0 < K < \infty$ and some $C = C(K, n, d, H)$ the estimate (3.10) is valid. Further for the same class of initial functions ρ_0 the limit (3.9) exists and coincides with $A_{n+1}\rho_0(x)$ where the operator A_{n+1} is defined by the relation (4.7).

(So in the terminology of Section 3 the operator A_n is the hydrodynamical operator of order n and the equation (3.7) is the hydrodynamical equation of order n for the semigroup (2.3).) The proof of the theorem is given in Sections 5–7.

[3] Following the construction in the proof of the thorem it is not difficult to evaluate s_0 explicitly. For example it is possible to let $s_0(n,1) = 2n$ if $n \ge 2$.

5. The solution of the equation. In this section we prove the statement of the theorem about the existence of the solution (4.10) of the hydrodynamical equation (3.7) with operators (4.7). Using the Fourier transform with respect to x

$$\hat{\rho}(\lambda,t) = \int_{\mathbb{R}^d} e^{i\lambda x} \rho(x,t)\, dx\ , \qquad \lambda \in \mathbb{R}^d \tag{5.1}$$

we can rewrite (at least formally) (3.7) as

$$\frac{\partial \hat{\rho}_n^\varepsilon(\lambda,t)}{\partial t} = \hat{S}_n^\varepsilon(\lambda)\, \hat{\rho}_n^\varepsilon(\lambda,t)\ , \qquad \lambda \in \mathbb{R}^d \tag{5.2}$$

where (see (4.6) and (4.7))

$$\hat{S}_n^\varepsilon(\lambda) = \sum_{m=1}^{n} i^m \varepsilon^{m-1} W_m(\lambda). \tag{5.3}$$

This equation with the initial condition

$$\hat{\rho}_0(\lambda) = \int_{\mathbb{R}^d} e^{i\lambda x} \rho_0(x) dx \tag{5.4}$$

has (at least formally) a solution

$$\hat{\rho}_n^\varepsilon(\lambda,t) = \hat{\rho}_0(\lambda)\, \exp\{t\hat{S}_n^\varepsilon(\lambda)\}. \tag{5.5}$$

Observe now that the condition of the theorem about signs of the functions W_n implies that for some constants $c > 0$

$$|\exp\{t\hat{S}_n^\varepsilon(\lambda)\}| \le \exp\{-ct\varepsilon|\lambda|^2\} \le 1. \tag{5.6}$$

So if the function ρ_0 is square integrable then the same is true for the functions $\hat{\rho}_n^\varepsilon(\cdot,t)$ defined by (5.5). In that case the previous formal considerations are rigorous justified and the hydrodynamical equation (3.7) has a unique solution

$$\rho_n^\varepsilon(x,t) = \rho_0(x) * V_n^\varepsilon(x,t) \tag{5.7}$$

where

$$V_n^\varepsilon(x,t) = (2\pi)^{-d} \int_{\mathbb{R}^d} e^{-i\lambda x}\, \exp\{t\hat{S}_n^\varepsilon(\lambda)\}\, dx \tag{5.8}$$

is the inverse Fourier transform of the function $\exp\{t\hat{S}_n^\varepsilon(\lambda)\}$ and we use the sign $*$ to denote convolutions of functions.

Observe now that because of the relation (5.6) the function $\exp\{t\hat{S}_n^\varepsilon(\lambda)\}$ is for any fixed t an element of the Schwartz space $\mathcal{S}(\mathbb{R}^d)$ of infinite differentiable functions of rapid decrease (see for example [4], §7.1). So the same is valid for the functions $V_n^\varepsilon(\cdot,t)$ and the convolution (5.7) is well defined for any uniformly bounded function ρ_0. It is clear that

$$\rho_n^{\varepsilon,\gamma}(x,t) = \left(\rho_0(x)\exp\{-\gamma|x|^2\}\right) * V_n^\varepsilon(x,t) \tag{5.9}$$

and it implies the existence of the limit (4.10) which coincides with (5.7). The previous formal calculations show that ρ_n^ε is a solution of the equation (3.7) in the sense of Schwartz distributions. But from the relation (5.7) it follows that also the function $\rho_n^\varepsilon(\cdot,t) \in \mathcal{S}(\mathbb{R}^d)$ for any $t > 0$ and is continuous in $t \geq 0$ for any fixed $x \in \mathbb{R}^d$. So the equation (3.7) is satisfied in a usual sense for any $x \in \mathbb{R}^d$, $t > 0$.

6. The derivation of the main estimate. In this section we will prove the most difficult statement of the theorem - the estimate (3.10). We shall use the following result about a decomposition of a unit (see [4], Theorem 1.4.6). There exists a non-negative infinite differentiable function $\phi(x), x \in \mathbb{R}^d$ with a support in a sphere $\{x \in \mathbb{R}^d : |x| \leq 1\}$ such that

$$\sum_{g\in\mathbb{Z}^d} \phi_g(x) \equiv 1, \qquad x \in \mathbb{R}^d \tag{6.1}$$

where

$$\phi_g(x) = \phi(x-g), \qquad x \in \mathbb{R}^d, \quad g \in \mathbb{Z}^d \tag{6.2}$$

is the shift of ϕ on g and $\mathbb{Z}^d$ is the d-dimensional integer lattice. We have a decomposition for the function (4.9):

$$\rho_0^\gamma(x) = \sum_{g\in\mathbb{Z}^d} \rho_0(x)\phi_g(x)\exp\{-\gamma|x|^2\}. \tag{6.3}$$

If we apply (5.7) with $\rho_0(x)$ changed on terms of the right part of (6.3) and then use the limit approach (4.10) we find that

$$\rho_n^\varepsilon(x,t) = \sum_{g\in\mathbb{Z}^d} (\rho_0(x)\phi_g(x)) * V_n^\varepsilon(x,t). \tag{6.4}$$

Turning to Fourier transforms we let

$$\psi_g(\lambda) = \int_{\mathbb{R}^d} e^{i\lambda x}\rho_0(x+g)\phi(x)dx, \ \lambda \in \mathbb{R}^d, \ g \in \mathbb{Z}^d. \tag{6.5}$$

Then

$$(6.6)\qquad \rho_n^\varepsilon(x,t) = \sum_{g\in\mathbb{Z}^d} (2\pi)^{-d} \int_{\mathbb{R}^d} e^{-i\lambda(x-g)}\psi_g(\lambda)\exp\{t\hat{S}_n^\varepsilon(\lambda)\}d\lambda.$$

On the other hand it follows from the definitions (2.1), (2.3), (2.5), (3.1) and the decomposition (6.1) that

$$(6.7)\qquad \begin{aligned} L_k\overline{\rho}_0^\varepsilon(x) &= \int_{\mathbb{R}^d} \overline{\rho}_0^\varepsilon(x-y)F^{*k}(dy) \\ &= \sum_{g\in\mathbb{Z}^d} \int_{\mathbb{R}^d} \phi_g(\varepsilon(x-y))\,\rho_0(\varepsilon(x-y))F^{*k}(dy). \end{aligned}$$

So turning to Fourier transforms and using also (3.2), (3.3) and (6.5) we find that

$$(6.8)\qquad \rho^\varepsilon(x,t) = \sum_{g\in\mathbb{Z}^d} (2\pi)^{-d} \int_{\mathbb{R}^d} e^{-i\lambda(x-g)}\,\psi_g(\lambda)(G(i\varepsilon\lambda))^{t\varepsilon^{-1}}\,d\lambda$$

where $G(h)$ is the generating function defined in (2.2). Now it follows from (6.6) and (6.8) that

$$(6.9)\qquad \rho^\varepsilon(x,t) - \rho_n^\varepsilon(x,t) = \sum_{g\in\mathbb{Z}^d} (2\pi)^{-d} B_g^\varepsilon(x,t)$$

where

$$(6.10)\qquad B_g^\varepsilon(x,t) = \int_{\mathbb{R}^d} e^{-i\lambda(x-g)}\psi_g(\lambda)((G(i\varepsilon\lambda))^{t\varepsilon^{-1}} - \exp\{t\hat{S}_n^\varepsilon(\lambda)\})\,d\lambda.$$

It is the starting formula for all the following estimates.

Let $M = (\kappa_{\{1,0,\dots,0\}},\dots,\kappa_{\{0,\dots,0,1\}}) \in \mathbb{R}^d$ be a vector of mean values induced by the probability measure F. In the first approximation particles of the system move with a constant velocity M. So it is natural to expect that $B_g^\varepsilon(x,t)$ if small if $|x-g-Mt|$ is large enough. Because of it we will estimate the quantity $|x-g-Mt|^{2v}B_g^\varepsilon(x,t)$ where v is a large enough integer which will be specified later. In the first place we observe that

$$(6.11)\qquad |x|^{2v} = \sum_{i_1,\dots,i_v=1,\dots,d} (x_{i_1})^2(x_{i_2})^2\dots(x_{i_v})^2.$$

So we obtain from (6.10) and (6.11) that

$$(6.12)\qquad \begin{aligned} &|x-g-Mt|^{2v}B_g^\varepsilon(x,t) \\ &= (-1)^v \sum_{i_1,\dots,i_v=1,\dots,d} \int_{\mathbb{R}^d} e^{-i\lambda(x-g-Mt)} \\ &\qquad \frac{\partial^{2v}}{\partial\lambda_{i_1}^2\dots\partial\lambda_{i_d}^2}\Big(\psi_g(\lambda)\left(G^\varepsilon(\lambda,t) - Q_n^\varepsilon(\lambda,t)\right)\Big)\,d\lambda \end{aligned}$$

where (see (5.3) and (4.6))

$$G^\varepsilon(\lambda,t) = (G(i\varepsilon\lambda))^{t\varepsilon^{-1}} e^{-it\lambda M}, \tag{6.13}$$

$$Q_n^\varepsilon(\lambda,t) = \exp\{t\hat{S}_n^\varepsilon(\lambda)\}e^{-it\lambda M} = \exp\left\{t\sum_{m=2}^{n} i^m\varepsilon^{m-1}W_m(\lambda)\right\}.$$

Differentiating the product inside the integral (6.12) we find that

$$\begin{aligned}(6.14)\quad & |x-g-Mt|^{2v}B_g^\varepsilon(x,t)\\ &= \sum_{R,R'\in\mathcal{R}:|R+R'|=2v} a_{R,R'}\int_{\mathbb{R}^d} e^{-i\lambda(x-g-Mt)}\cdot\\ &\qquad D_R(G^\varepsilon(\lambda,t)-Q_n^\varepsilon(\lambda,t))D_{R'}\psi_g(\lambda)d\lambda\end{aligned}$$

where the sum is taken over all pairs R, R' of multi-indexes such that their sum has the order $2v$ and $a_{R,R'}$ are some real coefficients. (It is not difficult to find these coefficients explicitly but it is not important for our purposes.) Now we want to estimate derivatives of the function Q_n^ε. For this purpose we use the following general fact. For any smooth enough function $f(\lambda)$, $\lambda\in\mathbb{R}^d$ and any multi-index $R\in\mathcal{R}$

$$\begin{aligned}(6.15)\quad & D_R\exp\{f(\lambda)\} =\\ &\left(\sum_{\substack{k,R_1,\dots,R_k\in\mathcal{R}:R_1+\dots+R_k=R\\ |R_i|\neq 0,i=1,\dots,k}}\prod_{i=1}^{k}R_i!D_{R_i}f(\lambda)\right)\exp\{f(\lambda)\},\end{aligned}$$

where the sum is taken over all representation of the multi-index R as a sum of other non-empty multi-indexes $R_1,\dots,R_k$ and k is arbitrary. For proving this identity it is enough to observe that if it is valid for some R the direct differentiation shows that it is also valid for any R' such that $|R'-R|=1$. We apply this identity to the function (compare (6.13))

$$f(\lambda) = q_{n,t}^\varepsilon(\lambda) = t\sum_{m=2}^{n} i^m\varepsilon^{m-1}W_m(\lambda). \tag{6.16}$$

Using the definition (4.6) we see that for every $R\in\mathcal{R}$

$$D_R q_{n,t}^\varepsilon(\lambda) = t\sum_{R'\in\mathcal{R}:R'\geq R,2\leq|R'|\leq n} i^{|R'|}((R'-R)!)^{-1}\kappa_{R'}\varepsilon^{|R'|-1}\lambda^{R'-R} \tag{6.17}$$

where $0!=1$ and $R'\geq R$ means that each component of the multi-index R' is not smaller than the corresponding component of R. (If $|R|>n$ we have an empty set of terms in (6.17) and we interpret the sum as zero.) From (6.17) it follows that for any γ, $0<\gamma\le 1$ and R, $0<|R|\le n$

$$(6.18)\qquad |D_R q^{\varepsilon}_{n,t}(\lambda)| \le \begin{cases} K_1 t\varepsilon^{n-1}|\lambda|^{n-|R|} & \text{if } \varepsilon|\lambda|\ge\gamma\,, \\ K_1 t\varepsilon^{|R|-1} & \text{if } \varepsilon|\lambda|<\gamma,\ |R|\ne 1\,, \\ K_1 t\varepsilon|\lambda| & \text{if } \varepsilon|\lambda|<\gamma,\ |R|=1\,, \end{cases}$$

where $K_1=K_1(\gamma,n)$ is a constant which depends on γ and n (and also as other constants below depends on semi-invariants $\kappa_{R'}$). Observe now that for any set $R_1,\dots,R_k$ of multi-indexes such that $R_1+\cdots+R_k=R$ we have

(6.19) if $\varepsilon|\lambda|\ge\gamma$ (remember that $0<\varepsilon\le 1$) then

$$\begin{aligned}\prod_{i=1}^{k}|D_{R_i}q^{\varepsilon}_{n,t}(\lambda)| &\le (K_1t)^k\varepsilon^{k(n-1)}|\lambda|^{kn-|R|} \\ &\le (K_1t)^k\varepsilon^{k(n-1)+(|R|-k)n}\gamma^{-(|R|-k)n}|\lambda|^{|R|(n-1)} \\ &\le K_2t^k(\varepsilon|\lambda|)^{|R|(n-1)}\end{aligned}$$

where $K_2=K_2(\gamma,n,|R|)$ is a new constant and

(6.20) if $\varepsilon|\lambda|<\gamma$ and we denote by p a number of indexes $i=1,\dots,k$ such that $|R_i|=1$ and let $s=k-p$ we have by using the estimates $s\le\left[\frac{|R|}{2}\right]$ and $p\le|R|$ that

$$\begin{aligned}\prod_{i=1}^{k}|D_{R_i}q^{\varepsilon}_{n,t}(\lambda)| &\le (K_1)^k(t\varepsilon)^s(t\varepsilon|\lambda|)^p \\ &\le K_3\left(1+(t\varepsilon)^{\left[\frac{|R|}{2}\right]}\right)\left(1+(t\varepsilon|\lambda|)^{|R|}\right)\,.\end{aligned}$$

where $K_3=K_3(\gamma,n)$ is a new constant.

It follows from (6.13) and (5.6) that

$$(6.21)\qquad |\,Q^{\varepsilon}_n(\lambda,t)\,| \le \exp\{-ct\varepsilon|\lambda|^2\}\le 1\,.$$

So we obtain from (6.15) and (6.19), (6.20), (6.21) that for any $R\in\mathcal{R}$

$$(6.22)\quad \left|D_R\,Q^{\varepsilon}_n(\lambda,t))\right| \quad\le$$

$$\begin{cases} K_4\,t^{|R|}\,(\varepsilon|\lambda|)^{|R|\,(n-1)}\exp\{-ct\varepsilon|\lambda|^2\} & \text{if } \varepsilon|\lambda|\ge\gamma \\ K_4\left(1+(t\varepsilon)^{\left[\frac{|R|}{2}\right]}\right)\left(1+(t\varepsilon|\lambda|)^{|R|}\right)\exp\{-ct\varepsilon|\lambda|^2\} & \text{if } \varepsilon|\lambda|<\gamma \end{cases}$$

where $K_4 = K_4(\gamma, n, |R|)$ is a new constant.

Now we want to estimate derivatives of the function $G^\varepsilon(\lambda,t)$. We can rewrite (6.13) as

$$G^\varepsilon(\lambda,t) = \left(G(i\varepsilon\lambda)\, e^{-i\varepsilon\lambda M}\right)^{t\varepsilon^{-1}} \tag{6.23}$$

From the definition (2.2) we find that for all $\lambda \in \mathbb{R}^d$ and $R \in \mathcal{R}$

$$\begin{aligned}(6.24)\quad &|D_R(G(i\varepsilon\lambda)\, e^{-i\varepsilon\lambda M})| \\ &= \varepsilon^{|R|}\left|\int_{\mathbb{R}^d}(x-M)^R \exp\{i\varepsilon\lambda(x-M)\}\, F(dx)\right| \le \varepsilon^{|R|}\, M_R\end{aligned}$$

where the absolute central moments

$$M_R = \int_{\mathbb{R}^d} |x-M|^R\, F(dx) \tag{6.25}$$

are finite because the condition (2.2). We will also use the following general fact (compare (6.15)). For any smooth enough function f and any integer $m \ge |R|$ where $R \in \mathcal{R}$

$$\begin{aligned}(6.26)\quad &D^R\left((f(\lambda))^m\right) = \\ &\sum_{k=1}^{|R|}\left(\sum_{\substack{R_1,\dots,R_k\in\mathcal{R}\\ R_1+\dots+R_k=R\\ |R_i|\ne 0,\, i=1,\dots,k}} \prod_{i=1}^{k} R_i!\, D_{R_i} f(\lambda)\right)\frac{m!}{(m-k)!}\,(f(\lambda))^{m-k},\end{aligned}$$

where the sum is taken over all representations of the multi-index R as a sum of other non-empty multi-indexes. Again it is not difficult to prove this identity by an induction with respect to $|R|$. Observe also that

$$\left|G(i\varepsilon\lambda)\, e^{-i\varepsilon\lambda M}\right| \le 1. \tag{6.27}$$

So from the relations (6.23) – (6.27) it follows that (remember again that $\varepsilon \le 1$)

$$|D_R\, G^\varepsilon(\lambda,t)| \le K_5\, |t|^{|R|}, \tag{6.28}$$

where $K_5 = K_5(|R|, n)$ is a constant.

For the case when $\varepsilon|\lambda|$ is small enough it is possible to obtain a better estimate. Observe that the function $\ell n G(i\lambda) - i\lambda M$ is an analytic function of λ in a sphere around the origin in which $G(i\lambda)$ does not vanish. Let

$$g_t^\varepsilon(\lambda) = \ell n G^\varepsilon(\lambda,t) = t\varepsilon^{-1}\ell n G(i\varepsilon\lambda) - it\lambda M. \tag{6.29}$$

Then using the analyticity we find that for some constants $\gamma_0' > 0$ and $K_6 = K_6(|R|)$ and any $R \in \mathcal{R}$

$$(6.30) \qquad |D_R\, g_t^\varepsilon(\lambda)| \le K_6 t\varepsilon^{|R|-1} \qquad \text{if} \quad \varepsilon|\lambda| < \gamma_0' .$$

Observe also that $g_t^\varepsilon(\lambda)$ and its first derivatives vanish at the point $\lambda = 0$. So for some constant K_7

$$(6.31) \qquad |D_R\, g_t^\varepsilon(\lambda)| \le K_7 t\varepsilon|\lambda| \qquad \text{if} \quad |R| = 1 \ \text{ and } \ \varepsilon|\lambda| < \gamma_0' .$$

The estimates (6.30), (6.31) are similar to the estimates (6.18) for the function $q_{n,t}^\varepsilon$ in the case $\varepsilon|\lambda| < \gamma$. Using the first two terms of the Taylor expansion for $g_t^\varepsilon(\lambda)$ we find that for some small enough $\gamma_0 > 0$ (we suppose that $\gamma_0 \le \gamma_0'$) and some $\bar{c} > 0$

$$(6.32) \qquad |G^\varepsilon(\lambda, t)| \le \exp\{-\bar{c}t\varepsilon|\lambda|^2|\} \qquad \text{if} \quad \varepsilon|\lambda| < \gamma_0 .$$

This estimate is similar to the estimate (6.21) for $Q_n^\varepsilon(\lambda, t)$. So we can repeat the calculations deriving (6.22) (for $\varepsilon|\lambda| < \gamma$) from (6.18) and (6.21) and conclude that for any $R \in \mathcal{R}$, if $\epsilon|\lambda| < \gamma_0$

$$(6.33) \quad |D_R\, G^\varepsilon(\lambda, t)| \le K_8 \left(1 + (t\varepsilon)^{[\frac{|R|}{2}]}\right) \left(1 + (t\varepsilon|\lambda|)^{|R|}\right) \exp\{-\bar{c}t\varepsilon|\lambda|^2\}$$

At last we can obtain even sharper estimates for the difference $G^\varepsilon(\lambda, t) - Q_n^\varepsilon(\lambda, t)$ in the case when $\varepsilon|\lambda|$ is small enough. In the first place we observe that $q_{n,t}^\varepsilon(\lambda)$ (see (6.16)) is a sum of terms of the Taylor expansion for $g_t^\varepsilon(\lambda)$ (see (6.29)) having orders smaller than $n+1$. So we obtain using (6.30) that for any $R \in \mathcal{R}$ and n

$$(6.34) \qquad \left| D_R \left(g_t^\varepsilon(\lambda) - q_{n,t}^\varepsilon(\lambda) \right) \right| \le K_9 t\varepsilon^n |\lambda|^{\max(n+1-|R|, 0)}$$

if $\varepsilon|\lambda| < \gamma_0$, where K_9 (and also constants $K_{10}, \ldots, K_{14}$ introduced later) is a constant depending on n and $|R|$. In the case $|R| = 0$ we obtain from (6.34) that

$$(6.35) \qquad \left| g_t^\varepsilon(\lambda) - q_{n,t}^\varepsilon(\lambda) \right| \le K_9 t\varepsilon^n |\lambda|^{n+1} \qquad \text{if} \quad \varepsilon|\lambda| < \gamma_0$$

It follows from (6.35) and (5.6) that if $|\lambda| < \alpha(t, \varepsilon)$

$$(6.36) \quad |G^\varepsilon(\lambda, t) - Q_n^\varepsilon(\lambda, t)| \le |Q_n^\varepsilon(\lambda, t)| \left| \exp\{g_t^\varepsilon(\lambda) - q_{n,t}^\varepsilon(\lambda)\} - 1 \right| \le K_{10} t\varepsilon^n |\lambda|^{n+1} \exp\{-ct\varepsilon|\lambda|^2\}$$

where

$$\alpha(t,\varepsilon) = \min\left(\gamma_0 \varepsilon^{-1}, (t\varepsilon^n)^{-\frac{1}{n+1}}\right) . \tag{6.37}$$

From the relation (6.15) we find that

$$\begin{aligned} D_R G^\varepsilon(\lambda,t) - D_R Q_n^\varepsilon(\lambda,t) = & \\ \sum_{\substack{k,R_1,\dots,R_k\in\mathcal{R}:R_1+\dots+R_k=R \\ |R_i|\neq 0, i=1,\dots,k}} & \left(\left(\prod_{i=1}^k R_i! D_{R_i} g_t^\varepsilon(\lambda)\right) G^\varepsilon(\lambda,t) - \right. \\ & \left. \left(\prod_{i=1}^k R_i! D_{R_i} q_{n,t}^\varepsilon(\lambda)\right) Q_n^\varepsilon(\lambda,t)\right) \end{aligned} \tag{6.38}$$

Now we shall estimate a separate term in the sum (6.38). Using the estimate (6.20), an analogous estimate for a product of $D_{R_d} g_t^\varepsilon(\lambda)$ following from (6.30) and (6.31) and the estimates (6.32), (6.34) and (6.36) we find that

$$\begin{aligned} &\left|\left(\prod_{i=1}^k R_i! D_{R_i} g_t^\varepsilon(\lambda)\right) G^\varepsilon(\lambda,t) - \left(\prod_{i=1}^k R_i! D_{R_i} q_{n,t}^\varepsilon(\lambda)\right) Q_n^\varepsilon(\lambda,t)\right| \\ &\le K_{11}\left\{\sum_{j=1}^k \left(\prod_{i=1}^{j-1} \left|D_{R_i} q_{n,t}^\varepsilon(\lambda)\right|\right)\left(\left|D_{R_j} g_t^\varepsilon(\lambda) - D_{R_j} q_{n,t}^\varepsilon(\lambda)\right|\right)\right. \\ &\qquad \left(\prod_{i=j+1}^k \left|D_{R_i} g_t^\varepsilon(\lambda)\right|\right) G^\varepsilon(\lambda,t) \\ &\qquad \left. + \left(\prod_{i=1}^k \left|D_{R_i} q_{n,t}^\varepsilon(\lambda)\right|\right)\left(\left|G^\varepsilon(\lambda,t) - Q_n^\varepsilon(\lambda,t)\right|\right)\right\} \\ &\le K_{12}\exp\{-\bar{c}t\varepsilon|\lambda|^2\}\left\{\sum_{j=1}^k \left(1 + (t\varepsilon)^{\left[\frac{|R|-|R_j|}{2}\right]}\right)\right. \\ &\qquad \left(1 + (t\varepsilon|\lambda|)^{|R|-|R_j|}\right) t\varepsilon^n |\lambda|^{\max(n+1-|R_j|,0)} \\ &\qquad \left. + \left(1 + (t\varepsilon)^{\left[\frac{|R|}{2}\right]}\right)\left(1 + (t\varepsilon|\lambda|)^{|R|}\right) t\varepsilon^n |\lambda|^{n+1} \exp\{-ct\varepsilon|\lambda|^2\}\right\} \\ &\le K_{13}\exp\{-\tilde{c}t\varepsilon|\lambda|^2\}\left(1 + (t\varepsilon)^{\left[\frac{|R|}{2}\right]}\right)\left(1 + (t\varepsilon|\lambda|)^{|R|}\right) t\varepsilon^n \left(1 + |\lambda|^{n+1}\right) \end{aligned} \tag{6.39}$$

if $|\lambda| < \alpha(t,\varepsilon)$, where $\tilde{c} = \min(c,\bar{c})$. Observe now that for some constant $K_{14} = K_{14}(|R|)$ for all t, ε, λ

$$\left(1 + (t\varepsilon|\lambda|)^{|R|}\right)\exp\{-\tilde{c}t\varepsilon|\lambda|^2\} \le K_{13}\exp\{-\frac{\tilde{c}}{2}t\varepsilon|\lambda|^2\} . \tag{6.40}$$

So it follows from (6.38), (6.39) and (6.40) that if $|\lambda| < \alpha(t,\varepsilon)$

$$(6.41)\quad | D_R G^\varepsilon(\lambda,t) - D_R Q_n^\varepsilon(\lambda,t) | \le K_{14} \exp\left\{ -\frac{\tilde{c}}{2} t\varepsilon |\lambda|^2 \right\} \left(1 + (t\varepsilon)^{[\frac{|R|}{2}]}\right) t\varepsilon^n \left(1 + |\lambda|^{n+1}\right)$$

An estimate of multipliers $D_R\psi_g(\lambda)$ used also in (6.14) is easier. Observe that a function (compare (6.5))

$$(6.42)\qquad \check{\psi}_g(x) = \rho_0(x+g)\varphi_0(x)$$

has a support in the sphere $\{|x| \le 1\}$ and is s time continuously differentiable. The derivative

$$(6.43)\qquad D_R\check{\psi}_g(x) = \sum_{R_1,R_2\in\mathcal{R}:\, R_1+R_2=R} (R_1!)\,(R_2!)\,\left(D_{R_1}\rho_0(x+g)\, D_{R_2}\phi_0(x)\right)$$

if $|R| \le s$ where the sum is taken over all pairs of multi-indexes R_1, R_2 such that $R_1 + R_2 = R$. So from the condition (4.11) of the theorem it follows that there exists a constant $C_1 = C_1(s)$ such that for all R such that $|R| \le s$ and all $\lambda \in \mathbb{R}^d$ and $g \in \mathbb{Z}^d$

$$(6.44)\qquad |D_R\check{\psi}_g(x)| \le C_1.$$

Using the definition (6.5) we see (compare (6.11), (6.12)) that for any integer β

$$(6.45)\quad |\lambda|^{2\beta} D_R\psi_g(\lambda) = \sum_{i_1,\dots,i_\beta=1,\dots,d} (-1)^\beta i^{|R|} \int_{\mathbb{R}^d} e^{i\lambda x} \frac{\partial^{2\beta}}{\partial x_{i_1}^2 \dots \partial x_{i_\beta}^2}\left(x^R\,\check{\psi}_g(x)\right) dx.$$

Because $\check{\psi}_g(x)$ is a finite-range function it follows from the relations (6.44), (6.45) and the relation (6.45) with $\beta = 0$ that for any $R \in \mathcal{R}$ and any β such that $2\beta \le s$ there exists a constant $C_2 = C_2(|R|,\beta)$ such that

$$(6.46)\qquad \left| D_R\,\psi_g(\lambda) l, \right| \le C_2 \left(1 + |\lambda|^{2\beta}\right)^{-1},\ \lambda \in \mathbb{R}^d,\ g\ \in \mathbb{Z}^d.$$

Now we are prepared for the estimation of the right part in (6.14). It is clear that there exists a constant $C_3 = C_3(v)$ such that

$$(6.47)\quad | x - g - Mt\,|^{2v} B_g^\varepsilon(x,t) \le C_3 \max_{\substack{R,\,R'\in\mathcal{R}\\ |R+R'|=2v}} \int_{\mathbb{R}^d} \left| D_R(G^\varepsilon(\lambda,t) - Q_n^\varepsilon(\lambda,t))\ , \right| \left| D_{R'}\psi_g(\lambda) \right| d\lambda.$$

We shall estimate integrals in (6.47) separately in different domains of integration. For any Borel $B \subset \mathbb{R}^d$ we let

$$I^g_{R,R'}(B) = \int_B |D_R(G^\varepsilon(\lambda,t) - Q^\varepsilon_n(\lambda,t))|\, |D_{R'}\psi_g(\lambda)| d\lambda. \tag{6.48}$$

If $s = s_0(n,d)$ is sufficiently large we can choose β such that

$$s \geq 2\beta \geq n + 1 + d. \tag{6.49}$$

For such β we find from (6.41) and (6.46) that for some constants $H_i = H_i(|R|, |R'|, n)$, $i = 1, 2, 3$

$$I^g_{R,R'}(\{\lambda \in \mathbb{R}^d : |\lambda| < \alpha(t,\varepsilon)\}) \tag{6.50}$$

$$\leq H_1(1 + (t\varepsilon)^{[\frac{|R|}{2}]})t\varepsilon^n \int_{\mathbb{R}^d} (1 + |\lambda|^{n+1})(1 + |\lambda|^{2\beta})^{-1} \exp\left\{-\frac{\tilde{c}}{2} t\varepsilon|\lambda|^2\right\} d\lambda$$

$$\leq H_2(1 + (t\varepsilon)^{[\frac{|R|}{2}]})t\varepsilon^n \min\left(\int_{\mathbb{R}^d} (1 + |\lambda|^{n+1})(1 + |\lambda|^{2\beta})^{-1} d\lambda\,, \int_{\mathbb{R}^d} \exp\left\{-\frac{\tilde{c}}{2} t\varepsilon|\lambda|^2\right\} d\lambda\right)$$

$$\leq H_3(1 + (t\varepsilon)^{[\frac{|R|}{2}]})t\varepsilon^n(1 + (t\varepsilon)^{d/2})^{-1}.$$

Further using the relations (6.22) (for $\gamma = \gamma_0$), (6.33) and (6.46) (with the condition (6.49)) and then (6.40) and observing that the following integral vanishes if $\alpha(t,\varepsilon) \neq (t\varepsilon^n)^{-1/(n+1)}$ we find that for some $H_i = H_i(|R|, |R'|, n)$, $i = 4, \ldots, 8$ and $c' = \min(c, \bar{c})/2$

$$I_{R,R'}\left(\left\{\lambda \in \mathbb{R}^d : \alpha(t,\varepsilon) \leq |\lambda| < \gamma_0\varepsilon^{-1}\right\}\right) \tag{6.51}$$

$$\leq \int_{\{|\lambda| : \alpha(t,\varepsilon) \leq |\lambda| < \gamma_0\varepsilon^{-1}\}} (|D_R G^\varepsilon(\lambda,t)| + |D_R Q^\varepsilon_n(\lambda,t)|) \left|D_{R'}\psi_g(\lambda)\right| d\lambda$$

$$\leq H_4\left(1 + (t\varepsilon)^{[\frac{|R|}{2}]}\right) \int_{\left\{|\lambda| \geq (t\varepsilon^n)^{-\frac{1}{n+1}}\right\}} \left(\left(1 + (t\varepsilon|\lambda|)^{|R|}\right) (\exp\{-ct\varepsilon|\lambda|^2\} + \exp\{-\bar{c}t\varepsilon|\lambda|^2\})\right) (1 + |\lambda|^{2\beta})^{-1} d\lambda$$

$$\leq H_5 \left(1+(t\varepsilon)^{[\frac{|R|}{2}]}\right) \int_{\left\{|\lambda|\geq (t\varepsilon^n)^{-\frac{1}{n+1}}\right\}} \left(1+|\lambda|^{2\beta}\right)^{-1} \exp\{-c't\varepsilon|\lambda|^2\}\, d\lambda$$

$$\leq H_6 \left(1+(t\varepsilon)^{[\frac{|R|}{2}]}\right) (t\varepsilon^n) \int_{\mathbb{R}^d} \left(1+|\lambda|^{2\beta}\right)^{-1} |\lambda|^{n+1} \exp\{-c't\varepsilon|\lambda|^2\}\, d\lambda$$

$$\leq H_7 \left(1+(t\varepsilon)^{[\frac{|R|}{2}]}\right) (t\varepsilon^n) \min\left(\int_{\mathbb{R}^d} \left(1+|\lambda|^{2\beta}\right)^{-1} |\lambda|^{n+1} d\lambda\ , \int_{\mathbb{R}^d} \exp\{-c't\varepsilon|\lambda|^2\}\, d\lambda\right)$$

$$\leq H_8 \left(1+(t\varepsilon)^{[\frac{|R|}{2}]}\right) (t\varepsilon^n) \left(1+(t\varepsilon)^{\frac{d}{2}}\right)^{-1} .$$

At last it follows from (6.22) (with $\gamma = \gamma_0$), (6.28) and (6.46) that if we additionally suppose that

$$s \geq 2\beta \geq |R|\,(n-1)+d \tag{6.52}$$

then for some $H_i = H_i(\,|R|\,,\,|R'|\,,n)\,,\ i=9,10$

$$I_{R,R'}\left(\left\{\lambda\in\mathbb{R}^d\ :\ |\lambda|\geq \gamma_0\varepsilon^{-1}\right\}\right) \tag{6.53}$$

$$\leq \int_{\{|\lambda|\,\geq\,\gamma_0\varepsilon^{-1}\}} \left(|\,D_R\, G^\varepsilon(\lambda,t)\,| + |\,D_R\, Q^\varepsilon_n(\lambda,t)\,|\right) \left|D_{R'}\,\psi_g(\lambda)\right|\, d\lambda$$

$$\leq H_9\, t^{|R|} \int_{\{|\lambda|\,\geq\,\gamma_0\varepsilon^{-1}\}} \left((\varepsilon|\lambda|)^{|R|(n-1)}+1\right) \left(|\lambda|^{2\beta}+1\right)^{-1} d\lambda$$

$$\leq H_{10}\, t^{|R|} \varepsilon^{-(2\beta-|R|(n-1)-d)} .$$

Remember now that we want to prove the estimate (3.10) under the condition $t \leq K\varepsilon^{-n}$. So if we suppose that

$$s \geq 2\beta \geq n\left(|R| - \left[\frac{|R|}{2}\right] - 1\right) + |R|(n-1) + d + \left[\frac{|R|}{2}\right] + n\ , \tag{6.54}$$

we can rewrite (6.53) as (compare (6.50) and (6.51))

$$I_{R,R'}\left(\left\{\lambda\in\mathbb{R}^d\ :\ |\lambda|\geq \gamma_0\varepsilon^{-1}\right\}\right) \leq H_{11} \left(1+(t\varepsilon)^{[\frac{|R|}{2}]}\right) (t\varepsilon^n) \left((t\varepsilon)^{\frac{d}{2}}+1\right)^{-1}, \tag{6.55}$$

if $t \leq K\varepsilon^{-n}$ where $H_{11} = H_{11}(|R|\,,n,K)$ is a new constant. Now from (6.47), (6.48), (6.50), (6.51) and (6.55) we obtain that under the condition $t \leq K\varepsilon^{-n}$

$$|\,x-g-Mt\,|^{2v} \left|B^\varepsilon_g(x,t)\right| \quad\leq\quad Ct\varepsilon^n\left(1+(t\varepsilon)^v\right)\left(1+(t\varepsilon)^{\frac{d}{2}}\right)^{-1} \tag{6.56}$$

where $C = C(n, K, v)$ is a constant and where we change the conditions (6.49), (6.52), (6.54) on the condition that β is chosen so that

$$s \geq 2\beta \geq 2\beta_0(n, d, v) \tag{6.57}$$

where $\beta_0(d, n, v)$ is a large enough value.

The estimate (6.56) will be used two times. In the first place we let $v = 0$ and find that if (6.57) is true for $v = 0$

$$\sum_{g\in\mathbb{Z}^d : |x-g-Mt| \leq (t\varepsilon)^{\frac{d}{2}}} \left| B_g^\varepsilon(x,t) \right| \leq Ct\varepsilon^n \left(1 + (t\varepsilon)^{\frac{d}{2}}\right)^{-1} \left(1 + (t\varepsilon)^{\frac{d}{2}}\right) \leq C_1\, t\varepsilon^n \tag{6.58}$$

if $t \leq K\varepsilon^{-n}$ where C_1 depends on K, n and d only. Further we let

$$v_0 = \begin{cases} \frac{d+1}{2} & \text{if } d \text{ is odd,} \\ \frac{d}{2} + 1 & \text{if } d \text{ is even.} \end{cases} \tag{6.59}$$

Then if (6.57) is true for $v = v_0$

$$\begin{aligned} \sum_{g\in\mathbb{Z}^d : |x-g-Mt| > (t\varepsilon)^{\frac{d}{2}}} \left| B_g(x,t) \right| &\leq Ct\varepsilon^n \left(1 + (t\varepsilon)^{v_0}\right) \left(1 + (t\varepsilon)^{\frac{d}{2}}\right)^{-1} \sum_{g\in\mathbb{Z}^d : |x-g-Mt| > (t\varepsilon)^{\frac{d}{2}}} |x - g - Mt|^{-2v_0} \\ &\leq C_2 t\varepsilon^n \left(1 + (t\varepsilon)^{\sigma_0}\right) \left(1 + (t\varepsilon)^{\frac{d}{2}}\right)^{-1} \min\left(1, (t\varepsilon)^{\frac{d}{2}(-2\sigma_0+d)}\right) \\ &\leq C_3 t\varepsilon^n, \qquad \text{if } \ t \leq K\varepsilon^{-n} \end{aligned} \tag{6.60}$$

where C_2, C_3 are new constants depending on k, n and d only. If we let

$$s_0(n, d) = \max\left(2\beta_0(n, d, 0)\,,\; 2\beta_0(n, d, v_0)\right) \tag{6.61}$$

we find that we can chose values of β such that (6.57) is valid both for $v = 0$ and $v = v_0$. So both estimates (6.59) and (6.60) are valid and because of the main relation (6.9) it implies the desired estimate (3.10).

7. The determination of hydrodynamic operators. Finally we want to prove that the operator A_{n+1} given by the relation (4.7) is the hydrodynamic operator of order $n + 1$ of our dynamics. It means that we have to check the relation (3.9). The proof of this relation is similar to the constructions of the previous section and so we describe it more briefly. From the relation (6.9) it follows that

$$\varepsilon^{-n}\left(\rho^{\varepsilon}(x,t)-\rho_n^{\varepsilon}(x,t)\right)=\sum_{g\in\mathbb{Z}^d}(2\pi)^{-d}\varepsilon^{-n}B_g^{\varepsilon}(x,t)\,. \tag{7.1}$$

Note that from the estimates (6.58) and (6.60) it follows that the series in the right part of (7.1) converges uniformly with respect to ε for each fixed t. We shall prove below that for any fixed $g\in\mathbb{Z}^d$ and t there exists the limit

$$\lim_{\varepsilon\to 0}(2\pi)^{-d}\varepsilon^{-n}B_g^{\varepsilon}(x,t)=t\,A_{n+1}\Big(\rho_0(x-Mt)\,\varphi_g(x-Mt)\Big) \tag{7.2}$$

Because of the linearity of the operator A_{n+1} and the identity (6.1) it implies that

$$\lim_{\varepsilon\to 0}\varepsilon^{-n}\left(\rho^{\varepsilon}(x,t)-\rho_n^{\varepsilon}(x,t)\right)=t\,A_{n+1}\left(\rho_0(x-Mt)\right) \tag{7.3}$$

Because the function ρ_0 is $(n+1)$-time continuously differentiable if the condition (4.11) of the theorem is true with $s=s_0(n,d)\geq n+1$ the relation (7.3) implies also the desired relation (3.9).

So we need to prove the relation (7.2). From the relation (6.12) with $v=0$ it follows that

$$B_g^{\varepsilon}(x,t)=\int_{\mathbb{R}^d}e^{-i\lambda(x-g-Mt)}\psi_g(\lambda)\left(G^{\varepsilon}(\lambda,t)-Q_n^{\varepsilon}(\lambda,t)\right)d\lambda\,. \tag{7.4}$$

On the other hand it follows from the relations (4.7), (4.6) and (6.5) that

$$\begin{aligned} &A_{n+1}\Big(\rho_0(x-Mt)\,\varphi_g(x-Mt)\Big)\\ &\qquad=\frac{i^{n+1}}{(2\pi)^d}\int_{\mathbb{R}^d}e^{-i(x-g-Mt)\lambda}\psi_g(\lambda)\,W_{n+1}(\lambda)\,d\lambda\,. \end{aligned} \tag{7.5}$$

So using the notations (6.48) we find that

$$\begin{aligned} &\left|(2\pi)^{-d}\varepsilon^{-n}B_g^{\varepsilon}(x,t)\;-\;tA_{n+1}(\rho_0(x-Mt)\,\varphi_g(x-Mt))\right|\\ &\leq(2\pi)^{-d}\left(\int_{\left\{|\lambda|\leq\varepsilon^{-\frac{1}{2(n+1)}}\right\}}|\psi_g(\lambda)|\right.\\ &\qquad\qquad\varepsilon^{-n}\left(G^{\varepsilon}(\lambda,t)\;-\;Q_n^{\varepsilon}(\lambda,t)\right)\;-\;i^{n+1}t\,W_{n+1}(\lambda)\,|\,d\lambda\\ &\qquad\qquad\left.+\,t\int_{\left\{|\lambda|>\varepsilon^{-\frac{1}{2(n+1)}}\right\}}|\psi_g(\lambda)|\,|W_{n+1}(\lambda)|\,d\lambda\right)\\ &\quad+\varepsilon^{-n}\,I_{0,0}^{g}\left\{\lambda\in\mathbb{R}^d\,:\,|\lambda|>\varepsilon^{-\frac{1}{2(n+1)}}\right\} \end{aligned} \tag{7.6}$$

We have to prove that each of the terms in the right part of (7.6) goes to zero when $\varepsilon \to 0$.

Remember that the function $q^{\varepsilon}_{n,t}(\lambda)$ is a sum of the first n terms of the Taylor expansion of $g^{\varepsilon}_t(\lambda)$ (see (6.16) and (6.29)). So using also the $(n+1)$-term of this expansion and an estimate for the remainder of the series implied by (6.30) we find that

$$(7.7) \qquad \begin{aligned} g^{\varepsilon}_t(\lambda) - q^{\varepsilon}_{n,t}(\lambda) &= i^{n+1} t \varepsilon^n W_{n+1}(\lambda) + \theta_1 C_1 \varepsilon^{n+1} |\lambda|^{n+2} \\ &= i^{n+1} t \varepsilon^n W_{n+1}(\lambda) + \theta_2 C_2 \varepsilon^{n+1/2} \end{aligned}$$

if $|\lambda| \le \varepsilon^{-\frac{1}{2(n+1)}}$, where here and in the following C_j, $j = 1, \ldots, 10$ are some constants depending on t and n and $\theta_i = \theta_i(n, \lambda, t)$ are such that $|\theta_i| \le 1$. It follows from (7.7) and the definition (4.6) that

$$(7.8) \qquad |g^{\varepsilon}_t(\lambda) - q^{\varepsilon}_{n,t}(\lambda)|^2 \le \theta_3 C_3 \varepsilon^{2n-1} \quad \text{if} \quad |\lambda| \le \varepsilon^{-\frac{1}{2(n+1)}}.$$

At last we observe that (see the definitions (6.16) and (4.6))

$$(7.9) \qquad \begin{aligned} |Q^{\varepsilon}_n(\lambda, t) - 1| &= |\exp\{q^{\varepsilon}_{n,t}(\lambda)\} - 1| \\ &\le |e^{C_4 \varepsilon |\lambda|^2} - 1| \\ &\le C_5 \varepsilon^{1-1/(n+2)} \quad \text{if} \quad |\lambda| \le \varepsilon^{-\frac{1}{2(n+1)}}. \end{aligned}$$

So it follows from (7.7), (7.8), (7.9) that

(7.10)
$$\begin{aligned} &\varepsilon^{-n}(G^{\varepsilon}(\lambda, t) - Q^{\varepsilon}_n(\lambda, t)) \\ &\quad = \varepsilon^{-n} Q^{\varepsilon}_n(\lambda, t)(\exp\{g^{\varepsilon}_t(\lambda) - q^{\varepsilon}_{n,t}(\lambda)\} - 1) \\ &\quad = \varepsilon^{-n}(1 + \theta_5 C_5 \varepsilon^{1-1/(n+2)}) \\ &\qquad (i^{n+1} t \varepsilon^n W_{n+1}(\lambda) + \theta_2 C_2 \varepsilon^{n+1/2} + \theta_6 C_6 \varepsilon^{2n-1}) \\ &\quad = i^{n+1} t W_{n+1}(\lambda)(1 + \theta_7 C_7 \varepsilon^{1/2}) \end{aligned}$$

if $|\lambda| \le \varepsilon^{-1/2(n+1)}$. Suppose now that the condition (6.52) is valid. Using also the estimate (6.46) for $R = \emptyset$ we find from (7.10) that

$$\int_{\{|\lambda| \le \varepsilon^{-1/2(n+1)}\}} |\psi_g(\lambda)| \, |\varepsilon^{-n}(G^{\varepsilon}(\lambda, t) - Q^{\varepsilon}_n(\lambda, t)) - i^{n+1} t W_{n+1}(\lambda)| d\lambda$$

$$(7.11) \qquad \le C_8 \varepsilon^{1/2} \int_{\mathbb{R}^d} |\psi_g(\lambda)| d\lambda \to 0 \quad \text{if} \quad \varepsilon \to 0.$$

It is clear from the definition (4.6) and the estimates (6.46) that if we suppose that $s_0(n, d) \ge n + 2 + d$ then

$$(7.12) \qquad \begin{aligned} &\int_{\{|\lambda| > \varepsilon^{-1/2(n+1)}\}} |\psi_g(\lambda)| \, |W_{n+1}(\lambda)| d\lambda \\ &\quad \le C_9 \int_{\{|\lambda| > \varepsilon^{-1/2(n+1)}\}} (1 + |\lambda|^{2\beta})^{-1} \, |\lambda|^{n+1} d\lambda \to 0 \quad \text{if} \quad \varepsilon \to 0. \end{aligned}$$

Repeating the consideration used in the estimate (6.50) we see that

$$(7.13)\qquad \varepsilon^{-n} I^{g}_{0,0}\left(\left\{\lambda \in \mathbb{R}^d : \varepsilon^{-1/2(n+1)} \leq |\lambda| < \alpha(t,\varepsilon)\right\}\right)$$

$$\leq C_{10} \int_{\{|\lambda| \geq \varepsilon^{-1/2(n+1)}\}} \left(1+|\lambda|^{n+1}\right)\left(1+|\lambda|^{2\beta}\right)^{-1} \exp\left\{-\frac{\tilde{c}}{2}|\lambda|^2\right\} d\lambda$$

which $\to 0$ if $\varepsilon \to 0$. At last the estimates (6.51) and (6.53) shows that

$$(7.14)\qquad \varepsilon^{-n} I^{g}_{0,0}\left(\left\{\lambda \in \mathbb{R}^d : \alpha(t,\varepsilon) \leq |\lambda| < \gamma_0 \varepsilon^{-1}\right\}\right) \to 0 \quad \text{if} \quad \varepsilon \to 0$$

and

$$(7.15)\qquad \varepsilon^{-n} I^{g}_{0,0}\left(\left\{\lambda \in \mathbb{R}^d : |\lambda| \geq \gamma_0 \varepsilon^{-1}\right\}\right) \to 0 \quad \text{if} \quad \varepsilon \to 0.$$

From the estimates (7.11)-(7.15) it follows that the right part of (7.6) goes to zero when $\varepsilon \to 0$. So the equality (7.2) is valid and as we have explained it implies the statement (3.9) of the theorem.

8. A cut-off approximation. The condition on the forms W_m, $m \geq 4$, m-even, introduced in the Main Theorem seems to be a very essential restriction. We know only one non-trivial class of situations when this condition is valid. In the paper [7] Shlosman proved that it is true for sums of values of ferromagnetic Gibbsian fields. In the case when this condition is not fulfilled the formal solution defined as the Fourier transform of the expression (5.5) cannot be interpreted even as a Schwartz distribution. It seems that such divergences are inherent to the problem. Nevertheless the following theorem shows that we can construct a good approximation to the original dynamics under a unique natural hypothesis about a non-degeneracy of the covariance matrix if we shall introduce a cut-off of the formal solution of the hydrodynamic equation at the domain of high frequencies.

Theorem. *Fix a probability measure F defining a semigroup of operators (2.3) and an integer $n \geq 1$. Let the form $W_2(\lambda)$ (see (4.6)) be positive for all $\lambda \neq 0$. Let $\chi(\lambda)$, $\lambda \in \mathbb{R}^d$ be an infinite differentiable function with a support on a sphere $\{|\lambda| \leq 1\}$ and such that $\chi(\lambda) \equiv 1$ for small enough λ. Let (compare (5.5))*

$$(8.1)\qquad \tilde{\rho}^{\varepsilon}_n(x,t) = \frac{1}{(2\pi)^d} \int_{\mathbb{R}^d} e^{-i\lambda x} \chi(\tilde{\gamma}^{-1}\varepsilon\lambda) \exp\{t\hat{S}^{\varepsilon}_n(\lambda)\}\hat{\rho}_0(\lambda) d\lambda.$$

Suppose that an initial condition $\rho_0(x)$ satisfies again the condition (4.11) with the same value of s_0. Then there exists a value $\tilde{\gamma} = \tilde{\gamma}(n) > 0$ such that for any K, $0 < K < \infty$ and some $C = C(K,n,d,H)$

$$(8.2)\qquad |\rho^{\varepsilon}(x,t) - \tilde{\rho}^{\varepsilon}_n(x,t)| \leq Ct\varepsilon^n \quad \text{if}\; t \leq K\varepsilon^{-n},\; x \in \mathbb{R}^d$$

(compare (3.10)).

Proof: The proof is similar to the constructions of sections 5 and 6, so we only note the main changes. The conditions of the main theorem on signs of forms was used only in the derivation of the estimate (5.6). Instead of it we observe that it follows from the condition of the theorem about the form W_2 that for some $c > 0$

$$W_2(\lambda) \leq -2c|\lambda|^2, \quad \lambda \in \mathbb{R}^d. \tag{8.3}$$

Further for $\lambda = (\lambda_1, \dots, \lambda_d) \in \mathbb{R}^d$

$$\begin{aligned} \Big|\sum_{m=3}^{n} i^m \varepsilon^{m-1} W_m(\lambda)\Big| &= \sum_{m=3}^{n} \left(\sum_{R \in \mathcal{R}:|R|=m} |\kappa_R| \right) \varepsilon^{m-1} \left(\max_{i=1,\dots,d} |\lambda_i| \right)^m \\ &\leq \varepsilon|\lambda|^2 \sum_{m=3}^{n} \left(\sum_{R \in \mathcal{R}:|R|=m} |\kappa_R| \right) (\varepsilon|\lambda|)^{m-2} \\ &\leq \varepsilon c|\lambda|^2 \quad \text{if} \quad \varepsilon|\lambda| \leq \tilde{\gamma} \end{aligned} \tag{8.4}$$

and $\tilde{\gamma} = \tilde{\gamma}(n)$ is small enough. We obtain from (8.3), (8.4) and (5.3) that

$$|\exp\{t\hat{S}_n^\varepsilon(\lambda)\}| \leq \exp\{-ct\varepsilon|\lambda|^2\} \quad \text{if} \quad \varepsilon|\lambda| \leq \tilde{\gamma} \tag{8.5}$$

and $\tilde{\gamma}$ is small enough.

Similar to (6.1)-(6.9) we find that

$$\rho^\varepsilon(x,t) - \tilde{\rho}_n^\varepsilon(x,t) = \sum_{g \in \mathbb{Z}^d} \tilde{B}_g^\varepsilon(x,t) \tag{8.6}$$

where $\tilde{B}_g^\varepsilon(x,t) =$

$$\int_{\mathbb{R}^d} e^{-i\lambda(x-g)} \psi_g(\lambda)((G(i\varepsilon\lambda))^{t\varepsilon^{-1}} - \chi(\tilde{\gamma}^{-1}\varepsilon\lambda)\exp\{t\hat{S}_n^\varepsilon(\lambda)\})d\lambda. \tag{8.7}$$

In (6.12) and the following relations it is necesary to replace $Q_n^\varepsilon(\lambda,t)$ (see (6.13)) on

$$\tilde{Q}_n^\varepsilon(\lambda,t) = \chi(\tilde{\gamma}^{-1}\varepsilon\lambda) Q_n^\varepsilon(\lambda,t). \tag{8.8}$$

Because

$$D_R \tilde{Q}_n^\varepsilon(\lambda,t) = \sum_{R_1,R_2 \in \mathcal{R}:R_1+R_2=R} R_1!R_2! D_{R_1}\chi(\tilde{\gamma}^{-1}\varepsilon\lambda) D_{R_2} Q_n^\varepsilon(\lambda,t) \tag{8.9}$$

the estimate (6.22) remains valid for $\tilde{Q}_n^\varepsilon(\lambda,t)$ with a changed constant K_4. A function $\chi(\lambda)$ is chosen such that for small enough γ_0

$$\tilde{Q}_n^\varepsilon(\lambda,t) = Q_n^\varepsilon(\lambda,t) \quad \text{if} \quad |\lambda| \leq \varepsilon^{-1}\gamma_0. \tag{8.10}$$

So we again have the estimate (6.41). All other considerations of section 6 carry over to the case of the function $\tilde{Q}_n^\varepsilon(\lambda,t)$.

9. Comments and generalizations. All the difficulties in the previous proofs are connected with the fact that under the condition (4.11) the Fourier transform of the initial function ρ_0 can be interpreted in general only as a Schwartz distribution. So we need the decomposition (6.1) and all the following tremendous estimates. The proofs become almost trivial if we suppose that ρ_0 is a square integrable smooth enough function. But in this case both the function ρ^ε and its hydrodynamic approximation ρ_n^ε go to zero when $t \to \infty$. Physically it means that we have a finite mass of a liquid which spreads over all the infinite space. In this case the estimate (3.10) is interesting for bounded t only. If we have

$$\rho_0(x) = \sum_{i=1}^{\infty} \rho_0^{T_i}(x) \tag{9.1}$$

where $\rho_0^T(x)$ is a periodic function with a period T then both functions ρ^ε and ρ_n^ε go to a constant when $t \to \infty$ and the estimate (3.10) is again non-trivial only for bounded t. A situation when a system has non-trivial hydrodynamic behavior in all moments and therefore we need the hydrodynamic expansion for its description can arise for example in the case

$$\rho_0(x) = \prod_{i=1}^{\infty} \rho_0^{T_i}(x) \tag{9.2}$$

where the periods $T_i \to \infty$ when $i \to \infty$. This case is relevant to an interesting physical situation when an initial density is a superposition of heterogenities of growing scales.

The results of the paper can be extended without any essential changes to the case when instead of independent random walks we consider independent random processes with independent increments and continuous time. In this case instead of semi-invariants (4.5) it is necessary to take semi-invariants of an increment during a unit of time. The study of the density of particles for some interacting Markov processes (see [5] for example) can be reduced to this case. For example in the case of one-dimensional symmetric exclusion processes we obtain increments which are sums of two independent Poisson random variables with the same parameters λ and jumps of lengths $+1$ and -1. In this case semi-invariants

$$\kappa_m = \begin{cases} 2\lambda & \text{if} \quad m \quad \text{is even,} \\ 0 & \text{if} \quad m \quad \text{is odd,} \end{cases} \tag{9.3}$$

and it is necessary to use a cut-off solution of hydrodynamic equations, described in section 8.

The hypothesis that the trajectories of particles are sums of independent jumps is not very essential. It seems that all the results can be extended to cases when these trajectories are sums of asymptotically weakly dependent random variables. In this case the semi-invariant κ_R can be replaced on limit values

$$\kappa_R = \lim_{t\to\infty} t^{-1}\kappa_R(0,t) \tag{9.4}$$

where $\kappa_R(0,t)$ is the semi-invariant of order R of the increment during a time interval $(0,t)$. The hypothesis about the independence of the trajectories of different particles is very essential. Even if it is possible to hope that the definition of hydrodynamic equations for such general cases is similar to definitions of section 3 these equations will not be linear in general and the study of the problem requires new ideas.

REFERENCES

[1] A. DeMasi, N. Ianiro, A. Pelligrinotti, E. Presutti *A Survey of the hydrodynamical behavior of many-particle systems* pagges 123 – 294 in Non-equilibrium Phenomena II; From Stochastics to Hydrodynamics Edited by J. L. Lebowitz, E. W. Montroll, North-Holland, Amsterdam. 1984

[2] R. L. Dobrushin *Caricatures of hydrodynamics,* pages 117–132 in 9th International Congress on Mathematical Physics 17-27 July 1988, Swansea, Wales, edited by B.Simon, A. Truman, I. M. Davies, Adam Higler, Bristol, 1989

[3] R. L. Dobrushin, R. Siegmund-Schultze *The hydrodynamic limit for systems of particles with independent evolution* Math. Nachr. **105** (1982) , 199–224

[4] L. Hörmander *The analysis of linear partial differential operators,* Vol. 1 Springer, New York, 1983

[5] T. M. Liggett *Interacting particle systems* Springer, New York, 1985

[6] C. B. Morrey *On the derivation of the equation of hydrodynamics from statistical mechanics* Comm. Pure Appl. Math. **8** (1955), 279–326

[7] S. Shlosman *Signs of the Ising model Ursell functions* Comm. Math. Phys. **102** (1986), 679–686

[8] F. Spitzer *Principles of random walk,* second edition, Springer, New York, 1964

[9] H. Spohn *Large scale dynamics of interacting particles,* Texts and Monographs in Physics, Springer Verlag, to appear

[10] Boldrighini, C., Dobrushin, R.L., Suhov, Yu. M., *Hydrodynamical limit for a degenerate model of classical statistical mechanics,* Uspekhi, Mathm. Nauk., **35**(1980), 152

R. Dobrushin
Inst. for Problems
of Information Transmission
Yermelovoi 19, Moscow 101447
USSR

F. Sokolovskii
Moscow State University
Leninskie gory
Moscow 119899
USSR

Making Money From Fair Games

R. DURRETT,[1] H. KESTEN,[2] AND G. LAWLER[3]

Suppose that a casino offers p fair games. Can we make money by playing the games (a) according to a fixed schedule, (b) using a strategy that depends on our wealth? The answers depend on the moments of the payoff distribution.

1. Introduction. The origin of this work is a letter from A. Spataru to Frank Spitzer. He wrote "In a study of a p-type branching process I was led naturally to consider a random walk $S_n = X_1 + \cdots + X_n$, where the steps X_n are independent, and $\mathcal{L}(X_n) \in \{\mathcal{L}(Y_1) \ldots \mathcal{L}(Y_p)\}$ where $\mathcal{L}(X)$ denotes the law of X and $Y_1, \ldots, Y_p$ are given random variables in L^1." After observing that several classical results generalized easily to the new set-up, he continued: "I would like to know whether the next statement is correct: If $E(X_n) = 0$ then $-\infty = \liminf S_n < \limsup S_n = \infty$."

Our first reaction was that this should be true and that furthermore the process we will call a "finitely inhomogeneous random walk" (or f.i.r.w.) should be recurrent in the sense that there is an $L < \infty$ so that $P(|S_n| \in [-L, L] \text{ i.o.}) = 1$. In support of this viewpoint we would like to observe that it is easy to prove the following result.

Theorem 0. *Suppose that $Y_1, \ldots, Y_p$ are integer valued and have $E(Y_i) = 0$. Let $i_1, i_2, \ldots$ be any stationary sequence taking values in $\{1, \ldots p\}$, let $X_1, X_2, \ldots$ be independent with $X_j = \mathcal{L}(Y_{i_j})$, and let $S_n = X_1 + \cdots + X_n$. Then $P(S_n = 0 \text{ i.o.}) = 1$.*

Proof. $X_1, X_2, \ldots$ is a stationary sequence of integer valued random variables with $E(X_i|\mathcal{I}) = 0$, where $\mathcal{I}$ is the shift-invariant σ-field, so the conclusion follows from a 'well-known result' that says partial sums of such sequences are recurrent. See Durrett [3], p.1011, for a proof and references to the literature.

Notice that the stationary sequence in Theorem 0 is arbitrary and that the conclusion $P(S_n = 0 \text{ i.o.}) = 1$ holds for almost every environment $i_1, i_2 \ldots$. The last observation should convince you that the answer to

---

[1]Partially supported by the National Science Foundation, the National Security Agency, and the Army Research Office through the Mathematical Sciences Institute at Cornell University.

[2]Partially supported by the National Science Foundation.

[3]Partially supported by the National Science Foundation. This work was done while this author was visiting Cornell partially supported by a Sloan fellowship and the Army Research Office through the Mathematical Sciences Institute at Cornell University.

Spataru's question is 'Yes,' so we turn now to the task of showing that it is 'No.' The inspiration for the first counterexample came from Feller's unfair "fair" games (see [4] or Exercise 17 on p.58 of Breiman [1]). Let $X_1, X_2, \ldots$ be i.i.d. with

$$P(X_i = 2^k - 1) = \frac{1}{2^k k(k+1)} \qquad k \geq 1$$

and $P(X_i = -1)$ defined to make the sum unity. Summing the telescoping series shows that $E(X_i + 1) = 1$, i.e., $E(X_i) = 0$. However, this fair game is unfavorable in the following sense: If $S_n = X_1 + \cdots + X_n$ then

$$S_n/(n/\log_2 n) \to -1 \text{ in probability.}$$

Being a random walk with mean 0 steps, S_n must be recurrent. To see how it returns to 0, notice that the probability of a jump of size 2^{k+1} is $1/2^{k+1}(k+1)(k+2)$, so the probability of a jump this large during $((k-1)2^{k-1}, k2^k]$ is about $1/4k$, and the Borel-Cantelli lemma implies that these big jumps will occur infinitely often. If a big jump occurs in the kth interval, it is likely that just before the big jump S_n is larger than

$$(-3/2)\, k2^k / \log_2(2^{k-1}) > -(.8)2^{k+1}, \qquad \text{for large } k$$

and hence the big jump carries us above 0. After the big jump the random walk will drift down and hit 0 (since negative steps have size -1).

Having identified how the last random walk returns to 0, it is not too hard to construct a f.i.r.w. that does not get there. We define an increasing sequence of constants a_k with $a_0 = 1$, let F_1 have mass $1/(a_k k^2)$ at a_k for odd k, let F_2 have mass $1/(a_k k^2)$ at a_k for even k, and then pick negative constants $c_i, i = 1, 2$, so that if the rest of the mass in the distribution F_i is placed at c_i, the mean will be 0. If $a_{k+1}/a_k \to \infty$ fast enough then it is not hard to show that the f.i.r.w. that uses F_1 when $i \in [a_{2j-2}, a_{2j-1})$ and uses F_2 when $i \in [a_{2j-1}, a_{2j})$ converges to $-\infty$ as $n \to \infty$. Fine tuning the last construction leads to

Example 1. Let $a_{j+1}/a_j \to \infty$. To define F_j for $1 \leq j \leq p$ put mass $1/(a_k^q k^2)$ at a_k if $k \equiv j \pmod p$. Put the rest of the mass at $c_j < 0$ to make the mean 0. Let $b_0 = 0$ and $b_k = b_{k-1} + a_k^q$ for $k \geq 1$. Let X_n have distribution F_j when $b_{k-1} < n \leq b_k$ if $k \equiv j \pmod p$. These distributions have finite qth moments. If $q < (2p-1)/p$ then letting $a_k = \exp(\lambda^k)$ and picking λ close to 1, we get a f.i.r.w. with $S_n \to -\infty$ almost surely. In fact, we will show in Section 2 that

$$\limsup_{n \to \infty} S_n / n^{1/q\lambda} = -\infty.$$

Remark. As the reader has probably noticed, we have just figured out how to lose money at a fair game, or to put things another way, we have designed a collection of "fair" lotteries that allow a gambling house or state government to make money.

As $p \to \infty$, $(2p-1)/p \to 2$. The next result shows that we can never construct an example with a finite second moment. Indeed, we will prove a stronger conclusion – even if the gambler chooses the game to play at time n according to an *adapted strategy* it is impossible to make money. To be precise let $\mathcal{F}_0$ be the information the gambler has at time 0 and suppose $S_0 \in \mathcal{F}_0$ (i.e, S_0 is $\mathcal{F}_0$ measurable). Let $X_n^i, 1 \le i \le p, n \ge 0$ be independent r.v.'s that are independent of $\mathcal{F}_0$, and suppose $P(X_n^i \le x) = F_i(x)$. To build the *controlled random walk, (c.r.w.)*, let $\psi_n \in \mathcal{F}_{n-1}$ be the choice of game at time n, let $S_n = S_{n-1} + X_n^{\psi_n}$, and let $\mathcal{F}_n = \mathcal{F}_{n-1} \vee \sigma(\psi_n, X_n^i, 1 \le i \le p)$, i.e., the smallest σ-field containing $\mathcal{F}_{n-1}$ and the indicated random variables.

Theorem 1. *If $F_1, \ldots, F_p$ are mean zero distributions with finite non-zero variances then the c.r.w. is recurrent. That is, there is an $L < \infty$ so that $P(S_n \in [-L, L]$ i.o.$) = 1$ for all adapted strategies.*

Having seen that it is possible by using 11 laws to get an example with finite 1.9th moment, it is natural to ask if this is possible with two laws. The answer is that when $p = 2$ you cannot do any better than Example 1.

Theorem 2. *Let S_n be a f.i.r.w. with $p = 2$. Assume that*

$$\int x\, dF_i(x) = 0 \quad \text{and} \quad \int |x|^{q_i}\, dF_i(x) < \infty.$$

If $\lim P(S_n > 0) = 1$ then $q_1 + q_2 \le 3$.

Looking at Theorem 2 and Example 1, the reader might guess that for p distributions the condition for the existence of a f.i.r.w. that goes off to infinity is $q_1 + \cdots + q_p \le (2p-1)$. However, we believe the right condition is much more complicated.

Example 2. Suppose $q_i < 2$ and let $r_i = (2 - q_i)/(q_i - 1)$. By generalizing the proof of Example 1 it is not hard to show that if

$$(\star) \qquad \sum_{\emptyset \ne S \subseteq \{1,\ldots,p\}} (|S| - 1) \prod_{i \in S} r_i > 1,$$

where $|S|$ denotes the number of points in S or equivalently,

$$(\star\star) \qquad \left[\frac{d}{dt} \prod_{i=1}^{p} (1 + t r_i)\right]_{t=1} > \prod_{i=1}^{p} (1 + r_i),$$

then we can find a f.i.r.w. in which $F_i, 1 \le i \le p$ has mean 0 and finite q_ith moment but $S_n \to \infty$ almost surely.

We conjecture that that there is a f.i.r.w. with $P(S_n > 0) \to 1$ if and only if $(\star)$ holds. In support of this, the reader should observe that

(a) when $p = 2$, $(\star)$ reduces to $r_1 r_2 > 1$ which is just another way of saying $q_1 + q_2 < 3$; (b) when all the r_i are equal to r condition $(\star\star)$ says $p(1+r)^{p-1} > (1+r)^p$, or $p > 1 + (2-q)/(q-1)$, and rearranging gives $q < (2p-1)/p$.

To complete the discussion we should mention the one paper that we know of that treats a related problem. Rogozin and Foss [7] considered an "oscillating random walk" Y on Z that takes steps according to one law F_1 when it is positive and according to a different law F_2 when it is negative, and they found necessary and sufficient conditions for the walk to be recurrent. Their conditions are difficult to check in general but are quite simple in the case in which each walk is *strongly attracted to a stable law.* That is, there is a sequence of constants b_n^i so that $P(S_n^i < b_n^i x)$ approaches a nondegenerate stable law, $G^i(x)$, the superscript indicating the random walk under consideration. If we let α_i be the index of the limiting stable law and let $a_1 = \alpha_1(1 - G_1(0))$ and $a_2 = \alpha_2 G_2(0)$ then their result can be stated as

Theorem 3. *If* $a_1 + a_2 > 1$ *then* Y *is recurrent while if* $a_1 + a_2 < 1$ Y *is not.*

To relate this to our results note that by 1 on page 159 in [7], $a \le 1$. Applying this result to $-S_n^i$ we see that $G_i(0) \le 1/\alpha_i$ and hence $a_i \ge \alpha_i - 1$. Combining the last observation with Theorem 3 we see that if Y is transient then $\alpha_1 + \alpha_2 \le 3$. When the F_i have mean 0, the oscillating random walk has

$$-\infty = \liminf Y_n < \limsup Y_n = \infty$$

since both of the random walks that make it up are recurrent. However, Theorem 3 shows that Y_n can be transient. To complete the picture we give an example of a transient f.i.r.w. that oscillates between $-\infty$ and ∞.

Example 3. For $k \ge 0$, let $a_k = \exp(\lambda^k)$ and let

$$F_1 \text{ have mass } \begin{cases} \dfrac{1}{(2k)^2 a_{2k} a_{2k+1}^{q-1}} & \text{at } a_{2k} \\[2ex] \dfrac{1}{(2k)^2 a_{2k+1}^{q}} & \text{at } -a_{2k+1}. \end{cases}$$

and

$$F_2 \text{ have mass } \begin{cases} \dfrac{1}{(2k+1)^2 a_{2k+1} a_{2k+2}^{q-1}} & \text{at } -a_{2k+1} \\[2ex] \dfrac{1}{(2k+1)^2 a_{2k+2}^{q}} & \text{at } a_{2k+2}. \end{cases}$$

The parts of the distributions that have been defined have mean 0, so put the rest of their mass at 0. Let $b_0 = 0$ and $b_k = b_{k-1} + a_k^q$ for $k \ge 1$. We let X_n have distribution F_1 when $b_{2k} < n \le b_{2k+1}$ and have distribution F_2 when $b_{2k+1} < n \le b_{2k+2}$. Here $k \ge 0$. We will show that if $q < 3/2$ and λ is close to 1, S_n is transient but

$$-\infty = \liminf S_n < \limsup S_n = \infty.$$

Notice that F_1 and F_2 have finite qth moments, and as in Example 1 our method fails to produce an example with $q \geq 3/2$. We do not know whether there is an example that has moments of order $q > 3/2$.

In Section 2 we give the calculations for Examples 1 and 2. In Section 3 we treat Example 3. In Section 4 we will prove Theorem 1. Section 4 is independent of Sections 2 and 3, and can be read now if the reader desires. The proof of Theorem 2 is too lengthy to be included here so it will be given in [5]. To motivate the reader to look at the details we leave him or her with the question: "How can the proof take advantage of the fact that only two distributions are used in the sequence?"

Section 2. Computations for Examples 1 and 2. The technique is standard — truncate then compute means and variances. Let $a_k = \exp(\lambda^k)$ with $\lambda > 1$, and $\bar{X}_n = X_n 1_{(|X_n| \leq a_{k-p})}$ for $b_{k-1} < n \leq b_k$.

$$P(\bar{X}_n \neq X_n) \leq \sum_{\ell \geq k} 1/(a_\ell^q \ell^2) \leq C/(a_k^q k^2),$$

since $a_{\ell+1}/a_\ell \geq 1 + \delta$ for some $\delta > 0$. Here and in what follows C is a constant whose value is not important and will change from line to line. The last estimate implies

$$P(\bar{X}_n \neq X_n \text{ for some } n \in [b_{k-1}, b_k)) \leq C/k^2.$$

Summing over k and using the Borel–Cantelli lemma gives $P(\bar{X}_n \neq X_n \text{ i.o.}) = 0$, so it suffices to investigate the behavior of $\bar{S}_n = \bar{X}_1 + \cdots + \bar{X}_n$.

Let $T_k = \bar{S}_{b_k}$. We will show that for large k, $T_k \leq -a_k/2k^2$ and use this to conclude that $\bar{S}_n \to -\infty$ almost surely. To see where the $-a_k/k^2$ comes from, notice that for $n \in (b_{k-1}, b_k]$ we have $E\bar{X}_n \leq -a_k^{1-q}/k^2$ so for $n = b_{k-1} + m$ with $0 < m \leq a_k^q$ we have

$$E\bar{S}_n \leq -a_k^q a_{k-1}^{1-q}/(k-1)^2 - m a_k^{1-q}/k^2 \leq -(a_{k-1} + m a_k^{1-q})/k^2 \tag{2.1}$$

To show that $\bar{S}_n$ is close to its expected value we estimate

$$\operatorname{var}(\bar{X}_n) \leq E(\bar{X}_n^2) \leq \max_i c_i^2 + \sum_{\ell \leq k-p} a_\ell^2/a_\ell^q \ell^2 \leq C a_{k-p}^{2-q}/k^2. \tag{2.2}$$

Letting $\Delta_m = \bar{S}_{b_{k-1}+m} - T_{k-1}$ and using the Hajek-Renyi inequality (Exercise 16 on p.434 in [6]) gives

$$\begin{aligned} &P\left(\Delta_m - m\mu_k \geq \frac{a_{k-1} + m a_k^{1-q}}{3k^2} \text{ for some } 0 < m \leq a_k^q\right) \\ &\leq \sum_{m=1}^{a_k^q} \frac{9k^2 a_{k-p}^{2-q}}{(a_{k-1} + m a_k^{1-q})^2} \end{aligned} \tag{2.3}$$

where $\mu_k = E\bar{X}_n$ for $n \in (b_{k-1}, b_k]$. Let $N = a_{k-1}a_k^{q-1}$. Breaking the sum into $m \le N$ and $m > N$, we see that the right-hand side of (2.3) is bounded by

$$N\frac{9k^2 a_{k-p}^{2-q}}{a_{k-1}^2} + \frac{9k^2 a_{k-p}^{2-q}}{a_k^{2-2q}} \sum_{m=N+1}^{\infty} m^{-2} \le Ck^2 \frac{a_k^{q-1} a_{k-p}^{2-q}}{a_{k-1}}.$$

If we can pick a_k so that

$$\sum_{k=1}^{\infty} k^2 a_k^{q-1} a_{k-p}^{2-q}/a_{k-1} < \infty, \tag{2.4}$$

then the Borel-Cantelli lemma guarantees that with probability one, for large k we have

$$\Delta_m \le -ma_k^{1-q}/k^2 + (a_{k-1} + ma_k^{1-q})/3k^2 \quad \text{for } 1 \le m \le a_k^q. \tag{2.5}$$

Using (2.5) with $m = a_k^q$ shows that for large k, $T_k - T_{k-1} \le (-2a_k + a_{k-1})/3k^2$. Since our a_k grow more than exponentially fast it follows that $T_k \le -a_k/2k^2$ for large k. The last conclusion and (2.5) imply that for large k

$$S_{b_{k-1}+m} \le -(a_{k-1} + ma_k^{1-q})/6k^2 \quad \text{for } 1 \le m \le a_k^q.$$

To get the rate cited in the introduction, we note that (i) $b_k \sim a_k^q$, (ii) for $n \in (b_{k-1}, b_k]$, $k \sim \log\lambda \log\log n$, (iii) use $-a_{k-1}/6k^2$ as an upper bound for $m \le N = a_{k-1}a_k^{q-1}$, and use $-ma_k^{1-q}/6k^2$ for $N < m \le a_k^q$

To investigate when (2.4) holds, we notice that if $a_k = \exp(\lambda^k)$ then

$$\begin{aligned} a_{k-p}^{2-q} a_k^{q-1}/a_{k-1} &= \exp((2-q)\lambda^{k-p} + (q-1)\lambda^k - \lambda^{k-1}) \\ &= \exp(\lambda^{k-p}((2-q) + (q-1)\lambda^p - \lambda^{p-1})). \end{aligned}$$

We would like to make

$$g(\lambda) \equiv (2-q) + (q-1)\lambda^p - \lambda^{p-1} < 0.$$

Differentiating gives

$$g'(\lambda) = p(q-1)\lambda^{p-1} - (p-1)\lambda^{p-2}.$$

$g(1) = 0$ and $g'(1) = pq - p - p + 1$ so if $q < (2p-1)/p$ we can find a $\lambda > 1$ that makes $g(\lambda) < 0$.

The first step in treating Example 2 is to describe it precisely. Extend q_j to $j \ge 1$ to be periodic ($q_j = q_{j-p}$). To define F_j for $1 \le j \le p$ put mass $1/a_k^{q_k}k^2$ at a_k if $k \equiv j \pmod p$. Put the rest of the mass at $-c_j$ to make the mean 0. Let $b_0 = 0$ and $b_k = b_{k-1} + a_k^{q_k}$ for $k \ge 1$. Let X_n

have distribution F_j when $b_{k-1} < n \le b_k$ if $k \equiv j \pmod p$. Defining $\bar{X}_n$ as before and repeating the previous argument shows that it is enough to consider $\bar{S}_n$. Let $T_j = \bar{S}_{b_j}$ and $\Delta_m = \bar{S}_{b_{k-1}+m} - T_{k-1}$. Letting $N = a_{k-1}a_k^{q_k}$ and repeating our analysis of (2.3) shows

$$(2.6) \qquad P(\Delta_m - m\mu_k \ge (a_{k-1} + m a_k^{1-q_k})/3k^2) \le Ck^2 a_k^{q_k-1} a_{k-p}^{2-q_k}/a_{k-1}.$$

Recalling the end of the treatment of Example 1, we see it is enough to show

$$(2.7) \qquad \sum_k Ck^2 a_{k-p}^{2-q_k} a_k^{q_k-1}/a_{k-1} < \infty.$$

If we let $a_k = \exp(\prod_{j=1}^k \lambda_j)$ then

$$\begin{aligned} a_{k-p}^{2-q_k} a_k^{q_k-1}/a_{k-1} &= \exp((2-q_k)\prod_{j=1}^{k-p}\lambda_j + (q_k-1)\prod_{j=1}^{k}\lambda_j - \prod_{j=1}^{k-1}\lambda_j) \\ &= \exp(\prod_{j=1}^{k-1}\lambda_j\{(2-q_k)\prod_{j=k-p+1}^{k-1}\lambda_j^{-1} + (q_k-1)\lambda_k - 1\}). \end{aligned}$$

We would like to pick $\lambda_j > 1$ periodic so that

$$(2.8) \qquad (2-q_k)\prod_{j=k-p+1}^{k-1}\lambda_k^{-1} + (q_k-1)\lambda_k - 1 < 0.$$

To do this we let $\lambda_j = 1 + \epsilon\alpha_j$ where $\alpha_j > 0$ is periodic and observe that if

$$(2.9) \qquad (q_k-1)\alpha_k - (2-q_k)\sum_{j=k-p+1}^{k-1}\alpha_j < 0,$$

then (2.8) holds for small ϵ.

To investigate when (2.9) holds, we let $r_k = (2-q_k)/(q_k-1)$ and use periodicity to rewrite the inequalities in terms of $\alpha_k, k = 1, \ldots, p$

$$(2.10) \qquad \alpha_k < r_k \sum_{j\in[1,p]-\{k\}} \alpha_j.$$

When $p = 2$ we have two inequalities $\alpha_1 < r_1\alpha_2$ and $\alpha_2 < r_2\alpha_1$. For this it is necessary that $r_1 r_2 > 1$. To see that the last condition is also sufficient let $\alpha_2 = 1$, $\alpha_1 = r_1 - \delta$ and pick δ small. To find the conditions under which the $p > 2$ inequalities in (2.10) are satisfied, we proceed by induction. Set $\alpha_p = r_p(\alpha_1 + \cdots + \alpha_{p-1})$. If with this choice of α_p the other $p-1$ conditions can be satisfied then by increasing α_1 a little we have p strict inequalities.

Substituting our choice of α_p into (2.10) gives

$$\alpha_k < r_k \left(\sum_{j\in[1,p)-\{k\}} \alpha_j + r_p \sum_{j<p} \alpha_j \right) \qquad \text{for } k \le p-1.$$

The last system can be rearranged to give a version of (2.10) with $p-1$ variables. Before we can do this we need to discard some trivial cases. If $r_k r_p > 1$ for some k then an example can be constructed using F_k and F_p. In the borderline case $r_k r_p = 1$, three laws suffice. Pick $j \in [1, p-1] - k$, let $\alpha_k = 1$, $\alpha_j < r_j$, $\alpha_p < r_p$, and set the other $\alpha_i = 0$. For our induction then we can suppose $r_k r_p < 1$ for all k and rearrange to get

$$\alpha_k < \frac{r_k(1+r_p)}{(1-r_k r_p)} \sum_{j\in[1,p-1]-\{k\}} \alpha_j. \tag{2.11}$$

To verify (⋆⋆) in the introduction, we need to show that

$$\left[\frac{d}{dt}\prod_{i=1}^{p}\left(1+t\frac{r_i(1+r_{p+1})}{1-r_i r_{p+1}}\right)\right]_{t=1} > \prod_{i=1}^{p}\left(1+\frac{r_i(1+r_{p+1})}{1-r_i r_{p+1}}\right) \tag{2.12}$$

is equivalent to

$$\left[\frac{d}{dt}\prod_{i=1}^{p+1}(1+tr_i)\right]_{t=1} > \prod_{i=1}^{p+1}(1+r_i). \tag{2.13}$$

Now (2.12) is equivalent to

$$\left[\frac{d}{dt}\prod_{i=1}^{p}\{1-r_i r_{p+1}+tr_i(1+r_{p+1})\}\right]_{t=1} > \prod_{i=1}^{p}(1+r_i), \tag{2.14}$$

and the left-hand side of (2.14) is

$$\begin{aligned}
&\sum_{i=1}^{p} r_i(1+r_{p+1}) \prod_{j\in[1,p]-\{i\}} \{1-r_j r_{p+1}+r_j(1+r_{p+1})\} \\
&\quad = \sum_{i=1}^{p} r_i \prod_{j\in[1,p+1]-\{i\}} (1+r_j) \\
&\quad = \left[\frac{d}{dt}\prod_{i=1}^{p+1}(1+tr_j)\right]_{t=1} - r_{p+1}\prod_{i=1}^{p+1}(1+r_j).
\end{aligned}$$

Thus (2.12) is equivalent to

$$\left[\frac{d}{dt}\prod_{i=1}^{p+1}(1+tr_i)\right]_{t=1} > (r_{p+1}+1)\prod_{i=1}^{p}(1+r_i),$$

which is (2.13).

Section 3. Computations for Example 3. As in the two previous examples, we let $\bar{X}_n = X_n 1_{(|X_n| \le a_j)}$ for $b_j < n \le b_{j+1}$ and the Borel–Cantelli lemma implies that it is enough to consider $\bar{S}_n = \bar{X}_1 + \dots \bar{X}_n$. Let $T_k = \bar{S}_{b_k}$. For $n \in (b_j, b_{j+1}]$

$$E\bar{X}_n \sim (-1)^j / j^2 a_{j+1}^{q-1}$$

since the only contribution comes from $P(\bar{X}_n = a_j)$, and as before

$$\mathrm{var}(\bar{X}_n) \le C a_j / j^2 a_{j+1}^{q-1}.$$

Recalling $a_j = \exp(\lambda^j)$, $b_{j+1} - b_j = a_{j+1}^q$, and summing we see

$$ET_{j+1} \sim (-1)^j a_{j+1}/j^2$$
$$\mathrm{var}(T_{j+1}) \le C a_j a_{j+1}/j^2$$

Now Chebyshev's inequality implies

$$\text{(3.1)} \qquad P(|T_{j+1} - ET_{j+1}| \ge ET_{j+1}/j) \le \frac{j^2 \, \mathrm{var}(T_{j+1})}{(ET_{j+1})^2} \le C j^4 a_j / a_{j+1},$$

which is summable, so the Borel–Cantelli lemma implies

$$\limsup_{n\to\infty} \bar{S}_n = \infty \qquad \liminf_{n\to\infty} \bar{S}_n = -\infty.$$

(3.1) shows that at time $n = b_{2k}$, $\bar{S}_n$ is close to $-a_{2k}/(2k-1)^2$. Let

$$\begin{aligned} \hat{X}_n &= X_n 1_{(|X_n| \le a_{j-1})} \qquad \text{for } n \in (b_j, b_{j+1}] \\ U_j(m) &= \hat{X}_{b_j+1} + \dots + \hat{X}_{b_j+m} \\ \tilde{X}_n &= X_n 1_{(|X_n| = a_j)} \qquad \text{for } n \in (b_j, b_{j+1}] \\ V_j(m) &= \tilde{X}_{b_j+1} + \dots + \tilde{X}_{b_j+m} \end{aligned}$$

Let $M = k^4 a_{2k} a_{2k+1}^{q-1}$. To prove that S_n is transient, we will show that the following four events have high probability (i.e., the probabilities of their complements are summable):

(3.2) There is a jump of size a_{2k} in $(b_{2k}, b_{2k} + M]$
(3.3) $|U_{2k}(m)| \le a_{2k}/8k^2$ for all $m \le M$
(3.4) $V_{2k}(m) > m/4k^2 a_{2k+1}^{q-1}$ for all $M \le m \le a_{2k}^q$
(3.5) $U_{2k}(m) > -(a_{2k} + m a_{2k+1}^{1-q})/4k^2$ for all $M \le m \le a_{2k}^q$

(3.1)–(3.5) imply that $|\bar{S}_{b_{2k}+m}| \ge a_{2k}/8k^2$ during $(b_{2k}, b_{2k+1}]$. The corresponding statements for $(b_{2k+1}, b_{2k+2}]$ can be proved in the same way and this gives the desired transience. It remains then to prove (3.2)–(3.5). Since $1-x \le e^{-x}$, we have

(3.2)
$$P(\text{ no jump of size } = a_{2k} \text{ in } (b_{2k}, b_{2k}+M]) \le \left(1 - \frac{1}{4k^2 a_{2k} a_{2k+1}^{q-1}}\right)^M \le e^{-k^2/4},$$

since $M = k^4 a_{2k} a_{2k+1}^{q-1}$ and $(1-x) \le e^{-x}$.

For $n \in (b_{2k}, b_{2k+1}]$, we have $E\hat{X}_n = 0$ and $\operatorname{var}(\hat{X}_n) \le C a_{2k-1}^{2-q}/k^2$, so Kolmogorov's inequality implies

(3.3)
$$\begin{aligned} P(|U_{2k}(m)| > a_{2k}/8k^2 \text{ for some } m \le M) &\le k^4 a_{2k} a_{2k+1}^{q-1}\, C \frac{a_{2k-1}^{2-q}}{k^2} \left(\frac{8k^2}{a_{2k}}\right)^2 \\ &= Ck^6 a_{2k-1}^{2-q} a_{2k+1}^{q-1}/a_{2k}. \end{aligned}$$

For $n \in (b_{2k}, b_{2k+1}]$, $\tilde{X}_n$ has mass $1/(4k^2 a_{2k} a_{2k+1}^{q-1})$ at a_{2k} and the rest of the mass at 0 so

$$E\tilde{X}_n = 1/(4k^2 a_{2k+1}^{q-1}), \qquad \operatorname{var}(\tilde{X}_n) \le a_{2k}/(4k^2 a_{2k+1}^{q-1}),$$

and using the Hajek-Renyi inequality gives

(3.4)
$$\begin{aligned} P(V_{2k}(m) \le m/4k^2 a_{2k+1}^{q-1} \text{ for some } M \le m \le a_{2k}^q) &\le \frac{M a_{2k}}{4k^2 a_{2k+1}^{q-1}} \left(\frac{M}{4k^2 a_{2k+1}^{q-1}}\right)^{-2} \\ &\quad + \sum_{m=M+1}^{\infty} \frac{a_{2k}}{4k^2 a_{2k+1}^{q-1}} \left(\frac{m}{4k^2 a_{2k+1}^{q-1}}\right)^{-2} \\ &\le C \frac{4k^2 a_{2k+1}^{q-1} a_{2k}}{M} = Ck^{-2}. \end{aligned}$$

Another use of the Hajek-Renyi inequality gives

(3.5)
$$\begin{aligned} P\left(U_{2k}(m) \le \frac{-(a_{2k} + m a_{2k+1}^{1-q})}{4k^2} \text{ for some } M \le m \le a_{2k}^q\right) &\le M \frac{a_{2k-1}^{2-q}}{k^2} \left(\frac{a_{2k}}{4k^2}\right)^{-2} \\ &\quad + \sum_{m=M+1}^{\infty} \frac{a_{2k-1}^{2-q}}{k^2} \left(\frac{a_{2k} + m a_{2k+1}^{1-q}}{4k^2}\right)^{-2} \\ &\le \frac{16k^6 a_{2k-1}^{2-q} a_{2k+1}^{q-1}}{a_{2k}} + \frac{a_{2k-1}^{2-q}}{k^2} \left(\frac{4k^2}{a_{2k+1}^{1-q}}\right)^2 M^{-1} \\ &= Ck^6 a_{2k-1}^{2-q} a_{2k+1}^{q-1}/a_{2k}. \end{aligned}$$

The bounds in (3.2) and (3.4) are summable. The bounds in (3.3) and (3.5), apart from an extra k^4, are the same as the summand in (2.4) with $p = 2$ and k replaced by $2k+1$. It follows as in Section 2 that if $q < 3/2$ we can pick $\lambda > 1$ to make the series summable. We have proved (3.2)–(3.5) and it follows that S_n is transient.

Section 4. Proof of Theorem 1. The key to the proof is the following fact that can be proved by embedding the c.r.w. into Brownian motion. Let $\tau = \inf\{j \geq 0 : S_j \geq 0\}$.

(4.1) Lemma. *There is a* $c > 0$ *depending on* $F_1, \dots F_p$ *but not on* n *or the adapted strategy so that*

$$P(\tau \leq n^2 | \mathcal{F}_0) \geq c \qquad \text{a.s. on } S_0 \geq -n.$$

We will prove (4.1) at the end of the section. An immediate consequence is

(4.2) Lemma. $P(\tau < \infty) = 1$.

Proof. Let $T_1 = S_0^2$. For $n \geq 1$, we stop if $S_{T_n} \geq 0$. If $S_{T_n} < 0$, let $T_{n+1} = T_n + (S_{T_n})^2$. $\hat{S}_m = S_{T_n+m}, m \geq 0$, is a c.r.w. with initial information $\mathcal{F}_{T_n}$, so (4.1) implies

$$P(S_m \geq 0 \text{ for some } m \in [T_n, T_{n+1}] | \mathcal{F}_{T_n}) \geq c$$

and the desired result follows from Theorem 9.5.2. in Chung [2].

Our c.r.w. is not a Markov chain but since the estimates we are about to derive depend on the starting point, it will be convenient to use E_x (resp. P_x) for expected value (resp. probability) for a c.r.w that starts from $S_0 = x$ and to write

$$G(x, A) = E_x \sum_{j=0}^{\tau-1} 1_{(S_n \in A)}.$$

Let

$$g_n = \sup_x \sup_\psi G(x, [-n, 0)) = \sup_{-n \leq x < 0} \sup_\psi G(x, [-n, 0)),$$

where $\sup_\psi$ indicates we are taking the supremum over all adapted strategies. Simple arguments similar to the one in (4.2) show that $g_n < \infty$. To get a bound on g_n notice that if $x \in [-n, 0)$,

$$G(x, [-n, 0)) \leq n^2 + P_x(\tau > n^2) g_n \leq n^2 + (1-c) g_n$$

by (4.1). Taking the sup over x and rearranging gives

$$(4.3) \qquad g_n \leq n^2/c.$$

Letting

$$H(x) = \sum_i (1 - F_i(x)) \geq \sup_i (1 - F_i(x)),$$

and considering the location of $S_{\tau-1}$ gives

$$P_x(S_\tau > L) = \int_{(-\infty,0)} G(x, dy) H(L-y) = \int_{(L,\infty)} G(x, [L-z, 0)) dH(z),$$

where the second equality comes from integrating by parts and changing variables $z = -y + L$. Using (4.3) now and recalling all the F_i have finite variance shows that if we pick L large enough the last expression is less than 1/2 and it follows that

$$(4.4) \qquad P_x(S_\tau \leq L) \geq 1/2 \qquad \text{for all } x < 0.$$

Applying (4.2) to $-S_n$ we see that S_n visits the negative half line infinitely often, so it follows from (4.4) and another application of the lemma from Chung that $P_x(S_n \in [0, L] \text{ i.o.}) = 1$ for all x.

Proof of (4.1). Let $\alpha = \inf_i \{\text{variance of } F_i\} > 0$ and $\nu = \alpha n^2/2p$, where p is the number of distributions. Let β_t be a Brownian motion. Brownian scaling implies that there is a constant $c > 0$ so that

$$(4.5) \qquad P_x(\beta_t > 0 \text{ for } t \in [\nu/2, \nu]) > 2c \qquad \text{for all } x \in [-n, 0].$$

We will now obtain (4.1) from (4.5) by "embedding" the c.r.w. in Brownian motion. We put the word embedding in quotes because we will build S_m and β_t at the same time. For $1 \leq k \leq p$, define three i.i.d. sequences indexed by $j \geq 1$: $X(k,j)$ with distribution F_k, Brownian motions $B(k,j;t)$ and stopping times $T(k,j)$ so that $B(k,j;T(k,j)) = X(k,j)$ and $ET(k,j) = EX(k,j)^2 = \int x^2 dF_k$. We construct the process S_m from these ingredients in the obvious way. If $k_{m+1} = \psi_m$, then we let $S_{m+1} - S_m$ be the first unused term in the k_{m+1}th sequence. That is, $X(k_{m+1}, j_{m+1})$, where $j_{m+1} = N(k_{m+1}, m) + 1$ and $N(k,m) = |\{l \leq m : k_l = k\}|$ is the number of terms that have been used in the kth sequence at time m. At the same time we extend our Brownian motion by setting $U_{m+1} = U_m + T(k_{m+1}, j_{m+1})$ and letting $\beta(U_m + t) = B(k_{m+1}, j_{m+1}; t)$ for $0 \leq t \leq T(k_{m+1}, j_{m+1})$.

Since the $T(k,j), j \geq 1$ are i.i.d. with $\alpha < ET(k,j) < \infty$, it follows that as $n \to \infty$

$$(4.6) \qquad P(T(k,1) + \cdots + T(k, n^2/p) < \nu) \to 0,$$

(recall $\nu = \alpha n^2/2p$) and

$$(4.7) \qquad P(T(k,j) \geq \nu/2 \text{ for some } j \in [1, n^2]) \to 0.$$

Now to construct $S_m, m \leq n^2$, we must use n^2/p terms from some sequence, so (4.6) implies $U_{n^2} > \nu$ with high probability. Since we can never use more than n^2 terms from any sequence, (4.7) implies that with high probability some U_m with $m \leq n^2$ is in $[\nu/2, \nu]$, and the desired result follows from (4.5).

REFERENCES

[1] L. Breiman, *Probability*, Addison Wesley, Reading, MA, 1968

[2] K.L. Chung, *A Course in Probability Theory* , second edition. Academic Press, New York, NY, 1974

[3] R. Durrett, *Oriented percolation in two dimensions*, Ann. Probab. **12** (1984), 999-1040

[4] W. Feller, *Note on the law of large numbers and "fair" games*, Ann. Math. Statist. **16** (1945), 301-304

[5] H. Kesten and G. Lawler, *A necessary condition for making money from fair games* , Preprint

[6] A. Renyi, *Probability Theory*, North-Holland, Amsterdam, 1970

[7] B.A. Rogozin and S.G. Foss, *Recurrency of an oscillating random walk*, Theor. Probab. Appl. **23** (1978), 155-162

R Durrett
Dept. of Mathematics
Cornell University
Ithaca, NY 14853

H. Kesten
Dept. of Mathematics
Cornell University
Ithaca, NY 14853

G. Lawler
Dept. of Mathematics
Duke University
Durham, NC 27706

Additive Functionals of Superdiffusion Processes

E.B.DYNKIN[1]

To every elliptic differential operator L in $\mathbb{R}^d$ (with time dependent coefficients) and to every $1 < \alpha \leq 2$ there corresponds a (non homogeneous) measure-valued Markov process X which we call the superdiffusion with parameters (L, α). A class $\mathcal{S}_\alpha$ of "exceptional sets" in $\mathbb{R}^{d+1}$ is introduced and an additive functional I^η of X is constructed for every finite measure η which does not charge any set of class $\mathcal{S}_\alpha$. The construction is based on a relationship between superdiffusions and a class of nonlinear parabolic partial differential equations and on results of Baras and Pierre about these equations.

1. Introduction. 1.1. Let D be a domain in $E = \mathbb{R}^d$ and let $Q = \mathbb{R} \times D$. A diffusion process in D with generator L is a Markov process $\xi = (\xi_t, \Pi_{r,x})$ with continuous paths and with the transition density $p(r, x; t, y)$ which is the fundamental solution of the equation

$$\dot{u} + Lu = 0 \text{ in } Q. \tag{1.1}$$

Here $\dot{u} = \partial u / \partial r$ and

$$L = \sum_{i,j} a_{ij} \frac{\partial}{\partial x_i} \frac{\partial}{\partial x_j} + \sum_i b_i \frac{\partial}{\partial x_i}$$

is a differential operator in Q with smooth coefficients subject to the condition: there exists a constant $\gamma > 0$ such that

$$\sum_{i,j} a_{ij}(r, x) u_i u_j \geq \gamma \sum_i u_i^2$$

for all $(r, x) \in Q$ and all $u_1, \dots, u_d \in \mathbb{R}$.

Denote by $\mathcal{B}$ the Borel σ-algebra in D, by M the set of all finite measures on $\mathcal{B}$ and by $\mathcal{M}$ the σ-algebra in M generated by the functions $f_B(\mu) = \mu(B)$, $B \in \mathcal{B}$. We put $Q_\Delta = \Delta \times D$ and we write $Q_\Delta = Q_{<t}$ if $\Delta = (-\infty, t)$. Let $\mathcal{B}_Q$ stand for the Borel σ-algebra in Q and let $\mathfrak{M}$ be the set of all measures μ on $(Q, \mathcal{B}_Q)$ such that $\mu(Q_{<t}) < \infty$ for every t.

[1]Partially supported by the National Science Foundation Grant DMS–8802667

A *superdiffusion process with parameters* (L,α) is a Markov process $(X_t, P_{r,\nu})$ in $(M,\mathcal{M})$ such that:

1.1.A. If $f^t(x)$ is a bounded continuous function on Q, then $\langle f^t, X_t\rangle$ is right continuous in t (writing $\langle v,\mu\rangle$ means the integral of v with respect to μ).

1.1.B. For every $r,t \in \mathbb{R}$, $\nu \in M$, $f \in \mathcal{B}$,

$$P_{r,\nu}\exp\langle -f, X_t\rangle = \exp\langle -v^r, \nu\rangle, \tag{1.3}$$

where $v^r(x) = v(r,x)$ is a solution of the integral equation

$$v(r,x) + \Pi_{r,x}\int_r^t v(s,\xi_s)^\alpha\, ds = \Pi_{r,x} f(\xi_t) \text{ for } r \le t. \tag{1.4}$$

The existence of a process with these properties for every $\alpha \in (1,2]$ has been proved in [6] and [12]. Moreover, according to [7] and [8], for every $\mu \in M$, there exists a probability measure P_μ such that

$$P_\mu \exp\langle -f, X_t\rangle = \exp\{-\int_{Q_{<t}} v(r,x)\mu(dr,dx)\}. \tag{1.3a}$$

(Note that $P_\mu = P_{r,\nu}$ if μ is the image of ν under the mapping $x \to (r,x)$ from D to Q.)

For every subdomain $\tilde D$ of D, we consider the restriction $\tilde L$ of L to $\tilde Q = \mathbb{R}\times\tilde D$ and we denote by $\tilde X$ the superdiffusion with parameters $(\tilde L,\alpha)$. It is shown in the Addendum to [8] that all these superdiffusions can be defined on the same sample space Ω in such a way that $\tilde X'_t(B) \le \tilde X_t(B)$ for all $t \in \mathbb{R}$, $B \subset \tilde D'$ if $\tilde D' \subset \tilde D$.

A superdiffusion X can be obtained by a passage to the limit from a branching particle system as the mass β of individual particles tends to 0 (see, *e.g.*, [7], [8]). Namely, the random measure X_t is the weak limit of the mass distribution

$$X_t^\beta(B) = \beta\sum_a 1_B(w_t^a),\ B \in \mathcal{B}$$

where w_t^a is the historical path of a particle which consists of its own trajectory and of the trajectories of all its ancestors. The initial probability distribution is given by the Poisson point process in Q with intensity measure μ.

In the last paragraph we assume that the process ξ can be started at any time $r \in \mathbb{R}$ from any point $x \in D$. Formally, ξ is defined on the space W whose elements are paths; a path w is a continuous mapping from an interval $[\alpha,\zeta)$, $-\infty < \alpha < \zeta \le +\infty$ to D. We add to the state space two extra elements ∂ and $\dagger$ and we put $w_t = \partial$ for $t < \alpha$, $w_t = \dagger$ for $t \ge \zeta$; $\xi_t(w) = w_t$ for all $t \in \mathbb{R}$. We denote by $\mathcal{F}(\Delta)$ the σ-algebra in W generated by ξ_t, $t \in \Delta$. The measure $\Pi_{r,x}$ is defined on $\mathcal{F}(\mathbb{R})$ and is concentrated on the set $\{\alpha = r, \xi_\alpha = x\}$. For every $C \in \mathcal{F}(\mathbb{R})$, $\Pi_{r,x}(C)$ is a Borel function in r, x and we set

$$\Pi_\mu(C) = \int \mu(dr, dx)\Pi_{r,x}(C).$$

1.2. An *additive functional* of ξ is a random measure A on $\mathbb{R}$ concentrated on $[\alpha, \zeta)$ such that, for every $r < t$, $A[r,t) \in \mathcal{F}[r,t)^*$ where the superscript * indicates the universal completion of a σ-algebra.

Suppose that ξ is a diffusion with generator L and that X is a superdiffusion with parameters (L, α). It is proved in [8] that to every additive functional A of ξ there corresponds an additive functional I_A of X such that, for every $r < t \in \mathbb{R}$, $\mu \in \mathfrak{M}$ and every interval Δ,

$$P_\mu \exp\{-I_A(\Delta)\} = \exp\langle -v, \mu\rangle, \tag{1.5}$$

where

$$v(r,x) + \Pi_{r,x} \int_r^\infty v(s, \xi_s)^\alpha ds = \Pi_{r,x} A(\Delta_{\geq r}); \tag{1.6}$$

here $\Delta_{\geq r} = \Delta \cap [r, \infty)$. [The corresponding functional of a branching particle system is defined by the formula

$$I_A(\Delta) = \beta \sum_a A(w^a, \Delta).]$$

By replacing A with λA and then taking the derivatives with respect to λ at $\lambda = 0$, we get from (1.5), (1.6) that

$$P_\mu I_A(\Delta) = \Pi_\mu A(\Delta_{\geq r}). \tag{1.7}$$

1.3. To every $f \in \mathcal{B}_Q$ and every measure σ on $\mathcal{B}_{\mathbb{R}}$ there corresponds an additive functional

$$A(B) = \int_B f(s, \xi_s)\sigma(ds) \tag{1.8}$$

of ξ. The corresponding functional of X is given by the formula

$$I_A(B) = \int_B \langle f^s, X_s\rangle \sigma(ds) \tag{1.9}$$

where $f^s(x) = f(s,x)$.

If A is an additive functional of ξ, then there exists a unique measure η on $(Q, \mathcal{B}_Q)$ such that

$$\Pi_{r,x} \int_{(r,\infty)} f(s, \xi_s) A(ds) = \int_Q p(r,x;s,y) f(s,y)\eta(ds,dy) \tag{1.10}$$

for all $r \in \mathbb{R}$, $x \in E$, $f \in \mathcal{B}_Q$ (we set $p(r,x;s,y) = 0$ for $r \geq s$). We call η *the characteristic measure of* A. It is known (see, *e.g.*, [2] and [3]) that:

(i) A σ-finite measure η is a characteristic measure if and only if it does not charge any ξ-polar set. [A Borel set B is ξ-polar if

$$\Pi_{r,x}\{(t,\xi_t) \text{ for some } t > r\} = 0 \tag{1.11}$$

for all $(r,x) \in Q$}.]

(ii) A Borel set B is ξ-polar if and only if the function

$$f(r,x) = \int_B k(r,x;s,y)\nu(ds,dy) \tag{1.12}$$

is unbounded for every measure ν unless $\nu(B) = 0$. Here

$$k(r,x;s,y) = k_{s-r}(y-x) \tag{1.13}$$

where

$$\begin{aligned} k_r(x) &= (2\pi r)^{-d/2} \exp\{-r/2 - |x|^2/2r\} \text{ for } r > 0, \\ k_r(x) &= 0 \text{ for } r \leq 0. \end{aligned} \tag{1.14}$$

We denote by $\mathcal{S}_\alpha$ the class of all Borel sets B in $S = \mathbb{R} \times E$ with the property: for every finite measure ν on S, function f defined by (1.12) belongs to $L^\alpha(S)$ only if $\nu(B) = 0$. The class $\mathcal{S}_\alpha$ can be also characterized by the condition $c_{k,m,\alpha'}(B) = 0$ where k is defined by (1.13), m is the Lebesgue measure, α' is the conjugate to α (i.e., $1/\alpha + 1/\alpha' = 1$) and $c_{k,m,\alpha'}$ is the capacity introduced in Section 5 of [18]. This is one of Choquet's capacities and therefore $B \in \mathcal{S}_\alpha$ if and only if $K \in \mathcal{S}_\alpha$ for all compact sets $K \subset B$.

Let $\mathfrak{M}_\alpha$ stand for the class of all measures $\mu \in \mathfrak{M}$ such that $\mu(dr,dx) = \rho(r,x)drdx$ with $\rho \in L^{\alpha'}(Q)$. The main result of this paper is given by the following theorem.

Theorem 1.1. *Let ξ be a diffusion with generator L and let X be the superdiffusion with parameters (L,α). To every finite measure η on $(Q,\mathcal{B}_Q)$ which charges no set of class $\mathcal{S}_\alpha$, there corresponds an additive functional I^η of X such that:*

1.3.A. *If $\mu \in \mathfrak{M}_\alpha$, then*

$$P_\mu \exp\{-I^\eta(\Delta)\} = \exp\langle -v, \mu\rangle \tag{1.15}$$

where v satisfies the equation

$$\begin{aligned} v(r,x) &+ \int_Q p(r,x;s,y)v(s,y)^\alpha \, ds\, dy \\ &= \int_{Q_\Delta} p(r,x;s,y)\eta(ds,dy) \text{ on } Q. \end{aligned} \tag{1.16}$$

1.3.B. For every $\mu \in \mathfrak{M}_\alpha$,

$$P_\mu I^\eta(\Delta) = \int_{Q\times Q_\Delta} \mu(dr, dx)p(r, x; s, y)\eta(ds, dy) \tag{1.17}$$

1.3.C. $I^{\eta'+\eta''} = I^{\eta'} + I^{\eta''}$ *and* $I^{c\eta} = cI^\eta$ *for any constant* $c \geq 0$.

1.3.D. *If* $\eta_n \uparrow \eta$, *then* $I^{\eta_n} \uparrow I^\eta P_\mu - a.s.$ *for all* $\mu \in M_\alpha$.

1.3.E. *If* $\eta(dr, dx) = \rho^r(x)drdx$, *then* $I^\eta(dt) = \langle \rho^t, X_t\rangle dt$.

1.3.F. *Let* $\phi^t(x) \geq 0$ *be a continuous function on* Q. *If* η *is concentrated on a compact set* $K \subset \langle \phi > 0\rangle$, *then, for every* $\mu \in \mathfrak{M}_\alpha$,

$$I^\eta(\mathbb{R}) = 0 \quad P_\mu\text{-a.s. on } \Omega_\phi = \{\omega : \langle \phi^t, X_t\rangle = 0 \text{ for all } t \in \mathbb{R}\}$$

The proof of Theorem 1.1 is based on results of Baras and Pierre [1] and on the following theorem proved in [10]:

Theorem 1.2. *Suppose that* D *is a bounded domain with smooth boundary and* $\Delta = (t_0, t)$. *Let* f *be a bounded Hölder continuous function on* $Q_\Delta = \Delta \times Q$ *and let* I_A *be the functional defined by* (1.9). *Then, for all* $\mu \in \mathfrak{M}$,

$$P_\mu \exp\{-I_A(\Delta)\} = \exp\langle -v, \mu\rangle \tag{1.18}$$

where $v(r, x)$ *is the classical solution of the boundary value problem*

$$\dot{v} + Lv - v^\alpha = -f \text{ in } Q_\Delta, \tag{1.19}$$

$$v(r, x) \to 0 \text{ as } r \uparrow t,\ v(r, x) \to 0 \text{ as } x \to a \in \partial D.$$

1.4. If the coefficients in L does not depend on time, then ξ and X are time homogeneous Markov processes (see Section 1.12 in [8]). An additive functional I of X is called *homogeneous* if $I(\Delta, \Theta_u\omega) = I(\Delta + u, \omega)$ for all $u > 0$, $\omega \in \Omega$.

Theorem 1.3. *Let* X *be a time homogeneous superdiffusion and let* $\eta(dr, dx) = dr\,\nu(dx)$. *Then* I^η *can be chosen homogeneous. If* $\mu(dx) = \rho(x)dx$ *with* $\rho \in L^{\alpha'}$, *then*

$$P_\mu \exp\{-I^\eta[0, t)\} = \exp\langle -v, \mu\rangle \tag{1.20}$$

where

$$v(x) + \int_D g(x, y)v(y)^\alpha dsdy = \int_D g_t(x, y)\nu(dy) \text{ on } D. \tag{1.21}$$

Here

$$g_t(x,y) = \int_0^t p(0,x;s,y)ds$$

and $g(x,y) = g_\infty(x,y)$ *is Green's function of* L *in* D.

It follows from Proposition 2.3 in [1] that a measure $\eta(dr,dx) = dr\nu(dx)$ does not charge sets of class $\mathcal{S}_\alpha$ if and only if ν belongs to the class $\mathcal{H}_\alpha(D)$ described in [9]. Therefore Theorems 1.1 and 1.3 imply Theorem 1.7 in [9].

1.5. The first results on additive functionals of superprocesses are due to Iscoe. In [15] he studied functionals of the form (1.9) which he called weighted occupation times and he established, in the case $L = \Delta, \alpha = 2$, the equations (1.18), (1.19). In [16], he introduced "local occupation times" which correspond, in our terminology, to $\eta = m \times \delta_c$ where m is the Lebesgue measure on $\mathbb{R}$ and δ_c is the Dirac's measure concentrated at a point $c \in \mathbb{R}^d$.

Additive functionals of more general superprocesses have been investigated in [4], [5] using the moment functions. This method is applicable only to $\alpha = 2$.

2. Proofs. 2.1. The Sobolev space $W_\alpha^k = W_\alpha^k(D)$ is a Banach space formed by all functions in $L^\alpha(D)$ which have weak derivatives up to order k belonging to $L^\alpha(D)$. We denote by $C_0^\infty(D)$ the class of all infinitely differentiable functions in D with compact supports and by $\mathring{W}_\alpha^k = \mathring{W}_\alpha^k(D)$ the closure of $C_0^\infty(D)$ in W_α^k.

Let $Q_\Delta = \Delta \times Q$ where $\Delta = (t_0, t)$ and D is a bounded domain with smooth boundary. The "parabolic" Sobolev spaces $W_\alpha^{1,2} = W_\alpha^{1,2}(Q_\Delta)$ consists of all functions $u \in L^\alpha(Q_\Delta)$ such that $\dot{u}$, $D_i u$, $i = 1, ..., d$ and $D_i D_j u$, $i,j = 1,...,d$ belong to $L^\alpha(Q_\Delta)$. The space $\mathring{W}_\alpha^{1,2}$ is the closure in $W_\alpha^{1,2}$ of the set of $u \in C^\infty(Q_\Delta)$ which satisfy the boundary conditions

$$u(r,x) \to 0 \text{ as } r \downarrow t_0,\ u(r,x) \to 0 \text{ as } x \to a \in \partial D. \tag{2.1}$$

Formula

$$F(s,y) = \int_{Q_\Delta} \varphi(r,x)\, dr\, dx\, p(r,x;s,y) \tag{2.2}$$

determines a continuous linear mapping from $L^\alpha(Q_\Delta)$ onto $\mathring{W}_\alpha^{1,2}$ and the inverse mapping (also continuous) is given by the formula

$$\varphi = \dot{F} - L^* F \tag{2.3}$$

where L^* is the formal adjoint operator for L (see [17], Chapter IV, Section 9).

We set $\eta \in \mathfrak{M}_\alpha$ if η is a finite measure on Q with a compact support and if "the heat potential"

$$h(r,x) = \int_Q p(r,x;s,y)\eta(ds,dy) \tag{2.4}$$

belongs to $L^\alpha(Q_\Delta)$ for every finite interval Δ. (This condition is satisfied if and only if $\langle F, \eta\rangle$ is a continuous functional on $\overset{\circ}{W}{}^{1,2}_{\alpha'}(\Delta)$.)

A function u on Q_Δ belongs to the space $L^1(\Delta, \overset{\circ}{W}{}^1_1)$ if $u^r(\cdot) = u(r,\cdot)$ belongs to $\overset{\circ}{W}{}^1_1$ for almost all $r \in \Delta$ and if

$$\int_\Delta \|u^r\|_{W^1_1} dr < \infty.$$

We define spaces $L^\infty(\Delta, L^1(D))$ and $L^\alpha(\Delta, L^\beta(D))$ similarly and we denote by $\mathcal{L}_\alpha(\Delta)$ the intersection of $L^1(\Delta, \overset{\circ}{W}{}^1_1(D))$, $L^\infty(\Delta, L^1(D))$ and $L^\alpha(Q_\Delta) = L^\alpha(\Delta, L^\alpha(D))$.

Writing "$Lv = f$ in $C_0^\infty(Q_\Delta)$" means

$$\int_{Q_\Delta} L^*\varphi(r,x)v(r,x)\text{drdx} = \int_{Q_\Delta} \varphi(r,x)f(r,x)drdx \text{ for all } \varphi \in C_0^\infty(Q_\Delta).$$

Writing "$Lv = \eta$ in $C_0^\infty(Q_\Delta)$" has an analogous meaning with $f(r,x)drdx$ replaced by $\eta(dr,dx)$.

Suppose that $\epsilon_n \to 0$ and that $\rho_n \in C^\infty(Q)$ are positive, $\int \rho_n drdx = 1$ and $\rho_n(r,x) = 0$ for $|r| + |x| > \epsilon_n$. We say that ρ_n is a *mollifier* if, in addition, $\|f_n - f\|_{L^p(Q_\Delta)} \to 0$ for every finite interval Δ, every $p < \infty$ and every $f \in L^p(Q_\Delta)$ (here f_n is the convolution of f and ρ_n). For instance, if $\gamma \in C_0^\infty(Q)$ and if

$$\gamma_\epsilon(r,x) = \epsilon^{-d-1}\gamma(r/\epsilon, x/\epsilon), \tag{2.5}$$

then $\rho_n = \gamma_{\beta_n}$ is a mollifier for every sequence $\beta_n \to 0$ (see e.g., [14], Lemma 7.2).

The following result easily follows from Lemma 3.5 in [1].

Lemma 2.1. *Let D be a domain with a smooth boundary and let $Q_\Delta = \Delta \times D$ where $\Delta = (t_0, t)$. Suppose that $\eta \in \mathfrak{M}_\alpha$. Then there exists a unique solution of the problem*

$$u \in \mathcal{L}_\alpha(\Delta),$$
$$\dot{u} + Lu - u^\alpha = -\eta \text{ in } C_0^\infty(Q_\Delta). \tag{2.6}$$

Moreover, if ρ_n is a mollifier and if

$$f_n(r,x) = \int \rho_n(r-s, x-y)\eta(ds,dy), \tag{2.7}$$

then u is the $L^\alpha(Q_\Delta)$-limit of functions u_n which are uniquely determined by the conditions

$$u_n \in \mathcal{L}_\alpha(\Delta),$$

$$\dot{u}_n + Lu_n - u_n^\alpha = -f_n \text{ in } C_0^\infty(Q_\Delta). \tag{2.8}$$

(The concluding part is not stated explicitly in [1] but it can be extracted from the proof of Lemma 3.5.)

We also need the following lemmas:

Lemma 2.2. *Let $i_t^r \in [0, +\infty]$ be given for every $r < t \in \mathbb{R}$ and let*

$$i_s^r + i_t^s = i_t^r \quad \text{for all } r < s < t, \tag{2.9}$$

$$i_{t-}^r(x) = i_t^r(x) \quad \text{for all } r < t \tag{2.10}$$

Then there exists a measure $I(dt)$ on $(\mathbb{R}, \mathcal{B})$ such that

$$I[r,t) = i_t^r \quad \text{for all } r < t \in \mathbb{R}. \tag{2.11}$$

Proof. Denote by $\mathcal{G}$ the class of all intervals $[r,t)$ such that $i_t^r < \infty$ and put $t \in G$ if t is an interior point of some $\Delta \in \mathcal{G}$. Clearly, G is open and $i_t^r < \infty$ for every $[r,t] \subset G$. Therefore there exists a measure $I(dt)$ on G such that (2.9) holds for all $[r,t) \subset G$.

Put $t \in H$ if $t \notin G$ but t belongs to an interval $\Delta \in \mathcal{G}$ (clearly, t is the left end point of Δ). The condition (2.9) holds for $[r,t) \subset G \cup H$ if we set $I\{t\} = i_{t+}^t$ for $t \in H$.

By setting $I(\Gamma) = \infty$ for all Γ which are not contained in $G \cup H$, we get a measure which satisfies condition (2.9).

Corollary. *If $I_\epsilon(dt)$, $\epsilon > 0$ is a family of measures on $(\mathbb{R}, \mathcal{B}_{\mathbb{R}})$ and if $I_\epsilon[r,t) \uparrow i_t^r$ as $\epsilon \downarrow 0$ for all $r < t \in \mathbb{R}$, then there exists a measure I such that $I[r,t) = i_t^r$ for all $r < t$.*

Indeed, i satisfies (2.9). It satisfies also (2.10) because i_t^r is upper semicontinuous and monotone increasing in t.

Lemma 2.3. *Suppose that J_ϵ, $\epsilon > 0$ is a family of additive functionals of a Markov process $\xi = (\xi_t, \Pi_{r,x})$ and let $\mathfrak{M}^\circ \subset \mathfrak{M}$. Suppose that, for every $r < t \in \mathbb{R}$ and every $\mu \in \mathfrak{M}^\circ$,*

$$J_\epsilon[r,t) \to j_t^r(\mu) \text{ in } \Pi_\mu\text{-probability} \tag{2.12}$$

and

$$j_{t_n}^r(\mu) \to j_t^r(\mu) \quad \Pi_\mu\text{-a.s. as } t_n \uparrow t. \tag{2.13}$$

Then there exists an additive functional I of ξ such that

$$J_\epsilon[r,t) \to I[r,t) \text{ in } \Pi_\mu\text{-probability} \tag{2.14}$$

for all $r < t \in \mathbb{R}$, $\mu \in \mathfrak{M}^o$. Moreover I depends linearly on i.

Proof. Put

$$i_t^r = \lim \operatorname{med} J_{1/n}[r,t) \tag{2.15}$$

where lim med means the medial limit of Mokobodzki (see, e.g.,the Appendix in [11]). By the fundamental properties of this limit:

(a) j satisfies (2.9);
(b) $i_t^r = j_t^r(\mu)$ Π_μ-a.s. for all $\mu \in \mathfrak{M}^o$;
(c) $i_t^r \in \mathcal{F}[r,t)^*$.

Clearly, i_{t-}^r satisfies (2.9) and (2.10) and, by (2.13) and (*b*), $i_{t-}^r(\mu) = i_t^r(\mu)$ $\Pi_{r,x} - a.s.$ By Lemma 2.3, for every ω, there exists a measure $I(\omega, dt)$ on $(\mathbb{R}, \mathcal{B}_{\mathbb{R}})$ such that $I\{[r,t),\omega\} = i_{t-}^r(\omega)$ for all $r < t \in \mathbb{R}$. Hence $I[r,t) \in \mathcal{F}[r,t)^*$ and $I[r,t) = j_t^r(\mu)$ $\Pi_\mu - a.s.$ for all $\mu \in \mathfrak{M}^o$.

2.2. Proof of Theorem 1.1. 1°. *Suppose that D is a bounded domain with a smooth boundary and $\eta \in \mathfrak{N}_\alpha$.* Let γ_ϵ be given by (2.5) and let

$$f_\epsilon^r(x) = \int \gamma_\epsilon(r-s, x-y)\eta(ds, dy). \tag{2.16}$$

Formula

$$J_\epsilon(\Delta) = \int_\Delta \langle f_\epsilon^s, X_s\rangle ds \tag{2.17}$$

defines a family of additive functionals of X. By Theorem 1.2,

(2.18)
$$P_\mu[\exp\{-\frac{1}{2}J_\beta[r,t)\} - \exp\{-\frac{1}{2}J_\epsilon[r,t)\}]^2 = \exp\langle -u_{\beta\beta}, \mu\rangle + \exp\langle -u_{\epsilon\epsilon}, \mu\rangle - 2\exp\langle -u_{\beta\epsilon}, \mu\rangle,$$

where $u_{\beta\epsilon}(r,x)$ is the classical solution of the boundary value problem (1.19) with f replaced by $f_{\beta\epsilon} = (f_\beta + f_\epsilon)/2$. Choose $t_0 < t$ and put $\Delta = (t_0, t)$. Clearly, $u_{\beta\epsilon} \in \mathcal{L}_\alpha(\Delta)$. If $\beta_n, \epsilon_n \to 0$, then $f_{\beta_n\epsilon_n}$ correspond by formula (2.7) to a mollifier $\rho_n = (\gamma_{\beta_n} + \gamma_{\epsilon_n})/2$ and therefore $u_{\beta_n\epsilon_n}$ converge in $L^\alpha(Q_\Delta)$ to the solution u of the problem (2.6). Hence $u_{\beta\epsilon} \to u$ in $L^\alpha(Q_\Delta)$ as $\beta, \epsilon \downarrow 0$.

Let $\varphi \in C_0^\infty(Q_\Delta)$ and let F be given by formula (2.2). The function $u_\epsilon = u_{\epsilon\epsilon}$ satisfies (1.19) with f replaced by f_ϵ and, by Green's formula and (2.3),

$$\int F f_\epsilon = -\int F(\dot{u}_\epsilon + Lu_\epsilon - u_\epsilon^\alpha) = \int \varphi u_\epsilon + \int F u_\epsilon^\alpha$$

(all integrals over Q_Δ). By passing to the limit as $\epsilon \to 0$, we get

(2.19)
$$u(r,x) + \int_{Q_{<t}} p(r,x;s,y)u(s,y)^\alpha ds dy = \int_{Q_{<t}} p(r,x;s,y)\eta(ds,dy) \text{ a.e. on } Q_\Delta = (t_0,t)\times D.$$

If $\mu \in M_\alpha$, then $\langle u_{\beta\beta},\mu\rangle, \langle u_{\epsilon\epsilon},\mu\rangle$ and $\langle u_{\beta\epsilon},\mu\rangle$ converge to $\langle u,\mu\rangle$ and therefore $J_\epsilon[r,t)$ converges in P_μ-probability to a limit $j_t^r(\mu)$ such that

$$P_\mu \exp\{-j_t^r(\mu)\} = \exp\langle -u,\mu\rangle. \tag{2.20}$$

It follows from (2.19) and (2.20) that

$$P_\mu j_t^r(\mu) = \int_{Q\times Q_\Delta} \mu(dr,dx)p(r,x;t,y)\eta(dt,dy). \tag{2.21}$$

Since $\eta \in \mathfrak{N}$, $\mu \in \mathfrak{M}_\alpha$, the right side is finite.

Let $\mu \in \mathfrak{M}_\alpha$. Then $j_s^r(\mu) \le j_t^r(\mu)$ P_μ-a.s. for all $r \le s < t$ and therefore $j_{t_n}^r(\mu) \uparrow \tilde{j}_t^r(\mu) \le j_t^r(\mu)$ P_μ-a.s. as $t_n \uparrow t$. By (2.21), $P_\mu \tilde{j}_t^r(\mu) = P_\mu j_t^r(\mu)$ and therefore $\tilde{j}_t^r(\mu) = j_t^r(\mu)$ P_μ-a.s.

By Lemma 2.3 (applied to the superprocess X), there exists an additive functional I^η of X such that

$$I^\eta[r,t) = \lim_{\epsilon\downarrow 0} J_\epsilon[r,t) \text{ in } P_\mu\text{-probability} \tag{2.22}$$

for every $r < t \in \mathbb{R}$ and every $\mu \in \mathfrak{M}_\alpha$. By (2.19),

$$v(r,x) = -\int_{Q_{<t}} p(r,x;s,y)u(s,y)^\alpha ds dy + \int_{Q_{<t}} p(r,x;s,y)\eta(ds,dy) \tag{2.23}$$

coincides with $u(r,x)$ a.e. on Q. This implies (1.16). Since $P_\mu\{I^\eta[r,t) \ne j_t^r(\mu)\} = 0$, (1.15) follows from (2.20). Hence I^η satisfies 1.3.A. Property 1.3.*B* is an implication of (2.21). The functional J_ϵ defined by (2.16), (2.17) depends linearly on η and so does function i determined by (2.15). Therefore 1.3.*C* follows from Lemma 2.3.

2°. *Let η be an arbitrary finite measure which does not charge any set of class S_α.* By Proposition 3.2 in [1], there exists a sequence $\eta_n \in \mathfrak{N}_\alpha$ such that $\eta_n \uparrow \eta$. By 1.3.*C*, the corresponding sequence I^{η_n} is monotone increasing and, by Corollary to Lemma 2.2, there exists an additive functional I^η of X such that

$$I^{\eta_n}[r,t) \uparrow I^\eta[r,t) \text{ for all } r < t \in \mathbb{R}. \tag{2.24}$$

We claim that I^η is defined by this formula uniquely up to P_μ-equivalence for all $\mu \in \mathfrak{M}_\alpha$. Indeed, let $\eta'_n \in \mathfrak{N}_\alpha$, $\eta'_n \uparrow \eta$. First we assume that $\eta'_n \le \eta_n$. Let $Y_n = I^{\eta_n}[r,t) \uparrow Y$, $Y'_n = Y_D^{\eta'_n}[r,t) \uparrow Y'$. By 1.3.A, for every $\mu \in \mathfrak{M}_\alpha$,

$$P_\mu \exp(-Y_n) = \exp\langle -v_n, \mu\rangle \tag{2.25}$$

where

$$\begin{aligned} v_n(r,x) &+ \int_Q p(r,x;s,y) v_n(s,y)^\alpha ds dy \\ &= \int_{Q_\Delta} p(r,x;s,y)\eta_n(ds,dy) \text{ in } Q. \end{aligned} \tag{2.26}$$

Clearly,

$$P_\mu e^{-Y} = e^{-\langle v,\mu\rangle} \tag{2.27}$$

where $v = \lim\, v_n$ satisfies (1.16). Analogously,

$$P_\mu e^{-Y'} = e^{-\langle v',\mu\rangle} \tag{2.28}$$

where $v' = \lim\, v'_n$ also satisfies (1.16). Well known bounds for $p(r,x;s,y)$ (see, *e.g.*, [13], Chapter 1, Section 8) imply that $\int_{(t_0,t)\times D} dr dx p(r,x;s,y)$ is bounded and therefore $\int_Q p(r,x;s,y)\eta(ds,dy) < \infty$ a.e. in Q for every finite measure η. Since $v' \le v$, we conclude that $v' = v$ *a.e* in Q. By 1.3.C, $Y' \le Y$ and, by (2.27), (2.28), $Y = Y'$ $P_\mu - a.s.$

Let now η'_n, $\eta''_n \in \mathfrak{N}_\alpha$ and $\eta'_n \uparrow \eta$, $\eta''_n \uparrow \eta$. Denote by η_n a measure defined by the formula $d\eta_n = a' \vee a'' d\eta$ where $a' = d\eta'/d\eta$, $a'' = d\eta''/d\eta$. Clearly, $\eta_n \in \mathfrak{N}_\alpha$ and $\eta_n \uparrow \eta$. Therefore

$$\lim\, I^{\eta'_n} = \lim\, I^{\eta''_n} = \lim\, I^{\eta_n} \quad P_\mu\text{- a.s.}$$

It is easy to see that I^η defined by (2.24) satisfy conditions 1.3.A, B, C.

3°. *If D is an arbitrary domain*, then we consider a sequence of bounded domains D_n with smooth boundaries such that $D^n \cup \partial D^n \subset D^{n+1}$ and $D^n \uparrow D$. Let X^n be the superdiffusion with parameters (L^n, α) where L^n is the restriction of L to $Q^n = \mathbb{R} \times D^n$. Suppose that η is a finite measure on D which charges no set of class $\mathcal{S}_\alpha$. Then the restriction η_n of η to D^n has an analogous property. The additive functional I^{η_n} of X^n constructed in 2° can be interpreted as a functional of X. Moreover $I^{\eta_n}(\Delta) \le I^{\eta_{n+1}}(\Delta)$ for all n and Δ. This follows from monotonicity of X_t^n in n, formulae (2.15), (2.16) and (2.17) and Lemma 2.3. By Corollary to Lemma 2.2, there exists an additive functional I^η of X which satisfies (2.24). Passing to the limit, we establish for I^η properties 1.3.A, B, C.

To prove 1.3.D, we note that, if $\eta_n \uparrow \eta$, then there exists an additive functional $\tilde{I}$ such that $I^{\eta_n} \uparrow \tilde{I} \le I^{\eta}$. It follows from 1.3.A that $P_\mu \exp\{-\tilde{I}(\Delta)\} = P_\mu \exp\{-I^\eta(\Delta)\}$ and therefore $\tilde{I} = I^\eta P_\mu - a.s.$

Clearly, 1.3.E holds for every f if it holds for continuous f with compact supports. For such f, the functions f_ϵ given by (2.16) converge uniformly to f as $\epsilon \downarrow 0$. Therefore $J_\epsilon(\Delta) \to \int_\Delta \langle f^s, \xi_s \rangle ds \; P_\mu - a.s.$ for all $\mu \in \mathfrak{M}_\alpha$, and 1.3.$E$ follows from (2.22).

We check 1.3.F. in the situation described in 1° (an extension to the general case is trivial). For sufficiently small $\epsilon > 0$, $\{f_\epsilon > 0\} \subset \{\phi > 0\}$ and therefore $J_\epsilon[r,t) = 0$. By (2.22), $I^\eta[r,t) = 0$ P_μ-a.s. on Ω_ϕ for every $\mu \in \mathfrak{M}_\alpha$.

2.3. Proof of Theorem 1.3. If $\eta(dr, dx) = dr\nu(dx)$, then f_ϵ determined by (2.16) does not depend on r and the additive functional J_ϵ given by (2.17) is homogeneous. Arguments in the Appendix to [11] show that the functional I^η constructed in Section 2.2 can also be chosen homogeneous. Formulae (1.20) and (1.21) follow from (1.15) and (1.16).

REFERENCES

[1] P. Baras, M. Pierre, Problems paraboliques semi-linéares avec donnees measures, Applicable Analysis **18** (1984), 111-149.

[2] J.L.Doob, Classical Potential Theory and Its Probabilistic Counterpart, Springer, New York/ Berlin/ Heidelberg/ Tokyo, 1984.

[3] E. B. Dynkin, Markov systems and their additive functionals, Ann. Probab. **5** (1977), 653-677.

[4] E. B. Dynkin, Representation for functionals of superprocesses by multiple stochastic integrals, with applications to self-intersection local times, Astérisque **157-158** (1988), 147-171.

[5] E. B. Dynkin, Superprocesses and their linear additive functionals, Trans. Amer. Math. Soc. 314 (1989), 255-282.

[6] E. B. Dynkin, Regular transition functions and regular superprocesses, Trans. Amer. Math. Soc. **316** (1989), 623-634.

[7] E. B. Dynkin, Branching particle systems and superprocesses (1989), Ann.Probab. to appear.

[8] E. B. Dynkin, Path processes and historical processes (1990), submitted to Probab. Th. Rel. Fields.

[9] E. B. Dynkin, A probabilistic approach to one class of nonlinear differential equations, submitted to Probab. Th. Rel. Fields.

[10] E. B. Dynkin, Superdiffusions and parabolic nonlinear differential equations, (1990), Preprint.

[11] E. B. Dynkin and R. K. Getoor, Additive functionals and entrance laws, J. Funct. Analysis **60** (1985), 221-265.

[12] P. J. Fitzsimmons, Construction and regularity of measure-valued Markov branching processes, Israel J. Math. **64** (1988), 337-361.

[13] A. Friedman, Partial Differential Equations of Parabolic Type, Prentice Hall, Englewood Cliffs, N. J., 1964.

[14] D. Gilbarg, N. S. Trudinger, Elliptic Partial Differential Equations of Second Order, Springer, Berlin/New York, 1977.

[15] I. Iscoe, A weighted occupation time for a class of measure-valued branching processes, Probab. Theory Related Fields **71** (1986), 85-116.

[16] I. Iscoe, Ergodic theory and local occupation time for measure-valued critical branching Brownian motion, Stochastics **18** (1986), 197-243.

[17] O. A. Ladyzenskaja, V. A. Solonnikov, N. N. Ural'ceva, Linear and quasilinear equations of parabolic type, Translations of Math. Mon, Amer. Math. Soc., Providence, 1968.

[18] N. G. Meyers, A theory of capacities for potentials of functions in Lebesgue classes, Math. Scand. **26** (1970), 255-292.

E. B. Dynkin
Department of Mathematics, White Hall,
Cornell University, Ithaca, N.Y. 14853

Interacting Systems, Stirrings, and Flows

T.E. HARRIS

Stirring processes are related both to certain interacting particle systems and to continuous stochastic flows. Examples are given including a flow generated by Moebius automorphisms of the unit disk.

1. Introduction. The pioneering paper [21] of Frank Spitzer gave new direction and impetus to the theory of interacting particle systems, and its influence is still apparent in the vigorous progress of that field today. The purpose of this note is to comment on a connection between certain interacting systems and stochastic flows of transformations, via "stirring processes" and their limiting behavior.

I am indebted to Paul Yang for helpful comments, and to Peter Baxendale for helpful comments and for acquainting me with the Lie theory for the Moebius functions.

2. Stirring Processes. As pointed out in Wang Lee's paper [18], one can regard the simple exclusion process in Z^d (see [21]), in the symmetric case, as a Poisson flow of permutations of Z^d. For example, take $d = 1$ and consider the case where the only jumps are to nearest neighbors. Let S be an arbitrary subset of Z^1, and put a particle at each point of S. Attach a Poisson point process $0 < T_{x1} < T_{x2} < \dots$ to each point $x \varepsilon Z^1$; the processes for different x's are independent, with the same intensity. At time T_{xi}, move the particle at x, if any, to $x+1$ and move the particle at $x+1$, if any, to x. If both x and $x+1$ are occupied, or if neither is, the counting measure for the set of particles does not change.

If we are interested only in the counting measure and not in individual particles, the process just described is the same as symmetric nearest-neighbor simple exclusion. It is an example of a stochastic stream of transformations operating on a base space, the result being a flow in the space, in this case Z^1. Other examples of particle systems in a countable base space operated by flows of transformations are given in Bertein and Galvez [6], Griffeath [9], and Harris [10].

Going to a continuous space, the "stirring processes" studied by Wang Lee in [18] have R^1 as the base space, and the transformations are translates of a Lebesgue-measure- preserving transformation. Such stirring processes have single- point paths of the jump- type, although they can be modified to have continuous paths (see [20], Remark 3.2).

By appropriate limiting procedures applied to stirring processes in R^d or other manifolds, we can arrive at continuous flows of transformations (sometimes homeomorphisms or diffeomorphisms) with diffusion paths. For example, we can arrive at any nice homogeneous isotropic flow in R^2 that preserves Lebesgue measure as a limit of stirrings with circular stream lines [11]. In this case there are infinitely many stirrings in each time interval.

Matsumoto and Shigekawa [20] have given general results about the convergence of stirrings to flows (see below.) The work of Watkins on convergence of random evolutions (e.g., [22], [23], and [24]) is related and contains a variety of examples. The processes treated by Kesten and Papanicolaou [15] are also related.

Kunita [16], [17] has given very general limit theorems for convergence of various types of approximating processes to flows of diffeomorphisms. In [16] the continuous process $\tilde{\psi}_t^\varepsilon$ in Sec. 3.9. is closely related to the stirrings discussed here, but more general in that the control variable corresponding to u in (2.1) may be a certain type of Markov process. Other convergence results have been given by a number of authors, e.g. Baxendale [3], Bismut [7], LeJan and Watanabe [19]. However, the class of stirrings seems to the author to be simple enough and of enough intuitive appeal to warrant some special consideration.

Here is the idea of the limiting procedure for stirrings, without details of rigor, and not in the most general setting. Let $A_u, u \varepsilon U$, be nice vector fields in R^d, given in component form as

$$A_u(x) = \left(A_u^1(x), A_u^2(x), \dots, A_u^d(x)\right). \tag{2.1}$$

In general U is an abstract set, but the points $u \varepsilon U$ may have a geometrical significance, as in the examples discussed below.

Let ν be a probability measure on the measurable subsets of U, and assume, as a centering condition,

$$\int_U A_u^i(x)\, d\nu = 0. \tag{2.2}$$

If (2.2) does not hold, centering may be achieved by modifying the process. There is then the question how much this distorts the original process. See, e.g., Sec. 5.

For $u \varepsilon U, x \varepsilon R^d$ let $\{Y(u,\tau,x), \tau \geq 0\}$ be the integral curve of (2.1) with initial position x, $Y_{un} = Y(u, 1/\sqrt{n}, \cdot)$. Let $0 < \sigma_1 < \sigma_2 < \dots$ be a Poisson point process with intensity n, and let $u_1, u_2, \dots$ be i.i.d. values of u, chosen according to the probability measure ν. Apply $Y_{u_1 n}$ to R^d at time σ_1, then $Y_{u_2 n}$ at time σ_2, etc. Let $X_n(t) = X_n(t,\cdot)$ be the composition $\dots \circ Y_{u_2 n} \circ Y_{u_1 n}$, including all $\sigma_i \leq t$; $X_n(t) =$ identity for $t < \sigma_1$. We call $\{X_n(t), t \geq 0\}$ a *stirring process.*

Let $n \to \infty$. According to [20], under rather general assumptions $\{X_n(t,x)\}$ has a limiting flow of diffeomorphisms $\{X_t(x)\}$. (The basic setting of [20] is a compact manifold, but much of the work carries over to R^d.) The motion of each finite set of points is a diffusion. The flow is determined uniquely by its two-point correlation function (or matrix),

$$\begin{aligned} a_{pq}(x,y) &\equiv \lim_{t\downarrow 0} E\{(X_t^p(x)-x^p)\ (X_t^q(y)-y^q)\}/t \\ &= \int_U A_u^p(x)\, A_u^q(y)\ d\nu, \end{aligned} \tag{2.3}$$

and the operator for the diffusion of one point,

$$\begin{aligned} Gf(x) &= \frac{1}{2}\sum_{pq} a_{pq}(x,x)\,\partial_p\partial_q f(x) \\ &+ \sum_p \left\{\frac{1}{2}\sum_k \partial/\partial x^k a_{pk}(x,y)\,|_{y\to x}\right\}\partial_p f(x). \end{aligned} \tag{2.4}$$

The coefficient of $\partial_p f(x)$ is the Itô drift of the pth coordinate. Although the Stratonovich drift of a diffusion in $R^d, d \geq 2$ is not uniquely determined by the diffusion operator, it does have a natural determination for stochastic flows, which in the present case turns out to be 0. Hence, the law of the inverse mapping $(X_t)^{-1}$, for fixed t, is the same as the law of X_t. See [17], Ch. 4.

In the special case where U is finite and $\nu(B) = \sum_B \nu_0(u)$, the flow can be generated by the finite system of stochastic differential equations

$$dX_t^p(x) = \sum_u (\nu_0(u))^{\frac{1}{2}} A_u^p(X_t(x))\, dW_{ut} + b^p(X_t(x))\, dt \tag{2.5}$$

where $\{W_{u\cdot}\}$ are independent standard Wiener processes, and $b^p(\cdot)$ is the coefficient of $\partial_p f(\cdot)$ in (2.4), $1 \leq p \leq d$.

3. Example of a Deterministic Stirring. Aref [2] considers an idealized model for the stirring of an incompressible planar fluid in the unit disk. Let D_0 (resp. D) denotes the open (resp. closed) unit disk in R^2, treated as the complex plane with points $z = (x,y) = x+iy$. Let $S^1 = \partial D$ be the unit circle. If $\zeta = \zeta_1 + i\ \zeta_2$ is a point not on S^1, representing the location of a stirring device, define a vector field by

$$\begin{aligned} &A_\zeta(z) = \left(A_\zeta^1(z), A_\zeta^2(z)\right) = A_\zeta^1(z) + i\ A_\zeta^2(z), \\ &A_\zeta^1(z) = (y-\zeta_2)/|z-\zeta|^2,\ A_\zeta^2(z) = -(x-\zeta_1)/|z-\zeta|^2, \\ &z \neq \zeta, \end{aligned} \tag{3.1}$$

and then, for $|\zeta| < 1$ put

$$V_\zeta(z) = A_\zeta(z) - A_{\frac{1}{\zeta^*}}(z), \quad \zeta* = \text{ complex conjugate of } \zeta,$$
(3.2)
$$z \neq \zeta, \ \frac{1}{\zeta*}; \quad |\zeta| < 1.$$

The velocity field V_ζ is the complex conjugate of an analytic function with poles at ζ and $\frac{1}{\zeta*}$, and has divergence and curl 0. It is tangent to S^1, so it determines an incompressible flow in $D\backslash\{\zeta\}$, corresponding to a fixed position for the stirring device. The stream lines are (non-concentric) circles of the form

$$|z-\zeta|^2 = \lambda^2 \left| z - \frac{1}{\zeta*} \right|^2, \qquad 0 \leq \lambda \leq |\zeta|. \tag{3.3}$$

As λ increases from 0 to $|\zeta|$, the center moves from ζ to 0, while the radius increases from 0 to 1.

In [2], stirring corresponds to letting ζ change with time. In the case treated in most detail, ζ stays fixed a non-random time T at a position ζ', then stays the same length of time at $-\zeta'$, then goes back to ζ' for a time T, etc. Computer studies in [2] show that suitable choices of T lead to chaotic behavior, even though no explicit randomness is introduced.

For a probabilist, it is natural to wonder what happens if we introduce randomness, for example by keeping ζ' and $-\zeta'$ fixed, but choosing between them at random (probability $\frac{1}{2}$.) We can realize the centering condition (2.2) by also randomizing the direction of the spin, choosing at random between the given velocity field and its negative. The poles at $\pm\ \zeta'$ must then be treated carefully.

We shall not go further into this. A modification of Aref's framework has been studied in a stochastic context by Jaekel [14], who analyzes a related model with one fixed pole plus a rigid Brownian motion of the plane, and obtains limiting results.

4. Moebius Stirrings. We now consider stirrings which are *automorphisms* of D_0; that is, 1:1 analytic mappings of D_0 onto itself. The possibilities are limited, since the composition group of automorphisms of D_0 consists of a certain subgroup (group operation = composition) $\mathcal{M}$ of the fractional linear (Moebius) functions. Each function in $\mathcal{M}$ can be expressed in the form (4.1) below, and likewise in the form (4.2). The parameters a, c, γ, β are complex. Again * = complex conjugation.

$$f(z) = \frac{az + c*}{cz + a*}, \qquad |a|^2 - |c|^2 = 1; \tag{4.1}$$

$$f(z) = \beta \frac{z-\gamma}{1-\gamma^* z}, \qquad |\beta| = 1, \ 0 \leq |\gamma| < 1. \tag{4.2}$$

Relevant information about the Moebius functions can be found in [1], Sec. 4.3; [5], Ch. 5; [13] I, Sec. 4.1 - 4.2 and II, 15.1.

Flows based on real fractional linear mappings have been studied by Carverhill [8], whose results are in a different direction from the ones considered here. Baxendale [4] has considered a general class of flows on certain matrix Lie groups; some of them can be specialized to give the limiting flows to be discussed here, as noted below.

The function on the right side of (4.1) can be associated with the matrix M:

$$M = \begin{vmatrix} a & c* \\ c & a* \end{vmatrix} \rightarrow H(M) : H(M; z) = \frac{az + c*}{cz + a*}; \ |a|^2 - |c|^2 = 1. \tag{4.3}$$

H is a homomorphism of the multiplicative group $\mathcal{G}$ of matrices of the given form onto $\mathcal{M}$. If g is in the Lie algebra $\boldsymbol{g}$ of $\mathcal{G}$, then

$$g = \begin{vmatrix} i\,\alpha & \delta \\ \delta* & -i\,\alpha \end{vmatrix}, \ \alpha \text{ real}, \ \delta \text{ complex}. \tag{4.4}$$

Then $\{H(e^{tg}), t \in R^1\}$ is a 1-parameter subgroup of $\mathcal{M}$, and produces a flow in D_0, corresponding to the vector field A_g:

$$(d/dt)\, H\left(e^{tg}; z\right)|_{t=0} = A_g(z) := -\delta^* z^2 + 2\alpha i z + \delta. \tag{4.5}$$

Each function $w_t(g; z)$ in the subgroup satisfies

$$dw_t(g; z)/dt = A_g(w_t(g; z)); \ w_0(g; z) = z. \tag{4.6}$$

We can then construct stirrings, as in Sec. 2, by putting a probability measure on the parameter g.

If $\delta = 0, w_t(g; z) = \exp(2i\alpha t) \cdot z$. Otherwise, the product of the roots of the quadratic in (4.5) is $-\delta/\delta*$, which shows that either

(I) Both roots (possibly identical) are on S^1 or

(II) one root is inside and the other outside S^1.

We first look at II, which occurs iff $|\delta| < |\alpha|$; a special case of I arises in Sec. 5. Let ζ be the root of smaller magnitude, the other being $\frac{1}{\zeta*}$.

The vector field of (4.5) then becomes, except for a real multiplier affecting only the time scale and direction,

$$\begin{aligned} A_\zeta(z) &= i(z - \zeta)(1 - z\zeta*)/\left(1 - |\zeta|^2\right) \\ &\equiv A_\zeta^1(z) + iA_\zeta^2(z), \ \zeta \neq 0; \quad A_0(z) = -iz. \end{aligned} \tag{4.7}$$

The special definition of A_0 is for centering purposes. Then w in (4.6) satisfies $dw/dt = A_\zeta(w)$, noting $dw/dt = -iw$ if $\zeta = 0$. The solution for $\zeta \neq 0$ is

$$w(t; \zeta; z) = \frac{\left[\exp(it) - |\zeta|^2\right] z + \zeta\,[1 - \exp(it)]}{\zeta^*\,[\exp(it) - 1]\, z + 1 - \exp(it)\,|\zeta|^2}. \tag{4.8}$$

The relation (put $w(t;\zeta;z) = w$)

(4.9) $$(w-\zeta)/(1-w\ \zeta*) = \exp(it)\ \ (z-\zeta)/(1-z\ \zeta*)$$

shows that the trajectories for a given ζ are the same set of circles in D as in Aref's example.

(4.10) **Definition.** We shall denote by $\mathcal{M}_1$ the subset of functions in $\mathcal{M}$ having the form (4.8) for some ζ, $0 \le |\zeta| < 1$ and $t\ \varepsilon\ [0, 2\pi)$, corresponding to II above.

For the restriction a_ζ of A_ζ to S^1, parametrized as $\{e^{i\phi}\}$, we have

(4.11) $$a_\zeta(\phi) = \left[1 + r^2 - 2r\cos(\phi - \theta)\right] / \left(1 - r^2\right), \ \ \zeta = re^{i\theta} \ \ \neq 0;$$

(4.12) $$a_0(\phi) = -1.$$

Define a probability measure for $\zeta = re^{i\theta}$ of the form

(4.13) $$d\nu = \mu(dr)\ \ d\theta/2\pi,$$

where the probability measure μ has its support in $[0, \rho_1]$, $0 < \rho_1 < 1$. The uniform law for θ is for sake of isotropy. To achieve centering (i.e., (2.2)), let μ' be any probability measure on $(0, \rho_1]$, and let

$$\mu = (1-\kappa)\,\Delta_{\{0\}} + \kappa\mu',$$

(4.14) $$\kappa = \frac{1}{\left\{1 + \int_0^1 (1 + r^2)\,\mu'(dr)\,/\,(1 - r^2)\right\}}.$$

Then (2.2) is satisfied.

Now consider the stirring process $\{Z_n(t,z), t \ge 0, z\varepsilon R^2\}$ produced by the vector fields $\{A_\zeta\}$. Let $w_n(\zeta;z) = w(1/\sqrt{n};\zeta;z)$ (see (4.8).) Since for fixed $t, Z_n(t,\cdot) \equiv Z_n(t)$ is a finite composition of functions in $\mathcal{M}_1$ of the form $w_n(\zeta_i;\cdot)$, it is in $\mathcal{M}$ (not necessarily in $\mathcal{M}_1$.) Note that $w_n(0;z) = \exp(-i/\sqrt{n}) \cdot z$.

Since $|\zeta| < 1$ (see above), the fields A_ζ are smooth in R^2. Given $1 < \rho_2 < \rho_3 < \infty$, we may multiply A_ζ by a smooth scalar function which is 1 for $|z| < \rho_2$ and 0 for $|z| > \rho_3$. Then (2.2) still holds for the modified field, which we still call A_ζ, and the new stirring, still denoted by $Z_n(t)$, is the same in D as the original one.

From the results of [20], as $n \to \infty$ $Z_n(t,z)$, considered as a random field in (t,z), convergences weakly to a stochastic flow of diffeomorphisms $\{Z(t,z)\}$ in R^2, and in fact the spatial-derivative fields also converge.

Until further notice, we consider Z as a flow in D. In this context, a function is said to be in a function class such as $\mathcal{M}$ if it coincides with the restriction to D of a function in the class.

(4.15). **Proposition**. *For fixed t_0, $Z(t_0, \cdot)$ is a.s. an automorphism of D_0, hence is in $\mathcal{M}$ restricted to D.*

Proof. For fixed t_0, consider $Z_n(t_0)$ as a random point in the space S of continuous mappings $D \to D$ with the usual Polish topology. Let $|f|$ be the supremum norm of $f \in S$. Given σ, $0 \le \sigma < 1$, the functions in (4.2) with $|\gamma| \le \sigma < 1$ are a compact subset of S, whence $\mathcal{M}$ is Borel.

We know $Z_n(t_0)$ converges in law to $Z(t_0)$, which is a.s. a homeomorphism of D. Then, according to a result of Skorohod, there exists a sequence (H_n) of S-valued random variables such that H_n has the same law as $Z_n(t_0)$ (hence $H_n \in \mathcal{M}$ a.s.) and (H_n) converges a.s. to an S-valued random variable H with the same law as $Z(t_0)$; $\sup_n |H_n|$ is a.s. finite, whence, from the theory of normal families, H is a.s. analytic in D_0.

Let γ_n be the root of H_n in D_0, and let β_n be the corresponding value of β in (4.2). Suppose, for some sample sequence, we have $\sup_n |\gamma_n| = 1$. Pick a subsequence (n') such that $\beta_{n'} \to \beta_0, \gamma_{n'} \to e^{i\theta_0}$. Then for each $z \in D_0$ we have $Z_{n'}(t_0, z) \to -\beta_0 \exp(i\theta_0)$. But $Z(t_0)$ is a.s. a homeomorphism, a contradiction, whence

$\sup |\gamma_n| < 1$ a.s. It follows that $Z(t_0) \in \mathcal{M}$ a.s. QED

For $z = e^{i\phi}$, we may write $Z\left(t, e^{i\phi}\right) = \exp\{i\Phi(t,\phi)\}$, where $\{\Phi(t,\phi)\}$ is a flow on S^1. From (2.3), (4.11), and (4.14) the correlation function of Φ is

$$
\begin{aligned}
B_0(\phi,\psi) &= \gamma_1 + \gamma_2 \cos(\phi - \psi), \\
\gamma_1 &= \int_0^1 \left[(1+r^2)/(1-r^2)\right]^2 d\mu, \\
\gamma_2 &= 2\int_0^1 \left[r/(1-r^2)\right]^2 d\mu.
\end{aligned}
\tag{4.16}
$$

The term γ_1 in (4.16) corresponds to an independent rigid Brownian rotation of S_1 superposed on the rest of the flow. Below we shall consider whether it can be eliminated by a more sensitive centering.

The hyperbolic metric is known to be well adapted to dealing with Moebius mappings. The hyperbolic element of arc is

$$
ds = \left(dx^2 + dy^2\right)^{\frac{1}{2}} / \left(1 - |z|^2\right). \tag{4.17}
$$

Hyperbolic curve lengths, distances, angles, and areas are left invariant by the mappings in $\mathcal{M}$. For a hyperbolic observer, the flow $Z(t)$ is rigid, in the sense that shapes and areas are unchanged. This implies that a small figure which returns near its starting point is approximately congruent, even in the Euclidean sense, to its original shape.

There are other evidences of limited randomness. For example, from the fact that Moebius mappings leave cross ratios invariant, there is a simple expression for $Z_t(z_4)$ in terms of $Z_t(z_i)$, $i = 1, 2, 3$.

Much more about the flow $\{Z(t,z)\}$ follows from specializing results of [4]. Here, we give only a simple direct proof of one result, which also follows from some of the ones in [4].

(4.18) **Proposition.** *Let D' and D'' be closed disks centered at 0 with respective radii $\rho' < 1$ and $\rho'' < 1$. Then*

$$P\{Z(t,D') \cap D'' = \emptyset \text{ for all sufficiently large } t\} = 1.$$

Proof. *The restriction of Z_t to S^1 may be written* $\exp\{i\ \Phi_t(\phi)\}$ *(see (4.16), where we will regard $\Phi_t(\phi)$ as a process in R^1, with period 2π in ϕ. From Cauchy's formula*

$$(4.19)\quad |Z_t(0)|^2 = (4\pi^2)^{-1} \iint_{0 \le \phi,\psi \le 2\pi} \exp\{i\ (\Phi_t(\phi) - \Phi_t(\psi))\}\, d\phi\ d\psi.$$

From the theory of flows in R^1 (see e.g. [12]; the restriction about the point spectrum in (1.1) of [12] is irrelevant here),

(4.20)

$$P\left\{\lim_{t\to\infty} (\Phi_t(\phi) - \Phi_t(\psi)) \text{ exists and is 0 or } 2\pi\right\} = 1, \quad |\phi - \psi| \le 2\pi.$$

From (4.19), (4.20), and Fubini's theorem, we see that $|Z_t(0)|^2 \to 1$. From the definition of the hyperbolic metric, the Euclidean diameter of the set $Z_t(D') \to 0$, since its hyperbolic diameter remains fixed. QED

5. Centering and Another One-Parameter Group. The functions $w(t;\zeta;z)$ of (4.8) satisfy

$$(5.1)\qquad w(t;\omega\zeta;\omega z) = \omega\, w(t;\zeta;z), \quad \omega \text{ complex}, \quad |\omega| = 1.$$

From this we deduce that for fixed t, $Z_n(t)$ of Sec. 4 has the same law as $\exp(-iN_t/\sqrt{n}) \cdot Z'_n(t)$, where the stirring $Z'_n(t)$ is constructed with the probability measure μ' for ζ (no centering; see (4.14)), with stirring intensity $n\kappa$ instead of n; N_t is Poisson, mean $nt(1-\kappa)$, independent of $Z'_n(t)$. Hence, an observer willing to twist his neck does not need the centering condition (2.2) in this case.

The centering determined by (4.14) was rather crude, in that it does not depend on ζ. Using a distribution μ' on $(0, \rho_1]$ for $|\zeta|$, we could apply a rotation $\exp\{-i(1+r^2)/(1-r^2)\sqrt{n}\}$ after each stirring $w_n(\zeta)$, $r = |\zeta|$.

Assuming a limiting flow exists, one might expect this to eliminate the rigid Brownian rotation mentioned below (4.16) because, from (4.11), for each ϕ,

$$\int_0^{2\pi} a_\zeta(\phi)\,\frac{d\theta}{2\pi} = \frac{(1+r^2)}{(1-r^2)}, \quad \zeta = re^{i\theta}. \tag{5.2}$$

Rather than examining this, let's try modifying the vector field A_ζ of (4.7), $\zeta \neq 0$, by subtracting $iz\left(1+r^2\right)/\left(1-r^2\right)$, which should have a similar effect. We obtain a new field U_ζ, $\zeta = re^{i\theta}$,

$$\begin{aligned} U_\zeta(z) &= A_\zeta(z) - i\frac{\left(1+r^2\right)z}{\left(1-r^2\right)} \\ &= \frac{-i\,r\exp(-i\theta)\ \left(z^2+\exp(2i\theta)\right)}{1-r^2}, \quad 0<r<1. \end{aligned} \tag{5.3}$$

The solution of $\frac{dw_t}{dt} = U_\zeta(w_t)$, $w_0 = z \ \varepsilon\ D$, is

$$w = \frac{z - i\exp(i\theta)\ S_t(\zeta)}{1 + i\,z\exp(-i\theta)\ S_t(\zeta)},$$

$$S_t(\zeta) = \tanh\left[\frac{rt}{(1-r^2)}\right]. \tag{5.4}$$

The functions on the right side of (5.4) form a group with t as an additive parameter, corresponding to the choice $\alpha = 0$ in (4.4). Since U_ζ vanishes at $\pm\ ie^{i\theta}$, we are in the regime I (see below (4.6)). The integral curve through $z\ \varepsilon\ D_0$ is the circular arc in D_0 through $z, ie^{i\theta}$, and $-ie^{i\theta}$ (straight line if those three points are collinear.) The trajectories approach $-ie^{i\theta}$ (resp. $ie^{i\theta}$) as $t \to \infty$ (resp. $-\infty$). Taking $\nu(d\zeta) = \mu'(dr)\frac{d\theta}{2\pi}$, where $\mu'(0,\rho_1] = 1$ as above, and using only $t \geq 0$, we see that (2.2) holds, and we again get a limiting flow. This time, the correlation function on S^1 has the form $B_1(\phi,\psi) = c\cos(\phi-\psi)$, $c = 2\int_0^1 [r/(1-r^2)]^2\mu'(dr)$.

REFERENCES

[1] L. V. Ahlfors, *Complex Analysis*, 3d Ed., McGraw-Hill, London. 1979.

[2] H. Aref, *Stirring by chaotic advection*, J. Fluid Mechanics **143** (1984), 1 -21.

[3] P. Baxendale, *Brownian motions in the diffeomorphism group*, I, Compositio Math. **53** (1984), 19 - 50.

[4] P. Baxendale, *Asymptotic behavior of stochastic flows of diffeomorphisms: two case studies*, Prob. Th. Rel. Fields **73** (1986), 51 - 85.

[5] A. Beardon, *A Primer on Riemann Surfaces*, London Math. Soc. Lectures Notes **78**, Cambridge 1984.

[6] F. Bertein and A. Galvez, *Une classe de systèmes de particules stables par association*, Z. Wahr. Verw. Geb. **41** (1977), 73 - 85.

[7] J.-M. Bismut, *Méchanique Aléatoire*, Springer Lecture Notes in Math. **866**, 1981.

[8] A. Carverhill, *Flows of stochastic dynamical systems: ergodic theory*, Stochastics **14** (1985), 273 - 318.

[9] D. Griffeath, *Additive and Cancellative Interacting Particle Systems*, Springer Lecture Notes in Math. **724**, 1979.

[10] T. Harris, *Additive set-valued Markov processes and graphical methods*, Ann. Prob. **6** (1978), 355 - 378.

[11] T. Harris, *Brownian motions on the homeomorphisms of the plane*, Ann. Prob. **9** (1981), 232 - 254.

[12] T. Harris, *Coalescing and non-coalescing flows in* R^1, Stoch. Proc. Appl. **17** (1984), 187 - 210.

[13] E. Hille, *Analytic Function Theory*, Vols. I & II, Blaisdell, NY, 1959.

[14] T. Jaekel, *Stochastic Flow with a Singular Vortex*, Doctoral Dissertation, University of Southern California, 1989.

[15] H. Kesten and G. Papanicolaou, *A limit theorem for turbulent diffusion*, Comm. Math. Phys. **65** (1979), 97 - 128.

[16] H. Kunita, *Lectures on Stochastic Flows and Applications*, Tata Inst., Bombay, 1986 and Springer-Verlag, New York.

[17] H. Kunita, *Stochastic Flows and Stochastic Differential Equations*, book in press.

[18] W. Lee, *Random Stirring of the Real Line*, Ann. Prob. **2** (1974), 580 - 592.

[19] Y. LeJan and S. Watanabe, *Stochastic flows of diffeomorphisms*, Proceedings of the Taniguchi Symposium 307 - 332, North Holland, New York, 1984.

[20] H. Matsumoto and I. Shigekawa, *Limit theorems for stochastic flows of diffeomorphisms of jump type*, Z. Wahr. Verw. Geb. **69** (1985), 507 - 540.

[21] F. Spitzer, *Interaction of Markov processes*, Advances in Math. **5** (1970), 246 - 290.

[22] J. Watkins, *A central limit problem in random evolutions*, Ann. Prob. **12** (1984), 480 - 513.

[23] J. Watkins, *Limit theorems for stationary random evolutions*, Stoch. Proc. Applic. **19** (1985), 189 - 224.

[24] J. Watkins, *A stochastic integral representation for random evolutions*, Ann. Prob. **13** (1985), 531 - 557.

T. E. Harris
Department of Mathematics DRB 306
University of Southern California
Los Angeles, CA 90089-1113

The One Dimensional Stochastic X-Y Model

RICHARD HOLLEY[1]

Consider infinitely many diffusions on an infinite number of circles, one at each integer, that are interacting with each other by means of a finite range potential. We prove that for any finite dimensional marginal of the process, the rate of convergence to the corresponding marginal of the GIBBS state for the potential is exponentially fast.

1. Introduction. Recently the author and Dan Stroock [1]0 have proved that for one dimensional stochastic ISING models with finite range interactions the rate of convergence to equilibrium of the semi-group acting on cylinder functions is exponentially fast starting from all initial configurations. Moreover the exponent is the one that would be expected from L^2 considerations. It is the goal of this paper to prove the analogous result for the stochastic X-Y model.

By the one dimensional stochastic X-Y model we mean the following MARKOV process. Let T denote the unit circle and Z denote the integers. The state space for the process is $\mathcal{S} = T^Z$. We use letters such as η, σ, and ω to denote elements of $\mathcal{S}$. The process will be a diffusion on each component of $\mathcal{S}$ with the drift coefficient in each component determined by the positions of the diffusions at nearby components. The interaction in the drift coefficients is given by a potential $\{\Phi_R : R \subset \mathbf{R}\}$, where each Φ_R is a C^1 function of $x_i, i \in R$, and we assume that $\Phi_R \equiv 0$ if $diam(R) > r$, where r is the range of the interaction. We also assume that $\Phi_R(\eta) = \Phi_{R+k}(\sigma)$ if σ and η are such that $\eta_x = \sigma_{x+k}$ for all $x \in R$. Let $\mathcal{D}$ denote the infinitely differentiable functions on $\mathcal{S}$ which depend on only a finite number of coordinates, and for $f \in \mathcal{D}$ define

$$Lf(\eta) = \sum_{k \in Z} e^{H_k(\eta)} \left(\frac{\partial}{\partial \eta_k} \left(e^{-H_k(\eta)} \frac{\partial}{\partial \eta_k} f(\eta) \right) \right),$$

where

$$H_k(\eta) = \sum_{R \ni k} \Phi_R(\eta).$$

It is known (see [8]) that there is a unique MARKOV process on $\mathcal{S}$ whose infinitesimal generator when restricted to $\mathcal{D}$ is given by L. This MARKOV

[1]Partially supported by the National Science Foundation.

process is the stochastic X-Y model with potential $\{\Phi_R\}$. We denote the semi-group of this process by $\{T_t : t \geq 0\}$.

Let $\mathcal{F}_\Lambda$ be the sigma algebra generated by $\{\omega_k : k \notin \Lambda\}$. The GIBBS state, μ, for the potential $\{\Phi_R\}$ is the unique (because we are in one dimension) measure on the BOREL subsets of $\mathcal{S}$ whose regular conditional probabilities given $\mathcal{F}_\Lambda$, Λ a finite subset of Z, have a density with respect to the product of normalized Lebesgue measures given by

$$\mu_\Lambda(\eta|\omega) = \frac{\exp(-\sum_{R\cap\Lambda\neq\emptyset}\Phi_R(\eta^\Lambda\omega))}{Z(\Lambda,\omega)}.$$

We use $\eta^\Lambda\omega$ to denote the configuration which is equal to η on Λ and equal to ω on Λ^c, and $Z(\Lambda,\omega)$ is the normalizing constant needed to make $\mu_\Lambda(\cdot,\omega)$ have integral one with respect to the product of normalized Lebesgue measures on T. We sometimes write $\mu(f)$ for the integral of f with respect to μ.

It is easily checked that for all ϕ, $\psi \in \mathcal{D}$

$$\int \phi L\psi \, d\mu = \int \psi L\phi \, d\mu.$$

Let $\mathbf{L}$ denote the generator of the semigroup of the stochastic X-Y model on $L^2(\mu)$. By combining the techniques in [1] with the methods of comparing generators found in section 5 of [6] it can be shown that 0 is an isolated point in the spectrum of $\mathbf{L}$. Let γ be the distance between 0 and the rest of the spectrum of $\mathbf{L}$. It is immediate from the spectral theorem that for all $f \in L^2(\mu)$

$$(1.1) \qquad ||T_tf(\cdot) - \mu(f)||_{L^2(\mu)} \leq e^{-\gamma t}||f - \mu(f)||_{L^2(\mu)}.$$

Our goal in this paper is to prove the following theorem. We use $||\cdot||_u$ to represent the uniform norm on continuous functions from $\mathcal{S}$ to $\mathbf{R}$.

Theorem 1.2. *Let $0 < \epsilon < \gamma$. Then for all $f \in \mathcal{D}$ there is an $A_f < \infty$ such that for all $t \geq 0$*

$$||T_tf(\cdot) - \mu(f)||_u \leq A_f e^{-\epsilon t}.$$

The results that we need in order to prove Theorem (1.2) involve logarithmic Sobolev inequalities, and the inequalities in [9] are strong enough for our purposes; however, recently Boguslaw Zegarlinski [12] has proved a logarithmic Sobolev inequality for GIBBS states on $\{-1,1\}^Z$ having finite range interactions. His proof is complicated somewhat by the need to use finite differences rather than derivatives. In the present setting Zegarlinski's proof is simpler and the resulting inequality will simplify the proof of Theorem (1.2). Thus we devote section 2 to a proof of the logarithmic Sobolev inequality for μ and L. Theorem (1.2) is then proved in section 3.

2. A Logarithmic Sobolev Inequality. Let $C^1(\mathcal{S})$ be the space of continuous, real valued functions on $\mathcal{S}$ which are continuously differentiable in each coordinate. For $f \in C^1(\mathcal{S})$ we define

$$\mathcal{E}(f,f) = \sum_{k\in Z} \int \left(\frac{\partial}{\partial \eta_k} f(\eta)\right)^2 \mu(d\eta).$$

It is not hard to check that for $f \in \mathcal{D}$,

$$\mathcal{E}(f,f) = -\int fLf\, d\mu. \tag{2.1}$$

The goal of this section is to prove the following theorem.

Theorem 2.2 (Zegarlinski). *Let μ be a* GIBBS *state as described in the introduction. Then there is an $\alpha < \infty$ such that for all $f \in C^1(\mathcal{S})$,*

$$\int f^2(\eta) \log(f^2(\eta))\, \mu(d\eta) \le \alpha\mathcal{E}(f,f) + \int f^2(\eta)\, \mu(d\eta) \log\left(\int f^2(\eta)\, \mu(d\eta)\right).$$

At high temperatures, i.e. weak interactions, Theorem (2.2) is known and a proof can be found in **[3]**. Our proof of Theorem (2.2) follows the outline of the proof given by Zegarlinski **[12]**; however, since we are dealing with differential operators instead of finite difference operators, the calculations are considerably easier.

We refer to a set $\Lambda \subset Z$ as an interval if it is a set of consecutive integers.

Lemma 2.3. *For every n there is an $\epsilon_n > 0$ such that if $\Lambda \subset Z$ consists of n disjoint intervals and $\eta \in \mathcal{S}$ then*

$$\epsilon_n \left\{ \int (f(\sigma) - \mu_\Lambda(f|\eta))^2 \mu_\Lambda(d\sigma|\eta) \right\} \le \sum_{k\in\Lambda} \int \left(\frac{\partial}{\partial \sigma_k} f(\sigma)\right)^2 \mu_\Lambda(d\sigma|\eta)..$$

Proof. From (1.1) it follows that for every $g \in \mathcal{D}$

$$\gamma \int (g(\sigma) - \mu(g))^2 \mu(d\sigma) \le \sum_k \int \left(\frac{\partial}{\partial \sigma_k} g(\sigma)\right)^2 \mu(d\sigma). \tag{2.4}$$

Given $f \in \mathcal{D}$, Λ a finite disjoint union of n intervals, and $\eta \in \mathcal{S}$, let $g(\sigma) = f(\sigma^\Lambda \eta)$. Then $g \in \mathcal{D}$ and since g depends only on the sites in Λ, (2.4) becomes

$$\gamma \int (f(\sigma^\Lambda \eta) - \mu(f(\cdot^\Lambda \eta)))^2 \mu(d\sigma) \le \sum_{k\in\Lambda} \int \left(\frac{\partial}{\partial \sigma_k} f(\sigma^\Lambda \eta)\right)^2 \mu(d\sigma). \tag{2.5}$$

Let ν_Λ be the marginal of μ on T^Λ. From the definition of μ one sees that there is a constant, $\delta_n > 0$ such for all η

$$\delta_n \le \frac{d\nu_\Lambda}{d\mu_\Lambda(\cdot|\eta)} \le \delta_n^{-1}.$$

Thus we may replace μ in (2.5) by $\mu_\Lambda(\cdot|\eta)$ if we replace γ by $\gamma\delta_n^2$. ■

Before stating the next lemmas we introduce some notation. In the following r is an even number that is at least as large as the range of the interaction, and L is a large even number which will be chosen later. For every integer k we define Λ_k to be the interval of integers $[\frac{k-1}{2}L + \frac{kr}{2}, \frac{k+1}{2}L + \frac{kr}{2}]$. Note that each Λ_k has length L and that there is gap of length r between Λ_k and Λ_{k+2}. Moreover, that gap is in the center of Λ_{k+1}. Set $\Gamma_0 = \cup_j \Lambda_{2j}$ and $\Gamma_1 = \cup_j \Lambda_{2j+1}$. If f is a bounded measurable function on $\mathcal{S}$ and $\Lambda \subset Z$, let

$$E_\Lambda[f](\eta) = \int f(\sigma)\,\mu_\Lambda(d\sigma|\eta).$$

Lemma 2.6. *There is an $\alpha_0 < \infty$ such that for all $f \in C^1(T^Z)$*

$$E_{\Gamma_i}[f^2\log(f^2)](\eta) \le \alpha_0 \sum_{k\in\Gamma_i} E_{\Gamma_i}[|\frac{\partial}{\partial\eta_k} f|^2](\eta) + E_{\Gamma_i}[f^2](\eta)\log(E_{\Gamma_i}[f^2](\eta)).$$

Proof. Let λ be normalized Lebesgue measure on T. Then there is a constant c_0 such that for all $f \in C^1(T)$

(2.7)

$$\int_T f^2(x)\log(f^2(x))\lambda(dx) \le c_0 \int_T \left(\frac{d}{dx}f(x)\right)^2 \lambda(dx) + \int_T f^2(x)\lambda(dx)\log\left(\int_T f^2(x)\lambda(dx)\right)$$

see [11]. By the results in [4], for any Λ_k as above,

$$\int_{T^{\Lambda_k}} f^2(\eta)\log(f^2(\eta))\lambda^{\Lambda_k}(d\eta) \le c_0 \sum_{n\in\Lambda_k} \int_{T^{\Lambda_k}} \left(\frac{\partial}{\partial\eta_n} f(\eta)\right)^2 \lambda^{\Lambda_k}(d\eta) + \int_{T^{\Lambda_k}} f^2(\eta)\lambda^{\Lambda_k}(d\eta)\log\left(\int_{T^{\Lambda_k}} f^2(\eta)\lambda^{\Lambda_k}(d\eta)\right).$$

Next note that since all of the Λ_k have the same length, there is a constant $c_1 < \infty$ such that for all η and all k

$$c_1^{-1} \le \frac{d\mu_{\Lambda_k}(\cdot|\eta)}{d\lambda^{\Lambda_k}} \le c_1.$$

Thus the above inequality holds if we replace λ^{Λ_k} by $\mu_{\Lambda_k}(\cdot|\eta)$ provided we replace c_0 by $c_0 c_1^2$. (see Lemma (2.5)) in [5]. Finally since the range of the interaction is less than or equal to r it follows that $\mu_{\Gamma_i}(\cdot|\eta)$ is a product of $\mu_{\Lambda_k}(\cdot|\eta)$'s and hence again by [4] the lemma follows with $\alpha_0 = c_0 c_1^2$. ■

Lemma 2.8. *For all* $f \in C^1(T^{\mathbf{Z}})$

$$E[f^2 \log(f^2)] \leq \alpha_0 \sum_{k \in \Gamma_i} E[|\frac{\partial}{\partial \eta_k} f|^2] + E[E_{\Gamma_i}[f^2](\eta) \log(E_{\Gamma_i}[f^2](\eta))].$$

Proof. This follows from Lemma (2.6) once one observes that

$$E[f^2 \log(f^2)] = E[E_{\Gamma_i}[f^2 \log(f^2)](\eta)].$$

■

Zegarlinski's idea is to iterate the conclusion of Lemma (2.8) alternating between Γ_0 and Γ_1. In the first iteration f^2 is replaced by $E_{\Gamma_0}[f^2]$. In the next iteration it is replaced by $E_{\Gamma_1}[E_{\Gamma_0}[f^2]]$ and so on. This means that f is replaced by $\sqrt{E_{\Gamma_0}[f^2]}$ and the question of differentiability arises in case $E_{\Gamma_0}[f^2]$ should happen to be zero. However, it is an easy exercise to show by a limiting argument that if Theorem (2.2) is true for all uniformly positive $f \in C^1(\mathcal{S})$ then it is true for all $f \in C^1(\mathcal{S})$. Thus we assume from now on that f is uniformly positive.

Since $\sqrt{E_{\Gamma_i}[f^2](\eta)}$ depends only on the η_k for $k \notin \Gamma_i$ it follows that for $k \in \Gamma_i$,

$$\frac{\partial}{\partial \eta_k} \sqrt{E_{\Gamma_i}[f^2](\eta)} = 0.$$

Before stating the next lemma we let $k \notin \Gamma_i$ and define functions $\psi_k^{(\eta)}(\cdot)$ by

$$\psi_k^{(\eta)}(\sigma) = \sum_{R \ni k} \frac{\partial}{\partial \eta_k} \phi_R(\sigma^{\Gamma_i}\eta) - E_{\Gamma_i}\left[\sum_{R \ni k} \frac{\partial}{\partial \eta_k} \phi_R\right](\eta).$$

Note that from the definitions of $E_{\Gamma_i}[f](\eta)$ and μ_{Γ_i} it follows that for any $f \in C^1(\mathcal{S})$ and $k \notin \Gamma_i$

$$\frac{\partial}{\partial \eta_k} E_{\Gamma_i}[f](\eta) = E_{\Gamma_i}[\frac{\partial}{\partial \eta_k} f](\eta) + E_{\Gamma_i}[\psi_k^{(\eta)} f](\eta). \tag{2.9}$$

Lemma 2.10. *For all* $f \in C^1(\mathcal{S})$ *which are uniformly positive*

$$\begin{aligned} &\left(\frac{\partial}{\partial \eta_k} \sqrt{E_{\Gamma_i}[f^2](\eta)}\right)^2 \\ &\leq 2E_{\Gamma_i}[|\frac{\partial}{\partial \eta_k} f|^2](\eta) + \frac{1}{2E_{\Gamma_i}[f^2](\eta)} \left(E_{\Gamma_i}[\psi_k^{(\eta)} f^2](\eta)\right)^2. \end{aligned} \tag{2.11}$$

Proof. From (2.9) we have

$$\frac{\partial}{\partial\eta_k}\sqrt{E_{\Gamma_i}[f^2](\eta)} = \frac{E_{\Gamma_i}[2f\frac{\partial}{\partial\eta_k}f](\eta) + E_{\Gamma_i}[\psi_k^{(\eta)}f^2](\eta)}{2\sqrt{E_{\Gamma_i}[f^2](\eta)}}.$$

Applying Schwartz' inequality to the first term in the numerator of the above expression, we arrive at

$$|\frac{\partial}{\partial\eta_k}\sqrt{E_{\Gamma_i}[f^2](\eta)}| \le \sqrt{E_{\Gamma_i}[|\frac{\partial}{\partial\eta_k}f|^2](\eta)} + \frac{1}{2\sqrt{E_{\Gamma_i}[f^2](\eta)}}E_{\Gamma_i}[\psi_k^{(\eta)}f^2](\eta).$$

The conclusion then follows from the inequality $(a+b)^2 \le 2a^2 + 2b^2$. ∎

The first term on the right side of (2.11) is the type of expression that we are looking for. We must still bound the second term. We do so with the help of the following lemma.

Lemma 2.12. *Let m be a probability measure and ψ and f be random variables. Assume that $E_m[\psi] = 0$. Then*

$$|E_m[\psi f^2]| \le 2\|\psi\|_\infty \|f\|_2 \sqrt{Var_m[f]}.$$

Proof.

$$\begin{aligned} E_m[\psi f^2] &= E_m[\psi(f^2 - (E_m[f])^2)] \le \|\psi\|_\infty E_m[|f^2 - E_m[f]^2|] \\ &\le \|\psi\|_\infty \left(E_m[(f + E_m[f])^2]\right)^{\frac{1}{2}} \left(E_m[(f - E_m[f])^2]\right)^{\frac{1}{2}} \\ &\le 2\|\psi\|_\infty \|f\|_2 \sqrt{Var_m[f]}. \end{aligned}$$

∎

Lemma 2.13. *Let $f \in C^1(S)$ be uniformly positive and let $k \notin \Gamma_i$. Let j be such that k is in the gap between Λ_j and Λ_{j+2}. Then*

$$\begin{aligned} &\frac{1}{2E_{\Gamma_i}[f^2](\eta)}\left(E_{\Gamma_i}[\psi_k^{(\eta)}f^2](\eta)\right)^2 \\ &\le 2\|\psi_k^{(\eta)}\|_\infty^2 \frac{1}{\epsilon_2} \sum_{n\in\Lambda_j\cup\Lambda_{j+2}} E_{\Gamma_i}\left[\left(\frac{\partial}{\partial\eta_n}f\right)^2\right](\eta). \end{aligned}$$

Here ϵ_2 is as in Lemma (2.3).

Proof.

(2.14)

$$\begin{aligned}
\left(E_{\Gamma_i}[\psi_k^{(\eta)} f^2](\eta)\right)^2 &= \left(E_{\Gamma_i}[E_{\Lambda_j\cup\Lambda_{j+2}}[\psi_k^{(\eta)} f^2]](\eta)\right)^2 \\
&\leq \left(E_{\Gamma_i}\left[2\|\psi_k^{(\eta)}\|_\infty \sqrt{E_{\Lambda_j\cup\Lambda_{j+2}}[f^2]}\sqrt{Var_{\Lambda_j\cup\Lambda j+2}[f]}\right](\eta)\right)^2 \\
&\leq 4\|\psi_k^{(\eta)}\|_\infty^2 E_{\Gamma_i}[E_{\Lambda_j\cup\Lambda_{j+2}}[f^2]]E_{\Gamma_i}[Var_{\Lambda_i\cup\Lambda_{i+2}}[f]](\eta) \\
&\leq 4\|\psi_k^{(\eta)}\|_\infty^2 E_{\Gamma_i}[f^2]E_{\Gamma_i}\left[\frac{1}{\epsilon_2}\sum_{n\in\Lambda_j\cup\Lambda_{j+2}} E_{\Lambda_j\cup\Lambda j+2}\left[\left(\frac{\partial}{\partial\eta_n}f\right)^2\right]\right](\eta) \\
&= 4\|\psi_k^{(\eta)}\|_\infty^2 E_{\Gamma_i}[f^2](\eta)\frac{1}{\epsilon_2}\sum_{n\in\Lambda_j\cup\Lambda_{j+2}} E_{\Gamma_i}\left[\left(\frac{\partial}{\partial\eta_n}f\right)^2\right](\eta).
\end{aligned}$$

In deriving (2.14) we have used Lemma (2.3). The lemma follows easily from (2.14). ■

Combining Lemmas (2.10) and (2.13) we see that for uniformly positive $f \in C^1(\mathcal{S})$ and $k \notin \Gamma_i$,

(2.15)

$$\begin{aligned}
&\left(\frac{\partial}{\partial\eta_k}\sqrt{E_{\Gamma_i}[f^2](\eta)}\right)^2 \\
&\leq 2E_{\Gamma_i}\left[\left(\frac{\partial}{\partial\eta_k}f\right)^2\right](\eta) + 2\|\psi_k^{(\eta)}\|_\infty^2\frac{1}{\epsilon_2}\sum_{n\in\Lambda_j\cup\Lambda_{j+2}} E_{\Gamma_i}\left[\left(\frac{\partial}{\partial\eta_n}f\right)^2\right](\eta).
\end{aligned}$$

Lemma 2.16. *There is a constant $C_0 < \infty$ such that for all uniformly positive $f \in C^1(T^Z)$*

$$\sum_{k\in Z}\left(\frac{\partial}{\partial\eta_k}\sqrt{E_{\Gamma_0}[f^2](\eta)}\right)^2 \leq C_0\sum_{k\in Z} E_{\Gamma_0}\left[\left(\frac{\partial}{\partial\eta_k}f\right)^2\right](\eta) \tag{2.17}$$

Proof. Since $\sup\{\|\psi_k^{(\eta)}\| : k \notin \Gamma_0, \eta \in \mathcal{S}\} < \infty$, this follows from (2.15). In fact we may take

$$C_0 = 2 + 4\sup\|\psi_k^{(\eta)}\|^2 r\frac{1}{\epsilon_2}.$$

■

The next lemma is a strengthening of the previous one for functions f of the form $f(\eta) = E_{\Gamma_i}[g](\eta)$ Note that such functions are $\mathcal{F}_{\Gamma_i}$ measurable, i.e. they depend only on the coordinates η_k for $k \notin \Gamma_i$.

Lemma 2.18. *For all sufficiently large L there is a $\kappa < 1$ such that for all uniformly positive $f \in C^1(T^Z)$ that are $\mathcal{F}_{\Gamma_{1-i}}$ measurable,*

$$\sum_{k\in Z}\left(\frac{\partial}{\partial\eta_k}\sqrt{E_{\Gamma_i}[f^2](\eta)}\right)^2 \le \kappa\sum_{k\in Z}E_{\Gamma_i}\left[\left(\frac{\partial}{\partial\eta_k}f\right)^2\right](\eta) \tag{2.19}$$

Proof. Note that the first term on the right side of (2.15) is 0 for $k \notin \Gamma_i$ and f that is $\mathcal{F}_{\Gamma_{1-i}}$ measurable. Thus we must deal with the second term. Recall that the second term came from Lemma (2.13) and consider the expression on the right side of the first line of (2.14). Since k is in the gap between Λ_j and Λ_{j+2}, it follows that the distance between k and Γ_{1-i} is at least $\frac{L-r}{2}$. We compute $E_{\Lambda_j\cup\Lambda_{j+2}}[\psi_k^{(\eta)}f^2](\eta)$ as follows. First let $\Omega = (\Lambda_j \cup \Lambda_{j+2}) \setminus \Gamma_{1-i}$, and note that $\mathcal{F}_{\Lambda_j\cup\Lambda_{j+2}} \subset \mathcal{F}_\Omega$. Also f is $\mathcal{F}_\Omega$ measurable. Thus

$$\begin{aligned}&E_{\Lambda_j\cup\Lambda_{j+2}}[\psi_k^{(\eta)}f^2](\eta)\\ &= E_{\Lambda_j\cup\Lambda_{j+2}}[E_\Omega[\psi_k^{(\eta)}f^2]](\eta)\\ &= E_{\Lambda_j\cup\Lambda_{j+2}}[E_\Omega[\psi_k^{(\eta)}]f^2](\eta).\end{aligned}$$

Now k is in the gap between Λ_j and Λ_{j+2}, and since the interaction has range r, $\psi_k^{(\eta)}$ as a function on $T^{\Lambda_j\cup\Lambda_{j+2}}$ depends only on the sites within a distance r of the gap between Λ_j and Λ_{j+2}. The sites in $\Lambda_j\cup\Lambda_{j+2}\setminus\Omega$ are at least a distance $\frac{L}{2}-\frac{3}{2}r$ from the sites in $\Lambda_j\cup\Lambda_{j+2}$ on which $\psi_k^{(\eta)}$ depends. Also $\mu_\Lambda(\cdot|\eta)$ is the conditional distribution for a MARKOV process on a compact state space whose transition function has a density with respect to Lebesgue measure which is uniformly positive and uniformly bounded. Hence the influence of the boundary conditions goes to zero exponentially fast. It follows that there is an $A < \infty$ and a $\delta < 1$ such that

$$\sup\{\|E_\Omega[\psi_k^{(\eta)}]\|_\infty : k \in Z, \eta \in \mathcal{S}\} \le A\delta^L.$$

Using this bound we see, just as in the proof of (2.15), that

$$\sum_{k\in Z}\left(\frac{\partial}{\partial\eta_k}\sqrt{E_{\Gamma_i}[f^2](\eta)}\right)^2 \le 4rA\delta^L\frac{1}{\epsilon_2}\sum_{k\in Z}E_{\Gamma_i}\left[\left(\frac{\partial}{\partial\eta_k}f\right)^2\right](\eta).$$

For L sufficiently large $\kappa \equiv 4rA\delta^L\frac{1}{\epsilon_2} < 1$. ∎

Now define $f_0 = f$, $f_1(\eta) = \sqrt{E_{\Gamma_0}[f_0^2](\eta)}$, $f_2(\eta) = \sqrt{E_{\Gamma_1}[f_1^2](\eta)}$, and in general $f_n(\eta) = \sqrt{E_{\Gamma_{(n-1)\bmod_2}}[f_{n-1}^2](\eta)}$, $n \ge 3$.

We assume from now on that L is large enough so that Lemma (2.18) holds.

Lemma 2.20. *Let $\kappa < 1$ be as in Lemma (2.18) and C_0 be as in Lemma (2.16). Then for all $f \in C^1(\mathcal{S})$ which are uniformly positive and for all $n \geq 1$,*

$$\begin{aligned}\sum_{k\in Z}\left(\frac{\partial}{\partial\eta_k}f_n(\eta)\right)^2 \\ \leq C_0\kappa^{n-1}\sum_{k\in Z}E_{\Gamma_{(n-1)mod_2}}E_{\Gamma_{(n-2)mod_2}}\cdots E_{\Gamma_0}\left[\left(\frac{\partial}{\partial\eta_k}f\right)^2\right](\eta)\end{aligned} \tag{2.21}$$

Proof. This follows easily from Lemmas (2.16) and (2.18) by induction. ■

Lemma 2.22. *For all $f \in C^1(\mathcal{S})$ which are uniformly positive, and for all $n \geq 0$,*

$$\begin{aligned}\int f^2(\eta)\log\left(f^2(\eta)\right)\,\mu(d\eta) \\ \leq \alpha_0 C_0\frac{1-\kappa^n}{1-\kappa}\mathcal{E}(f,f) + \int f_n^2(\eta)\log\left(f_n^2(\eta)\right)\,\mu(d\eta).\end{aligned} \tag{2.23}$$

Proof. The proof is by induction on n. For $n = 0$ there is nothing to do. Thus assume that the (2.23) is true for n, and consider

$$\int f_n^2(\eta)\log\left(f_n^2(\eta)\right)\,\mu(d\eta) = \int E_{\Gamma_{(n)mod_2}}[f_n^2(\cdot)\log\left(f_n^2(\cdot)\right)](\eta)\,\mu(d\eta).$$

Setting $i = (n)mod_2$ and applying Lemma (2.6) we have
(2.24)

$$\begin{aligned}&\int E_{\Gamma_i}[f_n^2(\cdot)\log\left(f_n^2(\cdot)\right)](\eta)\,\mu(d\eta) \\ &\leq \int\left(\alpha_0\sum_{k\in\Gamma_i}E_{\Gamma_i}[|\frac{\partial}{\partial\eta_k}f_n|^2](\eta) + E_{\Gamma_i}[f_n^2](\eta)\log(E_{\Gamma_i}[f_n^2](\eta))\right)\,\mu(d\eta) \\ &= \alpha_0\int\sum_{k\in Z}\left(\frac{\partial}{\partial\eta_k}f_n(\eta)\right)^2\,\mu(d\eta) + \int f_{n+1}^2(\eta)\log(f_{n+1}^2(\eta))\,\mu(d\eta).\end{aligned}$$

An application of Lemma (2.20) and the induction hypothesis completes the proof. ■

Theorem (2.2) will follow from Lemma (2.22) by letting n go to infinity once we show that $f_n^2 \to \int f^2(\eta)\,\mu(d\eta)$ in μ measure. That will be enough because f is bounded and thus

$$\int f_n^2(\eta)\log\left(f_n^2(\eta)\right)\,\mu(d\eta) \to \int f^2(\eta)\,\mu(d\eta)\log\left(\int f^2(\eta)\,\mu(d\eta)\right).$$

Lemma 2.25. *If $f \in C^1(T^Z)$ is uniformly positive and $\mathcal{E}(f,f) < \infty$, then $f_n^2 \to \int f^2(\eta)\,\mu(d\eta)$ in μ measure.*

Proof. Since $\int f_n^2(\eta)\,\mu(d\eta) = \int f^2(\eta)\,\mu(d\eta)$, it suffices to show that as $n \to \infty$, $Var_\mu(f_n) \to 0$. From Lemma (2.20) it follows that $\mathcal{E}(f_n, f_n) \to 0$, and from (2.4) we see that

$$Var_\mu(f_n) \le \frac{1}{\gamma}\mathcal{E}(f_n, f_n).$$

■

3. Proof of Theorem (1.2). Before beginning the proof we collect a few known facts that we will find useful.

Fact 3.1. For all $\delta > 0$ there is an $a < \infty$ such that for all $f \in \mathcal{D}$ and all sufficiently large t

$$\sup_\eta |T_s f(\eta) - T_s f(\eta')| \le e^{-\delta t}, \qquad 0 \le s \le t,$$

where

$$\eta'(k) = \begin{cases} \eta(k) & \text{if } |k| \le a \cdot t \\ 1 & \text{if } |k| > a \cdot t \end{cases}$$

The proof of this fact is just like the proof of Theorem (2.16) in [7] if one replaces Δ_k there by $\frac{\partial}{\partial \eta_k}$. The proof depends on the fact that for the processes that we are considering the interaction is all in the drift coefficient and the diffusion coefficient in each coordinate is constant (or at least independent of the other coordinates).

Fact 3.2. Let $\Lambda \subset Z$ be a finite set. Let $m_s^\Lambda(\cdot|\eta)$ denote the Λ marginal of the distribution of the process $\{\eta_t : t \ge 0\}$ at time s starting from initial configuration η. Then $m_1^\Lambda(\cdot|\eta)$ has a density

$$f^\Lambda(\cdot|\eta) \equiv \frac{dm_1^\Lambda(\cdot|\eta)}{d\mu^\Lambda}$$

with respect to μ^Λ, the Λ marginal of μ (see [8]) which satisfies

$$\|f^\Lambda(\cdot|\eta)\|_{L^\infty(\mu)} \le e^{\Gamma|\Lambda|} \tag{3.3}$$

for some $\Gamma < \infty$.

Since all of the interaction enters through the drift in each component, (3.3) can be seen by means of the Cameron Martin formula by thinking of the distribution of the process on T^Λ as being absolutely continuous with respect to the product of independent diffusions on T^Λ.

Lemma 3.4. *Let $q(t) = 1 + e^{\frac{4t}{\alpha}}$, where α is as in Theorem (2.2). Then for all $f \in L^2(\mu)$,*

$$||T_t f||_{L^{q(t)}(\mu)} \leq ||f||_{L^2(\mu)}.$$

Proof. This follows immediately from Theorem (2.2) and Gross's Theorem, which can be found as Theorem (6.1.14) in [2]. ■

Now let γ be as in (1.1), take δ in Fact (3.1) to be larger than γ, let a from Fact (3.1) be taken accordingly, and set $\Lambda_t = [-a \times t, a \times t] \cap Z$. Let $\phi \in \mathcal{D}$ be such that $\int \phi d\mu = 0$, and let $0 < \epsilon < \gamma$. We want to show that

$$\lim_{t \to \infty} e^{\epsilon t} ||T_t \phi||_u = 0.$$

If $\eta \in \mathcal{S}$ we denote the distribution of the process at time 1 starting from η by $T_1^* \delta_\eta$. Then letting $t = 1 + s + s^2$ we have

$$\begin{aligned} |T_t \phi(\eta)| &= \left| \int T_{s+s^2} \phi(\sigma) T_1^* \delta_\eta (d\sigma) \right| \\ &\leq \left| \int T_{s+s^2} \phi(\sigma') T_1^* \delta_\eta (d\sigma) \right| + e^{-\delta t}, \end{aligned}$$

where

$$\sigma'(k) = \begin{cases} \sigma(k) & \text{if } k \in \Lambda_t \\ 1 & \text{if } k \notin \Lambda_t. \end{cases}$$

Since $e^{\epsilon t} e^{-\delta t} \to 0$ as $t \to \infty$, it will suffice to show that

$$e^{\epsilon t} \sup_\eta \left| \int T_{s+s^2} \phi(\sigma') T_1^* \delta_\eta (d\sigma) \right| \to 0 \qquad \text{as } t \to \infty.$$

Let $q(s)$ be as in Lemma (3.4), f^{Λ_t} be as in Fact (3.2), and calculate as follows:

$$\begin{aligned} \left| \int T_{s+s^2} \phi(\sigma') T_1^* \delta_\eta (d\sigma) \right| &= \left| \int T_{s+s^2} \phi(\sigma') f^{\Lambda_t}(\sigma|\eta) \mu(d\sigma) \right| \\ &\leq ||T_{s+s^2} \phi(\cdot')||_{L^{q(s)}(\mu)} ||f^{\Lambda_t}(\cdot|\eta)||_{L^{q'(s)}(\mu)} \\ &\leq \left(||T_{s+s^2} \phi||_{L^{q(s)}(\mu)} + e^{-\delta t} \right) ||f^{\Lambda_t}(\cdot|\eta)||_{L^{q'(s)}(\mu)} \\ &\leq \left(||T_{s^2} \phi||_{L^2(\mu)} + e^{-\delta t} \right) ||f^{\Lambda_t}(\cdot|\eta)||_{L^{q'(s)}(\mu)} \\ &\leq \left(e^{-\gamma s^2} ||\phi||_{L^2(\mu)} + e^{-\delta t} \right) ||f^{\Lambda_t}(\cdot|\eta)||_{L^{q'(s)}(\mu)}. \end{aligned}$$

Since $\epsilon < \gamma < \delta$, it follows that $e^{\epsilon t} \left(e^{-\gamma s^2} ||\phi||_{L^2(\mu)} + e^{-\delta t} \right) \to 0$ as $t \to \infty$. Thus to complete the proof for Theorem (1.2) we need only show that $||f^{\Lambda_t}(\cdot|\eta)||_{L^{q'(s)}(\mu)}$ stays bounded as $t = 1 + s + s^2$ goes to infinity.

Lemma 3.5. $\lim_{t\to\infty}\sup_\eta \|f^{\Lambda_t}(\cdot|\eta)\|_{L^{q'(s)}(\mu)} = 1$.

Proof. From Holder's inequality if $\frac{1}{r} = \frac{\theta}{p} + \frac{1-\theta}{q}$ then

$$\|f^{\Lambda_t}(\cdot|\eta)\|_{L^r(\mu)} \leq \|f^{\Lambda_t}(\cdot|\eta)\|^{\theta}_{L^p(\mu)}\|f^{\Lambda_t}(\cdot|\eta)\|^{1-\theta}_{L^q(\mu)}.$$

We take $r = q'(s) = 1 + e^{\frac{-4s}{\alpha}}$, $p = 1$, $q = \infty$, and $\theta = \frac{1}{1+e^{-4s/\alpha}}$. From the definition of $f^{\Lambda_t}(\cdot|\eta)$ we have $\|f^{\Lambda_t}(\cdot|\eta)\|_{L^1(\mu)} = 1$. Thus

$$1 \leq \|f^{\Lambda_t}(\cdot|\eta)\|_{L^{q'(s)}(\mu)} \leq e^{\Gamma(2a\times t)(1-\theta)} = e^{\Gamma 2a(1+s+s^2)/(1+e^{\frac{4s}{\alpha}})}.$$

Since

$$\frac{(1+s+s^2)}{1+e^{\frac{4s}{\alpha}}} \to 0$$

as $t = s^2 + s + 1 \to \infty$, we have the desired result. ■

The reader will notice that the argument in this section will work in any dimension provided that the GIBBS state admits a logarithmic Sobolev inequality. Zegarlinski has also proved a logarithmic Sobolev inequality for GIBBS states at sufficiently high temperatures in dimensions greater than one [**13**].

REFERENCES

[1] M. Aizenman, and R. Holley *Rapid convergence to equilibrium of stochastic Ising models in the Dobrushin Shlosman regime* Percolation Theory and Ergodic Theory of Infinite Particle Systems, IMA Volumes in Mathematics and its Applications **8** (1987), 1-11

[2] J. D. Deuschel, and D. W. Stroock *Large Deviations* Academic Press, New York 1989

[3] J. D. Deuschel, and D. W. Stroock *Hypercontractivity and Spectral Gap of Symmetric Diffusions with Applications to the Stochastic Ising Model* to appear in J. Functional Analysis (1990)

[4] L. Gross *Logarithmic Sobolev inequalities* Am. J. Math. **97** (1976), 1061-1083

[5] R. Holley *One dimensional stochastic Ising models* Percolation Theory and Ergodic Theory of Infinite Particle Systems, IMA Volumes in Mathematics and its Applications **8** (1987), 187-202

[6] R. Holley *Convergence in L^2 of stochastic Ising models: jump processes and diffusions* Taniguchi Symp. on Stochastic Analysis at Katata 1982, 149-167

[7] R. Holley and D. W. Stroock *L^2 theory for the stochastic Ising model* Z. Wahr. Verw. Geb. **35** (1976), 87-101

[8] R. Holley and D. W. Stroock *Diffusions on an infinite dimensional torus* J. Functional Analysis **42** (1981), 29-63

[9] R. Holley and D. W. Stroock *Logarithmic Sobolev inequalities and stochastic Ising models* J. Statistical Physics **46** (1987), 1159-1194
[10] R. Holley and D. W. Stroock *Uniform and L^2 convergence in one dimensional stochastic Ising models* Commun. Math. Phys. **123** (1989), 85-93
[11] O. S. Rothaus *Diffusions in compact Riemannian manifolds and logarithmic Sobolev inequalities* J. Funct. Anal. **42** (1981), 102-109
[12] B. Zegarlinski *Log-Sobolev inequalities for infinite one dimensional lattice systems* preprint (1989)
[13] B. Zegarlinski *Dobrushin Uniqueness Theorem and Logarithmic Sobolev Inequalities* preprint (1989)

Richard Holley
Department of Mathematics
University of Colorado
Boulder, Colorado

Relations Between Solutions to a Discrete and Continuous Dirichlet Problem

HARRY KESTEN[1]

Consider the Dirichlet problem on the planar region W consisting of the unit disc minus a sector: $W = \{x = (x_1, x_2) : 0 < |x| < 1, -\alpha_2 < \text{argument}(x) < \alpha_1\}$ for some $0 < \alpha_1, \alpha_2 \le \pi$, $\alpha_1 + \alpha_2 > \pi$ (see Fig. 1). We denote by L_i, $i = 1, 2$, the line segment $\{x : \text{argument}(x) = (-1)^{i-1}\alpha_i$, $0 \le |x| \le 1\}$ in ∂W. Let φ be a piecewise continuous function on ∂W which vanishes on $L_1 \cup L_2$, and let $u(\varphi, \cdot)$ be the harmonic function on W with boundary values φ. Thus $\Delta\varphi = 0$. The discrete analogue, or the analogue for a simple random walk with step size h, is a function $u_h(\varphi, \cdot)$ on $h\mathbb{Z}^2$ which satisfies

$$\Delta_h u_h(\varphi, x) := \frac{1}{4}\sum_{i=1}^{2}[u_h(\varphi, x + he_i) + u_h(\varphi, x - he_i)] - u_h(\varphi, x) = 0$$

for $x \in h\mathbb{Z}^2 \cap W$, where e_i is the $i-th$ coordinate vector. Moreover $u_h(\varphi, \cdot)$ should be "close to φ on the discrete boundary of W". We prove an asymptotic relation (as $h \downarrow 0$) between $u_h(\varphi, hv)$ and $u(\varphi, hv)$ for fixed $v \in \mathbb{Z}^2$. This can be used to find the asymptotic behavior (as $n \to \infty$) of the probability of a simple random walk "starting at ∞" to hit the "triangular" region $\{x \in \mathbb{Z}^2 : |x| \le n$, $-\alpha_2 \le \text{argument}(x) \le \alpha_1\}$ near the origin. This kind of estimate is useful for a modified version of Diffusion–Limited Aggregation.

1. Introduction and first-order approximations. Frank Spitzer has investigated many fundamental properties of random walk. In particular he studied the potential theory of recurrent random walk. Most of Frank's results in this area can be found in his beautiful book [12]. Several of my own papers were inspired by Frank's work on random walks, and some were even written jointly with Frank. It is a pleasure to acknowledge here my indebtedness to Frank, and to present a paper which is based on his results about hitting probabilities of random walk on $\mathbb{Z}^2$.

Our investigations here are closely related to properties of hitting probabilities of large wedges for random walks. To be specific, let $S_n = S_0 + \sum_1^n X_i$ be simple symmetric random walk on $\mathbb{Z}^2$. That is, if e_i, $i = 1, 2$ denote the coordinate vectors in $\mathbb{R}^2$, then the X_j are i.i.d. with $P\{X_j =$

---

[1]Partially supported by a National Science Foundation grant to Cornell University
Key Words and Phrases: Dirichlet problem, Brownian motion, random walk, diffusion -limited aggregation, hitting probabilities.

$\pm e_i\} = 1/4$, $i = 1, 2$. S_0 is not necessarily equal to 0. We use P_v to denote the conditional distribution of $\{S_n\}_{n\geq 0}$, given $S_0 = v$; E_v is expectation with respect to P_v. For any set A,

$$(1.1) \qquad T(A) := \inf\{n \geq 0 : S_n \in A\}$$

is the hitting time of A. The hitting distribution (starting at u) is

$$H_A(u,v) = P_u\{S_{T(A)} = v\},\ v \in A.$$

(A brief word of caution. Our definitions here differ slightly from [11] and [12] in that $S_{T(A)} = S_0$ is allowed.) Our starting point is the following result of Frank's ([12, Theorem 14.1]): For any *finite* set A

$$\lim_{u\to\infty} H_A(u,v) = \mu_A(v)$$

exists and the limit is a probability measure on A, i.e.,

$$\sum_{v\in A} \mu_A(v) = 1.$$

This result is made to measure for the definition of the Diffusion–Limited Aggregation model of Witten and Sander [13]. This is a stochastic growth model. Its state at time n is a connected set A_n of $\mathbb{Z}^d$ which contains the origin and contains n lattice sites. A_{n+1} is formed from A_n by adding a single point $y_n \in \partial_1 A_n$. Here and in the sequel $\partial_1 B$, the boundary of B, is the set of vertices adjacent on $\mathbb{Z}^2$ to B, but not in B. For $d = 2$, once A_n has been found, one chooses y_n according to the distribution $\mu_{\partial_1 A_n}(\cdot)$ (see [6], [7] for more details and the definition when $d \geq 3$).

For the study of a modification of this model (cf. [8], [9]) we would like to have an asymptotic evaluation as $n \to \infty$ of $\mu_{B(n)}(v)$ or rather $\mu_{\partial_1 B(n)}(v)$ for fixed v and $B(n) := [-n, 0] \times \{0\}$ an interval on the x-axis (and $v \in B(n)$ or $v \in \partial_1 B(n)$, respectively). By means of time reversal this reduces to the special case $\tilde{\varphi} \equiv 1$ on $\partial W \backslash (L_1 \cup L_2)$, $\alpha_1 = \alpha_2 = \pi$, $h = n^{-1}$, of the following discrete Dirichlet problem. Let $0 \leq \alpha_1,\ \alpha_2 \leq \pi$, $\alpha_1 + \alpha_2 > \pi$,

$$(1.2) \qquad W = W(\alpha_1, \alpha_2) = \{x : 0 < |x| < 1,\ -\alpha_2 < \arg(x) < \alpha_1\}$$

(see Fig. 1). Here $|x|$ denotes the Euclidean norm of x and $\arg(x)$ stands for the argument of x. Unless otherwise stated we always choose $-\pi \leq -\alpha_2 \leq \arg(x) \leq \alpha_1 \leq \pi$. Let ∂W be the topological boundary of W and define

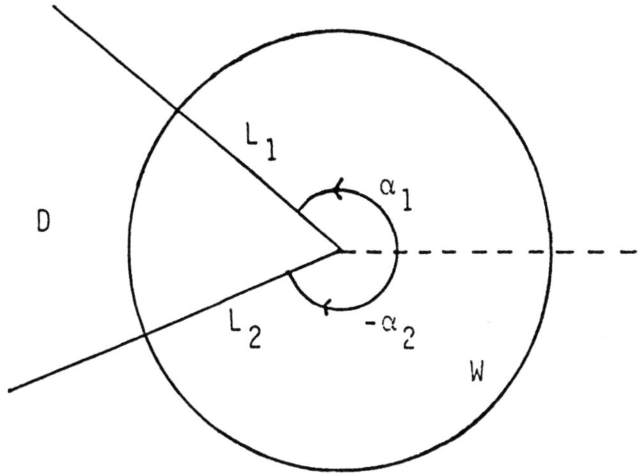

Fig. 1. The region W and the sector D.

(1.3) $\partial_h W = \{x : x \in h\mathbb{Z}^2,\ x \notin W \text{ but } x = y \pm he_i \text{ for some } y \in h\mathbb{Z}^2 \cap W, \text{ and } i = 1 \text{ or } 2\}.$

Now let $\tilde{\varphi} = \tilde{\varphi}_h$ be a function on $\partial_h W$ and define $u_h(\tilde{\varphi}, \cdot)$ on $h\mathbb{Z}^2 \cap (W \cup \partial_h W)$ as the unique solution to the discrete Dirichlet problem

(1.4) $\Delta_h u_h(\tilde{\varphi}, x) = 0,\ x \in h\mathbb{Z}^2 \cap W, \text{ and } u_h(\tilde{\varphi}, z) = \tilde{\varphi}(z), z \in \partial_h W,$

where the discrete Laplacian Δ_h is defined by

(1.5) $\Delta_h f(x) = \frac{1}{4}[f(x+he_1)+f(x-he_1)+f(x+he_2)+f(x-he_2)]-f(x).$

It is well known ([4, Sect. 2.2], [5, Exercises 18-21 in Ch. 1], [11, Theorem 1.4.5]) that

(1.6) $$u_h(\tilde{\varphi}, x) = E_{x/h}\{\tilde{\varphi}(hS_{T(h^{-1}\partial_h W)})\}.$$

What is the asymptotic behavior of $u_h(\tilde{\varphi}, hv)$ as $h \downarrow 0$ and $v \in \mathbb{Z}^2$ fixed (such that $hv \in W$ for small h)?

Of course the last question is not meaningful without suitable hypotheses on $\tilde{\varphi}_h$. Write (see Fig. 1)

(1.7) $$L_i = \{x : |x| \le 1, (-1)^{i-1} \arg(x) = \alpha_i\}, i = 1, 2,$$

for the two straight line segments in ∂W. Also set

$$(1.8)\qquad D = \{x : -\pi \le \arg(x) \le -\alpha_2 \text{ or } \alpha_1 \le \arg(x) \le \pi\}$$

(see Figs. 1 and 2). Throughout we shall make the following assumptions (1.9)-(1.13). There exists a finite set $Z \subset \partial W$ and a function φ on ∂W such that

$$(1.9)\qquad \varphi \text{ is continuous on } \partial W \backslash Z \text{ and bounded on all of } \partial W,$$

$$(1.10)\qquad \varphi(z) = 0 \text{ for } z \in L_1 \cup L_2 \quad \text{(see Fig. 1)},$$

$$(1.11)\qquad \text{For some } K < \infty, |\tilde{\varphi}_h(x)| \le K \text{ for } h > 0, x \in \partial_h W,$$

$$(1.12)\qquad \varphi_h(x) = 0 \text{ for } \quad x \in \partial_h W \cap D,$$

$$(1.13)\qquad \lim_{h\downarrow 0} \tilde{\varphi}_h(x_h) = \varphi(z) \text{ whenever } x_h \in \partial_h W \text{ converges to some } z \in \partial W \backslash Z.$$

Since $\{h^{1/2}S_{t/h}\}$ converges weakly to a two-dimensional Brownian motion $\{B(t)\}$([2, Theorem 10.1]; the two-dimensional case is entirely analogous to the one-dimensional case) it is reasonable in this setup to compare $u_h(\tilde{\varphi}, \cdot)$ with the function

$$(1.14)\qquad u(\varphi, x) = E_x\{\varphi(B(\tau))\},\ x \in W,$$

where E_x is conditional expectation given that the Brownian motion starts at x and

$$(1.15)\qquad \tau = \inf\{t \ge 0 : B(t) \notin W\}.$$

This is the unique bounded solution of the Dirichlet problem

$$(1.16)\qquad \Delta u(\varphi, x) = 0,\ x \in W, \text{ and}$$

$$\lim_{x \to z, x \in W} u(\varphi, x) = \varphi(z),\ z \in \partial W \backslash Z$$

(cf. [5, Ch. II.4-6]). It has been known for a long time ([3], also Lemma 1 in [8]) that

$$u_h(\tilde{\varphi}, x_h) \to u(\varphi, x) \text{ as } \quad h \downarrow 0,\ x_h \to x,$$

for fixed $x \in W$. This result was the impetus for a great deal of work by numerical analysis on the approximation of solutions to partial differential equations by solutions to a corresponding difference equation. While much in this direction is known (see for instance [10] and its references for results closely related to ours), the literature does not seem to contain sharp information on the behavior of $u_h(\tilde{\varphi}, x_h)$ when x_h tends to a corner of the domain as $h \downarrow 0$. The main content of our results here is when $x_h \to 0$ in the domain W. It must be said, though, that our boundary conditions are very restrictive (cf. Remark (ii) below.

We need some more notation. For any set A, write $d(x, A) = \inf\{|x-y| : y \in A\}$. Define (see Fig. 2)

$$\rho(x) = \rho(x, \alpha_1, \alpha_2) = \begin{cases} d(x, L_1) & \text{if } x_2 \geq 0 \\ d(x, L_2) & \text{if } x_2 < 0 \end{cases} \tag{1.17}$$

For $x \in W$, $\rho(x) = d(x, D)$. Note that $\rho(x) \leq |x|$.

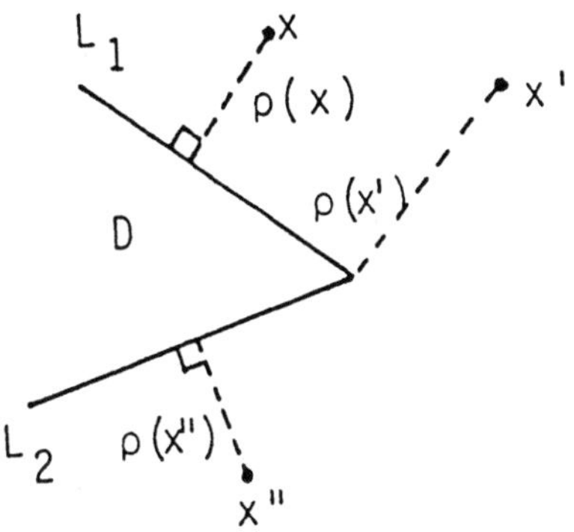

Fig. 2. Illustration of ρ for typical points x, x', x''. ρ is the length of the respective dashed segments.

Theorem 1. *Assume (1.9)-(1.13) hold, and*

$$0 < \alpha_1, \alpha_2 \leq \pi, \quad 2\alpha := \alpha_1 + \alpha_2 > \pi. \tag{1.18}$$

Then for each $v \in \mathbb{Z}^2 \backslash D$ there exists a $\lambda_1(v) = \lambda_1(v, \alpha_1, \alpha_2)$ (independent of φ, $\tilde{\varphi}$ and h), such that

$$\lim_{h \downarrow 0} h^{-\pi/(2\alpha)} |u_h(\tilde{\varphi}, hv) - \lambda_1(v) u(\varphi, hv)| = 0, \quad v \in \mathbb{Z}^2 \backslash D. \tag{1.19}$$

The constants λ_1 satisfy

$$\lambda_1(v) > 0, \tag{1.20}$$

$$\lambda_1(v) \to 1 \text{ as } |v| \to \infty, v \in \mathbb{Z}^2 \backslash D \text{ and such that } \rho(v/|v|) \geq \delta \text{ for some } \delta > 0. \tag{1.21}$$

Theorem 2. *Again assume (1.9)-(1.13) and (1.18) and let* $0 < \epsilon \le 1/8$ *be fixed. Then there exists a constant* $K_0 = K_0(\epsilon) < \infty$ *and a function* $\theta(\epsilon, h) \ge 0$ *such that*

$$\theta(\epsilon, h) \to 0 \text{ as } h \downarrow 0, \tag{1.22}$$

and for each $x \in h\mathbb{Z}^2 \backslash D$ *with* $|x| \le 1 - 2\epsilon$

$$\begin{aligned} |u_h(\tilde{\varphi}, x) - u(\varphi, x)| \\ \le K_0[h|x|^{\pi/(2\alpha)-1} + \frac{\rho(x)+h}{|x|}h^{\pi/(2\alpha)}] \\ + \theta(\epsilon, h)\{\rho(x) + h\}|x|^{\pi/(2\alpha)-1}. \end{aligned} \tag{1.23}$$

Corollary 1. *If* $\tilde{\varphi}(z) \ge 0$ *on* $\partial_h W$ *and* $\varphi > 0$ *on a set of strictly positive Lebesgue measure of* $\partial W \backslash L_1 \cup L_2$*, then for each fixed* $v \in \mathbb{Z}^2 \backslash D$ *one has, as* $h \downarrow 0$,

$$\begin{aligned} u_h(\tilde{\varphi}, hv) \sim h^{\pi/(2\alpha)}|v|^{\pi/(2\alpha)}\frac{2}{\pi}\lambda_1(v)\cos[\frac{\pi}{2\alpha}(\alpha - \alpha_1 + \arg(v))] \\ \cdot \int_{-\pi/2}^{\pi/2} \varphi(e^{i2\alpha\gamma/\pi - i(\alpha-\alpha_1)})\cos\gamma d\gamma, \end{aligned} \tag{1.24}$$

where $e^{i\zeta}$ *is short for* $(\cos\ \zeta,\ \sin\ \zeta) \in \mathbb{R}^2$. *In addition*

$$\frac{u_h(\tilde{\varphi}, x_h)}{u(\varphi, x_h)} \to 1 \text{ as } h \downarrow 0, \tag{1.25}$$

if for some $0 < \epsilon \le \frac{1}{8}$ *and* $\delta > 0$, $x_h \in h\mathbb{Z}^2$ *varies with* h *such that*

$$h^{-1}|x_h| \to \infty,\ x_h \in W \cap \{y : |y| \le 1 - 2\epsilon\},\ \rho(x_h/|x_h|) \ge \delta. \tag{1.26}$$

Corollary 2. *Let* U_n *be the "triangular region"*

$$\{v \in \mathbb{Z}^2 : |v| \le n, \arg(v) \in [-\pi, -\alpha_2] \cup [\alpha_1, \pi]\} = \{v \in D \cap \mathbb{Z}^2 : |v| \le n\}$$

(with D *as in (1.8); see Fig. 3). Also, let*

$$W_n = \{v \in \mathbb{Z}^2 : v \le n\} \backslash U_n = \{v \in \mathbb{Z}^2 \backslash D : |v| \le n\}.$$

If $v_n \in W_n$ *satisfies*

$$\begin{aligned} v_n \text{ is fixed, or } |v_n| \to \infty \text{ but } |v_n| \le n(1 - 2\epsilon) \\ \text{and } \rho(v_n/|v_n|) \ge \delta \text{ for some } 0 < \epsilon \le 1/8 \text{ and } \delta > 0, \end{aligned} \tag{1.27}$$

then

$$(1.28) \qquad P_{v_n}\{\arg(S_{T(\partial_1 W_n)}) \in \cdot \mid S_{T(\partial_1 W_n)} \notin D\}$$

converges weakly as $n \to \infty$. (In less precise terms, the conditional exit distribution for S. from W_n given that it leaves W_n through the circular arc of its boundary, has a weak limit.) In addition, for fixed $u \in \mathbb{Z}^2 \backslash D$,

$$(1.29) \qquad \lim_{|v| \to \infty} P_v\{S. \text{ hits } u \text{ before } U_n\} = \mu_{U_n \cup \{u\}}(u) \sim \kappa(u) n^{-\pi/(2\alpha)},$$

as $n \to \infty$, for some $\kappa(u) > 0$.

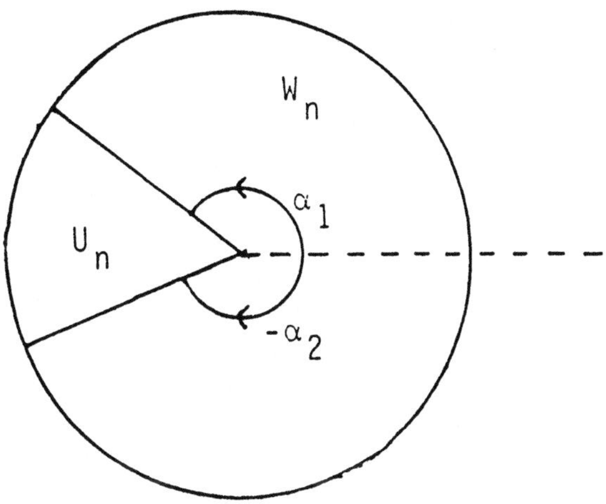

Fig. 3. The regions U_n and W_n in the disc of radius n.

Remarks. (i) For simplicity we have stated our results for $h \downarrow 0$ as a continuous variable. No significant changes are needed when h is restricted to a subsequence in the assumptions as well as conclusions.
(ii) Note that we have restricted ourselves to zero boundary values for φ on $L_1 \cup L_2$ (see (1.10)) as well as zero values for φ_h in D (see (1.12)). The situation becomes more complicated when φ is not constant on $L_1 \cup L_2$ (even for smooth φ).
(iii) For $\alpha_1 = \alpha_2 = \pi$, λ_1 can be calculated explicitly and $\lambda_1(v) \neq 1$ in general. As L. Wahlbin pointed out to us, this can be seen easily in general. (1.4) and (1.24) (with $\varphi(x) \equiv 1$ for $|x| = 1$) imply

(1.30)
$$\lambda_1(v)|v|^{\pi/(2\alpha)} \cos[\frac{\pi}{2\alpha}(\alpha_1 - \alpha + \arg(v))]$$
$$= \frac{1}{4}\sum_{i=1}^{2}\{\lambda_1(v+e_i)|v+e_i|^{\pi/(2\alpha)} \cos[\frac{\pi}{2\alpha}(\alpha_1 - \alpha + \arg(v+e_i))]$$
$$+\lambda_1(v-e_i)|v-e_i|^{\pi/(2\alpha)} \cos[\frac{\pi}{2\alpha}(\alpha_1 - \alpha + \arg(v-e_i))]\},$$

provided v, $v \pm e_1$ and $v \pm e_2 \notin D$. But (1.30) with $\lambda_1(\cdot)$ replaced by 1 throughout does not hold in general (e.g. when $v = 2e_1$ and $\alpha \to \pi$). Thus in general

$$\frac{u_h(\tilde{\varphi}, x_h)}{u(\varphi, x_h)} \text{ does not tend to } 1.$$

(iv) (1.29) improves results of Auer [1], who only estimated the probability in (1.29) up to factors (log n). For Diffusion-Limited Aggregation we shall use (1.29) with $\alpha_1 = \alpha_2 = \pi$ in [8] and [9].

2. Second order approximations and L-shaped regions. For applications to DLA it is more convenient to work with functions on $\mathbb{Z}^2$ rather than $h\mathbb{Z}^2$. One obtains such results from the previous section by taking $h = n^{-1}$ and multiplying the domain W by a factor n. The resulting statements have been summarized in Corollary 2. Of particular importance for DLA is the case $\alpha_1 = \alpha_2 = \pi$. For this case, Corollary 2 gives the behavior of the hitting probabilities of points near 0 before the (long) line segment $\{-j : 0 \le j \le n\}$ along the negative x-axis for a random walk starting at ∞ (see (1.29)). We will need similar information when the line segment is replaced by L-shaped objects. These are the unions of two line segments along the two coordinate axes. Actually, the way DLA has been formulated requires us to "fatten up" these objects by including all lattice points adjacent to them. We formulate our results for these fattened up objects, but expect that essentially the same results hold without fattening up.

The first result for L-shaped objects is an analogue of (1.29), the asymptotic behavior of hitting probabilities of points near the two tips (see (2.15), (2.16) and (2.19)). This result follows rather directly from Corollary 2. The next result is a second order result. It compares the hitting probabilities of the two tips of the L-shaped object (see (2.20)). This is a more delicate result which is based on a second order result in the setup of Section 1 for $\alpha_1 = \alpha_2 = \pi$. We state the result in the rescaled version and for the fattened-up line segment.

Define for $j = 0, 1$,

$$A_j = (-\infty, j] \times \{0\}, \tag{2.1}$$

$$B_j = \text{“boundary of } A_j\text{”} = (-\infty, j] \times \{-1, 1\} \cup \{(j+1, 0)\}, \tag{2.2}$$

$$\bar{A}_j = A_j \cup B_j \text{ (the “closure of A”)}, \tag{2.3}$$

$$W(j, n) = \{v \in \mathbb{Z}^2 : |v| \le n\} \backslash (A_j \cup B_j). \tag{2.4}$$

The "discrete circle of radius n" is

$$(2.5) \qquad \mathcal{C}(n) = \{v \in \mathbb{Z}^2 : |v| > n, \text{ but } v \text{ is an endpoint of an edge of } \mathbb{Z}^2, \text{ whose other endpoint, } v', \text{ satisfies } |v'| \le n\}.$$

Let φ be a continuous function on the unit circle and $\tilde{\varphi}_n$ a sequence of functions on $\mathcal{C}(n)$ such that

$$(2.6) \qquad \lim_{n\to\infty} \tilde{\varphi}(v(n)) = \varphi(e^{i\theta})$$

for any sequence $v(n) \in \mathcal{C}(n)$ with $\frac{1}{n}v(n) \to e^{i\theta}$ on the unit circle. Here again, $e^{i\theta}$ is short for the point $(\cos\theta, \sin\theta) \in \mathbb{R}^2$. In addition to the hitting times $T(\cdot)$ of (1.1) we introduce "return times":

$$(2.7) \qquad R(A) := \inf\{n \ge 1 : S_n \in A\}.$$

We now define for $j = 0, 1$

$$(2.8) \quad U_{j,n}(v) = U(v; j, n, \tilde{\varphi}_n) = E_v\{\tilde{\varphi}_n(S_{T(\mathcal{C}(n))})\ ;\ T(\mathcal{C}(n)) < R(\bar{A}_j)\}.$$

The reader should convince himself that except for the $\bar{A}_0$ instead of A_0, $U_{0,n}(v)$ is the same as $u_{1/n}(\tilde{\varphi}, v/n)$ of the previous section for $\alpha_1 = \alpha_2 = \pi$ (when D is just the negative x-axis). The restriction of the expectation to $\{T(\mathcal{C}(n)) < R(\bar{A}_j)\}$ takes the place of the requirement (1.12) that $\tilde{\varphi}$ vanish on $\partial W_h \cap D$. Also (2.6) is a version of (1.13) when $\alpha_1 = \alpha_2 = \pi$. In view of Theorem 1 and Corollary 1 it is therefore not surprising that for fixed v

$$(2.9) \qquad \lim_{n\to\infty} n^{1/2} U_{j,n}(v) \text{ exists.}$$

The required second order result takes the following form.

Theorem 3. *For each fixed $v \notin A_0$, there exists a finite constant $\lambda_2(v) = \lambda_2(v, \varphi)$ such that*

$$(2.10) \qquad \lim_{n\to\infty} n^{3/2}[U_{1,n}(v + e_1) - U_{0,n}(v)] = \lambda_2(v).$$

Moreover there exists a constant $\lambda_3(v) < \infty$ such that

$$(2.11) \qquad n^{3/2} \sum_{w\in\mathcal{C}(n)} |P_v\{T(\mathcal{C}(n)) < R(\bar{A}_0), S_{T(\mathcal{C}(n))} = w\} - P_v\{T(\mathcal{C}(n)) < R(\bar{A}_1), S_{T(\mathcal{C}(n))} = w\}| \le \lambda_3(v).$$

Next we define our L-shaped regions (see Fig. 4).

(2.12) $$L(k,\ell) = [0,k] \times \{0\} \cup \{0\} \times [0,\ell],$$

(2.13) $$\partial L(k,\ell) = [1,k] \times \{-1,+1\} \cup \{-1,+1\} \times [1,\ell] \cup \{(-1,0),(0,-1),(k+1,0),(0,\ell+1)\},$$

(2.14) $$\bar{L} = L \cup \partial L(k,\ell) = [0,k] \times \{-1,0,+1\} \cup \{-1,0,1\} \times [0,\ell] \cup \{(k+1,0),(0,\ell+1)\}.$$

(We should think of L, ∂L and $\bar{L}$, respectively, as the collection of lattice points in the indicated sets; to simplify notation we have not indicated that the right hand sides of (2.12)-(2.14) should be intersected with $\mathbb{Z}^2$. Also we write ∂L instead of $\partial_1 L$ in this section).

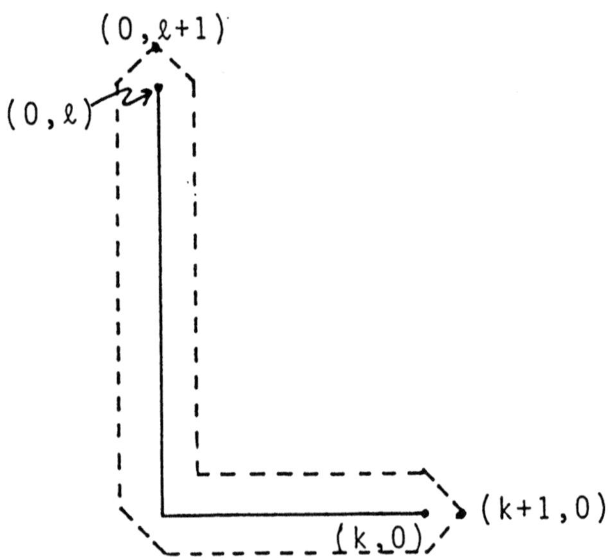

Fig. 4. Illustration of $L(k,\ell)$ (solid lines) and $\partial L(k,\ell)$ (dashed lines).

The hitting probabilities for the "tips" of $L(k,\ell)$ are written as

$$\pi_1(k,\ell) = \mu_{\partial L(k,\ell)}((k+1,0)), \text{ and}$$

$$\pi_2(k,\ell) = \mu_{\partial L(k,\ell)}((0,\ell+1)).$$

The results which we use for our variant of DLA take the following form.

Theorem 4. *For each fixed* $v \in \mathbb{Z}^2$ *and* $\beta \in [0,\infty]$ *there exist constants* $\lambda_4(v,\beta)$, $\lambda_5(v,\beta)$ *such that*

$$\lim (k+\ell)^{1/2} \mu_{\partial L(k,\ell)\cup\{v+ke_1\}}(v+ke_1) = \lambda_4(v,\beta) \tag{2.15}$$

and

$$\lim (k+\ell)^{1/2} \mu_{\partial L(k,\ell)\cup\{v+\ell e_2\}}(v+\ell e_2) = \lambda_5(v,\beta) \tag{2.16}$$

as

$$k,\ell \to \infty \text{ in such a way that } k/\ell \to \beta. \tag{2.17}$$

$$\lambda_4(e_1,\beta) > 0 \text{ and } \lambda_5(e_2,\beta) > 0 \tag{2.18}$$

so that (again under (2.17))

$$\begin{aligned} \lim (k+\ell)^{1/2}\pi_1(k,\ell) &= \lambda_4(e_1,\beta) > 0, \\ \lim (k+\ell)^{1/2}\pi_2(k,\ell) &= \lambda_5(e_2,\beta) > 0. \end{aligned} \tag{2.19}$$

There also exists a constant $C_1 > 0$ *such that*

$$\begin{aligned} &\mu_{\partial L(k,\ell)}((k+1,0)) - \mu_{\partial L(k,\ell)}((0,\ell+1)) \\ &= \pi_1(k,\ell) - \pi_2(k,\ell) \\ &= (C_1 + o(1))(k-\ell)/(k+\ell)^{3/2} \end{aligned} \tag{2.20}$$

as $k,\ell \to \infty$ *in such a way that*

$$k/\ell \to 1. \tag{2.21}$$

There exists a constant $a \in (0,1)$ *such that for all* k *and* ℓ *sufficiently large*

$$\frac{\mu_{\partial L(k,\ell)}(w)}{\mu_{\partial L(k,\ell)}(t)} \le 1-a \tag{2.22}$$

for all $w \in \partial L(k,\ell)\backslash\{(k+1,0),(0,\ell+1)\}$ *and*

$$t = \begin{cases} (k+1,0) & \text{if } \ell < k, \\ (0,\ell+1) & \text{if } \ell \ge k. \end{cases}$$

Finally, for all $\epsilon > 0$ *there exists an* $a(\epsilon) \in (0,1)$ *such that for sufficiently large* k *and* $\ell \ge (1+\epsilon)k$ *and all* $w \in \partial L(k,\ell)\backslash\{(0,\ell+1)\}$ *and* $t = (0,\ell+1)$

$$\frac{\mu_{\partial L(k,\ell)}(w)}{\mu_{\partial L(k,\ell)}(t)} \le 1-a(\epsilon). \tag{2.23}$$

(Note that $w = (k+1,0)$ is included in (2.23), but $w \notin \{(k+1,0),(0,\ell+1)\}$ in (2.22).

3. A few words about proofs. The proof of the first order estimates of Section 1 begins with the following completely general lemma.

Lemma 1. *Let R be a bounded subset of $\mathbb{R}^2$ and f a function defined and bounded on*

$$h\mathbb{Z}^2 \cap \{\text{points in } R \text{ or within distance } h \text{ of } R\}.$$

Let

$$\sigma = \sigma(R,h) = \inf\{r \geq 0 : hS_n \notin R\} \tag{3.1}$$

be the exit time from R for $hS_\cdot$. Then

$$E_v\{f(hS_\sigma) - f(hv)\} = E\{\sum_{n=0}^{\sigma-1} \Delta_h f(hS_n)\}. \tag{3.2}$$

(See (1.5) for Δ_h.)

(3.2) is a trivial consequence of the fact that

$$f(hS_{k\wedge\sigma}) - \sum_{n=0}^{(k\wedge\sigma)-1} \Delta_h f(hS_n),\ k \geq 0,$$

is a martingale (compare [7, Lemma 8]). The right hand side of (3.2) can be rewritten as

$$\sum_{w\in R} E_v\{\text{number of visits by } S_\cdot \text{ to } w \text{ before } \sigma\}\Delta_h f(hw). \tag{3.3}$$

We apply this with $f(\cdot) = u(\varphi,\cdot)$, $R = W \cap \{z : |z| \leq 1-\epsilon\}$. In this case we can estimate (3.3), as well as

$$E_v\{u(\varphi, hS_\sigma)\} - u_h(\tilde{\varphi}, hv),$$

at least for v's which are relatively far from $L_1 \cup L_2$. This leads to Theorem 2 and (1.25). This shows that relatively far from $L_1 \cup L_2$, $u_h(\tilde{\varphi},\cdot)$ behaves like $u(\varphi,\cdot)$. For starting points v not so far from $L_1 \cup L_2$ we just have to let the random walk run until it reaches the region where $u_h(\tilde{\varphi},\cdot)/u(\varphi,\cdot)$ is close to 1. This initial piece of the random walk is the source of the factor $\lambda_1(v)$ in Theorem 1.

As for Theorem 3, this is based on estimates for the difference between the harmonic measure of a large (discrete) circle for random walk, relative to the origin and relative to e_1 (see [11, Theorem 1.7.1]). The difficulty is that the random walk starting at je_1 has to avoid $\bar{A}_j$, $j = 0,1$. We circumvent this difficulty essentially by coupling these random walks until some stopping time T, when their positions will be S_T and $S_T + e_1$, with S_T relatively far away from $\bar{A}_0$. This will allow us to use the estimate on the difference of harmonic measures for some circle centered at S_T, relative to S_T and $S_T + e_1$, respectively, without having to avoid a set like $\bar{A}_j$.

REFERENCES

[1] Auer, P. (1989) Some hitting probabilities of random walk on $\mathbb{Z}^2$, preprint.

[2] Billingsley, P. (1968) *Convergence of probability measures*, John Wiley & Sons.

[3] Courant, R., Friedrichs, K. and Lewy, H. (1928) Über die partiellen Differenzengleichungen der mathematischen Physik, Math. Ann. **100**, 32-74

[4] Doyle, P. G. and Snell, J. L. (1984) *Random walks and electric networks*, Carus Monograph #22, Math. Assoc. America.

[5] Dynkin, E. B. and Yuskewitsch, A. A. (1969) *Sätze und Aufgaben über Markoffsche Prozesse*, Springer-Verlag.

[6] Kesten, H. (1987) How long are the arms in DLA?, J. Phys. A. **20**, L29-L33.

[7] Kesten, H. (1987) Hitting probabilities of random walks on $\mathbb{Z}^d$, Stoch. Proc. and their Appl. **25**, 165-184.

[8] Kesten, H. (1990) Relations between solutions to a discrete and continuous Dirichlet problem II, preprint.

[9] Kesten, H. (1991) Some caricatures of multiple contact diffusion-limited aggregation and the η-model, Proc. of London Math. Soc. Probability Conference, Durham, 1990.

[10] Laasonen, P. (1967) On the discretization error of the Dirichlet problem in a plane region with corners, Ann. Acad. Sci. Fenn. A I **408**, 2-16.

[11] Lawler, G. (1991) *Intersections of random walks*, Birkhäuser- Boston.

[12] Spitzer, F. (1964) *Principles of random walk*, D. van Nostrand Co.

[13] Witten, T. A. and Sander, L. M. (1981) Diffusion-limited aggregation, a kinetic phenomenon, Phys. Rev. Lett. **47**, 1400-1403.

H. Kesten
Department of Mathematics
White Hall
Cornell University
Ithaca, NY 14853 USA

On the Connected Components of the Complement of a Two-Dimensional Brownian Path

Jean-François Le Gall

1. Introduction. Motivated by a question of Mandelbrot [4, Chapter 25], Mountford [5] has investigated the properties of the connected components of the complement of a two-dimensional Brownian path. Let $B = (B_t, t \geq 0)$ be a Brownian motion in the complex plane $\mathbb{C}$, and $B[0,1] = \{B_t, 0 \leq t \leq 1\}$. For any u, v with $0 < u < v \leq \infty$, let $N_{[u,v)}$ denote the number of connected components of $\mathbb{C} \setminus B[0,1]$ with area in $[u, v)$. Mountford [5] has proved that for every $\lambda \in (0,1)$,

$$(1.a) \qquad x^2(\log x)^2 N_{[\pi(\lambda x)^2, \pi x^2)} \xrightarrow[x\to 0]{(Probability)} \frac{1}{2}(\lambda^{-2} - 1).$$

This result does not exactly answer Mandelbrot's question, which was to determine the limiting behavior of $N_{[u,\infty)}$ as u tends to 0. The point is that, since the convergence holds only in probability, it seems difficult to use (1.a) to get information about $N_{[u,\infty)}$.

The purpose of the present work is to sharpen Mountford's result by the following theorem.

Theorem 1.1. *For every $\varepsilon > 0$,*

$$\lim_{u\to 0} \sup_{v \in [(1+\varepsilon),\infty]} \left| \frac{(\log u)^2 N_{[u,v)}}{u^{-1} - v^{-1}} - 2\pi \right| = 0$$

almost surely. In particular,

$$\lim_{u\to 0} u(\log u)^2 N_{[u,\infty)} = 2\pi, \quad \text{a.s.}$$

The proof of Theorem 1.1, as well as the proof of (1.a) given in [5], depends heavily on estimates for the hitting probabilities of small disks. Such estimates were first derived by Spitzer [6]. For every $y \in \mathbb{C}$ and $\varepsilon > 0$, set

$$T_\varepsilon(y) = \inf\{s \geq 0, \ |B_s - y| \leq \varepsilon\}.$$

Lemma 1.2. (i) *For every* $y \in \mathbf{C} \setminus \{0\}$,

$$\lim_{\varepsilon \to 0} |\log \varepsilon| \, P[T_\varepsilon(y) \leq 1] = \pi \int_0^1 p_s(0, y) \, ds$$

where $p_s(0, y) = (2\pi s)^{-1} \exp(-|y|^2/2s)$.

(ii) *Set*

$$\Phi(y) = K\left(1 + \log_+ \frac{1}{|y|}\right) \exp\left(-\frac{|y|^2}{16}\right).$$

Then, provided that the constant $K > 0$ *is large enough, we have for every* $y, z \in \mathbf{C} \setminus \{0\}$, $y \neq z$, *for* $\varepsilon \in (0, 1/2)$,

$$|\log \varepsilon| \, P[T_\varepsilon(y) \leq 1] \leq \Phi(y),$$
$$|\log \varepsilon|^2 \, P[T_\varepsilon(y) \leq 1, \; T_\varepsilon(z) \leq 1] \leq \Phi(y)\Phi(z - y) + \Phi(z)\Phi(y - z).$$

Part (i) of Lemma 1.2 was proved in Spitzer [6] as Lemma 1. Part (ii) can easily be derived from part (i) (see [1, Lemma 2.1]).

Let us briefly explain the connection between Theorem 1.1 and hitting probabilities of disks.

We denote by $C(y)$ the connected component of $\mathbf{C} \setminus B(0,1)$ that contains y, and by $|A|$ the area of a measurable subset A of $\mathbf{C}$. Then, in order that $|C(y)| \leq \pi\varepsilon^2$, it is obviously necessary that $T_\varepsilon(y) \leq 1$. On the other hand, it is not hard to show that, conditionally on $\{T_\varepsilon(y) \leq 1\}$, we will have $|C(y)| \leq \pi\varepsilon^2$ outside a set of small probability. The reason is that if the Brownian path comes within a distance ε of y before time 1, then it will also come much closer with great probability. The previous heuristic arguments suggest that W_ε, defined as the total area of the components with area less than $\pi\varepsilon^2$, is close to the area of

$$\{y \in \mathbf{C}; \; T_\varepsilon(y) \leq 1\} =: S_\varepsilon(0, 1).$$

The set $S_\varepsilon(0,1)$ is the so-called Wiener sausage of radius ε associated with B, on the time interval $[0,1]$. It is known (see e.g. [2]) that

$$\lim_{\varepsilon \to 0} |\log \varepsilon| \, |S_\varepsilon(0, 1)| = \pi, \quad \text{a.s.}$$

(the convergence of expectations follows from Lemma 1.2). One may therefore expect:

$$W_\varepsilon \underset{\varepsilon \to 0}{\sim} \frac{\pi}{|\log \varepsilon|}.$$

It turns out that much more is true: for $\lambda \in (0, 1)$,

$$W_\varepsilon - W_{\lambda\varepsilon} \underset{\varepsilon \to 0}{\sim} \frac{\pi}{|\log \varepsilon|} - \frac{\pi}{|\log \lambda\varepsilon|} \underset{\varepsilon \to 0}{\sim} \frac{\pi |\log \lambda|}{|\log \varepsilon|^2}.$$

The latter result is more or less equivalent to (1.a). It provides some explanation for the factor $(\log x)^2$ in (1.a).

Estimates for the Wiener sausage play an important role in this paper. In particular, we use the fact that

$$E\big[m(S_\varepsilon(0,1))\big] = \frac{\pi}{|\log\varepsilon|} + O\Big(\frac{1}{|\log\varepsilon|^2}\Big) \tag{1.b}$$

which follows from Theorem 1 of Spitzer [7] (see [3] for refined expansions).

Section 2 contains some preliminary results. The basic ideas and techniques of this section are taken from Mountford [5], even if our estimates are slightly more precise than those of [5]. We have given complete proofs in order to make the paper self-contained, and also because we hope to use our sharp estimates in some future work. In Section 3, we use a new method to get the almost sure convergence in (1.a). This method was inspired by similar techniques which have been applied to the fluctuation of the Wiener sausage (see [2]). The proof of Theorem 1.1 is given at the end of Section 3.

2. Preliminary Estimates. Throughout this section, we consider a Brownian motion $(B_t,\ t \geq 0)$ with values in the complex plane $\mathbb{C}$, started at z_0 such that $|z_0| = 1$. For any $R > 1$, we set

$$T_R = \inf\{t \geq 0,\ |B_t| = R\}.$$

We denote by $C_R(0)$ the connected component of $\mathbb{C}\setminus B[0,T_R]$ that contains 0. The following basic lemma is due to Mountford [5].

Lemma 2.1. *There exists a constant $\beta > 0$ such that for any $R \geq 2$,*

$$P\big[C_R(0) \text{ is unbounded}\big] \leq R^{-\beta}.$$

Proof. Fix $\rho > 1$, and denote by $C_{(n)}(0)$ the connected component of $\mathbb{C}\setminus B\big[T_{\rho^n}, T_{\rho^{n+1}}\big]$ that contains 0. For any $n \geq 1$,

$$P\big[C_{\rho^n}(0) \text{ is unbounded}\big] \leq P\Big[\bigcap_{k=0}^{n-1}\{C_{(k)}(0) \text{ is unbounded}\}\Big].$$

However the strong Markov property shows that the events $\{C_{(k)}(0)$ is unbounded$\}$, $k = 0,1,\dots,n-1$ are independent. Furthermore, a scaling argument shows that they have the same probability c. Since the support of Wiener measure consists of all continuous functions starting from 0, it is also clear that $c < 1$. We get

$$P\big[C_{\rho^n}(0) \text{ is unbounded}\big] \leq c^n$$

and the statement of Lemma 2.1 follows easily. □

Remark. When $C_R(0)$ is bounded, $C_R(0)$ is obviously contained in the disk of radius R centered at 0. We shall use this simple fact on several occasions.

For any $r \in (0,1)$ set

$$\Gamma(r,R) = P\big[|C_R(0)| \le \pi r^2\big].$$

Note that it is easy to get good estimates for $\Gamma(r,R)$. Indeed,

$$\Gamma(r,R) \le P[T_r < T_R] = \frac{\log R}{\log R - \log r}.$$

On the other hand, we may get a lower bound on $\Gamma(r,R)$ by conditioning on $\{T_{r/2^n} < T_R\}$, applying the Markov property at time $T_{r/2^n}$, and using Lemma 2.1 together with a simple scaling argument. In this way, we get for every $n \ge 1$,

$$\Gamma(r,R) \ge P\big[T_{r/2^n} < T_R\big](1 - 2^{-n\beta}) = \frac{\log R}{\log R - \log r + n\log 2}(1 - 2^{-n\beta}).$$

If we take $n = [K \log\log R]$ with K large enough, we get that for any $M > 1$,

$$\Gamma(r,R) \ge \frac{\log R}{\log R - \log r + O(\log\log R)}\big(1 + O((\log R)^{-M})\big).$$

In what follows, we shall be interested in getting estimates as $R \to \infty$, uniformly in r. We shall always assume that $|\log r| \le (\log R)^{\alpha}$, for some $\alpha < 1$. We can then rewrite the previous estimates in the following more convenient way:

$$\text{(2.a)} \qquad \Gamma(r,R) \le 1 + \frac{\log r}{\log R} + O\big((\log R)^{2\alpha-2}\big),$$

$$\text{(2.b)} \qquad \Gamma(r,R) \ge 1 + \frac{\log r}{\log R} + O\left((\log R)^{2\alpha-2} + \frac{\log\log R}{\log R}\right).$$

Fix $\lambda \in (0,1)$ and set $Q(r,R) = \Gamma(r,R) - \Gamma(\lambda r,R)$. We are in fact interested in estimates for $Q(r,R)$. The simplest idea would be to use our lower and upper bounds on $\Gamma(r,R)$ to estimate $Q(r,R)$. This however does not lead to sufficiently good estimates, so that we will have to use a more sophisticated method again inspired by Mountford [5].

Lemma 2.2. *We can choose $K > 0$ so that, as $R \to \infty$,*

$$Q(r, R) = \frac{|\log \lambda|}{\log R} + O\big((\log R)^{-3/2} \log\log R\big)$$

uniformly for: $K \log\log R \leq |\log r| \leq (\log R)^{1/2}$.

Proof. The key idea is to compare $Q(r, R)$ and $Q(\lambda r, R)$ using a scaling transformation. We will then write, for an integer N suitably chosen,

$$\Gamma(r, R) - \Gamma(\lambda^n r, R) = \sum_{k=0}^{N-1} Q(\lambda^k r, R)$$

and finally apply the bounds (2.a), (2.b).

In what follows, we assume that, for some fixed $\alpha < 1$,

$$K \ \log\log R \leq |\log r| \leq (\log R)^\alpha$$

where the constant K is taken sufficiently large. We denote by $D(z, a)$ the disk of radius a centered at z. Notice that the event

$$A_r = \{\pi(\lambda r)^2 \leq |C_R(0)| \leq \pi r^2\}$$

is contained in $\{T_r < T_R\}$, so that Lemma 2.1, and the reasoning that leads to (2.b), imply

$$\text{(2.c)} \qquad P(A_r) = P\big(A_r \cap \{C_R(0) \subset D(0,1)\}\big) + O\big((\log R)^{-10}\big)$$

provided K has been chosen large enough.

We now want to estimate $Q(\lambda r, R) = P(A_{\lambda r})$. Notice that $A_{\lambda r}$ is trivially contained in the set $\{T_\lambda < T_R\}$. We then define $C'_R(0)$ as the connected component of $\mathbf{C} \setminus B[0, T_{\lambda R}]$ that contains 0. Set

$$A'_r = \{\pi(\lambda^2 r)^2 \leq |C'_R(0)| \leq \pi(\lambda r)^2\}.$$

Exactly as before,

$$\text{(2.d)} \qquad P(A'_r) = P\big(A'_r \cap \{C'_R(0) \subset D(0,\lambda)\}\big) + O\big((\log R)^{-10}\big).$$

Now, by the Markov property and scaling,

$$\text{(2.e)} \quad P\big[A'_r \cap \{C'_R(0) \subset D(0,\lambda)\}\big] \\ = P[T_\lambda < T_R] P\big[A_r \cap \{C_R(0) \subset D(0,1)\}\big].$$

The point is that, on the set $\{C'_R(0) \subset D(0,\lambda)\}$, $C'_R(0)$ is also the component of $\mathbf{C} \setminus B[T_\lambda, T_{\lambda r}]$ that contains 0, and we apply the Markov property at T_λ.

We now have to compare A'_r and $A_{\lambda r}$. The problem is that $C_R(0)$ may be smaller than $C'_R(0)$ because of the path between $T_{\lambda R}$ and T_R. However, by Lemma 2.1, we can choose $M > 0$ so large that $C'_R(0)$ is contained in $D(0,(\log R)^M)$ except on a set of probability $O((\log R)^{-10})$. It follows that, up to $O((\log R)^{10})$, $P(A_{\lambda r} \setminus A'_r)$ is bounded by

$$\begin{aligned}
&P\left[\inf_{[T_{\lambda R},T_R]} |B_s| \le (\log R)^M, \pi(\lambda^2 r)^2 \le |C_R(0)|\right] \\
&\le P[|C'_R(0)| \ge \pi(\lambda^2 r)^2]\, P\left[\inf_{[T_{\lambda R},T_R]} |B_s| \le (\log R)^M\right] \cdot \\
&\quad \times \left(1-\Gamma\left(\frac{\lambda^2 r}{(\log R)^M}, \frac{R}{(\log R)^M}\right)\right) \\
&= \frac{|\log\lambda|}{\log R - M\log\log R}(1-\Gamma(\lambda^2 r, \lambda R))\left(1-\Gamma\left(\frac{\lambda^2 r}{(\log R)^M}, \frac{R}{(\log R)^M}\right)\right) \\
&= O((\log R)^{2\alpha-3})
\end{aligned}$$

by (2.b). To get the first bound we have used the strong Markov property at the stopping time: $\inf\{t \ge T_{\lambda R}, |B_t| = (\log R)^M\}$, and then at $T_{\lambda R}$.

Similarly, $P(A'_r \setminus A_{\lambda r})$ is bounded, up to $O((\log R)^{-10})$, by:

$$\text{(2.f)} \quad P[A'_r]P\left[\inf_{[T_{\lambda R},T_R]} |B_u| < (\log R)^M\right] = P[A'_r]O((\log R)^{-1}) = O((\log R)^{\alpha-2})$$

Therefore we conclude that

$$\text{(2.g)} \qquad P(A'_r) = P(A_{\lambda r}) + O((\log R)^{\alpha-2}).$$

In view of (2.c), (2.d), (2.e) and (2.g), we obtain:

$$P(A_{\lambda r}) = P[T_\lambda < T_R]P(A_r) + O((\log R)^{\alpha-2})$$

uniformly for $K \log\log R \le |\log r| \le (\log R)^\alpha$. Equivalently,

$$P(A_{\lambda r}) = P(A_r)\left(1 + \frac{\log\lambda}{\log R}\right) + O((\log R)^{\alpha-2})$$

or

$$Q(\lambda r, R) = Q(r,R)\left(1+\frac{\log\lambda}{\log R}\right) + O((\log R)^{\alpha-2}).$$

Now let $N \ge 1$, and assume that $N|\log\lambda| \le \frac{1}{2}(\log R)^\alpha$. By the previous estimates, for any $k \in \{1,\dots,N\}$,

$$Q(\lambda^k r, R) = Q(r,R)\left(1+\frac{\log\lambda}{\log R}\right)^k + O(k(\log R)^{\alpha-2})$$

where the estimate $O\big(k(\log R)^{\alpha-2}\big)$ is uniform in k. It follows that:

$$\begin{aligned}\Gamma(r,R)-\Gamma(\lambda^N r,R) &= \sum_{k=0}^{N-1} Q(\lambda^k r,R)\\ &= \left(\frac{\log R}{|\log\lambda|}\right)\left(1-\left(1+\frac{\log\lambda}{\log R}\right)^N\right)Q(r,R)\\ &\quad + O\big(N^2(\log R)^{\alpha-2}\big)\end{aligned}$$

uniformly for $K \log\log R \le |\log r| \le \frac{1}{2}(\log R)^\alpha$. Furthermore,

$$1-\left(1+\frac{\log\lambda}{\log R}\right)^N = N\frac{|\log\lambda|}{\log R}+O\left(\frac{N^2}{(\log R)^2}\right)$$

and we conclude that

$$\Gamma(r,R)-\Gamma(\lambda^N r,R) = N\left(1+O\left(\frac{N}{\log R}\right)\right)Q(r,R)+O\big(N^2(\log R)^{\alpha-2}\big).$$

However the bounds (2.a), (2.b) give:

$$\Gamma(r,R)-\Gamma(\lambda^N r,R) = N\frac{|\log\lambda|}{\log R}+O\left((\log R)^{2\alpha-2}+\frac{\log\log R}{\log R}\right)$$

and it follows that

$$\begin{aligned}\left(1+O\left(\frac{N}{\log R}\right)\right)Q(r,R) &= \frac{|\log\lambda|}{\log R}\\ &\quad + O\left(N(\log R)^{\alpha-2}+\frac{1}{N}(\log R)^{2\alpha-2}+\frac{1}{N}\frac{\log\log R}{\log R}\right).\end{aligned}$$

At this point, choosing $N=[(\log R)^\gamma]$ with $\gamma>0$ small, we obtain that for any $\alpha<1$, there exists $\eta>0$ such that

$$Q(r,R)=\frac{|\log\lambda|}{\log R}+O\big((\log R)^{-1-\eta}\big)$$

uniformly for $K\log\log R \le |\log r| \le (\log R)^\alpha$.

We may now plug this estimate in (2.f), observing that $P[A'_r]=Q(\lambda r, \lambda R)$. For $\alpha \le 1/2$, we get $P(A'_r)=P(A_{\lambda r})+O\big((\log R)^{-2}\big)$, and the same calculations as previously give:

$$\begin{aligned}\left(1+O\left(\frac{N}{\log R}\right)\right)Q(r,R) &= \frac{|\log\lambda|}{\log R}\\ &\quad + O\left(N(\log R)^{-2}+\frac{1}{N}(\log R)^{2\alpha-2}+\frac{1}{N}\frac{\log\log R}{\log R}\right).\end{aligned}$$

Choosing $\alpha = 1/2$, $N = [c(\log R)^{1/2}]$ completes the proof. $\square$

Remark. One may wonder what happens when r is small in comparison with $1/R$. The previous arguments can easily be adapted to give:

$$\lim_{R\to\infty}\left(\sup_{0<r<(\log R)^{-K}}\left|\frac{(\log R+|\log r|)^2}{\log R}Q(r,R)-|\log\lambda|\right|\right)=0.$$

3. Main Results. In this section we assume that the Brownian motion $(B_t,\, t\geq 0)$ is started at 0. For every $y\in \mathbf{C}\setminus B[0,1]$, $C(y)$ denotes the component of $\mathbf{C}\setminus B[0,1]$ that contains y. We fix $\lambda\in(0,1)$ and for $\varepsilon>0$ we set

$$U_\varepsilon=\{y\in\mathbf{C}\setminus B[0,1];\pi(\lambda\varepsilon)^2\leq|C(y)|\leq\pi\varepsilon^2\}.$$

Theorem 3.1. *As ε goes to 0,*

$$E[|U_\varepsilon|]=\frac{\pi|\log\lambda|}{|\log\varepsilon|^2}+O\left(\frac{\log|\log\varepsilon|}{|\log\varepsilon|^{5/2}}\right).$$

Proof. For $\varepsilon>0$ small enough we define $\delta=\delta(\varepsilon)>\varepsilon$ by the condition:

$$\frac{\delta}{\exp(|\log\delta|^{1/4})}=\varepsilon.$$

Let y be such that $|y|\geq\delta$. Set

$$T_\delta(y)=\inf\{s\geq 0,\ |B_s-y|<\delta\}$$

and

$$R_\delta(y)=\inf\{s>T_\delta(y),\ |B_s-y|>(\log\delta)^{-4}\}.$$

Observe that $\{y\in U_\varepsilon\}\subset\{T_\delta(y)\leq 1\}$. We let $C_\delta(y)$ denote the connected component of $\mathbf{C}\setminus B[T_\delta(y),\,R_\delta(y)]$ that contains y. Then,

$$\tag{3.a}\begin{aligned}&|P[y\in U_\varepsilon]-P[T_\delta(y)\leq 1;\ \pi(\lambda\varepsilon)^2\leq|C_\delta(y)|\leq\pi\varepsilon^2]|\\&\leq P[T_\delta(y)\leq 1;\ \pi(\lambda\varepsilon)^2\leq|C_\delta(y)|\leq\pi\varepsilon^2;\ C(y)\neq C_\delta(y)]\\&\quad+P[T_\delta(y)\leq 1;\ \pi(\lambda\varepsilon)^2\leq|C(y)|\leq\pi\varepsilon^2;\ C(y)\neq C_\delta(y)].\end{aligned}$$

We first evaluate

$$P[T_\delta(y)\leq 1;\pi(\lambda\varepsilon)^2\leq|C_\delta(y)|\leq\pi\varepsilon^2]=P[T_\delta(y)\leq 1]Q\left(\frac{\varepsilon}{\delta},\frac{(\log\delta)^{-4}}{\delta}\right)$$

with the notation of the previous section. Here we have applied the Markov property at $T_\delta(y)$ and used a scaling transformation. Lemma 2.2 now implies

$$P\big[T_\delta(y) \le 1;\ \pi(\lambda\varepsilon)^2 \le |C_\delta(y)| \le \pi\varepsilon^2\big] \\ = P[T_\delta(y) \le 1]\bigg(\frac{|\log\lambda|}{|\log\delta|} + O\Big(\frac{\log|\log\delta|}{|\log\delta|^{3/2}}\Big)\bigg).$$

The next step is to bound the right hand side of (3.a). Let us consider the first term of the sum. We have

(3.b)

$$\begin{aligned} &P\big[T_\delta(y) \le 1;\ \pi(\lambda\varepsilon)^2 \le |C_\delta(y)| \le \pi\varepsilon^2; C(y) \ne C_\delta(y)\big] \\ &\le P\big[T_\delta(y) \le 1 \le R_\delta(y)\big] \\ &+ P\big[T_\delta(y) \le R_\delta(y) \le 1; \big(B[0,T_\delta(y)] \cup B[R_\delta(y),1]\big) \cap C_\delta(y) \ne \emptyset; \\ &\pi(\lambda\varepsilon)^2 \le |C_\delta(y)| \le \pi\varepsilon^2\big]. \end{aligned}$$

It is clear that

$$P\big[T_\delta(y) \le 1 \le R_\delta(y)\big] \le P\big[|B_1 - y| \le (\log\delta)^{-4}\big] \le (\log\delta)^{-8}\Psi(y)$$

for some integrable function Ψ. Next we observe that, by Lemma 2.1,

$$P\big[|C_\delta(y)| \le \pi\varepsilon^2; C_\delta(y) \cap (\mathbf{C} \setminus D(y,\delta)) \ne \emptyset\big] = O\big(|\log\delta|^{-10}\big)$$

uniformly in $y \in \mathbf{C}$. It follows that the second term of the right side of (3.b) is bounded by

$$\begin{aligned} &P[T_\delta(y) \le 1]O\big(|\log\delta|^{-10}\big) + P[T_\delta(y) \le 1; \\ &B\big[R_\delta(y), R_\delta(y)+1\big] \cap D(y,\delta) \ne \emptyset; \pi(\lambda\varepsilon)^2 \le |C_\delta(y)| \le \pi\varepsilon^2] \\ &= P[T_\delta(y) \le 1]\bigg(O\big(|\log\delta|^{-10}\big) \\ &+ Q\Big(\frac{\varepsilon}{\delta}, \frac{(\log\delta)^{-4}}{\delta}\Big)P\big[B[R_\delta(y), R_\delta(y)+1] \cap D(y,\delta) \ne \emptyset\big]\bigg) \end{aligned}$$

by the Markov property applied at $R_\delta(y)$ and then at $T_\delta(y)$. However, by Lemma 1.2,

$$P\big[B[R_\delta(y), R_\delta(y)+1] \cap D(y,\delta) \ne \emptyset\big] = O\Big(\frac{\log|\log\delta|}{|\log\delta|}\Big).$$

By Lemma 2.2 and Lemma 1.2 again, we conclude that the left side of (3.b) is bounded by

$$\frac{\log|\log\delta|}{|\log\delta|^3}\psi(y)$$

for some integrable function ψ.

It remains to bound the second term of the right side of (3.a). We observe that

(3.c)
$$\begin{aligned}
&P\big[T_\delta(y) \le 1; \pi(\lambda\varepsilon)^2 \le |C(y)| \le \pi\varepsilon^2; C(y) \ne C_\delta(y)\big] \\
&\quad \le P\big[T_\delta(y) \le 1 \le R_\delta(y)\big] + P\big[T_\delta(y) \le 1; \\
&\quad B\big[T_\delta(y), R_\delta(y)\big] \cap D(y,\varepsilon) \ne \emptyset; |C_\delta(y)| \ge \pi(\lambda\varepsilon)^2; C(y) \ne C_\delta(y)\big] \\
&\quad + P\big[T_\delta(y) \le 1; |C_\delta(y)| \ge \pi(\lambda\varepsilon)^2; B\big[R_\delta(y), R_\delta(y)+1\big] \cap D(y,\varepsilon) \ne \emptyset\big],
\end{aligned}$$

using the fact that on $\{|C(y)| \le \pi\varepsilon^2\}$, B must hit $D(y,\varepsilon)$ before time 1. The first term is bounded as previously. The third one is easily bounded by

$$\begin{aligned}
P[T_\delta(y) \le 1]\Big(1 - \Gamma\Big(\frac{\lambda\varepsilon}{\delta}, \frac{(\log\delta)^{-4}}{\delta}\Big)\Big) O\Big(\frac{\log|\log\delta|}{|\log\delta|}\Big) \\
= P[T_\delta(y) \le 1] O\Big(\frac{\log|\log\delta|}{|\log\delta|^{-7/4}}\Big)
\end{aligned}$$

by (2.b).

We now consider the second term of the right side of (3.c). Lemma 2.1 implies that on the set $\{B[T_\delta(y), R_\delta(y)] \cap D(y,\varepsilon) \ne \emptyset\}$, $C_\delta(y)$ is contained in $D(y,\delta)$ except on a set of probability $O(|\log\delta|^{-10})$. Therefore the second term is bounded by:

$$\begin{aligned}
&P\big[T_\delta(y) \le 1; |C_\delta(y)| \ge \pi(\lambda\varepsilon)^2; B[R_\delta(y), R_\delta(y)+1] \cap D(y,\delta) \ne \emptyset\big] \\
&\qquad + P[T_\delta(y) \le 1] O(|\log\delta|^{-10}) = P[T_\delta(y) \le 1] O\Big(\frac{\log|\log\delta|}{|\log\delta|^{7/4}}\Big)
\end{aligned}$$

as previously.

Putting together the previous estimates and using Lemma 1.2 we obtain that the left side of (3.c) is bounded by:

$$\Big(\frac{\log|\log\delta|}{|\log\delta|^{11/4}}\Big)\theta(y),$$

for some integrable function θ. Then (3.a) implies

$$\begin{aligned}
&\Big|P[y \in U_\varepsilon] - P[T_\delta(y) \le 1] Q\Big(\frac{\varepsilon}{\delta}, \frac{(\log\delta)^{-4}}{\delta}\Big)\Big| \\
&\qquad \le \Big(\frac{\log|\log\delta|}{|\log\delta|^{11/4}}\Big)(\psi(y) + \theta(y)).
\end{aligned}$$

Hence,

$$E[|U_\varepsilon|] = \int dy\, P[y \in U_\varepsilon]$$
$$= \left(\int dy\, P[T_\delta(y) \le 1]\right) Q\left(\frac{\varepsilon}{\delta}, \frac{(\log\delta)^{-4}}{\delta}\right) + O\left(\frac{\log|\log\delta|}{|\log\delta|^{11/4}}\right).$$

However, in the notation of the introduction,

$$\int dy\, P[T_\delta(y) \le 1] = E[m(S_\delta(0,1))] = \pi|\log\delta|^{-1} + O(|\log\delta|^{-2}).$$

Theorem 3.1 now follows from Lemma 2.2. □

Proposition 3.2. *There exists a constant K such that, for $\varepsilon \in (0, 1/4)$,*

$$\operatorname{var}|U_\varepsilon| \le K\,|\log\varepsilon|^{-11/2}.$$

Proof. Let B^1, B^2 be two independent Brownian motions with values in $\mathbf{C}$, started at 0. For every $y \in \mathbf{C} \setminus \big(B^1[0,1/2] \cup B^2[0,1/2]\big)$, denote by $C'(y)$ the connected component of $\mathbf{C} \setminus \big(B^1[0,1/2] \cup B^2[0,1/2]\big)$ that contains y. Set

$$U'_\varepsilon = \{y \in \mathbf{C} \setminus (B^1[0,1/2] \cup B^2[0,1/2]);\, \pi(\lambda\varepsilon)^2 \le |C'(y)| \le \pi\varepsilon^2\}.$$

Obviously $|U'_\varepsilon|$ and $|U_\varepsilon|$ are identically distributed so that $\operatorname{var}|U'_\varepsilon| = \operatorname{var}|U_\varepsilon|$.

For $i = 1, 2$, for every $y \in \mathbf{C} \setminus B^i[0,1/2]$, let $C^i(y)$ be the connected component of $\mathbf{C} \setminus B^i[0,1/2]$ that contains y, and

$$U^i_\varepsilon = \{y \in \mathbf{C} \setminus B^i[0,1/2];\, \pi(\lambda\varepsilon)^2 \le |C^i(y)| \le \pi\varepsilon^2\}.$$

The key idea of the proof is to show that $|U'_\varepsilon|$ is not too different from $|U^1_\varepsilon| + |U^2_\varepsilon|$.

Lemma 3.3. *As ε goes to 0, $E\big[(|U'_\varepsilon| - |U^1_\varepsilon| - |U^2_\varepsilon|)^2\big] = O(|\log\varepsilon|^{-11/2})$.*

Proof. We observe that

$$\begin{aligned} \big||U'_\varepsilon| - (|U^1_\varepsilon| + |U^2_\varepsilon|)\big| &\le \big||U'_\varepsilon| - |U^1_\varepsilon \cup U^2_\varepsilon|\big| + |U^1_\varepsilon \cap U^2_\varepsilon| \\ &\le |U'_\varepsilon \setminus (U^1_\varepsilon \cup U^2_\varepsilon)| + |U^1_\varepsilon \setminus U'_\varepsilon| + |U^2_\varepsilon \setminus U'_\varepsilon| + |U^1_\varepsilon \cap U^2_\varepsilon|. \end{aligned} \tag{3.d}$$

We will get bounds on the second moment of each of the four terms of the right side of (3.d). We start with:

$$E[|U^1_\varepsilon \setminus U'_\varepsilon|^2] = E\left[\int dy\, dz\, 1_{U^1_\varepsilon \setminus U'_\varepsilon}(y)\, 1_{U^1_\varepsilon \setminus U'_\varepsilon}(z)\right].$$

Let $\delta = \delta(\varepsilon)$ be as in the proof of Theorem 3.1. We claim that

(3.e)
$$E[|U_\varepsilon^1 \setminus U_\varepsilon'|^2] = E\Big[\int dydz\, 1_{\{C^1(y)\subset D(y,\delta);C^1(z)\subset D(z,\delta)\}} 1_{U_\varepsilon^1\setminus U_\varepsilon'}(y)\, 1_{U_\varepsilon^1\setminus U_\varepsilon'}(z)\Big] + O(|\log\varepsilon|^{-10}).$$

Indeed, writing $S_\varepsilon^1(0,a)$ for the Wiener sausage associated with B^1 on the time interval $[0,a]$,

$$E\Big[\int dydz\, 1_{U_\varepsilon^1}(y)\, 1_{U_\varepsilon^1}(z)\, 1_{\{C^1(y)\cap(\mathbf{C}\setminus D(y,\delta))\neq\emptyset\}}\Big]$$
$$\leq \int dy\, E\Big[1_{U_\varepsilon^1}(y)\, 1_{\{C^1(y)\cap(\mathbf{C}\setminus D(y,\delta))\neq\emptyset\}} |S_\varepsilon^1(0,1/2)|\Big]$$
$$\leq \Big(\int dy\, P\Big[y\in U_\varepsilon^1; C^1(y)\cap(\mathbf{C}\setminus D(y,\delta))\neq\emptyset\Big]^{1/2}\Big) E[|S_\varepsilon^1(0,1/2)|^2]^{1/2}$$

which may be bounded as in the proof of Theorem 3.1, using Lemma 2.1.

Now observe that, if $C^1(y)\subset D(y,\delta)$ and $y\in U_\varepsilon^1\setminus U_\varepsilon'$, then B^2 must hit $D(y,\delta)$ before time 1/2. It follows from (3.e) that:

$$E[|U_\varepsilon^1\setminus U_\varepsilon'|^2] \leq E\Big[\int dydz\, 1_{U_\varepsilon^1}(y)\, 1_{U_\varepsilon^1}(z)\, 1_{S_\delta^2}(y)\, 1_{S_\delta^2}(z)\Big] + O(|\log\varepsilon|^{-10})$$

(3.f)
$$= \int dydz\, P[y\in U_\varepsilon^1, z\in U_\varepsilon^1] P[y\in S_\delta^2, z\in S_\delta^2] + O(|\log\varepsilon|^{-10})$$

where for simplicity we write $S_\delta^2 = S_\delta^2(0,1/2)$. The problem is now to get a suitable bound on $P[y\in U_\varepsilon^1,\ z\in U_\varepsilon^1]$. Let $T_\delta^1(y)$, $R_\delta^1(y)$ be defined as previously with B replaced by B^1. By symmetry, we may restrict our attention to the case $\{T_\delta^1(y)\leq T_\delta^1(z)\}$. Suppose that $|z-y|\geq|\log\varepsilon|^{-3}$. Then, using the Markov property at time $T_\delta^1(z)$, we have for ε small,

$$P[y\in U_\varepsilon^1, z\in U_\varepsilon^1, T_\delta^1(y)\leq T_\delta^1(z)]$$
$$\leq P\Big[T_\delta^1(y)\leq T_\delta^1(z)\leq \frac{1}{2}, |C^1(y)|\geq \pi(\lambda\varepsilon)^2, |C^1(z)|\geq\pi(\lambda\varepsilon)^2\Big]$$
$$\leq P\Big[T_\delta^1(y)\leq T_\delta^1(z)\leq \frac{1}{2}, |\tilde{C}^1(y)|\geq\pi(\lambda\varepsilon)^2\Big]\Big(1-\Gamma\Big(\frac{\lambda\varepsilon}{\delta}, \frac{(\log\delta)^{-4}}{\delta}\Big)\Big)$$
$$+ P\Big[T_\delta^1(z)\leq\frac{1}{2}\leq R_\delta^1(z)\Big]$$
$$\leq K|\log\delta|^{-3/4} P\Big[T_\delta^1(y)\leq T_\delta^1(z)\leq\frac{1}{2}, |\tilde{C}^1(y)|\geq\pi(\lambda\varepsilon)^2\Big]$$
$$+ P\Big[T_\delta^1(z)\leq\frac{1}{2}\leq R_\delta^1(z)\Big]$$

where $\tilde{C}^1(y)$ is the connected component of $\mathbf{C} \setminus B[0, R^1_\delta(y)]$ that contains y. Clearly, $P\left[T^1_\delta(z) \le 1/2 \le R^1_\delta(z)\right] = O\left(|\log\delta|^{-8}\right)$. Therefore it suffices to bound:

$$P\left[T^1_\delta(y) \le T^1_\delta(z) \le \frac{1}{2}; |\tilde{C}^1(y)| \ge \pi(\lambda\varepsilon)^2\right]$$
$$\le P\left[T^1_\delta(y) \le \frac{1}{2}; |\tilde{C}^1(y)| \ge \pi(\lambda\varepsilon)^2\right] \frac{\Phi\left(\frac{z-y}{2}\right)}{|\log\delta|}$$
$$\le K'\,\Phi(y)\,\Phi\left(\frac{z-y}{2}\right)|\log\varepsilon|^{-11/4}.$$

The first bound uses the Markov property at time $R^1_\delta(y)$ and Lemma 1.2. The second one follows from Lemma 1.2, the Markov property at time $T^1_\delta(y)$ and (2.b).

We conclude that, for some constant $\bar{K}$, for any y, z such that $|z-y| \ge |\log\varepsilon|^{-3}$,

$$P\left[y \in U^1_\varepsilon, z \in U^1_\varepsilon, T^1_\delta(y) \le T^1_\delta(z)\right]$$
$$\le \bar{K}|\log\varepsilon|^{-7/2}\Phi(y)\Phi\left(\frac{z-y}{2}\right) + O\left(|\log\varepsilon|^{-8}\right).$$

On the other hand, by Lemma 1.2,

$$P\left[y \in S^2_\delta, z \in S^2_\delta\right] \le |\log\delta|^{-2}\left(\Phi(y)\Phi(z-y) + \Phi(z)\Phi(y-z)\right).$$

Putting together the previous estimates leads to:

$$E\left[|U^1_\varepsilon \setminus U'_\varepsilon|^2\right] = O\left(|\log\varepsilon|^{-11/2}\right). \tag{3.g}$$

Obviously the same bound holds for $E[|U^2_\varepsilon \setminus U'_\varepsilon|^2]$. It is also easy to check that $E[|U^1_\varepsilon \cap U^2_\varepsilon|^2]$ satisfies the same bound. Indeed, it is obvious that:

$$E\left[|U^1_\varepsilon \cap U^2_\varepsilon|^2\right] \le E\left[\int dy\,dz\, 1_{U^1_\varepsilon}(y)\, 1_{U^1_\varepsilon}(z)\, 1_{S^2_\varepsilon}(y)\, 1_{S^2_\varepsilon}(z)\right]$$

so that it suffices to use the previous estimates.

It only remains to bound

$$E\left[|U'_\varepsilon \setminus (U^1_\varepsilon \cup U^2_\varepsilon)|^2\right] = E\left[\int dydz\, 1_{U'_\varepsilon\setminus(U^1_\varepsilon\cup U^2_\varepsilon)}(y)\, 1_{U'_\varepsilon\setminus(U^1_\varepsilon\cup U^2_\varepsilon)}(z)\right]$$
$$\le E\left[\int dydz\, 1_{U'_\varepsilon\setminus(U^1_\varepsilon\cup U^2_\varepsilon)}(y)\, 1_{U'_\varepsilon\setminus(U^1_\varepsilon\cup U^2_\varepsilon)}(z)\, 1_{\{C'(y)\subset D(y,\delta)\}}\cdot\right.$$
$$\left.\cdot\, 1_{\{C'(z)\subset D(z,\delta)\}}\right] + O\left(|\log\varepsilon|^{-10}\right).$$

by the same arguments as previously. Then,

$$1_{\{C'(y)\subset D(y,\delta)\}}\, 1_{U'_\varepsilon\setminus(U^1_\varepsilon\cup U^2_\varepsilon)}(y) \\ \leq 1_{S^1_\delta}(y)\, 1_{S^2_\delta}(y)\, 1_{\{|C^1(y)|\geq\pi(\lambda\varepsilon)^2\}}\, 1_{\{|C^2(y)|\geq\pi(\lambda\varepsilon)^2\}}$$

so that

$$E\big[|U'_\varepsilon\setminus(U^1_\varepsilon\cup U^2_\varepsilon)|^2\big]\leq\int dydz\, P\big[y\in S^1_\delta, z\in S^1_\delta, \\ |C^1(y)|\geq\pi(\lambda\varepsilon)^2, |C^1(z)|\geq\pi(\lambda\varepsilon)^2\big]^2+O\big(|\log\varepsilon|^{-10}\big).$$

However, we have just proved that:

$$P\big[y\in S^1_\delta, z\in S^1_\delta, |C^1(y)|\geq\pi(\lambda\varepsilon)^2, |C^1(z)|\geq\pi(\lambda\varepsilon)^2\big] \\ \leq\bar{K}\,|\log\varepsilon|^{-7/2}(\Phi(y)\Phi(z-y)+\Phi(z)\Phi(y-z))+O\big(|\log\varepsilon|^{-8}\big).$$

We conclude that:

$$E\big[|U'_\varepsilon\setminus(U^1_\varepsilon\cup U^2_\varepsilon)|^2\big]=O\big(|\log\varepsilon|^{-7}\big).$$

This completes the proof of Lemma 3.3. □

We may now complete the proof of Proposition 3.2. Set $h(\varepsilon) = (\mathrm{var}|U_\varepsilon|)^{1/2} = (\mathrm{var}|U'_\varepsilon|)^{1/2}$. Notice that $|U^1_\varepsilon|$ and $|U^2_\varepsilon|$ are independent, and that $|U^1_\varepsilon| \overset{(d)}{=} \frac{1}{2}|U_{\varepsilon\sqrt{2}}|$ by a scaling argument. Then Lemma 3.3 implies

$$(\mathrm{var}|U_\varepsilon|)^{1/2}\leq\big(\mathrm{var}|(U^1_\varepsilon|+|U^2_\varepsilon|)\big)^{1/2}+O\big(|\log\varepsilon|^{-11/4}\big) \\ =2^{-1/2}\big(\mathrm{var}|U_{\varepsilon\sqrt{2}}|\big)^{1/2}+O\big(|\log\varepsilon|^{-11/4}\big).$$

Therefore

$$h(\varepsilon)\leq 2^{-1/2}h(\varepsilon\sqrt{2})+O\big(|\log\varepsilon|^{-11/4}\big).$$

Set $g(\varepsilon)=|\log\varepsilon|^{11/4}h(\varepsilon)$. It follows that for any $\rho\in(2^{-1/2},1)$, for ε small, $g(\varepsilon)\leq\rho\, g(\varepsilon\sqrt{2})+O(1)$. Since g is clearly bounded on any interval $(\eta,1/4)$ ($\eta>0$), the previous bound implies that g is bounded on $(0,1/4)$. □

Remark. By being more careful in the previous proof, we could have improved significantly the bounds of Proposition 3.2 and Lemma 3.3. We shall leave this refinement to the reader.

Proof of Theorem 1.1. Proposition 3.2 implies that

$$\sum_{n=1}^{\infty}E\big[\big((\log\lambda^n)^2|U_{\lambda^n}|-E[(\log\lambda^n)^2|U_{\lambda^n}|]\big)^2\big]<\infty$$

and therefore

$$(\log \lambda^n)^2 |U_{\lambda^n}| - E\big[(\log \lambda^n)^2 |U_{\lambda^n}|\big] \xrightarrow[n\to\infty]{a.s.} 0.$$

Hence, by Theorem 3.1

$$(\log \lambda^n)^2 |U_{\lambda^n}| \xrightarrow[n\to\infty]{a.s.} \pi |\log \lambda|. \tag{3.h}$$

Let $\bar{N}_{[u,v)}$ be the number of connected components with area in $[\pi u^2, \pi v^2)$. Note that:

$$(\pi \lambda^{2n})^{-1} |U_{\lambda^n}| \le \bar{N}_{[\lambda^{n+1},\lambda^n)} \le (\pi \lambda^{2(n+1)})^{-1} |U_{\lambda^n}|$$

so that (3.h) implies that, a.s.,

$$\begin{aligned} |\log \lambda| &\le \liminf \lambda^{2n} (\log \lambda^n)^2 \, \bar{N}_{[\lambda^{n+1},\lambda^n)} \\ &\le \limsup \lambda^{2n} (\log \lambda^n)^2 \bar{N}_{[\lambda^{n+1},\lambda^n)} \le \frac{|\log \lambda|}{\lambda^2}. \end{aligned}$$

Fix an integer $p \ge 1$ and set $\lambda' = \lambda^{1/p}$. Since

$$\bar{N}_{[\lambda^{n+1},\lambda^n)} = \sum_{i=1}^{p} \bar{N}_{[\lambda'^{np+i},\lambda'^{np+i-1})},$$

we get

$$\begin{aligned} \frac{|\log \lambda|}{p} \sum_{i=1}^{p} \lambda^{-\frac{2i-2}{p}} &\le \liminf \lambda^{2n} (\log \lambda^n)^2 \, \bar{N}_{[\lambda^{n+1},\lambda^n)} \\ &\le \limsup \lambda^{2n} (\log \lambda^n)^2 \, \bar{N}_{[\lambda^{n+1},\lambda^n)} \le \frac{|\log \lambda|}{p} \sum_{i=1}^{p} \lambda^{-\frac{2i}{p}}. \end{aligned}$$

Choosing p large we conclude that:

$$\lambda^{2n} (\log \lambda^n)^2 \, \bar{N}_{[\lambda^{n+1},\lambda^n)} \xrightarrow[n\to\infty]{a.s.} |\log \lambda| \int_0^1 \lambda^{-2s} ds = \frac{1}{2}(\lambda^{-2} - 1).$$

Then a simple monotonicity argument gives:

$$x^2 (\log x)^2 \, \bar{N}_{[x,\alpha x)} \xrightarrow[x\to 0]{a.s.} \frac{1}{2}(1 - \alpha^{-2})$$

for every $\alpha > 1$. Equivalently,

$$u (\log u)^2 \, N_{[u,\alpha u)} \xrightarrow[u\to 0]{a.s.} 2\pi \left(1 - \frac{1}{\alpha}\right)$$

in the notation of the introduction. Theorem 1.1 now follows smoothly. For $v > (1+\varepsilon)u$, we can approximate $N_{[u,v)}$ by

$$\sum_{j=0}^{p} N_{[\alpha^j u, \alpha^{j+1} u)}$$

for some $p \in \mathbb{N}$ and $\alpha > 1$, and observe that

$$u(\log u)^2 \sum_{j=0}^{p} N_{[\alpha^j u, \alpha^{j+1} u)} \xrightarrow[u \to 0]{a.s.} 2\pi(1 - \alpha^{-p-1})$$

for every $p \geq 0$, including $p = +\infty$.

Acknowledgment. I thank H. Kesten for several useful comments on the first version of this work.

REFERENCES

[1] J.F. Le Gall, *Sur la saucisse de Wiener et les points multiples du mouvement brownien,* Ann. Probab. **14** (1986), 1219–1244.

[2] J.F. Le Gall, *Fluctuation results for the Wiener sausage,* Ann. Probab. **16** (1988), 991–1018.

[3] J.F. Le Gall, *Wiener sausage and self-intersection local times,* J. Funct. Anal. **88** (1990), 299–341.

[4] B.B. Mandelbrot, *The Fractal Geometry of Nature,* W.H. Freeman and Co., New York, 1983.

[5] T.S. Mountford, *On the asymptotic number of small components created by planar Brownian motion,* Stochastics **28** (1989), 177–188.

[6] F. Spitzer, *Some theorems concerning 2-dimensional Brownian motion,* Trans. Amer. Math. Soc. **87** (1958), 187–197.

[7] F. Spitzer, *Electrostatic capacity, heat flow and Brownian motion,* Z. Wahrsch. verw. Gebiete **4** (1964), 110–121.

Laboratoire de Probabilités,
Université Paris VI
4, Place Jussieu
75230 Paris Cedex 05, France

The Periodic Threshold Contact Process

Thomas M. Liggett[1]

We consider the periodic threshold contact process with period 2 in one dimension with parameters λ and μ. This process dies out if $\lambda + \mu + 2 > 4\lambda\mu$. We obtain a sufficient condition for its survival, which is satisfied by $(\lambda, \mu) = (2.17, 2.18)$, $(2.00, 2.37)$, and $(1.50, 3.62)$, for example. These results were motivated by recent work of Cox and Durrett on the threshold voter model.

1. Introduction. My interest in interacting particle systems began when Chuck Stone handed me a preprint of Frank Spitzer's 1970 paper "Interaction of Markov Processes," with the comment that I might find something worthwhile in it. That paper, and several others which Frank wrote during the late sixties and early seventies, stimulated the development of a field that has since become one of the more active and important in probability theory. I owe a large personal debt to Frank, since about half of my publications, as well as a large part of my book, have resulted directly from the models he developed, and the questions he asked about them.

This paper deals with a version of the contact process, which is not one of the models whose study Frank started. (That distinction goes to Ted Harris.) Nevertheless, there have been a number of important connections between Frank's work and the contact process. For example, my proof with Dick Holley in 1978 that the basic contact process survives if $\lambda \geq 2$ was inspired by Frank's observation in his 1977 paper that reversible "contact-like" processes have renewal measures as their invariant measures. That suggested that one might be able to prove interesting things about the contact process by starting it off with a cleverly chosen renewal measure. A second example is my proof in 1981, again with Dick Holley, that the critical value of the d-dimensional contact process is asymptotic to $1/2d$ for large d. It was accomplished by comparing the contact process with a generalization of the smoothing process, which Frank introduced in his

[1] Research supported in part by NSF Grant DMS 86-01800.

1981 paper.

The periodic threshold contact process is the Markov process on $\{0,1\}^Z$ with the following transitions:

$1 \to 0$ at site k at rate 1,

$0 \to 1$ at site k at rate λ if k is odd and there is a one at either site $k-1$ or $k+1$,

and

$0 \to 1$ at site k at rate μ if k is even and there is a one at either site $k-1$ or$k+1$.

The process is said to survive if it has an invariant measure other than the pointmass on the configuration which is identically zero, and is said to die out otherwise. In this paper, we will first give a simple proof of extinction of the process if λ or μ is small, and will then apply the techniques developed in Holley and Liggett (1978) and Liggett (1991) to obtain sufficient conditions for survival of the process.

Our motivation for this project comes from some recent work of Cox and Durrett. In their paper, they study a process called the threshold voter model. It has the following transitions:

$0 \to 1$ at site k at rate 1 if there is a one at any site in $\{k-N,\dots,k+N\}$,

and

$1 \to 0$ at site k at rate 1 if there is a zero at any site in $\{k-N,\dots,k+N\}$.

Here N is a positive integer, which is called the range of the interaction. The usual one dimensional voter model has only the trivial invariant measures which concentrate on the identically zero and identically one configurations (see Chapter V of Liggett (1985), for example). Cox and Durrett found that in contrast to this, the threshold voter model can have a nontrivial invariant measure as well, which is obtained by starting the process with the product measure with density 1/2, and taking the limit as $t \to \infty$. When this limit is nontrivial, the process is said to survive, and otherwise it is said to die out.

It is natural to try to determine exactly for which N the threshold voter model survives. It is easy to show that it dies out if $N = 1$, and Cox and Durrett conjecture that it survives for all $N \geq 2$. They prove survival for large N by comparison with the threshold contact process. The comparison is such that survival of the threshold contact process with integer parameters λ and μ implies survival of the periodic threshold voter model for $N = \lambda + \mu - 1$. Previously, the best sufficient condition for survival of the threshold contact process was $\lambda, \mu \geq 4$. It was obtained via a comparison with the usual contact process, using the result in Holley and Liggett (1978). This gives survival of the threshold voter model for $N \geq 7$. The improved results for the periodic threshold contact process that are stated below give survival of the threshold voter model for $N \geq 4$. It appears likely that the remaining cases $N = 2, 3$ will have to be handled directly, rather than by comparison with the threshold contact process.

In order to state the theorem, let β be the middle root of the cubic polynomial

$$(\lambda\beta - \mu)(\lambda\beta + \mu)^2 + \beta(2 + \mu + \lambda)^2[(2 + \lambda)(1 + 2\mu) - \beta(2 + \mu)(1 + 2\lambda)],$$

which always has three positive real roots, as we shall see later. Then define

$$f = \frac{\lambda\beta + \mu}{2 + \mu + \lambda}.$$

Theorem. (a) *The periodic threshold contact process dies out if* $\lambda + \mu + 2 > 4\lambda\mu$.

(b) *The homogeneous threshold contact process (with* $\lambda = \mu$*) survives if* $\lambda \geq \lambda_1$*, where* λ_1 *is the largest root of the cubic* $\lambda^3 - \lambda^2 - 3\lambda + 1$ ($\lambda_1 = 2.17+$).

(c) *The periodic threshold contact process survives if the following inequalities are satisfied:*

$$f^2 + 2\mu f + \beta - 2\mu \geq 0, \tag{1.1}$$
$$f^2 + 2\lambda\beta f + \beta - 2\lambda\beta^2 \geq 0, \tag{1.2}$$
$$\mu f + \mu\lambda\beta - \beta\mu - \beta \geq 0, \tag{1.3}$$
$$\lambda f + \mu\lambda - \lambda - 1 \geq 0, \tag{1.4}$$

and

$$(\mu^2 + \beta + 2\mu f - 2\mu)(\beta\lambda^2 + 1 + 2\lambda f - 2\beta\lambda) \geq (\mu + \beta\lambda)^2. \tag{1.5}$$

The three parts of this theorem are proved in Sections 2, 4 and 5 respectively. Section 3 is devoted to the restatement in the present context of some results from Liggett (1991) that are needed for the proofs of parts (b) and (c) of the theorem. While part (b) is a special case of part (c), we prove this separately since the required computations are much simpler than in the periodic case.

The following table lists some (λ, μ) pairs which satisfy (1.1)–(1.5). For each λ listed, condition (1.5) fails if the corresponding value of μ is reduced by .01. The value of β is given (rounded to four decimal places) to facilitate the checking of the required inequalities.

λ	μ	β
2.17	2.18	1.0012
2.10	2.25	1.0180
2.00	2.37	1.0448
1.90	2.52	1.0754
1.80	2.70	1.1096
1.70	2.93	1.1488
1.60	3.22	1.1933
1.50	3.62	1.2458
1.40	4.17	1.3064
1.30	5.03	1.3800
1.20	6.50	1.4692
1.15	7.73	1.5219

We will conclude the introduction by mentioning two results that help to highlight certain features of our theorem. If we let λ_c denote the critical value of the homogeneous threshold contact process in which $\lambda = \mu$, then the theorem asserts that $1 \leq \lambda_c \leq 2.18$. Numerical work by Dickman and Burschka (1988) provides the estimate $\lambda_c = 1.74$, so that our upper bound appears to be quite good.

Secondly, our theorem implies that the asymptotes of the true survival region for the periodic threshold contact process lie between .25 and 1.15. In particular, the asymptotes are not the coordinate axes. This contrasts with one of the results proved in Liggett (1991): The periodic contact process with transitions

$$1 \to 0 \text{ at site } k \text{ at rate } 1,$$

$$0 \to 1 \text{ at site } k \text{ at rate } \lambda\eta(k-1) + \mu\eta(k+1) \text{ if } k \text{ is odd,}$$

and

$$0 \to 1 \text{ at site } k \text{ at rate } \mu\eta(k-1) + \lambda\eta(k+1) \text{ if } k \text{ is even,}$$

dies out if $\lambda\mu \leq 1$ and survives if $\lambda\mu \geq 4$. (Here η denotes the configuration in $\{0,1\}^Z$.)

2. Extinction. In this section, we give the (easy) proof of part (a) of the theorem. Let ν be the upper invariant measure of the process. Since ν is invariant, $\int \Omega f \, d\nu = 0$ for all cylinder functions f. Applying this to $f(\eta) = \eta(k)$ gives

$$\nu\{\eta(k) = 1\} = \lambda[\nu\{\eta(k) = 0\} - \nu\{\eta(k-1) = \eta(k) = \eta(k+1) = 0\}] \tag{2.1}$$

for odd k, and

$$\nu\{\eta(k) = 1\} = \mu[\nu\{\eta(k) = 0\} - \nu\{\eta(k-1) = \eta(k) = \eta(k+1) = 0\}] \tag{2.2}$$

for even k. Applying it to $f(\eta) = \eta(k)\eta(k+1)$ gives

$$\begin{aligned} 2\nu\{\eta(k) = 1, \eta(k+1) = 1\} =& \lambda\nu\{\eta(k) = 0, \eta(k+1) = 1\} \\ &+ \mu\nu\{\eta(k) = 1, \eta(k+1) = 0\}] \end{aligned} \tag{2.3}$$

for odd k, and

(2.4)
$$2\nu\{\eta(k)=1,\eta(k+1)=1\} = \mu\nu\{\eta(k)=0,\eta(k+1)=1\} + \lambda\nu\{\eta(k)=1,\eta(k+1)=0\}]$$

for even k. Let $\alpha = \nu\{\eta(k)=1\}$ for odd k, and $\beta = \nu\{\eta(k)=1\}$ for even k. Solving (2.3) and (2.4) gives

$$\nu\{\eta(k)=1,\eta(k+1)=1\} = \frac{\lambda\beta+\mu\alpha}{2+\mu+\lambda}$$

for all k. Using this in equation (2.1) yields

$$\begin{aligned}\lambda(2+\lambda+\mu)\nu\{\eta(k-1)=1,\eta(k)=0,\eta(k+1)=1\}\\ = 2\lambda\beta(2+\mu)-\alpha(2+\lambda+\mu+2\lambda\mu)\end{aligned}$$

for odd k, and hence it follows that

(2.5)
$$2\lambda\beta(2+\mu) \geq \alpha(2+\lambda+\mu+2\lambda\mu).$$

Using (2.2) in place of (2.1) gives

(2.6)
$$2\mu\alpha(2+\lambda) \geq \beta(2+\lambda+\mu+2\lambda\mu).$$

If α and β are strictly positive, (2.5) and (2.6) combine to imply that

$$0 \leq (4\lambda\mu-\lambda-\mu-2)(\lambda+\mu+2).$$

This completes the proof of part (a) of the theorem.

3. The Inhomogeneous Threshold Contact Process. In this section, we will state some results analogous to those in Liggett (1991) in the context of the threshold contact process. These will be used in the next two sections. The proofs will be largely omitted, since they are virtually identical to the proofs in the case of the inhomogeneous contact process. To make them look as much as possible like the statements in the earlier paper, and to prepare for other potential applications, we formulate them for a general spatially inhomogeneous threshold contact process, with the following transitions:

$1 \to 0$ at site k at rate 1,

and

$0 \to 1$ at site k at rate $\lambda(k)$ if there is a one at either site $k-1$ or $k+1$.

The following is the analogue of Theorem 5 of Liggett (1991).

Theorem 3.1. *Suppose that there are functions $\alpha(i) > 0$ and $\phi(k,l) \geq 0$, $-\infty < k < l < \infty$, which satisfy*

$$\sum_{k \leq i < l} \phi(k,l) = 1, \tag{3.2}$$

$$\alpha(i) = \sum_{k>i} \phi(i,k) = \sum_{k<i} \phi(k,i), \tag{3.3}$$

$$0 = \lambda(i+1)\alpha(i) + \lambda(i)\alpha(i+1) - [2 + \lambda(i) + \lambda(i+1)]\phi(i,i+1), \tag{3.4}$$

and

$$\begin{aligned} 0 = {} & \lambda(i-1)\phi(i-2,i+1) + \lambda(i+1)\phi(i-1,i+2) \\ & + \frac{\phi(i-1,i)\phi(i,i+1)}{\alpha(i)} - [2+\lambda(i)]\phi(i-1,i+1) \end{aligned} \tag{3.5}$$

for all i,

$$\begin{aligned} 0 = {} & \lambda(k)\phi(k-1,l) + \lambda(l)\phi(k,l+1) + \sum_{j=k+1}^{l-1} \frac{1}{\alpha(j)}\phi(k,j)\phi(j,l) \\ & -[2 + \lambda(k+1) + \lambda(l-1)]\phi(k,l) \end{aligned} \tag{3.6}$$

for $k < l - 2$, and

$$\phi(k,l-1)\phi(k-1,l) \geq \phi(k,l)\phi(k-1,l-1) \tag{3.7}$$

for $k < l - 1$. Then the inhomogeneous threshold contact process survives.

The main step in the proof of Theorem 3.1 is the derivation of a certain identity, which is the analogue of Proposition 3.7 of Liggett (1991), and which is proved in the same way. To state it, suppose α and ϕ satisfy (3.2) and (3.3). Let ν be the distribution of the inhomogeneous renewal process associated with α and ϕ in the following way:

$$\begin{aligned} & \nu\{\eta : \eta(k_j) = 1 \text{ for } 1 \leq j \leq n, \eta(k) = 0 \text{ for all other } k_1 < k < k_n\} \\ & \qquad = \alpha(k_1) \prod_{i=1}^{n-1} \frac{\phi(k_i, k_{i+1})}{\alpha(k_i)} \end{aligned} \tag{3.8}$$

for $n \geq 1$ and $k_1 < k_2 < \cdots < k_n$. Identities (3.2) and (3.3) are needed for consistency, and in order to guarantee that ν concentrates on configurations with infinitely many ones in both directions. Fix a finite subset A of Z, and define

$$R(k) = \nu\{\eta : \eta(j) = 0 \text{ for all } j \in A \cap (k,\infty) \mid \eta(k) = 1\} \tag{3.9}$$

and

$$L(k) = \nu\{\eta : \eta(j) = 0 \text{ for all } j \in A \cap (-\infty, k) \mid \eta(k) = 1\} \tag{3.10}$$

for $k \in Z$. Take ν to be the initial distribution of the process, and let ν_t be the distribution at time t. Then

(3.11)

$$\begin{aligned}
&\frac{d}{dt}\nu_t\{\eta : \eta(j) = 0 \text{ for all } j \in A\}\Big|_{t=0} \\
&\quad = \sum_{k<l;k,l\notin A} L(k)R(l)\frac{d}{dt}\nu_t\{\eta : \eta(k) = \eta(l) = 1, \\
&\qquad\qquad\qquad \eta(j) = 0 \text{ for } k < j < l\}\Big|_{t=0} \\
&\quad - \sum_{j\notin A} L(j)R(j)\frac{d}{dt}\nu_t\{\eta : \eta(j) = 1\}\Big|_{t=0} \\
&\quad - \sum_{j\notin A} \lambda(j)\phi(j-1, j+1)[L(j) - L(j-1)1_{A^c}(j-1)][R(j) \\
&\qquad\qquad - R(j+1)1_{A^c}(j+1)] \\
&\quad - \sum_{j\notin A} \lambda(j)\alpha(j-1)[L(j) - L(j-1)1_{A^c}(j-1][R(j-1) - R(j)] \\
&\quad - \sum_{j\notin A} \lambda(j)\alpha(j+1)[R(j) - R(j+1)1_{A^c}(j+1)][L(j+1) - L(j)].
\end{aligned}$$

In the first sum, k is restricted to be to the left of the rightmost point in A and l is restricted to be to the right of the leftmost point in A. In the second sum, j is restricted to be between the leftmost and rightmost points in A.

The proof of Theorem 3.1 is an easy consequence of identity (3.11). Briefly, (3.4), (3.5) and (3.6) imply that the first two terms on the right of (3.11) are zero, while (3.7) implies that the remaining terms are nonpositive. Therefore, the left side of (3.11) is nonpositive. By duality, it follows that $\nu_t\{\eta : \eta(j) = 0 \text{ for all } j \in A\}$ is decreasing in t for all A, and is hence bounded above by its value at $t = 0$. Therefore, the process survives.

4. Survival in the Homogeneous Case. In this section, we will prove part (b) of the theorem by verifying the assumptions of Theorem 3.1 in case $\lambda = \mu \geq \lambda_1$, the largest root of the cubic $\lambda^3 - \lambda^2 - 3\lambda + 1$. The computations are simpler than in the periodic case, and in fact, we can solve explicitly for the required α and (the generating function of) ϕ. Let $\alpha(i) \equiv \alpha$, and define $\phi(n) = \phi(i, i+n)$. Then equations (3.2)–(3.6) become

$$\sum_{n=1}^{\infty} n\phi(n) = 1, \tag{4.1}$$

$$\sum_{n=1}^{\infty} \phi(n) = \alpha, \tag{4.2}$$

$$(1+\lambda)\phi(1) = \lambda\alpha, \tag{4.3}$$

$$(2+\lambda)\phi(2) = 2\lambda\phi(3) + \frac{\phi^2(1)}{\alpha}, \tag{4.4}$$

and

$$(2+2\lambda)\phi(n) = 2\lambda\phi(n+1) + \frac{1}{\alpha}\sum_{k=1}^{n-1}\phi(k)\phi(n-k) \tag{4.5}$$

for $n \geq 3$.

To solve these equations, introduce the generating function

$$G(s) = \frac{1}{\alpha}\sum_{n=1}^{\infty}\phi(n)s^n.$$

Provided $\phi(n)$ is absolutely summable, equations (4.1)–(4.5) can be written as $G(1) = 1$, $G'(1) = 1/\alpha$, $G'(0) = \lambda/(1+\lambda)$, and

$$\begin{aligned} 0 = {} & G^2(s) - 2[1+\lambda-\lambda s^{-1}]G(s) + 2\lambda s \\ & - [2\lambda^2/(1+\lambda)] + [\lambda s(s-2)\phi(2)/\alpha]. \end{aligned} \tag{4.6}$$

Using $G(1) = 1$ gives

$$\frac{\phi(2)}{\alpha} = \frac{\lambda-1}{\lambda(\lambda+1)},$$

so that (4.6) becomes

$$\begin{aligned} 0 = {} & (\lambda+1)G^2(s) - 2(\lambda+1)[1+\lambda-\lambda s^{-1}]G(s) \\ & + (\lambda-1)s^2 + 2s(\lambda^2+1) - 2\lambda^2. \end{aligned} \tag{4.7}$$

Solving the quadratic for $G(s)$ and doing some algebra leads to

$$sG(s) = -\lambda + (1+\lambda)s + (1-s)\sqrt{\lambda^2 - 2\lambda s - \frac{\lambda-1}{\lambda+1}s^2}. \tag{4.8}$$

Note that the expression inside the square root is ≥ 0 for all $0 \leq s \leq 1$ if and only if $\lambda \geq \lambda_1$. Furthermore, $G'(1) < \infty$, so that $\alpha > 0$. We still need to check that $\phi(n)$ is positive, and that (3.7) is satisfied. It does not seem very easy to do so directly from expression (4.8). Therefore, we will use a different approach to solve (4.1)–(4.5) which will give these properties automatically. By continuity, we may as well assume now that $\lambda > \lambda_1$.

Define $f(t,n)$ for $t \geq 0$ and $n \geq 1$ by $f(t,1) \equiv \lambda/(\lambda+1)$, $f(0,n) = 0$ for $n \geq 2$,

$$\frac{d}{dt}f(t,2) = 2\lambda f(t,3) + f^2(t,1) - (2+\lambda)f(t,2), \tag{4.9}$$

and

$$\frac{d}{dt}f(t,n) = 2\lambda f(t,n+1) + \sum_{k=1}^{n-1} f(t,k)f(t,n-k) - (2+2\lambda)f(t,n) \tag{4.10}$$

for $n \geq 3$. From the form of this evolution, it is easy to see that $f(t,n)$ is nonnegative and increasing in t for each $n \geq 2$. We will construct the solution of (4.1)–(4.5) from the limit of $f(t,n)$ as $t \to \infty$.

Let

$$F(t) = \sum_{n=1}^{\infty} f(t,n), \quad F^*(t) = \sum_{n=1}^{\infty} nf(t,n), \quad \text{and } F^{**}(t) = \sum_{n=1}^{\infty} n(n-1)f(t,n).$$

Then these functions satisfy the following differential equations:

$$\frac{d}{dt}F(t) = [1-F(t)]^2 + 2f(t,1) - 1 - \lambda f(t,2), \tag{4.11}$$

$$\frac{d}{dt}F^*(t) = 2[1-F(t)][\lambda - F^*(t)], \tag{4.12}$$

and

$$\frac{d}{dt}F^{**}(t) = 2[\lambda - F^*(t)]^2 + 2\lambda f(t,2) + 4\lambda - 2\lambda^2 + 2[F(t)-1][2\lambda + F^{**}(t)]. \tag{4.13}$$

Since F, F^*, and F^{**} are increasing in t, the right hand sides of (4.11)–(4.13) are nonnegative. Combining the nonnegativity statements for (4.11) and (4.13) gives

$$[1-F(t)]^2 - 2[1-F(t)][2\lambda + F^{**}(t)] + [\lambda - F^*(t)]^2 \geq \frac{\lambda^3 - \lambda^2 - 3\lambda + 1}{\lambda + 1} \tag{4.14}$$

for all $t \geq 0$. Since $\lambda > \lambda_1$, the right side is strictly positive, and hence $F(t) = 1$ and $F^*(t) = \lambda$ cannot occur simultaneously. On the other hand, the nonnegativity statement for (4.12) implies that $F(t) = 1$ and $F^*(t) = \lambda$ must occur simultaneously, if at all.

Therefore, we conclude that $F(t) < 1$ and $F^*(t) < \lambda$ for all t, and hence the limits

$$f(n) = \lim_{t\to\infty} f(t,n)$$

are finite and satisfy

$$\sum_{n=1}^{\infty} f(n) \le 1 \quad \text{and} \quad \sum_{n=1}^{\infty} n f(n) \le \lambda. \tag{4.15}$$

The right side of (4.12) tends to zero as $t \to \infty$, so at least one of the inequalities in (4.15) must be an equality. To decide which one, note that

$$F(t) - F^*(t) + \frac{1}{2}F^{**}(t) = \sum_{n=1}^{\infty} \frac{(n-1)(n-2)}{2} f(t,n) \tag{4.16}$$

is also increasing in t. It follows by taking the appropriate linear combination of the right sides of (4.11)–(4.13) and using $F(t) < 1$ that

$$[(1 - F(t)) - (\lambda - F^*(t))]^2 \ge \frac{\lambda^3 - \lambda^2 - 3\lambda + 1}{\lambda + 1} > 0$$

for all $t \ge 0$. So the quantity $1 - F(t) - \lambda + F^*(t)$ never changes sign. It is negative at $t = 0$, so it is always negative. Hence $1 - F(t) < \lambda - F^*(t)$ for all $t \ge 0$.

Therefore, we conclude that the first inequality in (4.15) is an equality. Since the right hand side of (4.9) and (4.10) also tend to zero as $t \to \infty$, it follows that $\phi(n) = cf(n)$ for a properly chosen constant c solves (4.1)–(4.5).

We still need to check (3.7), which in the present homogeneous setting becomes $f^2(n) \le f(n-1)f(n+1)$ for all $n \ge 2$. To do so, we will check that this logconvexity property holds for $f(t,n)$ for all t, and then pass to the limit. The logconvexity property is trivially true at $t = 0$, and it is preserved by the evolution, since the derivatives of the relevant differences can be written in the following form:

$$\begin{aligned}\frac{d}{dt}[f(t,1)f(t,3) - f^2(t,2)] = 2\lambda[f(t,1)f(t,4) - f(t,2)f(t,3)]& \\ + 2f(t,3)[f(t,1) - \lambda f(t,2)]& \\ - 2(2+\lambda)[f(t,1)f(t,3) - f^2(t,2)],&\end{aligned}$$

$$\begin{aligned}\frac{d}{dt}[f(t,2)f(t,4) - f^2(t,3)] &= 2\lambda[f(t,2)f(t,5) - f(t,3)f(t,4)] \\ &+ \lambda f(t,2)f(t,4) \\ &+ \frac{[f^2(t,2) - f(t,1)f(t,3)]^2 + f^2(t,1)[f(t,2)f(t,4) - f^2(t,3)]}{f(t,2)} \\ &- 2(2+2\lambda)[f(t,2)f(t,4) - f^2(t,3)],\end{aligned}$$

and

$$\begin{aligned}\frac{d}{dt}[f(t,n-1)f(t,n+1)-f^2(t,n)] &= 2\lambda[f(t,n-1)f(t,n+2)\\ &- f(t,n)f(t,n+1)]\\ &+ \sum_{k=1}^{n-2}\frac{[f(t,n)f(t,k)-f(t,n-1)f(t,k+1)]}{f(t,n-1)}.\\ &\cdot[f(t,n)f(t,n-k-1)-f(t,n-1)f(t,n-k)]\\ &+ \frac{f(t,n-1)f(t,n+1)-f^2(t,n)}{f(t,n-1)}\sum_{k=1}^{n-2}f(t,k)f(t,n-k-1)\\ &- 2(2+2\lambda)[f(t,n-1)f(t,n+1)-f^2(t,n)]\end{aligned}$$

for $n \geq 4$. To complete the proof, we need to check that $f(t,1)-\lambda f(t,2) \geq 0$ for all t, since that expression occurs on the right of the first expression above. But this follows by passing to the limit in (4.11) and using monotonicity:

$$\lambda f(t,2) \leq \lim_{s\to\infty}\lambda f(s,2) = \frac{\lambda-1}{\lambda+1} < \frac{\lambda}{\lambda+1} = f(t,1). \tag{4.17}$$

5. Survival in the Periodic Case. In this section, we will prove part (c) of the theorem. We assume throughout that (1.1)–(1.5) are satisfied. In verifying the conditions of Theorem 3.1, we will look for $\alpha(i)$ and $\phi(k,l)$ which are periodic of period 2 and for which $\phi(i,i+2n+1)$ is independent of i. So, let

$$\begin{gathered}\alpha=\alpha(2i+1),\quad \beta=\alpha(2i),\\ f(n)=\phi(i,i+2n+1) \text{ for } n\geq 0, \text{ and}\\ g(n)=\phi(2i,2i+2n) \text{ and } h(n)=\phi(2i+1,2i+2n+1) \text{ for } n\geq 1.\end{gathered}$$

Then equations (3.2)–(3.6) become

$$\sum_{n=0}^{\infty}(2n+1)f(n)+\sum_{n=1}^{\infty}n[g(n)+h(n)]=1, \tag{5.1}$$

$$\alpha=\sum_{n=0}^{\infty}f(n)+\sum_{n=1}^{\infty}h(n),\quad \beta=\sum_{n=0}^{\infty}f(n)+\sum_{n=1}^{\infty}g(n), \tag{5.2}$$

$$(2+\mu+\lambda)f(0)=\alpha\mu+\beta\lambda, \tag{5.3}$$

$$(5.4)\quad (2+\lambda)g(1) = 2\mu f(1) + f^2(0)/\alpha, \quad (2+\mu)h(1) = 2\lambda f(1) + f^2(0)/\beta,$$

and for $n \geq 2$,

$$(5.5)\qquad (2+\mu+\lambda)f(n-1) = \lambda g(n) + \mu h(n) + \frac{1}{\alpha}\sum_{k=1}^{n-1} f(k-1)h(n-k) + \frac{1}{\beta}\sum_{k=1}^{n-1} f(k-1)g(n-k),$$

$$(5.6)\quad (2+2\mu)h(n) = 2\lambda f(n) + \frac{1}{\alpha}\sum_{k=1}^{n-1} h(k)h(n-k) + \frac{1}{\beta}\sum_{k=0}^{n-1} f(k)f(n-k-1),$$

and

$$(5.7)\quad (2+2\lambda)g(n) = 2\mu f(n) + \frac{1}{\beta}\sum_{k=1}^{n-1} g(k)g(n-k) + \frac{1}{\alpha}\sum_{k=0}^{n-1} f(k)f(n-k-1).$$

Note that if we can solve all of the equations above except (5.1), and the left side of (5.1) is finite for that solution, then we can solve the entire system of equations by multiplying the α, β, $f(n)$, $g(n)$ and $h(n)$ of the given solution by an appropriate constant. Therefore, we might as well take $\alpha = 1$ and solve (5.2)–(5.7).

Our next task is to determine β. Add equation (5.3), the first equation in (5.4), and (5.5) and (5.7) for $n \geq 2$, using (5.2) to obtain

$$(5.8)\qquad \mu[2f(0) + h(1)] = 2\mu - \beta.$$

Similarly, we see that

$$(5.9)\qquad \lambda[2f(0) + g(1)] = 2\beta\lambda - 1.$$

Then (5.3), (5.4), (5.8) and (5.9) provide five equations in the five unknowns β, $f(0)$, $f(1)$, $g(1)$ and $h(1)$. Solving these equations gives β as a root of the cubic

$$(5.10)\qquad P(\beta) = (\lambda\beta - \mu)(\lambda\beta + \mu)^2 + \beta(2+\mu+\lambda)^2[(2+\lambda)(1+2\mu) - \beta(2+\mu)(1+2\lambda)].$$

Note that $P(0) < 0$, $P(1)$ has the same sign as $\mu - \lambda$, and $P(\mu/\lambda)$ has the same sign as $\lambda - \mu$. Therefore, P always has three positive real roots, with the middle one between 1 and μ/λ. We will fix β to be that root. This choice is suggested by the fact that if $\mu = \lambda$, we know that we should choose $\beta = 1$.

Now that we have determined β, we can in principle solve recursively for $f(n)$, $g(n)$, and $h(n)$ by first using (5.3), (5.8), (5.9) and (5.4), and then using (5.5), (5.6) and (5.7) for each n, as a (nonsingular) system of three equations in the three unknowns $f(n)$, $g(n)$, $h(n)$. It is probably impossible to determine from this procedure when the resulting solution is positive and satisfies (3.7). Therefore, we use a modification of the technique we used in the homogeneous case.

Fix the values of α and β as indicated above, let

$$f(t,0) \equiv f = \frac{\lambda\beta + \mu}{2 + \mu + \lambda}, \tag{5.11}$$

and define an evolution $\{(f(t,n), g(t,n), h(t,n)),\ t \geq 0,\ n \geq 1\}$ by

$$f(0,n) = g(0,n) = h(0,n) = 0 \quad \text{for} \quad n \geq 1,$$

$$\begin{aligned} \frac{d}{dt} g(t,1) &= 2\mu f(t,1) + \mu h(t,1) + f^2(t,0) \\ &\quad + 2\mu f(t,0) + \beta - 2\mu - (2 + \lambda) g(t,1), \end{aligned} \tag{5.12}$$

$$\begin{aligned} \frac{d}{dt} h(t,1) &= 2\lambda f(t,1) + \lambda g(t,1) + f^2(t,0)/\beta \\ &\quad + 2\lambda f(t,0) + 1 - 2\beta\lambda - (2 + \mu) h(t,1), \end{aligned} \tag{5.13}$$

and for $n \geq 2$,

$$\begin{aligned} &\frac{d}{dt} f(t,n-1) = \lambda g(t,n) + \mu h(t,n) \\ &\quad + \sum_{k=1}^{n-1} f(t,k-1) h(t,n-k) + \frac{1}{\beta} \sum_{k=1}^{n-1} f(t,k-1) g(t,n-k) \\ &\quad - (2 + \mu + \lambda) f(t,n-1), \end{aligned} \tag{5.14}$$

$$\begin{aligned} \frac{d}{dt} h(t,n) &= 2\lambda f(t,n) + \sum_{k=1}^{n-1} h(t,k) h(t,n-k) \\ &\quad + \frac{1}{\beta} \sum_{k=0}^{n-1} f(t,k) f(t,n-k-1) - (2 + 2\mu) h(t,n), \end{aligned} \tag{5.15}$$

and

$$\begin{aligned} \frac{d}{dt} g(t,n) &= 2\mu f(t,n) + \sum_{k=0}^{n-1} f(t,k) f(t,n-k-1) \\ &\quad + \frac{1}{\beta} \sum_{k=1}^{n-1} g(t,k) g(t,n-k) - (2 + 2\lambda) g(t,n). \end{aligned} \tag{5.16}$$

Note that the right sides of (5.12) and (5.13) are not direct analogues of (4.9), and are not the first choices one would make in attempting to solve (5.4). However, it appears that the modification is necessary in order to make the rest of the proof work. (The more "natural" choices will be used later in verifying (3.7).)

All the coefficients of the unknown functions on the right side of (5.12)–(5.16) are positive, except the last one in each case, and that one corresponds to the function whose derivative occurs on the left side. Furthermore, the inhomogeneous terms in (5.12) and (5.13) are nonnegative by assumptions (1.1) and (1.2). Therefore $f(t,n)$, $g(t,n)$ and $h(t,n)$ are nonnegative and increasing in t for every n. The solution to (5.2)–(5.7) will be constructed from the limits as $t \to \infty$ of these functions, so we next need to show that these limits are finite.

For this purpose, let

$$F(t) = \sum_{n=0}^{\infty} f(t,n), \; G(t) = \sum_{n=1}^{\infty} g(t,n), \; H(t) = \sum_{n=1}^{\infty} h(t,n),$$

$$F^*(t) = \sum_{n=0}^{\infty} nf(t,n), \; G^*(t) = \sum_{n=1}^{\infty} ng(t,n), \; H^*(t) = \sum_{n=1}^{\infty} nh(t,n),$$

and

$$F^{**}(t) = \sum_{n=0}^{\infty} n(n-1)f(t,n), \quad G^{**}(t) = \sum_{n=1}^{\infty} n(n-1)g(t,n),$$

$$H^{**}(t) = \sum_{n=1}^{\infty} n(n-1)h(t,n).$$

Summing the appropriate subsets of (5.12)–(5.16) after multiplying by the appropriate function of n, leads to the following differential equations for these functions:

(5.17)
$$\begin{aligned}\frac{d}{dt}F(t) = {} & \lambda G(t) + \mu H(t) + F(t)H(t) \\ & + \frac{1}{\beta}F(t)G(t) - (2+\mu+\lambda)F(t) \\ & + (2+\mu+\lambda)f(t,0) - \lambda g(t,1) - \mu h(t,1),\end{aligned}$$

(5.18)
$$\begin{aligned}\frac{d}{dt}H(t) = {} & 2\lambda F(t) + H^2(t) + \frac{1}{\beta}F^2(t) \\ & - (2+2\mu)H(t) + \mu h(t,1) + \lambda g(t,1) + 1 - 2\beta\lambda,\end{aligned}$$

(5.19)
$$\begin{aligned}\frac{d}{dt}G(t) = {} & 2\mu F(t) + F^2(t) + \frac{1}{\beta}G^2(t) \\ & - (2+2\lambda)G(t) + \lambda g(t,1) + \mu h(t,1) + \beta - 2\mu,\end{aligned}$$

(5.20)
$$\frac{d}{dt}F^*(t) = \lambda G^*(t) + \mu H^*(t) - \lambda G(t) - \mu H(t) + F^*(t)[H(t) + \beta^{-1}G(t)] + F(t)[H^*(t) + \beta^{-1}G^*(t)] - (2+\mu+\lambda)F^*(t),$$

(5.21)
$$\frac{d}{dt}H^*(t) = 2\lambda F^*(t) + 2H^*(t)H(t) + \frac{1}{\beta}F(t)[2F^*(t) + F(t)] - (2+2\mu)H^*(t) + \mu h(t,1) + \lambda g(t,1) + 2\lambda f(t,0) + 1 - 2\beta\lambda,$$

(5.22)
$$\frac{d}{dt}G^*(t) = 2\mu F^*(t) + F(t)[2F^*(t) + F(t)] + \frac{2}{\beta}G^*(t)G(t) - (2+2\lambda)G^*(t) + \mu h(t,1) + \lambda g(t,1) + 2\mu f(t,0) + \beta - 2\mu,$$

(5.23)
$$\frac{d}{dt}F^{**}(t) = G^{**}(t)[\lambda + \beta^{-1}F(t)] + H^{**}(t)[\mu + F(t)] + F^{**}(t)[H(t) + \beta^{-1}G(t) - (2+\mu+\lambda)] - 2\lambda G^*(t) - 2\mu H^*(t) + 2F^*(t)H^*(t) + 2\beta^{-1}F^*(t)G^*(t) + 2\lambda G(t) + 2\mu H(t),$$

(5.24)
$$\frac{d}{dt}H^{**}(t) = 2F^{**}(t)[\lambda + \beta^{-1}F(t)] + 2H^{**}(t)[H(t) - 1 - \mu] + 2[H^*(t)]^2 + 2\beta^{-1}F^*(t)[F^*(t) + 2F(t)],$$

(5.25)
$$\frac{d}{dt}G^{**}(t) = 2F^{**}(t)[\mu + F(t)] + 2G^{**}(t)[\beta^{-1}G(t) - 1 - \lambda] + 2\beta^{-1}[G^*(t)]^2 + 2F^*(t)[F^*(t) + 2F(t)].$$

The functions whose derivatives appear on the left sides of (5.17)–(5.25) are all increasing in t, so the right sides of (5.17)–(5.25) are nonnegative. Adding (5.17) to (5.18) and rearranging (using (5.11)) gives

(5.26)
$$\frac{d}{dt}[F(t) + H(t)] = [1 - F(t) - H(t)]^2 + [1 - F(t) - H(t)][\mu + F(t)] - [\beta - F(t) - G(t)][\lambda + \beta^{-1}F(t)],$$

while adding (5.17) to (5.19) gives

$$\frac{d}{dt}[F(t)+G(t)] = \beta^{-1}[\beta - F(t) - G(t)]^2 - [1 - F(t) - H(t)][\mu + F(t)] + [\beta - F(t) - G(t)][\lambda + \beta^{-1}F(t)]. \tag{5.27}$$

Note that

$$P\left(\frac{2+\lambda}{\lambda}\right) < 0 < P\left(\frac{\mu}{2+\mu}\right),$$

so that

$$\frac{\mu}{2+\mu} < \beta < \frac{2+\lambda}{\lambda}. \tag{5.28}$$

Therefore $F(0)+H(0) = f < 1$ and $F(0)+G(0) = f < \beta$. It then follows from (5.26) and (5.27) that the vector $(1 - F(t) - H(t),\ \beta - F(t) - G(t))$ starts in the first quadrant and can only leave it through the origin. On the other hand, using (5.26) and (5.27) again, we see that while it is in the first quadrant,

$$v(t) = 1 - F(t) - H(t) + \beta - F(t) - G(t)$$

satisfies

$$\frac{v^2(t)}{\beta+1} \le -v'(t) \le \frac{(\beta \vee 1)v^2(t)}{\beta}.$$

Therefore, $v(t)$ is always positive and tends to zero as $t \to \infty$. It follows that

$$\lim_{t\to\infty}[F(t)+G(t)] = \beta \quad \text{and} \quad \lim_{t\to\infty}[F(t)+H(t)] = 1. \tag{5.29}$$

Since these limits are finite, we can pass to the limit in (5.17)–(5.19) and in (5.12)–(5.16), to conclude that the limits as $t \to \infty$ of $F(t)$, $G(t)$, $H(t)$, $f(t,n)$, $g(t,n)$ and $h(t,n)$ satisfy the identities obtained by setting the right sides of these equations equal to zero. We will denote these limits by F, G, H, $f(n)$, $g(n)$ and $h(n)$ respectively. The limits satisfy (5.2), (5.3), and (5.5)–(5.7). (The advantage of using this evolution instead of the one in which (5.12) and (5.13) are replaced by (5.28) and (5.29) is that one can prove (5.29) from (5.17)–(5.19) alone, without using (5.20)–(5.25).)

It remains to check (5.4), and to check that the left side of (5.11) is finite for our solution. In order to do both, we will show that $F^*(t)$, $G^*(t)$ and $H^*(t)$ remain bounded as $t \to \infty$. We will then be able to pass to the limit in (5.20)–(5.22). Add the right sides of (5.21), (5.22) and (5.17) to twice (5.20), using the fact that these expressions are nonnegative and that $F(t)+G(t)\le\beta$ and $F(t)+H(t)\le 1$ for all t to conclude that

$$\begin{aligned}(1+\beta^{-1})F^2(t)&+F(t)H(t)+\beta^{-1}F(t)G(t)\\&-(2+\lambda+\mu)F(t)-\lambda G(t)-\mu H(t)\\&+2(\lambda+\mu)f+\mu h(t,1)+\lambda g(t,1)+1+\beta-\beta\lambda-\mu\ge 0\end{aligned}$$

for all t. Passing to the limit in t and using (5.29) and the definition of f gives

$$\mu h(1)+\lambda g(1)\ge 4f-1-\beta. \tag{5.30}$$

Setting the limit of the right side of (5.17) equal to zero and using (5.29) and the definition of f gives

$$(1+\beta^{-1})F^2+2(\mu+\lambda)F+\mu h(1)+\lambda g(1)-2(\lambda\beta+\mu)=0.$$

Let ϕ be the positive solution of the quadratic equation

$$(1+\beta^{-1})\phi^2+2(\mu+\lambda)\phi-1-\beta-2(\mu+\lambda)f=0. \tag{5.31}$$

Using (5.30) and the definition of f, and comparing the two quadratics above, we conclude that $F\le\phi$.

Now let $U(t)=F^*(t)+G^*(t)$ and $V(t)=F^*(t)+H^*(t)$. By the analogues of (4.16) for G and H,

$$2F^{**}(t)+G^{**}(t)+H^{**}(t)-2G^*(t)+2G(t)$$

and

$$2F^{**}(t)+G^{**}(t)+H^{**}(t)-2H^*(t)+2H(t)$$

are both increasing in t. Applying this observation to the appropriate linear combinations of the right sides of (5.18)–(5.25) and using $F(t)+H(t)\le 1$ and $F(t)+G(t)\le\beta$ yields respectively the inequalities

$$[V(t)-\mu]^2+\beta^{-1}[U(t)+\beta-G(t)]^2\ge\mu^2+\beta+2\mu f-2\mu[F(t)+H(t)] \tag{5.32}$$

and

$$[V(t)-H(t)+1]^2+\beta^{-1}[U(t)-\lambda\beta]^2\ge\beta\lambda^2+1+2\lambda f-2\lambda[F(t)+G(t)]. \tag{5.33}$$

These inequalities can be interpreted as saying that the point $(V(t),\ \beta^{-1/2}U(t))$, which starts at $(0,0)$, always stays outside two (moving) circles which have their centers in the fourth and second quadrants respectively. This point will be trapped by the circles, and hence will remain bounded, provided that these circles intersect at all times — see the following figure.

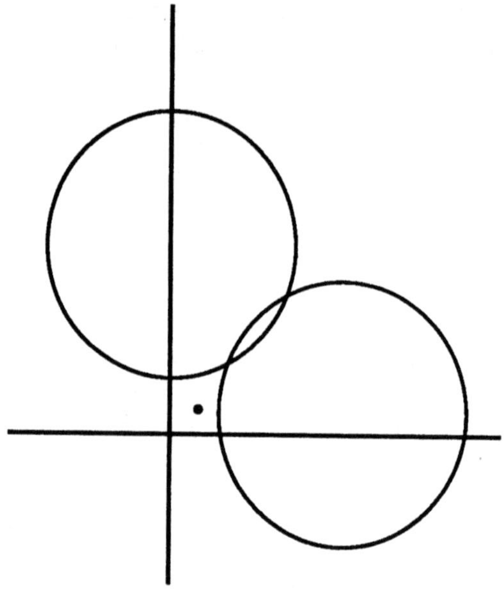

These circles intersect provided that the sum of their radii exceeds the distance between their centers, which in this case is the condition

$$(5.34)\qquad \begin{aligned}&\sqrt{\mu^2+\beta+2\mu[f-F(t)-H(t)]}\\ &\quad+\sqrt{\beta\lambda^2+1+2\lambda[f-F(t)-G(t)]}\\ &\quad>\sqrt{[\mu+1-H(t)]^2+\beta^{-1}[\beta\lambda+\beta-G(t)]^2}.\end{aligned}$$

To check that this holds for all t, regard the left side of (5.34) minus the right side of (5.34) as a function of (F, G, H) in the region determined by $f \le F \le \phi$, $0 \le G \le \beta - F$, and $0 \le H \le 1 - F$. We know that these inequalities are satisfied at all times by $(F(t), G(t), H(t))$. This function is clearly decreasing in F, and it is decreasing in G and H by assumptions (1.4) and (1.3) respectively, as can be checked by computing the relevant derivatives. Thus (5.34) will be satisfied for all t if it is satisfied when $F = \phi$, $G = \beta - F$ and $H = 1 - F$, i.e. if

$$\begin{aligned}\sqrt{\mu^2+\beta+2\mu(f-1)}+\sqrt{\beta\lambda^2+1-2\lambda(f-\beta)}\\ \ge\sqrt{(\mu+\phi)^2+\beta^{-1}(\beta\lambda+\phi)^2}.\end{aligned}$$

But using (5.31), it easy to check that this is just assumption (1.5) in disguise. Hence $U(t)$ and $V(t)$ remain bounded as $t \to \infty$.

Since $F^*(t)$, $G^*(t)$ and $H^*(t)$ have finite limits as $t \to \infty$, we can pass to the limit (5.20)–(5.22), to conclude that these limits satisfy the equations obtained from these by setting the right sides equal to zero. Adding (5.21), (5.22) and (5.17) to twice (5.20) and passing to the limit gives

$$(5.35)\qquad 0=\mu h(1)+\lambda g(1)+2(\mu+\lambda)f+1+\beta-2\mu-2\beta\lambda.$$

On the other hand, taking λ (5.12)–μ (5.13) and passing to the limit leads to

$$(5.36) \qquad 0=(2+\mu+\lambda)(\mu h(1)-\lambda g(1))+f^2(0)(\lambda-\beta^{-1}\mu) \\ +\ \lambda\beta-\mu-2\mu\lambda+2\beta\lambda\mu.$$

Noting that the statement $P(\beta)=0$ can be written as

$$f^2(0)(\lambda-\beta^{-1}\mu)=\beta(2+\mu)(1+2\lambda)-(2+\lambda)(1+2\mu),$$

one can use (5.35) and (5.36) to solve for $h(1)$ and $g(1)$, and hence to show that they satisfy (5.8) and (5.9). Passing to the limit in (5.12) and (5.13) then gives (5.4). Thus, we have solved (5.2)–(5.7) as required.

We still need to verify that our solution satisfies (3.7), which in the present context becomes

$$(5.37) \qquad f(n-1)f(n)\geq g(n)h(n),\ g(n)h(n+1)\geq f^2(n), \\ \text{and } h(n)g(n+1)\geq f^2(n)$$

for $n\geq 1$. Unfortunately, the evolution (5.12)–(5.16) does not preserve these inequalities. Therefore, we define a new evolution which does. Let $\{(f^*(t,n-1),\ g^*(t,n),\ h^*(t,n)),\ t\geq 0,\ n\geq 1\}$ have the same initial values as $\{(f(t,n-1),\ g(t,n),\ h(t,n)),\ t\geq 0,\ n\geq 1\}$ and satisfy

$$(5.38) \qquad \frac{d}{dt}g^*(t,1)=2\mu f^*(t,1)+f^{*2}(t,0)-(2+\lambda)g^*(t,1),$$

$$(5.39) \qquad \frac{d}{dt}h^*(t,1)=2\lambda f^*(t,1)+f^{*2}(t,0)/\beta-(2+\mu)h^*(t,1),$$

and (5.14)–(5.16) for $n\geq 2$ (with f, g, and h replaced by f^*, g^*, and h^*).

Note that

$$\mu h(t,1)+2\mu f(t,0)+\beta-2\mu\leq 0$$

and

$$\lambda g(t,1)+2\lambda f(t,0)+1-2\beta\lambda\leq 0,$$

since the two quantities on the left are increasing in t and tend to zero by (5.8) and (5.9). Therefore, comparing the two evolutions and using the fact that $(f(n),\ g(n),\ h(n))$ is a fixed point for both evolutions, we see that

$$f(t,n)\leq f^*(t,n)\leq f(n),\ g(t,n)\leq g^*(t,n)\leq g(n),$$

and

$$h(t,n)\leq h^*(t,n)\leq h(n)$$

for all n and t. It follows that f^*, g^*, and h^* have the same limits as f, g, and h. Therefore, it suffices to prove that the second evolution preserves the inequalities in (5.37). This verification will be omitted, since it is the same as the corresponding proof in the homogeneous case which is given at the end of Section 4. The analogue of (4.17) is

$$\lambda g(1) + \mu h(1) \leq 2f(0).$$

To check this, use (5.8), (5.9) and (5.11) to reduce it to

$$2(\lambda\beta + \mu) \leq (\beta + 1)(2 + \lambda + \mu),$$

which is a consequence of (5.28).

Acknowledgment. I would like to thank Rick Durrett for telling me about his results with Ted Cox on the threshold voter model, and for asking the questions which led to this paper. He also provided me with the Dickman and Burschka reference.

REFERENCES

[1] J.T. Cox and R. Durrett, *Nonlinear voter models*, this volume, 1991.

[2] R. Dickman and M.A. Burschka, *Nonequilibrium critical poisoning in a single species model,* Phys. Letters A **127** (1988), 132–137.

[3] R. Holley and T.M. Liggett, *The survival of contact processes,* Ann. Prob. **6** (1978), 198–206.

[4] R. Holley and T.M. Liggett, *Generalized potlatch and smoothing processes,* Z. Wahr. verw. Gebiete **55** (1981), 165–195.

[5] T.M. Liggett, *Interacting Particle Systems,* Springer-Verlag, NY, 1985.

[6] T.M. Liggett, *Spatially inhomogeneous contact processes.* In *Spatial Stochastic Processes,* A Festschrift in honor of the Seventieth Birthday of Ted Harris, Birkhäuser, Boston, 1991.

[7] F. Spitzer, *Interaction of Markov Processes,* Adv. Math. **5** (1970), 246–290.

[8] F. Spitzer, *Stochastic time evolution of one dimensional interacting particle systems,* Bull. Amer. Math. Soc. **83** (1977), 880–890.

[9] F. Spitzer, *Infinite systems with locally interacting components,* Ann. Prob. **9** (1981), 349–364.

Department of Mathematics
University of California, Los Angeles
Los Angeles, CA 90024

Bounds on the Critical Exponent of Self-Avoiding Polygons

NEAL MADRAS *

We obtain rigorous upper bounds on the critical exponent for self-avoiding polygons using probabilistic methods based on renewal processes and random walks.

1. Introduction. An *N-step self-avoiding walk* (SAW) is an ordered sequence of *distinct* points $(\omega_0, \ldots, \omega_N)$ of the d-dimensional integer lattice $\mathbf{Z}^d$ $(d \geq 2)$ such that consecutive points are unit distance apart. For a fixed nonzero point z in $\mathbf{Z}^d$, let $S_N(z)$ be the set of N-step SAWs with $\omega_0 = 0$ and $\omega_N = z$. If z is a nearest neighbour of 0, then a SAW in $S_N(z)$ may be viewed as a *self-avoiding polygon*, that is, a walk on the lattice which begins and ends at the same point and is otherwise self-avoiding. A more precise definition will be given below.

The main result of this paper is a set of rigorous dimension-dependent upper bounds on the critical exponent for self-avoiding polygons (see equations (14) - (16) below). Before we get to the statement of the bounds, we will discuss some background and history of the problem.

The SAW is a model of considerable interest in theoretical physics. In physical chemistry, it serves as a model of a flexible linear polymer chain in solution (Whittington [19]). In particular, self-avoiding polygons in three dimensions have been used to model closed chains of DNA and other molecules. In statistical mechanics, the SAW exhibits scaling behaviour and critical phenomena analogous to that seen in the Ising model and other systems.

My own interest in this problem began while I was a graduate student at Cornell. At Frank Spitzer's suggestion, I attended a statistical mechanics course given by Michael Fisher, where some properties of the self-avoiding walk were discussed. In spite of its apparent simplicity, very little is known rigorously about this model. Two fundamental open questions were mentioned in Frank's book *Principles of Random Walk* [17, p. 105-6 and p. 168], but virtually no progress has been made on either of them in the years since.

Let c_N denote the number of N-step SAWs with $\omega_0 = 0$, and let $c_N(z)$ denote the number of SAWs in $S_N(z)$. Hammersley ([6],[7]) proved that

* Supported by a University Research Fellowship from NSERC of Canada.

there exists a constant $\mu > 1$ such that

$$\lim_{N\to\infty} (c_N)^{1/N} = \mu \tag{1}$$

and that for each fixed nonzero $z = (z^{(1)}, \ldots, z^{(d)})$ in $\mathbf{Z}^d$,

$$\lim_{N\to\infty} (c_N(z))^{1/N} = \mu. \tag{2}$$

(In (2), N is restricted to be of the same parity as $|z^{(1)}| + \cdots + |z^{(d)}|$. This will be implicitly adopted as a convention for the remainder of the paper.)

The above results rely on the following well-known elementary lemma.

Lemma 1.1 (Subadditivity) *Suppose $\{h_n\}$ is a "subadditive" sequence of real numbers, that is*

$$h_{n+m} \le h_n + h_m \text{ for all } m, n \ge 1. \tag{3}$$

Then there exists a constant H such that

$$\lim_{n\to\infty} \frac{h_n}{n} = H; \tag{4}$$

in fact,

$$H = \inf_{n\ge 1} \frac{h_n}{n}. \tag{5}$$

Since every $(N+M)$-step SAW ω can be split into an N-step SAW $(\omega_0, \ldots, \omega_N)$ and an M-step SAW $(\omega_N, \ldots, \omega_{N+M})$ which uniquely determine ω, it follows that

$$c_{N+M} \le c_N c_M. \tag{6}$$

Taking logarithms of (6) allows us to apply the subadditivity lemma, obtaining (1) with $\mu = \exp H$; as a bonus, (5) gives the bound

$$\mu^N \le c_N \text{ for all } N. \tag{7}$$

A similar argument can be applied to polygons, but with all of the inequalities reversed. Before discussing this, let us define our terms precisely. A *bond* is a line segment (undirected) joining two points of $\mathbf{Z}^d$ that are unit distance apart. For even $N \ge 4$, an *N-step self-avoiding polygon* is a set $\mathcal{P}$ of N bonds with the following property: there exists a corresponding SAW $(\omega_0, \ldots, \omega_{N-1})$ having $|\omega_{N-1} - \omega_0| = 1$ such that $\mathcal{P}$ consists of precisely the bond joining ω_{N-1} to ω_0 and the $N-1$ bonds joining ω_{i-1} to ω_i $(i = 1, \ldots, N-1)$. (Observe that each N-step self-avoiding polygon has exactly $2N$ corresponding SAWs.) Thus, we can think of an N-step self-avoiding polygon as a self-avoiding closed curve consisting of N bonds, with neither the starting point nor the orientation specified. Two N-step self-avoiding polygons are said to be *equivalent up to translation* if there is a vector v in $\mathbf{Z}^d$ such that translation by v defines a bijection from the set of bonds of one polygon to the set of bonds of the other polygon. Let u_N be the number of distinct eqivalence classes up to translation of N-step self-avoiding polygons. (Informally, u_N is the number of N-step polygons modulo translation.) Then for any nearest neighbour e of the origin,

$$2Nu_N = 2dc_{N-1}(e) \text{ for all even } N \geq 4. \tag{8}$$

(E.g.: for $d = 2$: $u_4 = 1$, $u_6 = 2$, $c_3((1,0)) = 2$, and $c_5((1,0)) = 6$.)

One can join an N-step polygon to an M-step polygon to get an $(N + M)$-step polygon (see Hammersley [7] for details), and so

$$u_{N+M} \geq u_N u_M/(d-1) \text{ for all even } N, M \geq 4. \tag{9}$$

Observe that the direction of the inequality in (9) is opposite that of (6). By subadditivity, $\lim_{N\to\infty}(u_N/(d-1))^{1/N}$ exists and equals $\sup_{N\geq 1}(u_N/(d-1))^{1/N}$; Hammersley [7] showed that both equal μ. This gives us the upper bound

$$u_N \leq (d-1)\mu^N \tag{10}$$

and, from (8),

$$c_N(e) \leq N\mu^N \text{ if } |e| = 1. \tag{11}$$

The actual scaling behaviour of SAWs is believed to be as follows:

$$c_N \approx \mu^N N^{\gamma-1} \tag{12}$$

$$c_N(z) \approx \mu^N N^{\alpha_{sing}-2} \text{ for fixed nonzero } z \tag{13}$$

as $N \to \infty$. Here γ and α_{sing} are *critical exponents* which, unlike μ, are believed to be universal among all lattices of the same dimension d. We can summarize (7) and (11) by the inequalities $\gamma \geq 1$ and $\alpha_{sing} \leq 3$. The value of γ is believed to be exactly 43/32 in two dimensions (Nienhuis [13], [14]), approximately 1.16 in three dimensions (Guttmann [5]), Le Guillou and Zinn-Justin [11]), 1 in four dimensions (with logarithmic corrections to c_N), and 1 in more than four dimensions. Slade [16] proved that γ equals 1 in sufficiently high dimensions, but aside from this, virtually nothing has been proven rigorously. It is not rigorously known that the scaling behaviours (12) and (13) are correct; in fact, it is a difficult open problem just to prove that γ is *finite*.

We are in a slightly better position when it comes to α_{sing}, since there is a lower bound on the number of polygons which essentially says $\alpha_{sing} \geq -6d - 6$ (see Theorem 3 of Kesten [10]). As for upper bounds, however, the only rigorous result in the literature is $\alpha_{sing} \leq 3$, as described above. The value of $\alpha_{sing} - 2$ is believed to be exactly -1.5 in two dimensions (Nienhuis [13]), approximately -1.77 in three dimensions (Guttmann [5], Madras and Sokal [12], Le Guillou and Zinn-Justin [11]), and $-d/2$ in four or more dimensions. (These values depend on the conjectured "hyperscaling relation" $\alpha_{sing} - 2 = -d\nu$, where $N^{2\nu}$ is the scaling behavior of $|\omega_N - \omega_0|^2$ for a randomly chosen N-step SAW.)

To get a feeling for these quantities, consider ordinary simple random walk in $\mathbf{Z}^d$. Then $c_N = (2d)^N$, so $\mu = 2d$ and $\gamma = 1$. Also, the probability that the random walk is at a given point z after N steps scales like $N^{-d/2}$, so $c_N(z) \approx (2d)^N N^{-d/2}$, that is, $\alpha_{sing} - 2 = -d/2$ for simple random walk. These are the "mean field exponents" for self-avoiding walk, and should be the actual exponents in high dimensions, where self-avoidance is a less serious constraint. In low dimensions, however, it should be harder for a SAW to return to a specified site, and we expect $\alpha_{sing} - 2$ to be *less* than $-d/2$. The above example also indicates why α_{sing} is a more interesting quantity than μ: the latter depends on the details of the lattice, but the former depends only on the dimension.

In this paper we will prove the following bounds on $\alpha_{sing} - 3$, the scaling exponent for u_N (see (8) and (13)):

$$\alpha_{sing} - 3 \begin{cases} \leq -1/2 & \text{if } d = 2; \\ \leq -1 & \text{if } d = 3; \\ < -1 & \text{if } d > 3. \end{cases} \tag{14}$$

Since the existence of scaling exponents has not been rigorously proven, we must say exactly what we mean by (14). The simplest description is in terms of the generating function $U(x)$ of polygons:

$$U(x) := \sum_{N=1}^{\infty} \left(\frac{u_N}{\mu^N}\right) x^N \tag{15}$$

The bound (10) says that as x increases to 1, $U(x)$ cannot diverge faster than $O((1-x)^{-1})$. We will prove that there exists a constant A, depending on d, such that

$$U(x) \leq \sum_{N=1}^{\infty} A N^{-(d-1)/2} x^N \quad \text{for all } 0 \leq x \leq 1. \tag{16}$$

Thus as x increases to 1, $U(x)$ has at most a square root divergence in two dimensions, at most a logarithmic divergence in three dimensions, and is bounded above three dimensions. This is one interpretation of (14); more general interpretations will be given in Section 3. Of course, these bounds are still far from the conjectured values; in particular, $U(1)$ is expected to be finite in every dimension.

Two theorems of classical probability theory play important roles in this paper. One is the Basic Renewal Theorem (see, for example, Feller [4], p. 330). The other is the following bound for r-dimensional random walk (Spitzer [17], Proposition 7.6). Let $\{Y_i\}$ be an independent, identically distributed sequence of $\mathbf{Z}^r$-valued random variables, satisfying the condition

$$\text{(17)} \qquad \Pr\{Y_i \in Q\} < 1 \text{ for every hyperplane } Q$$

(this includes hyperplanes not passing through the origin). Then there exists a finite constant C (independent of k) such that

$$\text{(18)} \qquad \Pr\{Y_1 + \cdots + Y_k = 0\} \leq Ck^{-r/2}.$$

This can be viewed as a "mean field" bound, since the right-hand side is the exact rate when Y corresponds to the ordinary nearest-neighbour random walk in $\mathbf{Z}^r$. In our case, r will equal $d-1$, and the condition (17) will be obvious because every point of $\mathbf{Z}^{d-1}$ will be a possible value of Y_i.

The paper is organized as follows. Section 2 defines "bridges", a special subset of SAWs, and shows how they bring renewal theory and random walks into the picture. Section 3 shows how to relate bridges to self-avoiding polygons, and contains the proof of the main results. Finally, Section 4 discusses some related work of J.T. Chayes and L. Chayes.

2. Bridges. Bridges form an important subset of SAWs that is frequently useful in the rigorous analysis of SAWs. An *N-step bridge* is a SAW $(\omega_0, \ldots, \omega_N)$ such that $\omega_0 = 0$ and $\omega_0^{(1)} < \omega_i^{(1)} \leq \omega_N^{(1)}$ for all $i = 1, \ldots, N$. Let b_N be the number of N-step bridges. For convenience, we define $b_0 = 1$.

We will now discuss the decomposition of bridges. Given an N-step bridge ω, suppose that there exists an i $(0 < i \leq N)$ such that

$$\text{(19)} \qquad \omega_j^{(1)} \leq \omega_i^{(1)} < \omega_k^{(1)} \text{ for all } j = 0, \ldots, i \text{ and } k = i+1, \ldots, N.$$

Then ω can be decomposed into two smaller bridges, $(\omega_0, \ldots, \omega_i)$ and $(\omega_i, \ldots, \omega_N)$. Conversely, since any two bridges can be concatenated to make a third bridge, it is clear that $b_{N+M} \geq b_N b_M$ (c.f. (9)). Once again, subadditivity says that $\lim_{N\to\infty}(b_N)^{1/N}$ exists and equals $\sup_{N\geq 1}(b_N)^{1/N}$, and as should be no surprise, both equal μ (it follows, for example, from Hammersley and Welsh [8]). Thus we obtain the upper bound

$$\text{(20)} \qquad b_N \leq \mu^N \text{ for all } N.$$

We say that an N-step bridge is "irreducible" if the only i $(0 < i \leq N)$ for which (19) holds is $i = N$. Let I_k denote the number of irreducible k-step bridges. Given an N-step bridge ω, let k be the smallest index i for which (19) holds. Then $(\omega_0, \ldots, \omega_k)$ is an irreducible bridge. It is now straightforward to see that

$$b_N = \sum_{k=1}^{N} I_k b_{N-k}. \tag{21}$$

This may be transformed into a probabilistic renewal equation as follows. Let $p_k = I_k \mu^{-k}$ (for $k \geq 1$) and $a_N = b_N \mu^{-N}$ (for $N \geq 0$). (Observe that $a_0 = 1$, and $p_k \leq a_k \leq 1$ by (20).) Then multiplying (19) by μ^{-N} yields

$$a_N = \sum_{k=1}^{N} p_k a_{N-k}. \tag{22}$$

Kesten [9] wrote down these equations and proved that $\sum_{k=1}^{\infty} p_k = 1$. Therefore, if we have an i.i.d. sequence of random variables $X_1, X_2, \ldots$ with common distribution $\Pr\{X = k\} = p_k$, then

$$a_N = \Pr\{X_1 + \ldots + X_k = N \text{ for some } k \geq 0\}. \tag{23}$$

Remark. The Basic Renewal Theorem implies that $\lim_{N\to\infty} b_N/\mu^N$ exists and equals $(\sum_{k=1}^{\infty} k p_k)^{-1}$. This limit is believed to be 0, but proving it is an open problem.

The preceding paragraph focused on the first coordinates of the points of a bridge. To obtain more detailed information, we will also want to consider the remaining $d-1$ coordinates. For z in $\mathbf{Z}^{d-1}$, let $b_n(z)$ (respectively, $I_n(z)$) denote the number of n-step bridges (respectively, irreducible bridges) with $(\omega_n^{(2)}, \ldots, \omega_n^{(d)}) = z$. Then we have the following refinement of (21):

$$b_N(z) = \sum_{k=1}^{N} \sum_{v \in \mathbf{Z}^{d-1}} I_k(v) b_{N-k}(z - v). \tag{24}$$

Now define the random vector (X, Y) such that X takes values in $\{1, 2, \ldots\}$ and Y takes values in $\mathbf{Z}^{d-1}$, and

$$\Pr\{(X,Y) = (k,z)\} = \frac{I_k(z)}{\mu^k} \text{ for } k \geq 1, z \in \mathbf{Z}^{d-1}. \tag{25}$$

Observe that the marginal distribution of X is the same as in the preceding paragraph. Let $\{(X_n, Y_n) : n \geq 1\}$ be a sequence of independent copies of (X, Y). Then

$$\frac{b_N(z)}{\mu^N} = \Pr\{X_1 + \ldots + X_k = N \text{ and } Y_1 + \ldots + Y_k = z \text{ for some } k\}. \tag{26}$$

(This follows by iteration of (24), obtaining

$$b_N(z) = \sum_{j=1}^{N} \left[\sum \prod_{i=1}^{j} I_{n_i}(v_i) \right]$$

where the inner sum is over all $n_1, \ldots, n_j \geq 1$ and all $v_1, \ldots, v_j$ in $\mathbf{Z}^{d-1}$ such that $n_1 + \ldots + n_j = N$ and $v_1 + \ldots + v_j = z$.)

We remark that this multidimensional renewal process can be thought of as the following "bridge-building" process. Choose a random sequence of "pieces" $\omega^{[1]}$, $\omega^{[2]}, \ldots$ independently from the common distribution which assigns probability μ^{-m} to each m-step irreducible bridge. Let X_i be the number of steps in the i-th piece, and let Y_i be the vector in $\mathbf{Z}^{d-1}$ consisting of the second through d-th coordinates of the last point of the i-th piece. Then the joint distribution of (X_i, Y_i) is given by (25). One can now build an "infinite bridge" ω by concatenating the pieces in order. Observe that $b_n(z)/\mu^n$ is the probability that $\omega_i^{(1)} \leq \omega_n^{(1)} < \omega_j^{(1)}$ for all $i < n$ and all $j > n$ *and* that $(\omega_n^{(2)}, \ldots, \omega_n^{(d)}) = z$.

3. Proof of the Main Result. In this section we will prove the inequality (16), as an immediate corollary of the following two results.

Lemma 3.1. *$u_N \leq b_N(0)d/2$ for all even $N \geq 4$.*

Proposition 3.2. *Let $\{q_N\}$ be a non-increasing sequence of nonnegative real numbers. Then there exists a constant C (depending only on the dimension d) such that*

$$\sum_{N=1}^{\infty} q_N \left(\frac{b_N(0)}{\mu^N} \right) \leq \sum_{N=1}^{\infty} C q_N N^{-(d-1)/2}. \tag{27}$$

Proof of Lemma 3.1: There is a simple geometric picture behind this lemma: Cut the polygon at a site with minimal $x^{(1)}$ coordinate and at a site with maximal $x^{(1)}$ coordinate. Reflect one of the two pieces through the hyperplane of the maximal $x^{(1)}$ coordinate. If we do this properly, the result (after a translation) is a bridge with *both* endpoints on the $x^{(1)}$ axis; moreover, this transformation will be one-to-one.

Now for the details. For even $N \geq 4$, let S'_N denote the set of $(N-1)$-step SAWs with $\omega_0 = 0$, $|\omega_{N-1}| = 1$, $\omega_1 = (1, 0, \ldots, 0)$, and $\omega_m^{(1)} \geq 0$ for $m = 0, \ldots, N-1$. Then the cardinality $|S'_N|$ of this set is greater than or equal to the number of self-avoiding polygons (modulo translation) which do not lie in a hyperplane $x_1 =$ constant; therefore, by symmetry,

$$|S'_N| \geq \frac{2}{d} u_N. \tag{28}$$

Now, given a SAW ω in S'_N, define a new SAW v as follows. Let

$$M = \max\{\omega_i^{(1)} : 0 \le i \le N-1\}$$

and choose subscripts $r < s$ such that $\omega_r^{(1)} = 0$, $\omega_s^{(1)} = M$, and $0 < \omega_i^{(1)} < M$ for every $i = r+1, \ldots, s-1$. For $u \in \mathbf{Z}^d$, let $T_M(u)$ denote the reflection of u in the hyperplane $x^{(1)} = M$:

$$T_M(u^{(1)}, u^{(2)}, \ldots, u^{(d)}) = (2M - u^{(1)}, u^{(2)}, \ldots, u^{(d)}).$$

Define

$$\psi_i = \begin{cases} \omega_{r+i} & \text{if } 0 \le i \le s-r; \\ T_M(\omega_{r+i}) & \text{if } s-r < i \le N-r-1; \\ T_M(\omega_{r+i-N}) & \text{if } N-r-1 < i \le N \end{cases}$$

and then set $v_i = \psi_i - \psi_0$ for $0 \le i \le N$. Observe that $v = (v_0, \ldots, v_N)$ is a SAW (since $\psi_i^{(1)} \ge M$ for all $i \ge s-r$). In fact, v is a bridge with both endpoints on the $x^{(1)}$-axis ($v_N = (2M, 0, \ldots, 0)$). It is not hard to see that such a bridge could have come from at most one SAW in S'_N, and therefore the transformation $\omega \mapsto v$ is one-to-one. The lemma follows from this observation and equation (28). ■

Proof of Proposition 3.2: Let $\{q_N\}$ be a non-increasing sequence of nonnegative real numbers. Let $1\{\ldots\}$ denote the indicator function of the event $\{\ldots\}$. Recall that the random variable X is always greater than or equal to 1, so if $X_1 + \cdots + X_k = N$ for some k, then k must be less than or equal to N. Applying this observation to (26), we obtain

$$\begin{aligned} &\sum_{N=1}^{\infty} q_N \left(\frac{b_N(0)}{\mu^N}\right) \\ &\quad = E\Big(\sum_{N=1}^{\infty}\sum_{k=1}^{N} q_N 1\{X_1 + \cdots + X_k = N \text{ and } Y_1 + \cdots + Y_k = 0\}\Big) \\ &\quad \le E\Big(\sum_{k=1}^{\infty}\sum_{N=k}^{\infty} q_k 1\{X_1 + \cdots + X_k = N \text{ and } Y_1 + \cdots + Y_k = 0\}\Big) \\ &\quad = E\Big(\sum_{k=1}^{\infty} q_k 1\{Y_1 + \cdots + Y_k = 0\}\Big) \\ &\quad \le \sum_{k=1}^{\infty} C q_k k^{-(d-1)/2} \end{aligned}$$

where the last inequality follows from (18). This completes the proof of Proposition 3.2. ■

We are now ready to prove equation (16) as well as some more general interpretations of the bounds (14) for α_{sing}.

Corollary 3.3. *If* $0 \le x \le 1$, *then*

$$U(x) \le \sum_{N=1}^{\infty} \left(\frac{Cd}{2}\right) N^{-(d-1)/2} x^N.$$

Proof: Use Lemma 3.1 and Proposition 3.2 with $q_N = x^N$. ∎

Corollary 3.4.

$$\sum_{N=1}^{M} \frac{u_N}{\mu^N} = \begin{cases} O(M^{1/2}) & \text{if } d = 2; \\ O(\log M) & \text{if } d = 3; \\ O(1) & \text{if } d > 3. \end{cases}$$

(Here M is an arbitrary large integer.)

Proof: Given M, set $q_N = 1$ if $N \le M$ and $q_N = 0$ otherwise. ∎

Corollary 3.5. *If* $\{q_N\}$ *is a nonincreasing nonnegative sequence for which the sum* $\sum_{N=1}^{\infty} q_N N^{-(d-1)/2}$ *converges, then* $\sum_{N=1}^{\infty} q_N(u_N/\mu^N)$ *also converges.*

Corollary 3.6. *(d=2) For* $0 \le x < 1$,

$$\sum_{N=1}^{\infty} \left(\frac{u_N}{\mu^N}\right) N^{1/2} x^N = O\left(\frac{1}{1-x}\right).$$

Proof: Fix x. Define $f(a) = a^{1/2}x^a$ for $a > 0$. f is maximized at $a_0 = -(2\log x)^{-1}$, and $f'(a) < 0$ for $a > a_0$. So if we put

$$q_N = \begin{cases} f(a_0) & \text{if } N \le a_0; \\ f(N) & \text{if } N > a_0, \end{cases}$$

then $\{q_N\}$ is a nonincreasing positive sequence. The result follows from an application of Proposition 3.2 and some routine estimates, which are left as an exercise for the interested reader. ∎

The above results are not strong enough to give a bound such as $u_N/\mu^N = O(N^{-(d-1)/2})$, although this would follow (at least for $d \ge 3$) from any one of several "mild" conditions on the sequence $\{u_N/\mu^N\}$, such as monotonicity. Unfortunately, it is not easy to check the regularity conditions of this sequence. We do know that the N-th root of the N-th term approaches 1 (by (2) and (8)), but this is not enough to let us say much. For example, it is consistent with the above corollaries that the lim sup of this sequence is positive. Fortunately, it is not hard to rule out this possibility:

Corollary 3.7.

$$\lim_{N\to\infty} \frac{u_N}{\mu^N} = \lim_{N\to\infty} \frac{b_N(0)}{\mu^N} = 0. \tag{29}$$

Proof: By Lemma 3.1, it suffices to prove the second equality. Let J_N denote the number of N-step bridges ω having ω_N on the $x^{(1)}$-axis, with the following modified irreducibility property: there does not exist an i with $0 < i < N$ such that ω_i is on the $x^{(1)}$-axis AND (19) holds. Then we have the following analogue of (21):

$$b_N(0) = \sum_{k=1}^{N} J_k b_{N-k}(0). \tag{30}$$

Put $\bar{p}_n = J_n/\mu^n$ and $\bar{a}_n = b_n(0)/\mu^n$. Then $\bar{a}_n \le 1$, so $\sum_{k=1}^{\infty} \bar{p}_k \le 1$ (c.f. Section 4 of Kesten [9]). We now apply the Basic Renewal Theorem [4]. If $\sum \bar{p}_k < 1$, then $\sum \bar{a}_N < \infty$, so in particular $\lim \bar{a}_N = 0$. If $\sum \bar{p}_k = 1$, then $\lim \bar{a}_N$ exists; by Proposition 3.2, this limit cannot be positive. This proves the corollary. ∎

Remark. For a different proof of Corollary 3.7, first observe that $b_N(z)/\mu^N$ is the probability that the random walk in $\mathbf{Z}^d$ with jumps (X_i, Y_i) (started from the origin) ever hits the point (N, z). Then use Proposition 25.3 of Spitzer [17] to conclude that

$$\lim_{|N|+|z|\to\infty} \frac{b_N(z)}{\mu^N} = 0.$$

4. Bridges at Noncritical Temperatures. In this section we will briefly describe how some results of J.T. Chayes and L. Chayes ([2],[3]) fit into the preceding probabilistic framework. It is hoped that this discussion may give probabilists a more intuitive understanding of their results.

Let $b_{n,L}$ (respectively, $I_{n,L}$) denote the number of n-step bridges (respectively, irreducible bridges) with $\omega_n^{(1)} = L$. (Note that bridges are called "cylinder walks" by Chayes and Chayes.) For β real, define

$$\begin{aligned} \mathcal{H}_L(\beta) &= \sum_{k=0}^{\infty} b_{k,L} e^{-\beta k} \\ C_L(\beta) &= \sum_{k=1}^{\infty} I_{k,L} e^{-\beta k}. \end{aligned} \tag{31}$$

We shall consider a fixed β with $e^\beta > \mu$. In particular, $\mathcal{H}_L$ and C_L are both finite. Since $\mathcal{H}_{L+M} \ge \mathcal{H}_L \mathcal{H}_M$, the limit

$$m(\beta) = \lim_{L\to\infty} \left(-\frac{\log \mathcal{H}_L(\beta)}{L} \right)$$

exists, and

$$\mathcal{H}_L(\beta) \le e^{-m(\beta)L} \text{ for all } L.$$

Suppressing the β, let $a_L = \mathcal{H}_L e^{mL}$ and $p_L = \mathcal{C}_L e^{mL}$. (Observe that $a_0 = 1$ and $a_L \leq 1$.) Since

$$\mathcal{H}_L = \sum_{k=1}^{L} \mathcal{C}_k \mathcal{H}_{L-k},$$

we see that equation (22) holds. Proposition 4.1 of Chayes and Chayes [3] says that $\sum_L p_L = 1$ and $\sum_L Lp_L < \infty$. Thus $\{p_L\}$ defines a probability distribution with a finite mean. Let $\{X_i\}$ be an independent, identically distributed sequence drawn from this distribution; then equation (23) holds, and

$$\lim_{L\to\infty} a_L = \frac{1}{E(X_1)} = \frac{1}{\sum_L Lp_L} > 0. \tag{32}$$

Now define

$$m_c(\beta) = \liminf_{L\to\infty} \left(-\frac{\log \mathcal{C}_L(\beta)}{L} \right).$$

Let us assume that

$$m_c(\beta) > m(\beta). \tag{33}$$

This means that $\limsup (p_L)^{1/L} < 1$; therefore there exists a $t > 0$ such that $E(e^{tX})$ is finite. This condition implies exponential convergence in (32) ([15], p. 107); i.e., there exists a $\Delta(\beta) > 0$ such that

$$|\mathcal{H}_L e^{mL} - \frac{1}{\sum_k kp_k}| < e^{-\Delta L} \text{ for all } L.$$

This is Theorem 3.3(1) of Chayes and Chayes [3].

For z in $\mathbf{Z}^{d-1}$, let $b_{n,L}(z)$ (respectively, $I_{n,L}(z)$) be the number of n-step bridges (respectively, irreducible bridges) with $\omega_n = (L, z)$. Let $H_{(L,z)}(\beta)$ and $C_{(L,z)}(\beta)$ denote their respective generating functions (analogously to (31) above). Then

$$H_{(L,z)} = \sum_{k=1}^{L} \sum_{v \in \mathbf{Z}^{d-1}} C_{(k,v)} H_{(L-k, z-v)}.$$

Consider the random walk in $\mathbf{Z}^d$ with jumps (X, Y) having distribution

$$\Pr\{(X,Y) = (L,z)\} = C_{(L,z)} e^{mL}.$$

Observe that this random walk has positive, finite drift in the $x^{(1)}$ direction. Analogously to equation (26), we see that $H_{(L,z)} e^{mL}$ is the probability that this random walk ever hits the point (L, z). With the assumption (33), Theorem 3.3(2) of Chayes and Chayes [3] says that for large L, this probability factors into $(E(X))^{-1}$ (the limiting probability of hitting the hyperplane $x^{(1)} = L$) times a normal density in z with variance proportional to L. (See also Section 5.4 of Chayes and Chayes [2], as well as Chayes and Chayes [1].) This would be an immediate consequence of known probabilistic theorems (e.g. Theorem 3.2 of Stam [18]*) if Y were known to have finite second moments.

In fact, assumption (33) implies that Y has an *exponential* moment, which can be shown as follows. Since m_c is continuous at β (by concavity; see [2]), one can choose $s > 0$ such that $\beta - s > \log \mu$ and $m_c(\beta - s) > m(\beta)$. Then

$$\begin{aligned} E(e^{s|Y|}) &= \sum_{z,L} e^{s|z|} C_{(L,z)}(\beta) e^{m(\beta)L} \\ &\leq \sum_{z,L,k} e^{sk} I_{k,L}(z) e^{-\beta k} e^{m(\beta)L} \\ &= \sum_{L} C_L(\beta - s) e^{m(\beta)L} < \infty. \end{aligned}$$

It would be interesting to see if this probabilistic framework can be used to simplify the proof of the main result of [3], which is that (33) holds for all $\beta > \log \mu$.

REFERENCES

[1] J.T. Chayes and L. Chayes, *Random tubes as a model of pair correlations*, Contemporary Mathematics **41** (1985) 11-41.

[2] J.T. Chayes and L. Chayes, *Percolation and random media*, in *Critical Phenomena, Random Systems and Gauge Theories; Les Houches, Session XLIII*, K. Osterwalder and R. Stora (eds.). Elsevier, Amsterdam, Oxford, New York, 1986.

[3] J.T. Chayes and L. Chayes, *Ornstein-Zernike behavior for self-avoiding walks at all non-critical temperatures*, Commun. Math. Phys. **105** (1986), 221-238.

[4] W. Feller, *An Introduction to Probability Theory and Its Applications*, Vol. I (3rd edition), Wiley, New York, 1968.

[5] A.J. Guttmann, *On the critical behaviour of self-avoiding walks*, J. Phys. A **20** (1987), 1839-1854.

[6] J.M. Hammersley, *Percolation processes. II. The connective constant*, Proc. Camb. Phil. Soc. **53** (1957), 642-645.

[7] J.M. Hammersley, *The number of polygons on a lattice*, Proc. Camb. Phil. Soc. **57** (1961), 516-523.

[8] J.M. Hammersley and D.J.A. Welsh, *Further results on the rate of convergence to the connective constant of the hypercubical lattice*, Quart. J. Math. Oxford Ser. 2 **13** (1962), 108-110.

---

* I am grateful to H. Kesten for pointing out this reference.

[9] H. Kesten, *On the number of self-avoiding walks*, J. Math. Phys. **4** (1963), 960-969.
[10] H. Kesten, *On the number of self-avoiding walks II*, J. Math. Phys. **5** (1964), 1128-1137.
[11] J.C. Le Guillou and J. Zinn-Justin, *Accurate critical exponents from field theory* , J. Phys. France **50** (1989), 1365-1370.
[12] N. Madras and A.D. Sokal, *The pivot algorithm: a highly efficient Monte Carlo algorithm for the self-avoiding walk*, J. Stat. Phys. **50** (1988), 109-186.
[13] B. Nienhuis, *Exact critical exponents of $O(n)$ models in two dimensions*, Phys. Rev. Lett. **49** (1982), 1062-1065.
[14] B. Nienhuis, *Critical behavior of two-dimensional spin models and charge asymmetry in the Coulomb gas*, J. Stat. Phys. **34** (1984), 731-761.
[15] E. Nummelin, *General Irreducible Markov Chains and Non-negative Operators*, Cambridge University Press, Cambridge, 1984.
[16] G. Slade, *The scaling limit of self-avoiding random walk in high dimensions*, Ann. Prob. **17** (1989), 91-107.
[17] F. Spitzer, *Principles of Random Walk*, Springer, New York, 1976.
[18] A.J. Stam, *Renewal theory in r dimensions*, Comp. Math. **23** (1971), 1-13.
[19] S.G. Whittington, *Statistical mechanics of polymer solutions and polymer adsorption*, Adv. Chem. Phys. **51** (1982), 1-48.

Neal Madras
Department of Mathematics
York University
4700 Keele Street
North York, Ontario M3J 1P3
Canada

Spitzer's Formula Involving Capacity

SIDNEY C. PORT

Let X_t be a strongly transient Lévy process on a second countable locally compact Abelian group and let B be a bounded Borel set. Set $E_B(t,A) = \int P_x\,(T_B \le t,\, X_{T_B} \in A)\,dx$ where $T_B = \inf\{t > 0 : X_t \in B\}$. We give an asymptotic expansion to third order for $E_B\,(t,A)$. If X_t is a strongly transient strictly stable process on R^d we give an expansion to order 4 of $E_B\,(t,R^d)$.

1. Introduction. Let $\mathcal{G}$ be a second countable locally compact Abelian group. Throughout this paper X_t will be a Lévy process on $\mathcal{G}$. Let $P_x(\cdot)$ be the law of the process when $X_0 = x$ and let E_x be expectation relative to P_x. Recall that a point x is a possible value iff $P_0(N + x) > 0$ for all open sets N. The collection of possible values is a semigroup. We will always assume that $\mathcal{G}$ is the closed group generated by this semigroup. The process $-X_t$ is called the dual process (to X_t). Quantities relative to this process are denoted by $\wedge$, e.g. $\widehat{P}_x(\cdot)$ is the law of the process $-X_t$ with initial point x. A quantity Q relative to the dual process is often referred to as *co* $- Q$.

Let

$$G(x, A) = \int_0^\infty P_x(X_t \in A)dt.$$

The process is called transient iff $G(x, A) < \infty$ for all x and all bounded Borel sets A. The process is called strongly transient iff

$$\int_{\mathcal{G}} G(x, dy)G(y, A) < \infty$$

for all x and all bounded Borel sets A.

Let B be a Borel set and let

$$T_B = \inf\{t > 0 : X_t \in B\}(= \infty \quad \text{if} \quad X_t \notin B \quad \text{for all} \quad t > 0)$$

be the hitting time of B. Set $\varphi_B(x) = P_x(T_B < \infty)$ and $\hat{\varphi}_B(x) = \widehat{P}_x(T_B < \infty)$.

A potential theory was introduced for all transient Lévy processes in [PS 71]. For B a bounded Borel set there is a unique measure μ_B supported on $\overline{B}$ such that $\mu_B G(x, dy) = \int_{\mathcal{G}} \mu_B(dz)G(z, dy)$ is absolutely continuous w.r.t. Haar measure having $\hat{\varphi}_B(x)$ as density. Similarly there is a unique measure $\hat{\mu}_B$ supported on $\overline{B}$ such that $\hat{\mu}_B\widehat{G}(dy)$ has φ_B as density. The two measures μ_B and $\hat{\mu}_B$ have a common total mass $C(B)$ called the capacity of B and the measures μ_B and $\hat{\mu}_B$ are called respectively the capacitary and co-capacitary measure of B.

If $G(x, dy)$ is absolutely continuous w.r.t. Haar measure then there is a version of the density $g(y-x)$ that is an excessive function in x and a co-excessive function in y. For this density

$$\int_{\overline{B}} \mu_B(dx) g(y-x) = \hat{\varphi}_B(y)$$

and

$$\int_{\overline{B}} \hat{\mu}_B(dx) \hat{g}(y-x) = \varphi_B(y)$$

for all y. Let γ be a measure and write

$$g\gamma(y) = \int_{k^d} g(x-y)\gamma(dx).$$

Then in the case when $G(x, dy)$ is absolutely continuous

$$\hat{\varphi}_B(y) = \hat{g}\mu_B(y), \quad \varphi_B(y) = g\hat{\mu}_B(y) \tag{1.1}$$

where $\hat{g}(x) = g(-x)$.

Remark. The terminology used here is that used in [PS 71] and is at odds with the terminology used in the Markov process literature until just recently [G 90]. In the Markov process literature μ_B would be called the co-capacitary measure and $\hat{\mu}_B$ the capacitary measure.

A particular Lévy process on d-dimensional Euclidian space R^d is the Brownian motion process. In 1964 Spitzer, [S 64], investigated the behavior as $t \to \infty$ of the quantity

$$E_B(t) = \int_{R^d} P_x(T_B \leq t)\, dx \tag{1.2}$$

for Brownian motion processes on R^d, $d \geqq 2$. There is a substantial difference between the behavior of $E_B(t)$ for recurrent Brownian motions ($d \leq 2$) and transient Brownian motions ($d \geq 3$). For transient Brownian motions Spitzer showed that

$$E_B(t) = C(B)t + o(t). \tag{1.3}$$

This formula has subsequently become known as *Spitzer's formula*. Spitzer also investigated the next order term in the expansion of $E_B(t)$. Here there is a difference between weakly transient Brownian motions ($d = 3, 4$) and strongly transient Brownian motions ($d \geq 5$). For the former $E_B(t) - tC(B) \uparrow \infty$, and Spitzer determined that for these Brownian motions

$$E_B(t) - tC(B) = g(t) + O(1), \tag{1.4}$$

where for explicit numerical constants k_1 and $k_2, g(t) = k_1\sqrt{t}$ for $d = 3$ and $g(t) = k_2 \log(t)$, $d = 4$. For strongly transient Brownian motions Spitzer showed

$$E_B(t) - tC(B) = \int_{R^d} \hat{\varphi}_B(x)\varphi_B(x)\,dx + o(1). \tag{1.5}$$

Spitzer also determined the first two terms in the expansion of $E_B(t)$ for Brownian motion in R^2.

Suppose B is a bounded set. Consider the process $X_t + B$. This, for X_t a Lévy process, is called the Lévy sausage. The quantity $E_B(t)$ is just the expected volume of the sausage swept out by time t.

Suppose particles are distributed onto R^d as a point process with Lebesgue measure as intensity. Then allow the particles to move independently according to the law of X_t. Then $E_B(t)$ is the expected number of distinct particles to hit B by time t.

Spitzer's work on $E_B(t)$ stimulated further work over the years. In (G 65) Getoor, using Hunt's potential theory, showed that (1.3) held for a large class of Hunt processes that included all transient stable process. He showed additionally that analogs of (1.4) held for all weakly transient isotropic stable processes and that (1.5) held for all strongly transient isotropic stable processes. Port and Stone [PS 71] extended Spitzer's results to a very large class of Lévy processes. In particular, they showed that (1.5) held for every strongly transient Lévy process. Analogs of Spitzer's results for transient random walks can be found in [P 66]. Getoor and Stephen just recently developed a capacity theory valid for a very large class of Markov processes and showed that (1.3) was valid, where here,

$$E_B(t) = \int P_x(T_B \leq t)\gamma(dx)$$

for γ an excessive measure. This new capacity theory is fully treated in the new monograph by Getoor [G 90].

The interpretation of $E_B(t)$ in terms of an infinite system of particles leads naturally to the quantity

$$E_B(t, A) = \int_{\mathcal{G}} P_x(T_B \leq t, X_{T_B} \in A)\,dx.$$

Asymptotic expansions to order 2 were given for this quantity for a large class of Lévy processes by Port and Stone in [PS 71]. In particular they showed that for every strongly transient Lévy process

$$E_B(t, A) = t\mu_B(A) + \int_{R^d} \hat{\varphi}_B(x)P_x(T_B < \infty, X_{T_B} \in A)\,dx + o(1). \tag{1.7}$$

Even in [S 64] interest is expressed in higher order terms in the expansion of $E_B(t)$. Kac conjectured that for Brownian motion in R^3 the $O(1)$ term could be replaced by $k_3 + o(1)$ where k_3 was an explicitly given constant. This conjecture remained unsettled until just recently. LeGall [L 88] obtained the third order expansions of $E_B(t)$ for all Brownian motions in R^d, $d \geq 3$ and independently Port [P 90] obtained such expansions for all transient stable processes.

In this paper we will only be concerned with strongly transient processes. Our purpose is twofold. First, we will show that one can extend the expansion in (1.7) to a third order expansion for a very large class of strongly transient Lévy processes on $\mathcal{G}$. These will include all those processes for which the semigroup of possible values is actually $\mathcal{G}$ itself. Secondly, we will show that for a large class of strongly transient stable processes on R^d we can improve the third order expansion of $E_B(t)$ given in [P 90] to a fourth order expansion.

We will always assume our processes satisfy the following.

Condition A. There is a compact set K such that

$$\overline{\lim_{t\to\infty}}\; P_0(X_t \in K)^{1/t} = 1.$$

Under Condition A it follows (see Page 25 of [PS 71]) that if ψ is a continuous function with compact support having integral 1, then

$$E_x\psi(X_{t+s})/E_0\psi(X_t) \to 1 \tag{1.8}$$

uniformly in x on compact subsets of $\mathcal{G}$, and uniformly in s on compact subsets of $[0,\infty)$.

Fix $\psi \geq 0$, continuous with compact support having integral 1, and let

$$p(t) = \int_{R^d} E_x\psi(X_t)\psi(x)\, dx.$$

Then (1.8) shows that uniformly in s on compacts, $p(t+s)/p(t) \to 1$. Let $r(t) = \int_s^\infty p(s)\, ds$. Then uniformly in s on compacts

$$r(t+s)/r(t) \to 1. \tag{1.9}$$

We will also use the following.

Condition B. $Sup_{t>0}\, r(t)/r(2t) < \infty$.

Note that condition B is satisfied if $p(t)$ is a regularly varying function.

A Lévy process is called *singular* iff $X_t - X_0$ has a purely singular distribution for all $t > 0$. Otherwise, the process is *non-singular*.

Standard Conditions. Henceforth X_t will be a strongly transient Lévy process satisfying Conditions A and B and $r(t)$ will be the above function. Also B will be a relatively compact set which if the process is singular satisfies the additional requirement that $P_x(T_{\overline{B}} = T_{B^o}) = 1$ a.e. The function f will be a continuous non-negative function with compact support if the process is singular and any bounded non-negative Borel function with compact support if the process is non-singular. We will call these requirements the *standard conditions.*

2. Notation

$$E_B^t f = \int_{\mathcal{G}} E_x[f(X_{T_B}); T_B \le t]\, dx.$$

$$H_B f(x) = E_x[f(X_{T_B}); T_B < \infty].$$

$$\pi_B(ds, dz) = \int_{\mathcal{G}} \hat{\varphi}_B(x) P_x(T_B \in ds,\ X_{T_B} \in dz)\, dx.$$

$$\gamma_B(ds) = \mu_B(ds,\ \mathcal{G}).$$

$$P^t\psi(x) = E_x\psi(X_t)$$

$$\mu_B f = \int_{\mathcal{G}} f(x)\mu_B(dx).$$

$$H_B G = \int_{\mathcal{G}} \mu_B(dx) G(x, \cdot)$$

$$P_{\hat{\varphi}_B}(\cdot) = \int_{\mathcal{G}} \hat{\varphi}_B(x) P_x(\cdot).$$

$$E_{\hat{\varphi}_B}\psi = \int_{\mathcal{G}} \hat{\varphi}_B(x) E_x\psi\, dx.$$

3. Expansion of $E_B^t f$.

Theorem 1. *Under the standard conditions on X_t, B, and f,*

(3.1)
$$E_B^t f = t(\mu_B f) + \int_{\mathcal{G}} H_B f(x)\hat{\varphi}_B(x)\, dx - C(B)(\mu_B f)\int_t^\infty r(s)\, ds + o\left(\int_t^\infty r(s)\, ds\right).$$

To prove Theorem 1 we will need some facts from [PS 71].

Theorem 12.1 of [PS 71] shows that uniformly in x on compacts

$$P^t H_B f(x) = (\mu_B f) r(t) + o(r(t)). \tag{3.2}$$

Since $\sup_x P^t H_B f(x) \le \|f\|_\infty$ and $r(t)$ is bounded and decreasing it follows from (3.2) that there is a constant k_1 such that for C compact

$$\sum_{x \in C} P^t H_B f(x) \le k_1 r(t). \tag{3.3}$$

Theorem 14.3 of [PS 71] shows that

$$\int_{\overline{B}} \pi_B((t,t+h],dy)f(y) = h(\mu_B f)r(t) + o(r(t)). \tag{3.4}$$

It follows from (3.4) that there is a constant k_2 such that

$$\int_{\overline{B}} \pi_B((t,t+h],dy)f(y) \leq k_2 h r(t). \tag{3.5}$$

Theorem 11.1 of [PS 71] shows

$$t(\mu_B f) = \int_{\mathcal{G}} \hat{P}_x(T_B = \infty)E_x[f(X_{T_B}); T_B \leq t]\, dx. \tag{3.6}$$

Lemma 3.1.
(3.7)

$$\begin{aligned} E_B^t f - t(\mu_B f) - \int_{\mathcal{G}} \hat{\varphi}_B(x) H_B f(x)\, dx \\ = -\int_{\mathcal{G}} P_{\hat{\varphi}_B}(X_t \in dy) H_B f(y) + \int_{\overline{B}} \int_0^t \pi_B(ds,dz) P^{t-s} H_B f(z). \end{aligned}$$

Proof. Using (3.6) we see that

$$E_B^t f - t(\mu_B f) = \int_{\overline{B}} \pi_B([0,t],dz)f(z). \tag{3.8}$$

But
(3.9)

$$\int_{\overline{B}} \pi_B([0,t],dz)f(z) = \int_{\overline{B}} \pi_B([0,\infty),dz)f(z) - \int_{\overline{B}} \pi_B((t,\infty),dz)f(z).$$

However

(3.10)

$$\begin{aligned} \int_{\overline{B}} \pi_B((t,\infty),dz)f(z) &= P_{\hat{\varphi}_B}^t H_B f - E_{\hat{\varphi}_B}[H_B f(X_t); T_B \leq t] \\ &= P_{\hat{\varphi}_B}^t H_B f - \int_{\overline{B}} \int_0^t \pi_B(ds,dz) P^{t-s} H_B f(z). \end{aligned}$$

Combining (3.8) – (3.10) we find that (3.7) holds. □

Lemma 3.2.

$$\int_{\overline{B}} \int_0^t \pi_B(ds,dz) P^{t-s}\, H_B f(z) = O(r(t)).$$

Proof. Using (3.3) we see that

$$\int_{\overline{B}}\int_0^{t/2} \pi_B(ds,dz)P^{t-s}H_Bf(z)$$
$$\leq \int_0^{t/2} \gamma_B(ds)k_1\|f\|_\infty r(t/2) \leq k_1\gamma_B((0,\infty))r(t/2).$$

Since $\sup_t r(t/2)/r(t) < \infty$ it follows that

$$\int_{\overline{B}}\int_0^{t/2} \pi_B(ds,dz)P^{t-s}H_Bf(z) = O(r(t)). \tag{3.11}$$

Now using (3.3) and (3.5) we see that

$$\int_{\overline{B}}\int_{t/2}^{t} \pi_B(ds,dz)P^{t-s}H_Bf(z)$$
$$\leq \|f\|_\infty k_1 \sum_{j=1}^{\infty}\int_{t/2+j-1}^{t}/2+j\gamma_B(ds)r\left(\frac{t}{2}-j\right)$$
$$\leq \|f\|_\infty k_1k_2 r(t/2)\sum_{j=0}^{\infty} r(j) = O(r(t)).$$

Thus using this estimate and (3.11) we see that Lemma 3.2 holds. □

Lemma 3.3.

$$P^t_{\varphi_B}H_Bf \sim C(B)(\mu_Bf)\int_t^\infty r(s)\,ds.$$

Proof. Note that

$$P^t_{\varphi_B}H_Bf = \int_{\overline{B}}\mu_BG(dx)P^tH_Bf(x) = \int_{\overline{B}}\mu_B(dz)\int_t^\infty dsP^sH_Bf(z)ds.$$

By (3.2)

$$\int_{\overline{B}}\mu_B(dz)P^sH_Bf(z) \sim C(B)(\mu_Bf)r(s),$$

so

$$\int_t^\infty ds\int_{\overline{B}}\mu_B(dz)P^sH_Bf(z) \sim C(B)(\mu_Bf)\int_t^\infty r(s)\,ds.$$

Thus Lemma 3.3 holds. □

Proof of Theorem 1. Observe that for any n,

$$\frac{1}{r(t)}\int_t^\infty r(s)\,ds \geq \frac{1}{r(t)}\sum_{j=0}^{n} r(t+j).$$

Hence

$$\lim_{t\to\infty} \frac{1}{r(t)} \int_t^\infty r(s)\, ds \geq n+1,$$

so

$$\lim_{t\to\infty} \frac{r(t)}{\int_t^\infty r(s) ds} \leq \frac{1}{n+1}.$$

Thus

$$r(t)/\int_t^\infty r(s)\, ds \to 0.$$

Using this fact, (3.7), and Lemma's 3.2 and 3.3 we find that (3.1) holds.□

4. Expansion of $E_B(t)$ for Strictly Stable Processes. Throughout this section X_t will be a stable process on R^d. Recall that the log characteristic function of $X_1 - X_0$ is of the form

$$\psi(\theta) = -i\theta \cdot b - \lambda|\theta|^\alpha \int_{R^d} W_\alpha(\theta, \xi)\mu(d\xi),$$

where $\lambda > 0$, μ a probability measure on the unit sphere, $0 < \alpha \leq 2$, and

$$W_\alpha(\theta,\xi) = \left[1 - i\tan\left(\frac{\pi\alpha}{2}\right)\operatorname{sgn}(\theta\cdot\xi)\right]\left|\frac{\theta}{|\theta|}\cdot\xi\right|^\alpha, \qquad \alpha \neq 1$$

$$W_1(\theta,\xi) = \left(\frac{\theta}{|\theta|}\cdot\xi\right) + i\frac{2}{\pi}\left(\frac{\theta}{|\theta|}\cdot\xi\right)\log(|\theta\cdot\xi|), \qquad \alpha = 1.$$

Our basic assumption that R^d is the closed group generated by the possible values of X_t imply that μ is not supported on a hyperplane through 0. This in turn implies that $X_t - X_0$ has a bounded c^∞ density $p(t,x)$ for each $t > 0$. Let $\dot{p}(1,0)$ be the derivative of $p(1,\cdot)$ at 0 and let $\dot{p}(1,0)z = \sum_i^d \frac{\partial p}{\partial x_i}(1,0)z_i$.

A stable process is called *strictly stable* iff $X_t - X_0$ and $t^{-\frac{1}{\alpha}}(X_1 - X_0)$ have the same distribution for all $t > 0$. In that case $p(t,x)$ satisfies the scaling property

$$p(t,x) = t^{-d/\alpha} p(1, t^{-1/\alpha}x). \tag{4.1}$$

The vector b is called the *drift*. A stable process is strictly stable iff either $\alpha \neq 1$ and $b = 0$ or $\alpha = 1$ and $\int_{R^d} x\mu(dx) = 0$.

Suppose the process X_t is strictly stable. Then (4.1) shows that

$$\lim_{t\to\infty} t^{d/\alpha} p(t,x) = p(1,0)$$

the convergence being uniform in x on compacts. It follows that

$$\lim_{t\to\infty} r(t) t^{\frac{d}{\alpha}-1} = \left(\frac{d}{\alpha} - 1\right)^{-1} p(1,0). \tag{4.2}$$

Consequently, if $p(1,0) > 0$ then $r(t)$ satisfies Conditions A and B. It is known [T 67] that all strictly stable processes except those with $\alpha < 1$ and μ supported on a hemisphere have $p(1,0) > 0$. In particular, in one dimension all strictly stable processes except those that are subordinators or whose negative is a subordinator satisfy the conditions on $r(t)$.

It follows from (4.3) that a strictly stable process with $p(1,0) > 0$ is strongly transient iff $\alpha < d/2$.

Taking $f = 1$ in (3.1) shows

$$E_B(t) = tC(B) + P_{\hat{\varphi}_B}(T_B < \infty) - C(B)^2 \int_t^\infty r(s)\,ds + o\left(\int_t^\infty r(s)\,ds\right).$$

We will call this a third order expansion of $E_B(t)$. Using (4.2) we can write this expansion as

$$\begin{aligned} E_B(t) &= tC(B) + P_{\hat{\varphi}_B}(T_B < \infty) \\ &\quad - C(B)^2 p(1,0)\left[\left(\frac{d}{\alpha}-2\right)\left(\frac{d}{\alpha}-1\right)\right]^{-1} t^{2-d/\alpha} + O(t^{2-d/\alpha}). \end{aligned}$$

In [P 90] a better expansion was given, namely

$$\begin{aligned} E_B(t) &= tC(B) + P_{\hat{\varphi}_B}(T_B < \infty) \\ &\quad - C(B)^2 p(1,0)\left[\left(\frac{d}{\alpha}-2\right)\left(\frac{d}{\alpha}-1\right)\right]^{-1} t^{2-\frac{d}{\alpha}} \\ &\quad - \int_B\int_B \dot{p}(1,0)(y-x)\hat{\mu}_B(dy)\mu_B(dx)t^{2-\frac{d+1}{\alpha}} + O\left(t^{1-\frac{d}{\alpha}}\right). \end{aligned} \tag{4.3}$$

Our result here is to improve (4.3) by showing we can replace the term $O(t^{1-d/\alpha})$ by an exact expression of order $t^{1-d/\alpha}$ plus a term of order $o^{1-d/alpha}$. More precisely, the following holds.

Theorem 2. *Let X_t be a strictly stable process with exponent $\alpha < d/2$ and $p(1,0) > 0$. If $\alpha < 1$*

(4.4)

$$\begin{aligned} E_B(t) &= tC(B) + \int \hat{\varphi}_B(x)\varphi_B(x)\,dx \\ &\quad - C(B)^2 p(1,0)\left[\left(\frac{d}{\alpha}-2\right)\left(\frac{d}{\alpha}-1\right)\right]^{-1} t^{2-\frac{d}{\alpha}} \\ &\quad + \left(\frac{d}{\alpha}-1\right)^{-1} P(1,0)\left[C(B)\int \hat{\varphi}_B(x)\varphi_B(x)\,dx\right] t^{1-\frac{d}{\alpha}} + o(t^{1-\frac{d}{\alpha}}). \end{aligned}$$

If $1 \le \alpha < 2$,

(4.5)

$$E_B(t) = tC(B) + \int \hat{\varphi}_B(x)\varphi_B(x)\, dx - C(B)^2 p(1,0)\left[\left(\frac{d}{\alpha}-2\right)\left(\frac{d}{\alpha}-1\right)\right]^{-1} t^{2-\frac{d}{\alpha}} - \left[\int\int \hat{\mu}_B(dx)\mu_B(dy)\dot{p}(1,0)(y-x)\right] t^{2-\frac{d+1}{\alpha}} + \left(\frac{d}{\alpha}-1\right)^{-1} P(1,0)\left[C(B)\int \hat{\varphi}_B(x)\varphi_B(x)\, dx\right] t^{1-\frac{d}{\alpha}} + o(t^{1-\frac{d}{\alpha}}).$$

If $\alpha = 2$

(4.6)

$$E_B(t) = tC(B) + \int \hat{\varphi}_B(x)\varphi_B(x)\, dx - C(B)^2(4\pi)^{-\frac{d}{2}}\left[\left(\frac{d}{2}-1\right)\left(\frac{d}{2}-2\right)\right]^{-1} t^{2-\frac{d}{2}} - \left[(4\pi)^{-\frac{d}{2}}\int\int \mu_B(dx)\mu_B(dy)\frac{|y-x|^2}{4}\left[\left(\frac{d}{2}\right)\left(\frac{d}{2}-1\right)\right]^{-1} - C(B)\int \varphi_B(x)^2\, dx\right] t^{1-\frac{d}{2}} + o(t^{1-\frac{d}{2}}).$$

Remark. In the case when X_t is isotropic an explicit formula for $p(1,0)$ and $\int \varphi_B^2(x)\, dx$ can be found in [P 90]. Also in this case $\dot{p}(1,0) = 0$.

The proof of Theorem 2 will be carried out via two lemmas.

Lemma 4.1. *Let*

$$c_j = \left[\left(\frac{d+j}{\alpha}-1\right)\left(\frac{d+j}{2}-2\right)\right]^{-1}, \qquad j = 0,1,2.$$

Let

$$\gamma_1 = \int_{\overline{B}}\int_{\overline{B}} \dot{p}(1,0)(y-x)\hat{\mu}_B(dy)\mu_B(dx)$$

and let

$$\gamma_2 = (4\pi)^{-\frac{d}{2}} d(d-2)\int_{\overline{B}}\int_{\overline{B}} |y-x|^2 \mu_B(dy)\mu_B(dx).$$

Set

$$G_1(t) = \int_{R^d} P_{\hat{\varphi}_B}(X_t \in dy)\varphi_B(y).$$

If $\alpha < 2$,

$$G_1(t) = C(B)c_0 p(1,0)t^{2-\frac{d}{\alpha}} + c_1\gamma_1 t^{2-\frac{d+1}{\alpha}} + o(t^{1-\frac{d}{2}}). \tag{4.7}$$

If $\alpha = 2$

$$(4.8) \qquad G_1(t) = C(B)^2(4\pi)^{-d/2}t^{2-\frac{d}{2}} + \gamma_2 t^{1-\frac{d}{2}} + o\left(t^{1-\frac{d}{2}}\right).$$

Proof. Using (1.1) and the fact that $g(x) = \int_0^\infty p(t,x)\,dt$ it easily follows that

$$(4.9) \qquad G_1(t) = \int_{\overline{B}}\int_{\overline{B}} \mu_B(dx)\hat{\mu}_B(dy)\int_t^\infty (u-t)p(u,y-x)\,du.$$

Using (4.1), a Taylor expansion shows

$$(4.10) \qquad p(u,z) = u^{-\frac{d}{\alpha}}\left[p(1,0) + u^{-\frac{1}{\alpha}}\dot{p}(1,0)z + \varepsilon_1\right]$$

where for some constant k_3, $|\varepsilon_1| \le k_3|z|^2u^{-\frac{2}{\alpha}}$.

Using (4.9) and (4.10) we find

$$(4.11) \qquad G_1(t) = C(B)^2c_0p(1,0)t^{2-\frac{d}{\alpha}} + c_1\gamma_1 t^{2-\frac{d+1}{\alpha}} + O\left(t^{2-\frac{d+2}{\alpha}}\right).$$

If $\alpha < 2$ then $2 - \frac{d+2}{\alpha} < 1 - \frac{d}{\alpha}$ so $O\left(t^{2-\frac{d+2}{\alpha}}\right) = o\left(t^{1-\frac{d}{\alpha}}\right)$. Thus (4.7) follows from (4.11). If $d = 2$,

$$p(u,z) = (4\pi u)^{-\frac{d}{2}}e^{-\frac{|z|^2}{4u}} = (4\pi u)^{-\frac{d}{2}}\left[1 - \frac{|z|^2}{4u} + O(|z|^4u^{-2})\right].$$

Using this and (4.9) we find that (4.8) holds. □

Lemma 4.2. *Let X_t be any strongly transient Lévy process satisfying the standard conditions. Then*

$$(4.12) \qquad \int_{\overline{B}}\int_0^t \pi_B(ds,dz)P^{t-s}\varphi_B(z) = 2C(B)P_{\hat{\varphi}_B}(T_B < \infty)r(t) + o(r(t)).$$

Proof. By (3.2)

$$\lim_{t_0\to\infty}\left[\lim_{t\to\infty}\frac{1}{r(t)}\int_{\overline{B}}\int_0^{t_0}\pi_B(ds,dz)P^{t-s}\varphi_B(z)\right]$$
$$= \lim_{t_0\to\infty} C(B)P_{\hat{\varphi}_B}(T_B \le t_0) = C(B)P_{\hat{\varphi}_B}(T_B < \infty).$$

Thus

(4.13)

$$\lim_{t_0\to\infty}\lim_{t\to\infty}\frac{1}{r(t)}\int_{\overline{B}}\int_0^{t_0}\pi_B(ds,dz)P^{t-s}\varphi_B(z) = C(B)P_{\hat{\varphi}_B}(T_B < \infty).$$

Now by (3.3)

$$\overline{\lim_{t_0\to\infty}} \frac{1}{r(t)} \int_{\overline{B}} \int_{t_0}^{t/2} \pi_B(ds, dz) P^{t-s} \varphi_B(z) \leq k_1 \sup_t \frac{r(t/2)}{r(t)} \gamma_B((t_0, \infty))$$

so

$$\lim_{t_0\to\infty} \overline{\lim_{t\to\infty}} \frac{1}{r(t)} \int_{\overline{B}} \int_{t_0}^{t/2} \pi_B(ds, dz) P^{t-s} \varphi_B(z) = 0. \tag{4.14}$$

Next observe that by (3.3) and (3.5)

$$\begin{aligned}
\int_{\overline{B}} \int_{t/2}^{t-t_0} \pi_B(ds, dz) P^{t-s} \varphi_B(z) &= \int_{\overline{B}} \int_0^{\frac{t}{2}-t_0} \pi_B(ds, dz) P^{\frac{t}{2}-t_0} \varphi_B(z) \\
&\leq k_1 \int_0^{\frac{t}{2}-t_0} \gamma_B(ds) r\left(\frac{t}{2} - s\right) \\
&\leq k_1 \sum_{j=1}^{\frac{t}{2}-t_0} \gamma_B\left(\left(\frac{t}{2} + j - 1, \frac{t}{2} + j\right)\right) r\left(\frac{t}{2} - 1\right) \\
&\quad + k_1 \gamma_B\left(\left(\frac{t}{2} - t_0, \frac{t}{2} - t_0 + 1\right)\right) r(t_0) \\
&\leq k_1 k_2 r\left(\frac{t}{2}\right) \left[\sum_{j\geq t_0-1}^{\infty} r(j) + r(t_0)\right].
\end{aligned}$$

Thus

$$\lim_{t_0\to\infty} \overline{\lim_{t}} \int_{\overline{B}} \int_{t/2}^{t-t_0} \pi_B(ds, dz) P^{t-z} \varphi_B(z) = 0. \tag{4.15}$$

Let

$$G'(t) = \int_{\overline{B}} \int_{t-t_0}^{t} \pi_B(ds, dz) P^{t-s} \varphi_B(z).$$

Then

$$\begin{aligned}
G'(t) &= \int_{\overline{B}} \int_0^{t_0} \pi(ds + t - t_0, dz) P^{t_0-s} \varphi_B(z) \\
&\leq \int_{\overline{B}} \sum_{j=1}^{n} \pi_B\left[\left(t - t_0 + t_0 \frac{j-1}{n}, t - t_0 + \frac{jt_0}{n}\right], dz\right] P^{t_0 - \frac{jt_0}{n}} \varphi_B(z).
\end{aligned}$$

Using (3.2) we see that

$$\overline{\lim_{t\to\infty}} \frac{1}{r(t)} G'(t) \leq \frac{t_0 C(B)}{n} \sum_{j=0}^{n} \int_{\overline{B}} \overline{\mu}_B(dz) P^{\frac{jt_0}{n}} \varphi_B(z). \tag{4.16}$$

As $n \to \infty$ the RHS of (4.16) converges to

$$\int_{\overline{B}} \hat{\mu}_B(dz) \int_0^{t_0} P^s \varphi_B(z) ds.$$

Thus

$$\overline{\lim_{t\to\infty}} \frac{1}{R(t)} G'(t) \leq C(B) \int_{\overline{B}} \mu_B(dz) \int_0^{t_0} P^s \varphi_B(z) ds. \tag{4.17}$$

A similar computation shows

$$\underline{\lim_{t\to\infty}} \frac{1}{r(t)} G'(t) \geq C(B) \int_{\overline{B}} \overline{\mu}_B(dz) \int_0^{t_0} P^s \varphi_B(z) ds. \tag{4.18}$$

Using (4.17) and (4.18), we find

$$\lim_{t_0\to\infty} \left[\lim_{t\to\infty} \frac{1}{r(t)} G'(t) \right] = C(B) \int_{\overline{B}} \mu_B(dz) G\varphi_B(z). \tag{4.19}$$

But

$$\int_{\overline{B}} \mu_B(dz) G\varphi_B(z) = \int_{R^d} \hat{\varphi}_B(x) \varphi_B(x)\, dx$$

so (4.17) shows

$$\lim_{t_0\to\infty} \left[\lim_{t\to\infty} G'(t) r(t) \right] = C(B) P_{\hat{\varphi}_B}(T_B < \infty). \tag{4.18}$$

Equation (4.12) now follows from (4.13) – (4.15) and (4.18). □

Proof of Theorem 2. Recall that $r(t) \sim p(1,0) \left(\frac{d}{\alpha} - 1\right)^{-1} t^{1-\frac{d}{\alpha}}$. Lemma 3.1 shows

(4.19)

$$E_B(t) - tC(B) - P_{\hat{\varphi}_B}(T_B < \infty)$$
$$= - \int_{\overline{B}} P_{\hat{\varphi}_B}(X_t \in dy) \varphi_B(y) + \int_{\overline{B}} \int_0^t \pi_B(ds, dz) P^{t-s} \varphi_B(z).$$

The assertions of the theorem now follows from (4.19) and Lemmas 4.1 and 4.2 plus the fact that for $\alpha < 1$ $t^{2-\frac{d+1}{\alpha}} = o(t^{1-\frac{d}{\alpha}})$ and for $1 \leq \alpha < 2$, $t^{1-\frac{d+2}{\alpha}} = o(t^{1-\frac{d}{\alpha}})$ and for $1 \leq \alpha < 2$, $t^{2-\frac{d+2}{\alpha}} = o(t^{1-\frac{d}{\alpha}})$. □

5. Expansion of $E_B(t)$ for Stable Processes with Drift. Let $f(x)$ be the density of $X_1 + b$ where X_t is a stable process with drift b. Then

$$p(t, x) = t^{-\frac{d}{\alpha}} f\left(t^{1-\frac{1}{\alpha}} \left(b + \frac{x}{t}\right)\right). \tag{5.1}$$

We will first consider these processes in one dimension. Condition A is satisfied iff the measure μ does not assign probability 0 to either the point set $\{1\}$ or $\{-1\}$. Henceforth we assume this is the case. Then as shown in [P 90] the process is strongly transient iff $\alpha < 1/2$.

Theorem 3. *Let X_t be a stable process with drift $b \neq 0$ on R^1. Assume $\alpha < 1/2$ and $\mu(\{1\}) \neq 0$, $\mu(\{-1\}) = 0$. Then*

$$\begin{aligned} E_B(t) &= tC(B) + P_{\hat{\varphi}_B}(T_B < \infty) \\ &\quad - C(B)^2 \left[\left(\frac{1}{\alpha} - 2\right)\left(\frac{1}{\alpha} - 1\right)\right]^{-1} f(0)t^{2-\frac{1}{\alpha}} \\ &\quad + 2C(B)P_{\hat{\varphi}_B}(T_B < \infty)\left(\frac{1}{\alpha} - 1\right)^{-1} f(0)t^{1-\frac{1}{\alpha}} + o(t^{1-\frac{1}{\alpha}}). \end{aligned} \tag{5.2}$$

Proof. The assumption that $\mu(\{1\}) \neq 0$ and $\mu(\{-1\}) \neq 0$ implies $f(0) > 0$. Using (5.1) we find that uniformly in x on compacts $p(t,x) \sim t^{-\frac{1}{\alpha}} f(0)$. Thus $r(t) \sim \left(\frac{1}{\alpha} - 1\right)^{-1} f(0)t^{-\frac{1}{\alpha}}$. The proof of (5.2) now proceeds along much the same lines as the proof of Theorem 2. Lemma 3.3 shows that

$$\int_B \int_0^t \pi_B(ds, dz)P^{t-s}\varphi_B(z) \sim 2C(B)P_{\hat{\varphi}_B}(T_B < \infty)\left(\frac{1}{\alpha} - 1\right)^{-1} f(0)t^{1-\frac{1}{\alpha}}.$$

A Taylor series expansion shows that

$$P^t_{\hat{\varphi}_B}\varphi_B = \left[\left(\frac{2}{\alpha} - 1\right)\left(\frac{1}{\alpha} - 1\right)\right]^{-1} f(0)t^{2-\frac{1}{\alpha}} + o(t^{1-1/\alpha}).$$

Equation (5.2) now follows from these two facts via Lemma 3.1. □

Remark. The quantities $f(0)$ and $P_{\hat{\varphi}_B}(T_B < \infty)$ can be explicitly computed. See [P 90].

We now consider processes with drift in dimension $d \geq 2$. Here we must distinguish between processes with $\alpha < 1$, $\alpha = 1$, $1 < \alpha < 2$, and $\alpha = 2$.

If $\alpha < 1$ then $t^{1-\frac{1}{\alpha}} \to 0$ as $t \to \infty$. Thus by (5.1) $t^{\frac{d}{\alpha}}p(t,x) \to f(0)$, uniformly in x on compacts. If $f(0) > 0$ then $r(t) \sim \left(\frac{d}{\alpha} - 1\right)^{-1} f(0)t^{1-\frac{d}{\alpha}}$ and the process is strongly transient. Taylor [T 67] showed that for $\alpha < 1$, $f(0) > 0$ iff μ is not supported on a hemisphere.

Theorem 4. *Let X_t be a stable process with drift on R^d and assume μ is not concentrated on a hemisphere. Assume*

$$\frac{n-1}{n} \leq \alpha < \frac{n}{n+1}.$$

Let

$$e_j = \sum_{i_1=1}^{d} \cdots \sum_{i_j=1}^{d} \frac{\partial^j}{\partial x_{i_1} \cdots \partial x_{i_j}} f(0) b_{i_1} \cdots b_{i_j}$$

and let

$$D_j = \left[\left(\frac{d+j}{\alpha} - j - 2 \right) \left(\frac{d+j}{\alpha} - j - 1 \right) \right]^{-1} c_j .$$

Then

$$\begin{aligned} E_B(t) &= tC(B) + P_{\hat{\varphi}_B}(T_B < \infty) - \sum_{j=0}^{n-1} D_j C(B)^j t^{2+j-\frac{d+j}{\alpha}} \\ &+ \left(\frac{d}{\alpha} - 1 \right)^{-1} f(0) C(B) P_{\hat{\varphi}_B}(T_B < \infty) t^{1-\frac{d}{\alpha}} + o\left(t^{1-\frac{d}{\alpha}} \right). \end{aligned} \tag{5.3}$$

Proof. Lemma 3.3 shows that

$$\int_{\overline{B}} \int_0^t \pi_B(ds, dz) P^{t-s} \varphi_B(z) \sim 2C(B) P_{\hat{\varphi}_B}(T_B < \infty) \left(\frac{1}{\alpha} - 1 \right)^{-1} f(0) t^{1-\frac{d}{\alpha}}.$$

A Taylor series expansion shows

$$P_{\hat{\varphi}_B} \varphi_B = \sum_{j=0}^{n-1} D_j C(B)^j t^{2+j-\frac{j+d}{2}} + o(t^{1-d/\alpha}).$$

Equation (5.3) now follows from these two facts by Lemma 3.1. □

If $1 < \alpha < 2$ $t^{1-\frac{1}{\alpha}} \to \infty$ as $t \to \infty$ and the behavior of $p(t, x)$ depends on how $f(x)$ behaves as $x \to \infty$ along the direction b. In general we do not have very much information on this matter. Results in [PT 69] show that $f(x) = O(|x|^{-(1+\alpha)})$ which suffices to show that for $d \geq 2$ all such processes are strongly transient. For isotropic f very precise information is possible. Proposition A13 of [P 90] gives an asymptotic series for $f(x)$ as $x \to \infty$. Let

(5.4)

$$B_n = (-1)^{n-1} \cdot \frac{\Gamma\left(\frac{n\alpha}{2} + 1\right) \Gamma\left(\frac{n\alpha + d}{2}\right) 4^{\frac{\alpha+d}{2}} \left(1 + \tan^2\left(\frac{\pi\alpha}{4}\right)\right)^{\frac{n}{2}} \sin\left(n\frac{\pi\alpha}{4} + n \tan^{-1}\left(\frac{\pi\alpha}{4}\right)\right)}{\pi n! (4\pi)^{d/2}}$$

Then

$$f(x) \sim \sum_{n=1}^{\infty} B_n x^{-(d+\alpha n)}, \; x \to \infty. \tag{5.5}$$

Using (5.1) and (5.5) we find

$$r(t) \sim B_1 b^{-(d+\alpha)}(d+\alpha-1)^{-1}t^{2-d-\alpha}. \tag{5.6}$$

and

$$\int_t^\infty (u-t)p(u,z)\,du = \sum_{j=1}^n B_j' t^{j+2-d-\alpha j} + O(t^{n+3-d-\alpha(n+1)}) \tag{5.7}$$

where

$$B_j' = [(d+j\alpha-j-1)(d+j\alpha-j-2)]^{-1}B_j.$$

Using (5.6) and (5.7) we find via Lemma's 3.1 and 3.3 that the following holds.

Theorem 4. *Let $1<\alpha<2$ and assume that X_1+b is isotropic. Let n be such that $\frac{n+1}{n} \le \alpha < \frac{n}{n-1}$. Then*

$$E_B(t) = tC(B) + \int_{R^d} \varphi_B^2(x)\,dx - \sum_{j=1}^n B_j' b^{-(d+\alpha j)} C(B)^j t^{j+2-d-\alpha j}$$
$$+ 2C(B)P_{\hat{\varphi}_B}(T_B<\infty)B_1 b^{-(d+\alpha)}(d+\alpha-1)^{-1}t^{2-d-\alpha} + o(t^{2-d-\alpha}).$$

REFERENCES

[G 65] R.K. Getoor, *Some Asymptotic Formulas Involving Capacity,* Z. Wharscheinlichkeits Theorie verw. Geb. **4** (1965), 248-252.

[G 90] R.K. Getoor, *Excessive Measures,* Birkhäuser, Boston 1990.

[L 88] J.F. LeGall, *Sur Une Conjecture de M. Kac,* Prob. Th. Rel, Fields, **78** (1988), 389-402.

[P 66] S.C. Port, *Limit Theorems Involving Capacities,* J. Math. Mechanics, **15** (1966), 805-832.

[P 90] S.C. Port, *Asymptotic Expansions for the Expected Volume of a Stable Sausage,* Ann. Prob. (To appear 1990).

[PS 71] S.C. Port and C. Stone, *Infinitely Divisible Processes and Their Potential Theory,* (First Part) Ann. Inst. Fourier, **21** (1971), 157-275.

[PT 69] W.E. Pruitt and S.J. Taylor, *The Potential Kernel and Hitting Probabilities of the General Stable Process in R^n*. Trans. Amer. Math. Soc., **46** (1969), 299-321.

[S 64] F. Spitzer, *Electrostatic Capacity, Heat Flow, and Brownian Motion,* Z. Wharscheinlichkeits Theorie verw. Geb., **3** (1964), 110-121.

[T 67] S.J. Taylor, *Sample Path Properties of a Transient Stable Process,* J. Math. Mech., **16** (1967), 1229-1246.

Dept. of Mathematics
University of California, Los Angeles
Los Angeles, CA 90024

An Integral Test for Subordinators

WILLIAM E. PRUITT[1]

An integral test is obtained to determine whether $P\{X_t \leq \beta_t \text{ i.o.}\}$ is 0 or 1 where $t^{-1}\beta_t$ is nondecreasing and X_t is a subordinator belonging to a large class which contains the stable subordinators. This generalizes the results of Breiman for the stable subordinators.

1. Introduction. Let X_t, $t \geq 0$, be a subordinator, i.e., a real valued nondecreasing Lévy process. The problem of interest here is the rate of escape of the process from 0. This can be measured in different ways. One method is to find, if possible, a function β_t, defined for small t, such that

$$\liminf_{t \to 0} \frac{X_t}{\beta_t} = 1 \quad \text{a.s.} \tag{1.1}$$

Another method that gives more information is to find a function h, depending on the distribution of X_t and the function β_t, such that

$$P\{X_t \leq \beta_t \text{ i.o. as } t \to 0\} = \begin{matrix} 0 \\ 1 \end{matrix} \quad \text{according as} \quad \int h(t)dt \begin{matrix} < \\ = \end{matrix} \infty. \tag{1.2}$$

For example, (1.1) is not sharp enough to answer the question of whether X_t gets below β_t infinitely often for small t.

Our goal in the present paper is to give an integral test in the form (1.2), which we believe to be completely general, although we have only been able to prove it under an additional hypothesis. We will compare it with Breiman's test [1] for stable subordinators and in the final section we will give a couple of examples to illustrate the applicability.

The first integral test obtained was the Kolmogorov-Erdös test which considered the large values of Brownian motion. This supplemented the information given by the law of the iterated logarithm. Integral tests more closely related to the present work were obtained for the small values of the modulus of Brownian motion in dimensions $d \geq 3$ by Dvoretzky and Erdös [2] and for $d = 2$ by Spitzer [5]. A result for the small values of nondecreasing random walk that is very closely related to the theorem given here appears in [4].

2. The Stable Case. We start with a brief description of Breiman's result [1] for stable subordinators. This will serve as an introduction to the

[1]Research supported in part by N.S.F. Grants DMS-8603437 and DMS-8902581

type of result we want for the more general subordinators. The Laplace transform of the process is given by

$$Ee^{-\lambda X_t} = \exp\left(mt\int(e^{-\lambda x}-1)\nu(dx)\right),$$

with $\nu(dx) = x^{-(1+\alpha)}dx$, $0 < x < \infty$. $\alpha \in (0,1)$ is the usual parameter of the stable subordinator and $m > 0$ is a scale parameter. Let

$$\mu = (\Gamma(1-\alpha)m)^{1/(1-\alpha)}(1-\alpha)/\alpha.$$

The test integral is

$$\text{(2.1)} \qquad \int_0 \frac{1}{t}\left(\frac{t^{1/\alpha}}{\beta_t}\right)^{\alpha/2(1-\alpha)} \exp\left(-\mu\left(\frac{t^{1/\alpha}}{\beta_t}\right)^{\alpha/(1-\alpha)}\right)dt,$$

i.e., the probability in (1.2) is 0 if this integral converges and 1 if it diverges. The function β_t is assumed to satisfy

$$\text{(2.2)} \qquad \beta_t t^{-1/\alpha} \text{ is nondecreasing.}$$

Note that if we define

$$\text{(2.3)} \qquad \xi_t(k,\epsilon) = \ell_2\frac{1}{t} + \frac{3}{2}\ell_3\frac{1}{t} + \ell_4\frac{1}{t} + \cdots + \ell_{k-1}\frac{1}{t} + (1+\epsilon)\ell_k\frac{1}{t},$$

where ℓ_k is the logarithm iterated k times and let

$$\beta_t(k,\epsilon) = \mu^{(1-\alpha)/\alpha}t^{1/\alpha}(\xi_t(k,\epsilon))^{-(1-\alpha)/\alpha},$$

then we may apply Breiman's test to obtain for any $k \geq 4$ and any $\epsilon > 0$,

$$P\{X_t \leq \beta_t(k,0) \text{ i.o. as } t \to 0\} = 1,$$
$$P\{X_t \leq \beta_t(k,\epsilon) \text{ i.o. as } t \to 0\} = 0.$$

These last results are typical of the usefulness of the test in providing very fine information about the behavior of the sample paths of the subordinator. We will obtain comparable results in the more general situation in the Corollary below.

3. The Integral Test. Before stating the theorem, we will make a few comments about assumptions. One needs some regularity on the function β_t in order to have a reasonable test. Our assumption will be that

$$\text{(3.1)} \qquad \frac{\beta_t}{t} \text{ is nondecreasing.}$$

This is quite useful in the proof and it is milder than the usual assumptions of this type including Breiman's assumption (2.2).

The other assumption that we will make concerns the Lévy measure, ν, of the subordinator whose Laplace transform is given by

$$Ee^{-\lambda X_t} = \exp\left(-bt\lambda + t\int(e^{-\lambda x} - 1)\nu(dx)\right).$$

For X_t to be a subordinator we must have $b \geq 0$, ν supported on $[0, \infty)$, and $\int(x \wedge 1)\nu(dx) < \infty$. The additional assumption is

$$\liminf_{x\to\infty} \frac{\int_{y\leq x} y^2\nu(dy)}{x^2\nu((x,\infty))} > 0. \tag{3.2}$$

This condition is motivated by Feller's condition for stochastic compactness of normalized sums of i.i.d. random variables. It is needed to obtain the probability estimates that are required. In case of a stable subordinator of index α, the ratio in (3.2) is equal to $\alpha/(2-\alpha)$ for all x. Thus (3.2) is also a relatively mild assumption.

Theorem. *Assume that X_t is a nondegenerate subordinator which satisfies* (3.2) *and β_t a function defined for small t which satisfies* (3.1). *Then*

$$P\{X_t \leq \beta_t \text{ i.o. as } t \to 0\} = \begin{matrix} 0 \\ 1 \end{matrix} \text{ according as } \int_0 \frac{1}{t}\log\left(\frac{3}{p_t}\right) p_t\, dt \begin{matrix} < \\ = \end{matrix} \infty, \tag{3.3}$$

where $p_t = P\{X_t \leq \beta_t\}$. The convergent case is valid without the assumption (3.2).

Of course, this result is not useful unless one has good estimates of p_t. These are available in [3] under the assumption (3.2). To describe these estimates, we introduce the functions, defined for $\lambda > 0$,

$$g(\lambda) = b + \int_0^\infty xe^{-\lambda x}\nu(dx),\ R(\lambda) = \int_0^\infty (1 - e^{-\lambda x}(1 + \lambda x))\nu(dx).$$

Then g is decreasing and R is increasing. One solves the equation

$$g(\lambda_t) = \frac{\beta_t}{t},$$

and then

$$p_t \approx \frac{1}{(tR(\lambda_t))^{1/2}}e^{-tR(\lambda_t)}. \tag{3.4}$$

(We use the symbol $\approx$ to mean that the ratio of the two sides is bounded above and below by positive finite constants.) It is also shown in [3] that $p_t \to 0$ iff $tR(\lambda_t) \to \infty$ and, in this case

$$\log(1/p_t) \sim tR(\lambda_t) \tag{3.5}$$

even without assuming (3.2). Using (3.4), we see that under the assumption (3.2) the integral criterion in (3.3) is equivalent to

$$\int_0 \frac{1}{t}(tR(\lambda_t))^{1/2} e^{-tR(\lambda_t)} dt \ \substack{< \\ =} \ \infty.$$

This form leads immediately to the following

Corollary. *Define*

$$\xi_t(k,\epsilon) = \ell_2 \frac{1}{t} + \frac{3}{2}\ell_3 \frac{1}{t} + \ell_4 \frac{1}{t} + \cdots + \ell_{k-1}\frac{1}{t} + (1+\epsilon)\ell_k \frac{1}{t},$$

where ℓ_k is the logarithm iterated k times and then define $\lambda_t(k,\epsilon)$ by solving

$$R(\lambda_t(k,\epsilon)) = \frac{\xi_t(k,\epsilon)}{t}$$

and let

$$\beta_t(k,\epsilon) = tg(\lambda_t(k,\epsilon)).$$

Then, for any $k \geq 4$ and any $\epsilon > 0$,

$$P\{X_t \leq \beta_t(k,0) \text{ i.o. as } t \to 0\} = 1,$$
$$P\{X_t \leq \beta_t(k,\epsilon) \text{ i.o. as } t \to 0\} = 0.$$

This corollary gives the strengthened versions of (1.1) that are analogous to those obtained for the law of the iterated logarithm for Brownian motion by using the Kolmogorov-Erdös test and to those given above for stable subordinators. We will use the corollary in the examples in the next section.

Proof of the convergent case of (3.3). First we want to observe that we may make a couple of assumptions to rule out special cases. If $\nu(0,\infty) < \infty$, then X_t is a compound Poisson process with a drift if $b > 0$. Thus we will have $X_t = bt$ for t less than the first jump of the process. This means that the probability in (3.3) will be 0 if $\beta_t < bt$ for small t and 1 if $\beta_t \geq bt$ for small t. On the other hand, p_t will be 0 for small t in the first case and near 1 for small t in the second so the integral will converge or diverge accordingly. Thus we may assume that $\nu(0,\infty) = \infty$.

The other special case is when

$$\liminf_{t\to 0} tR(\lambda_t) < \infty. \tag{3.6}$$

If there is a sequence $t_k \to 0$ such that $t_k R(\lambda_{t_k}) \le M$, then for $t_k \le t \le 2t_k$, we have $tR(\lambda_t) \le 2t_k R(\lambda_{t_k}) \le 2M$ since λ_t is nonincreasing by (3.1). Then by Lemma 5.2 in [3], we will have p_t bounded below on these intervals and so for some $c > 0$,

$$\int_{t_k}^{2t_k} \frac{1}{t} \log\left(\frac{3}{p_t}\right) p_t\, dt \ge c \int_{t_k}^{2t_k} \frac{1}{t}\, dt = c \ \log 2;$$

thus the integral in (3.3) must diverge. But since

$$\limsup_{t\to 0} p_t > 0,$$

we must have $P\{X_t \le \beta_t \text{ i.o. as } t \to 0\} = 1$ by Blumenthal's $0-1$ law. Thus we may also assume that $tR(\lambda_t) \to \infty$ as $t \to 0$. Note that this implies that $\lambda_t \to \infty$.

We next observe that by (3.5) we may rewrite the test integral as $\int R(\lambda_t) p_t dt$ since we are assuming now that $tR(\lambda_t) \to \infty$. We define a decreasing sequence of times $\{t_k\}$ by

$$\int_{t_k}^{1} R(\lambda_t)\, dt = k.$$

This is possible since $R(\lambda_t) \ge t^{-1}$ for small t. Furthermore, $t_k \to 0$ since for $t \ge \epsilon$ we must have $R(\lambda_t) \le R(\lambda_\epsilon)$ so that the integral of $R(\lambda_t)$ over $(\epsilon, 1)$ is finite. Next, we consider $t_k \le s < t \le t_{k-1}$ and define

$$q_{st} = P\{X_t - X_s \le \beta_t - \beta_s\}.$$

If μ_{st} is defined by

$$g(\mu_{st}) = \frac{\beta_t - \beta_s}{t-s},$$

then we have $\mu_{st} \le \lambda_t$, since by (3.1)

$$\frac{\beta_t - \beta_s}{t-s} \ge \frac{\beta_t}{t}.$$

Thus

$$(t-s)R(\mu_{st}) \le (t-s)R(\lambda_t) \le \int_s^t R(\lambda_u)du \le \int_{t_k}^{t_{k-1}} R(\lambda_u)du = 1.$$

This means that

$$q_{st} \geq c > 0 \tag{3.7}$$

by Lemma 5.2 in [3]. We will use this key fact in two ways. First note that for $t_k \leq s < t \leq t_{k-1}$,

$$p_t = P\{X_t \leq \beta_t\} \geq P\{X_s \leq \beta_s, X_t - X_s \leq \beta_t - \beta_s\} = p_s q_{st} \geq c p_s.$$

In particular, we have

$$c p_{t_k} \leq p_t \leq c^{-1} p_{t_{k-1}} \quad \text{when} \quad t_k \leq t \leq t_{k-1},$$

and so

$$c p_{t_k} \leq \int_{t_k}^{t_{k-1}} R(\lambda_t) p_t \, dt \leq c^{-1} p_{t_{k-1}}.$$

Thus the integral criterion is equivalent to the convergence or divergence of the series $\sum p_{t_k}$.

Thus to complete the proof of the Theorem, we need to prove the supplementary condition for Borel-Cantelli in the divergent case and to fill in between the subsequence in the convergent case. The former is too long to include here; it is similar to the random walk proof given in [4]. For the latter, it turns out that it is too crude to fill in by monotonicity and one needs a "maximal" lemma. It is clearly enough to prove

$$P\left\{\inf_{t_k \leq t \leq t_{k-1}} \frac{X_t}{\beta_t} \leq 1\right\} \leq c^{-1} p_{t_{k-1}}. \tag{3.8}$$

To prove (3.8), let $t_k = s_1 < s_2 < \cdots < s_n = t_{k-1}$, take $\epsilon > 0$, and define

$$A_j = \left\{\frac{X_{s_j}}{\beta_{s_j}} < 1 + \epsilon, \frac{X_{s_i}}{\beta_{s_i}} \geq 1 + \epsilon, \ i < j\right\}.$$

Then, by (3.7)

$$\begin{aligned} P\left\{\min_j \frac{X_{s_j}}{\beta_{s_j}} < 1 + \epsilon\right\} &= \sum P(A_j) \\ &= \sum q_{s_j, t_{k-1}}^{-1} P\{A_j; X_{t_{k-1}} - X_{s_j} \leq \beta_{t_{k-1}} - \beta_{s_j}\} \\ &\leq c^{-1} \sum P\{X_{t_{k-1}} \leq (1+\epsilon)\beta_{t_{k-1}}; A_j\} \\ &\leq c^{-1} P\left\{\frac{X_{t_{k-1}}}{\beta_{t_{k-1}}} \leq 1 + \epsilon\right\}. \end{aligned}$$

By refining the partition in such a way that the mesh length goes to zero, we may conclude (using the right continuity of X_t and the monotonicity of β_t) that

$$P\left\{\inf_{t_k \le t \le t_{k-1}} \frac{X_t}{\beta_t} \le 1\right\} \le P\left\{\inf_{t_k \le t \le t_{k-1}} \frac{X_t}{\beta_t} < 1+\epsilon\right\}$$
$$\le c^{-1} P\left\{\frac{X_{t_{k-1}}}{\beta_{t_{k-1}}} \le 1+\epsilon\right\}.$$

Then (3.8) follows by letting $\epsilon \to 0$.

4. Examples. We will look at two examples. The first has a one-sided Lévy measure that is fairly close to the Lévy measure of a Cauchy process but modified to give the necessary integrability condition, $\int x\nu(dx) < \infty$, to be a subordinator. The second is similar but with the Lévy measure concentrated at the points 2^{-k} instead of being spread evenly. This is a sort of continuous time version (for small time) of Feller's St. Petersberg game.

Example 1. Let X_t be a subordinator with $b = 0$ and Lévy measure

$$d\nu(x) = \frac{dx}{x^2 \log^2(\frac{e}{x})}, \quad 0 < x \le 1.$$

Then fairly routine asymptotics gives

$$g(\lambda) = \frac{1}{\log\lambda} + \frac{C}{\log^2\lambda} + O\left(\frac{1}{\log^3\lambda}\right), \quad \lambda \to \infty,$$

where

$$C = \int_1^\infty e^{-v}\frac{dv}{v} - 1 - \int_0^1 (1-e^{-v})\frac{dv}{v} = -C_1 - 1,$$

and C_1 is Euler's constant. Also,

$$R(\lambda) = \frac{\lambda}{\log^2\lambda} + O\left(\frac{\lambda}{\log^3\lambda}\right), \quad \lambda \to \infty.$$

To apply the results of the Corollary, we want to solve $tR(\lambda_t) = \xi_t(k,\epsilon)$. Thus we want

$$\log\frac{1}{t} + \log\xi_t = \log R(\lambda_t) = \log\lambda_t - 2\log\log\lambda_t + O\left(\frac{1}{\log\lambda_t}\right), \tag{4.1}$$

which leads to

$$\log\lambda_t - 2\log\log\lambda_t = \log\frac{1}{t} + \log\xi_t + O\left(\frac{1}{\log t^{-1}}\right) \tag{4.2}$$

and taking logs,

$$\log\log\lambda_t + O\left(\frac{\log\log\lambda_t}{\log\lambda_t}\right) = \log\log\frac{1}{t} + O\left(\frac{\log\log\log t^{-1}}{\log t^{-1}}\right).$$

Since $\log\lambda_t \sim \log t^{-1}$ by (4.1), we may use this in (4.2) to obtain

$$\log\lambda_t = \log\frac{1}{t} + 2\log\log\frac{1}{t} + \log\xi_t + O\left(\frac{\log\log t^{-1}}{\log t^{-1}}\right).$$

To evaluate $\beta_t(k,\epsilon)$ we only need to know $\log\lambda_t$. Using the expansion for g leads to

$$\beta_t(k,\epsilon) = t\,\Big(\frac{1}{\log t^{-1}} - \frac{2\log\log t^{-1}}{\log^2 t^{-1}} - \frac{\log\xi_t(k,\epsilon)}{\log^2 t^{-1}} + \frac{C}{\log^2 t^{-1}} + O\left(\frac{(\log\log t^{-1})^2}{\log^3 t^{-1}}\right)\Big).$$

Since

$$\log\xi_t(k+1,\epsilon) - \log\xi_t(k,\epsilon) \sim \begin{cases} -\epsilon\frac{\ell_k t^{-1}}{\ell_2 t^{-1}} & \text{if} \quad \epsilon > 0, \\ \frac{\ell_{k+1} t^{-1}}{\ell_2 t^{-1}} & \text{if} \quad \epsilon = 0, \end{cases}$$

we see that the difference between $\beta_t(k,\epsilon)$ and $\beta_t(k+1,\epsilon)$ is larger than the error term so that we may drop the error term. Thus, if we let

$$\beta_t(k,\epsilon) = t\left(\frac{1}{\log t^{-1}} - \frac{2\log\log t^{-1}}{\log^2 t^{-1}} - \frac{\log\xi_t(k,\epsilon)}{\log^2 t^{-1}} + \frac{C}{\log^2 t^{-1}}\right),$$

then we have for any $k \geq 4$ and any $\epsilon > 0$,

$$P\{X_t \leq \beta_t(k,0) \text{ i.o. as } t \to 0\} = 1,$$
$$P\{X_t \leq \beta_t(k,\epsilon) \text{ i.o. as } t \to 0\} = 0.$$

Note that the term $C\log^{-2} t^{-1}$ cannot be dropped as it is larger than the difference between $\beta_t(k,\epsilon)$ and $\beta_t(k+1,\epsilon)$. Thus we have very sharp estimates on the behavior of the small values of the process X_t for small t.

Example 2. Let X_t be a subordinator with $b = 0$ and Lévy measure ν such that

$$\nu\left(\frac{1}{2^k}\right) = \frac{2^k}{k(k+1)}, \quad k = 1, 2, \ldots .$$

The asymptotics are very much like the first example except that the logs are to the base 2 and the coefficients are more interesting. We write Log for log to the base 2. Then

$$g(\lambda) = \frac{1}{\operatorname{Log}\lambda} + \frac{C(\lambda)}{\operatorname{Log}^2\lambda} + O\left(\frac{1}{\operatorname{Log}^3\lambda}\right), \quad \lambda \to \infty,$$
$$R(\lambda) = (D(\lambda) - C(\lambda))\frac{\lambda}{\operatorname{Log}^2\lambda} + O\left(\frac{\lambda}{\operatorname{Log}^3\lambda}\right), \quad \lambda \to \infty,$$

where C, D are defined for $\lambda > 0$ by

$$C(\lambda) = \sum_{k=1}^{\infty} \exp(-\lambda 2^k) + \sum_{k=1}^{\infty} (\exp(-\lambda 2^{1-k}) - 1) - \text{Log } \lambda^{-1},$$

$$\begin{aligned} D(\lambda) = & \frac{1}{\lambda} \sum_{k=1}^{\infty} (1 - \exp(-\lambda 2^k)) 2^{-k} \\ & + \frac{1}{\lambda} \sum_{k=1}^{\infty} (1 - \exp(-\lambda 2^{1-k}) - \lambda 2^{1-k}) 2^{k-1} - \text{Log } \frac{1}{\lambda}. \end{aligned}$$

The functions C, D satisfy

$$C(2\lambda) = C(\lambda), \quad D(2\lambda) = D(\lambda), \qquad \text{for all } \lambda > 0.$$

Then if

$$\beta_t(k, \epsilon) = t \left(\frac{1}{\text{Log } t^{-1}} - \frac{2\text{LogLog } t^{-1}}{\text{Log}^2 t^{-1}} - \frac{\text{Log } \xi_t(k, \epsilon)}{\text{Log}^2 t^{-1}} - \frac{h(\lambda_t(k, \epsilon))}{\text{Log}^2 t^{-1}} \right),$$

where $\lambda_t(k, \epsilon)$ is the solution of $R(\lambda_t(k, \epsilon)) = t^{-1} \xi_t(k, \epsilon)$ and

$$h(\lambda) = -\text{Log}(D(\lambda) - C(\lambda)) - C(\lambda),$$

we again have

$$P\{X_t \le \beta_t(k, 0) \text{ i.o. as } t \to 0\} = 1,$$

$$P\{X_t \le \beta_t(k, \epsilon) \text{ i.o. as } t \to 0\} = 0.$$

The function $\beta_t(k, \epsilon)$ is thus very similar to the one above with the exception that the logs are base 2 and the constant C has been replaced by $h(\lambda_t(k, \epsilon))$. The former is since we used 2^k instead of e^k in the definition of ν and the latter seems to happen when the limit of the ratio in (3.2) does not exist. The functions C, D and h all turn out to be very nearly constant. h also must satisfy $h(2\lambda) = h(\lambda)$ since C and D do and $h(\lambda) = 0.803979\ldots$ for all λ with the next three digits varying between 632 and 973. Thus the constant C in the previous example has been replaced by a coefficient that varies but only a little.

REFERENCES

[1] L. Breiman, *A delicate law of the iterated logarithm for non-decreasing stable processes*, Ann. Math. Statist. **39** (1968), 1818-1824 (Correction **41** (1970) 1126)

[2] A. Dvoretzky and P. Erdös, *Some problems on random walk in space*, Proc. Second Berkeley Symp. Math. Statist. Probab., 353-367, Univ. California Press, Berkeley, CA, 1951

[3] N.C. Jain and W.E. Pruitt, *Lower tail probability estimates for subordinators and nondecreasing random walks*, Ann. Probab. **15** (1987), 75-101

[4] W.E. Pruitt, *The rate of escape of random walk*, Ann. Probab. **18** (1990), 1417-1461.

[5] F. Spitzer, *Some theorems concerning 2-dimensional Brownian motion*, Trans. Amer. Math. Soc. **87** (1958), 187-197

William E. Pruitt
School of Mathematics
University of Minnesota
Minneapolis, MN 55455

Microcanonical Distributions, Gibbs States, and the Equivalence of Ensembles

DANIEL W. STROOCK AND OFER ZEITOUNI

0. Introduction. *Consider an ideal gas of n identical particles which have achieved equilibrium subject only to the constraint that their average energy is some specified constant.* For the purposes of this discussion, we will interpret this sentence as saying that we have

i) a Polish space E (the *phase space* of the individual particles),

ii) a σ-finite measure λ (the *Liouville measure* for the dynamical system governing the motion of the individual particles) on $(E, \mathcal{B}_E)$ ($\mathcal{B}_E$ is the BOREL field over E),

iii) a function $U : E \longrightarrow [0, \infty)$ (the *energy*),

and that we are looking at is a regular version Λ_n of the conditional probability under λ^n given that $\frac{1}{n}\sum_{m=1}^{n} U(x_m) = 1$.

In statistical mechanics (see, for example [7] or [8]), Λ_n is called the *microcanonical distribution* of the system, and the GIBBS given wisdom is that, at least when n is very large, the marginal distribution of a typical individual particle under Λ_n is well approximated by the *Gibbs' state* γ *for* U, where $\gamma \in \mathbf{M}_1(E)$ (the space of probability measures on $(E, \mathcal{B}_E)$) has the form

$$\gamma(dx) = \frac{e^{-\beta U(x)}}{Z_\beta}\, \lambda(dx).$$

In fact, the prediction goes on to assert the *equivalence of ensembles* in the sense that, for any $d \in \mathbb{Z}^+$ and $n >> d$, subsets of d particles will be *nearly independent* under Λ_n. That is, GIBBS predicts that the distribution of $\mathbf{x} \in E^n \longmapsto \mathbf{x}_{(d)} = (x_1, \ldots, x_d) \in E^d$ will be well approximated by γ^d.

The veracity of GIBBS' wisdom is easy to check in many reasonable examples. The easiest such example is the one discussed by E. BOREL in [1]. Namely, take $E = \mathbb{R}$, λ equal to LEBESGUE's measure, and $U(x) = x^2$. It is then clear that Λ_n is the normalized uniform surface measure on the sphere

The first author acknowledges support from the grant NSF DMS–8611487 and both authors were partially supported by the grant DAAL 03–86–K–0171

$\mathbf{S}^{n-1}(n^{1/2})$ and that, for $n \geq 2$, the marginal distribution $\mu_n \in \mathbf{M}_1(E)$ of each particle has density (with respect to λ) f_n given by

$$f_n(\xi) = \frac{\omega_{n-2}}{n^{1/2}\omega_{n-1}} \left(1 - \frac{\xi^2}{n}\right)^{\frac{n-3}{2}} \mathbf{1}_{(-1,1)}\left(\frac{\xi}{n^{1/2}}\right),$$

where ω_{m-1} has been used to denote the total surface area of the unit sphere $\mathbf{S}^{m-1}(1)$ in $\mathbb{R}^m$. Recalling that

$$\omega_{m-1} = \frac{2\pi^{m/2}}{\Gamma(m/2)},$$

where Γ is EULER's Gamma–function, and applying STIRLING's formula, one quickly sees that

$$\frac{\omega_{n-2}}{n^{1/2}\omega_{n-1}} \longrightarrow \frac{1}{\sqrt{2\pi}} \quad \text{as} \quad n \to \infty;$$

and as a consequence, one finds that $\mu_n \longrightarrow \gamma$ in variation, where γ is the GIBBS' (GAUSS, MAXWELL–BOLTZMANN) state for U with $\beta = \frac{1}{2}$. (The history of this computation is rather murky. In a good deal of the mathematical literature, in [9] for instance, it is attributed to H. POINCARÉ. However, in spite of their best scholarly efforts, DIACONIS and FREEDMAN say in [4] that they were unable to discover any sense in which POINCARÉ was even involved, and they come to the conclusion that the credit should go to E. BOREL. On the other hand, KAC asserts on p. 259 of [5] that MAXWELL, not BOREL, was the first. Finally, just to thoroughly muddy the waters, we will here thank E. CARLEN for pointing out to us that, in 1866, MEHLER [10] seems to have eclipsed everybody. Although MEHLER did not have statistical mechanics in mind, his treatment in other respects is the most profound. Indeed, he anticipates there exactly the sort of role which GAUSS measure plays in the type of infinite dimensional analysis on which MCKEAN reports in [9]!) Given the preceding, it is an easy matter to verify that the d-dimensional marginal distributions of Λ_n converge (again in variation) to γ^d. In fact, all that one has to do is remark that, for any $\boldsymbol{\theta} \in \mathbf{S}^{d-1}$, $\mathbf{x} \in \mathbb{R}^n \longmapsto \boldsymbol{\theta} \cdot \mathbf{x}_{(d)}$ has the same distribution as $\mathbf{x} \in \mathbb{R}^n \longmapsto x_1$ and then apply a little elementary FOURIER analysis to deduce the desired convergence. In other words, GIBBS' prediction is completely correct in this setting. (In [4], DIACONIS and FREEDMAN give a much sharper statement of these results.)

Before dropping the preceding example, it will be useful to point out that the same conclusion would have been reached even if we had replaced the strict conditioning in the true microcanonical distribution Λ_n with the more leniently conditioned *approximate microcanonical distribution*

$$\Lambda_{n,\delta} = \lambda^n\left(\cdot\,\Big|\,\left|\frac{|\mathbf{x}|}{n} - 1\right| \leq \delta\right), \qquad \delta \in (0,1),$$

and, at the end, passed to the limit as $\delta \searrow 0$. To see this, simply repeat the sort of compution which was performed above. This observation is important to us because, in more general circumstances, it will not be clear that a smooth version of the regular conditional distribution will exist; and, therefore, it will not be possible to speak with confidence about the true microcanonical distribution.

The goal of this article is to understand GIBBS' prediction from a more general mathematical standpoint. We are, by no means, the first ones to make such an attempt. Already in **[6]**, A. KHINCHIN used central limit theory to repeat BOREL's calculation for a reasonably general class of U's. As distinguished from KHINCHIN, we will be basing our analysis on the *theory of large deviations*, but here again we are far from the first! Indeed, D. RUELLE's groundbreaking article **[11]** and the lectures by O. LANFORD **[7]** not only establish a variant on the sort of connection which we are seeking, they do so in a setting which is physically more realistic. In addition, the ideas which they developed there have come to play an extremely important role in the subsequent development of large deviation theory (cf. Chapter III of **[3]** and the comments in Remark 1.20 below). From the point of view adopted here, the key to the connection between GIBBS' prediction and the theory of large deviations lies in the interpretation of the problem in terms of the *empirical distribution*

$$\mathbf{x} \in E^n \longmapsto \mathbf{L}_n(\mathbf{x}) \equiv \frac{1}{n}\sum_{k=1}^{n} \delta_{x_k} \in \mathbf{M}_1(E). \tag{0.1}$$

That is, $\mathbf{L}_n(\mathbf{x})$ is the measure which assigns to a set Γ the fraction of those k's from $\{1,\ldots,n\}$ for which x_k is an element of Γ. Using, as we will throughout, the notation

$$\langle \phi, \mu\rangle \equiv \int_E \phi\, d\mu \quad \text{for measurable } \phi : E \longrightarrow \mathbb{R} \text{ and } \mu \in \mathbf{M}_1(E),$$

notice that the conditioning which defines the microcanonical distribution becomes $\langle U, \mathbf{L}_n(\mathbf{x})\rangle = 1$ and that GIBBS' prediction would be a consequence of

$$\lim_{n\to\infty} \lambda^n\left(\mathbf{L}_n \notin G\,\Big|\,\langle U, \mathbf{L}_n\rangle = 1\right) = 0 \tag{0.2}$$

for every open neighborhood G (in $\mathbf{M}_1(E)$) of the GIBBS' state γ. Indeed, although the strength of (0.2) depends on the topology which one puts on $\mathbf{M}_1(E)$ (cf. the discussion in Section 1 for an explanation of the topology adopted by us), no matter what that topology may be, (0.2) leads immediately to the statement that

$$\lim_{n\to\infty} \int_{E^n} \phi(x_1)\,\Lambda_n(d\mathbf{x}) = \int_E \phi\, d\gamma$$

so long as ϕ is a function for which $\mu \in \mathbf{M}_1(E) \longmapsto \langle \phi, \mu \rangle$ is continuous; and statements about the d-dimensional marginals follow from consideration of the empirical distribution of d-tuples (cf. (1.24) below).

In Section 1, we present abstract statements of results (cf. Theorems 1.4 and 1.27 and Corollary 1.35) which cover a rather general class of conditionings based on functions (not just linear ones) of the empirical distribution. Actually, because of the problems with regular versions of conditional probabilities alluded to above, we do not deal with the true microcanical distribution but, instead, with approximations to it (cf. (1.21)). Section 2 contains two applications of the results in Section 1. The first of these (cf. Theorem 2.10) is to the classical case of ideal gases (i.e., when the conditioning is determined by a linear function of the empirical distribution), and the second (cf. Theorems 2.16 and 2.21) is to gasses in which there is *quadratic interaction.*

Finally, it is a pleasure to acknowledge the guidance to the relavent literature which we have received from PERSI DIACONIS and JOHN LEWIS. Even if we have not gotten everything precisely right, we are a lot closer than we would have been without their help.

1. The Main Results. Let $\mathbf{M}_1(E)$ be the space of BOREL probability measures on the Polish space E, and denote by $\mathcal{B}_{\mathbf{M}_1(E)}$ the smallest σ-algebra over $\mathbf{M}_1(E)$ with respect to which $\nu \in \mathbf{M}_1(E) \longmapsto \langle \phi, \nu \rangle$ is measurable for every bounded $\phi \in C(E;\mathbb{R})$ (i.e., the BOREL field when $\mathbf{M}_1(E)$ is given what probabilists call the *weak topology*). When talking about quantities which are *measurable on* $\mathbf{M}_1(E)$, we will always mean that they are $\mathcal{B}_{\mathbf{M}_1(E)}$-measurable. On the other hand, when talking about quantities on $\mathbf{M}_1(E)$ which are *continuous*, it will be important for us to consider the *τ-topology* (*strong topology*) on $\mathbf{M}_1(E)$. That is, the weakest topology with respect to which $\nu \in \mathbf{M}_1(E) \longmapsto \langle \phi, \nu \rangle$ is continuous for every $\phi \in B(E;\mathbb{R})$ (the space of bounded, $\mathbb{R}$-valued, $\mathcal{B}_E$-measurable functions on E). In keeping with the notation used in [3], we will use $\Gamma^{\circ\,\tau}$ and $\overline{\Gamma}^{\tau}$

to denote, respectively, the interior and closure of a set $\Gamma \subseteq \mathbf{M}_1(E)$ with respect to the τ-topology.

In what follows, λ will be a fixed σ-finite measure λ on $(E, \mathcal{B}_E)$, $\Psi : \mathbf{M}_1(E) \longrightarrow (-\infty, \infty]$ will be a $\mathcal{B}_{\mathbf{M}_1(E)}$- measurable, lower semi-τ-continuous function which is bounded below, and $\zeta : E \longrightarrow \mathbb{R}$ will be a $\mathcal{B}_E$-measurable function with the properties that

$$
\begin{aligned}
&\overline{\lambda} \in \mathbf{M}_1(E) \quad \text{where} \quad \overline{\lambda}(dx) \equiv e^{-\zeta(x)}\,\lambda(dx) \\
&M \equiv \sup\left\{ \|\zeta\|_{L^p(\nu)} : \Psi(\nu) \le 1 \right\} < \infty \quad \text{for some} \quad p \in (1,\infty).
\end{aligned} \tag{1.1}
$$

Notice that

$$
\ell \equiv \sup\left\{ \int_E \frac{d\nu}{d\lambda} \log^- \frac{d\nu}{d\lambda}\, d\lambda : \nu << \lambda \text{ and } \Psi(\nu) \le 1 \right\} < \infty,
$$

where $\log^-$ denotes the negative part of the logarithm function. Indeed, since

$$
\log \frac{d\nu}{d\lambda} = \log \frac{d\nu}{d\overline{\lambda}} - \zeta \quad \text{whenever} \quad \nu << \lambda,
$$

we see, from (1.1), that, when $\nu << \lambda$ and $\Psi(\nu) \le 1$,

$$
\int_E \frac{d\nu}{d\lambda} \log^- \frac{d\nu}{d\lambda}\, d\lambda \le \int_E \overline{f} \log^- \overline{f}\, d\overline{\lambda} + \|\zeta\|_{L^1(\nu)} \le \frac{1}{e} + M,
$$

where $\overline{f} = \frac{d\nu}{d\overline{\lambda}}$, and we have used $t \log t \ge -\frac{1}{e}$ for $t \in [0,\infty)$. (It is for estimates of this sort that one wants to replace the possibly infinite measure λ by probability measure $\overline{\lambda}$.) Thus, we can unambiguously define the *entropy function* $\mathbf{H}_\Psi : \mathbf{M}_1(E) \longrightarrow [-\ell, \infty]$ by the formula

$$
\mathbf{H}_\Psi(\nu) = \begin{cases} \int_E \frac{d\nu}{d\lambda} \log \frac{d\nu}{d\lambda}\, d\lambda & \text{if} \quad \nu << \lambda \text{ and } \Psi(\nu) \le 1 \\ \infty & \text{otherwise.} \end{cases} \tag{1.2}
$$

(Notice that our definition of entropy differs from the physical convention by a minus sign. That is, our entropy decreases.) In fact, if, for any pair of μ, $\nu \in \mathbf{M}_1(E)$, the *relative entropy function* is defined by

$$
\mathbf{H}(\nu|\mu) = \begin{cases} \int_E \frac{d\nu}{d\mu} \log \frac{d\nu}{d\mu}\, d\mu & \text{if} \quad \nu << \mu \\ \infty & \text{otherwise,} \end{cases}
$$

then

$$
\mathbf{H}_\Psi(\nu) = \begin{cases} \mathbf{H}(\nu|\overline{\lambda}) - \langle \zeta, \nu \rangle & \text{if} \quad \Psi(\nu) \le 1 \\ \infty & \text{otherwise.} \end{cases} \tag{1.3}
$$

In particular (cf. Theorem 3.2.21 in [**3**]), since $\mathbf{H}(\cdot\,|\,\overline{\lambda})$ is a non-negative function whose level sets are sequentially compact in the τ-topology on $\mathbf{M}_1(E)$, (1.3) together with (1.1) allow us to conclude that $\mathbf{H}_\Psi$ is bounded below and has sequentially τ-compact level sets.

Theorem 1.4. *Referring to the preceding, assume that the condition in (1.1) holds. In addition, assume that*

$$m_{\Psi} \equiv \inf\left\{\mathbf{H}_{\Psi}(\nu) : \Psi(\nu) \leq 0\right\} < \infty. \tag{1.5}$$

(Implicit in (1.5) is the assumption that $\Psi(\nu) \leq 0$ for some $\nu \in \mathbf{M}_1(E)$.) Then

$$\mathfrak{G}(\Psi) \equiv \left\{\mu : \Psi(\mu) \leq 0 \text{ and } \mathbf{H}_{\Psi}(\mu) = m_{\Psi}\right\} \tag{1.6}$$

is a non-empty set which is sequentially τ-compact. In fact, when $\{\nu : \Psi(\nu) \leq 0 \text{ and } \mathbf{H}_{\Psi}(\nu) < \infty\}$ is convex, $\mathfrak{G}(\Psi)$ contains precisely one element. Next, define $\mathbf{x} \in E^n \longmapsto \mathbf{L}_n(\mathbf{x}) \in \mathbf{M}_1(E)$ as in (0.1), and make the assumption that, for every $\epsilon > 0$,

$$\lim_{n\to\infty} \mu^n\left(\left|\Psi(\mathbf{L}_n)\right| \geq \epsilon\right) = 0 \quad \textit{whenever } \Psi(\mu) = 0 \text{ and } \mathbf{H}_{\Psi}(\mu) < \infty. \tag{1.7}$$

Then, for every $\delta \in (0,1)$, there is an $N_\delta \in \mathbb{Z}^+$ such that

$$\lambda^n\left(\Psi(\mathbf{L}_n) \leq \delta\right) \in (0,\infty) \quad \textit{for all} \quad n \geq N_\delta;$$

for every $\Gamma \in \mathcal{B}_{\mathbf{M}_1(E)}$,

$$\begin{aligned} -\inf&\left\{\mathbf{H}_{\Psi}(\nu) - m_{\Psi} : \nu \in \Gamma^{\circ\tau} \text{ and } \Psi(\nu) \leq 0\right\} \\ &\leq \varliminf_{\delta\searrow 0} \varliminf_{n\to\infty} \frac{1}{n} \log\left[\lambda^n\left(\mathbf{L}_n \in \Gamma \middle| \Psi(\mathbf{L}_n) \leq \delta\right)\right] \\ &\leq \varlimsup_{\delta\searrow 0} \varlimsup_{n\to\infty} \frac{1}{n} \log\left[\lambda^n\left(\mathbf{L}_n \in \Gamma \middle| \Psi(\mathbf{L}_n) \leq \delta\right)\right] \\ &\leq -\inf\left\{\mathbf{H}_{\Psi}(\nu) - m_{\Psi} : \nu \in \overline{\Gamma}^{\tau} \text{ and } \Psi(\nu) \leq 0\right\}; \end{aligned} \tag{1.8}$$

and therefore, for every $G \in \mathbf{M}_1(E)$ which is a τ-open neighborhood of $\mathfrak{G}(\Psi)$,

$$\begin{aligned} &\varlimsup_{\delta\searrow 0} \varlimsup_{n\to\infty} \frac{1}{n} \log\left[\lambda^n\left(\mathbf{L}_n \notin G \middle| \Psi(\mathbf{L}_n) \leq \delta\right)\right] \\ &\quad\leq -\inf\left\{\mathbf{H}_{\Psi}(\nu) - m_{\Psi} : \nu \notin G \text{ and } \Psi(\nu) \leq 0\right\} < 0. \end{aligned} \tag{1.9}$$

Finally, if, in addition,

$$\mathfrak{G}(\Psi) \subseteq \{\mu : \Psi(\mu) = 0\}, \tag{1.10}$$

then, for every $\delta \in (0,1)$, *there is an* $N_\delta \in \mathbb{Z}^+$ *for which*

$$\lambda^n\Big(\big|\Psi(\mathbf{L}_n)\big| \le \delta\Big) \in (0,\infty) \quad \text{for all} \quad n \ge N_\delta, \tag{1.11}$$

and (1.8) *and* (1.9) *can be replaced, respectively, by*

$$\begin{aligned} &-\inf\Big\{\mathbf{H}_\Psi(\nu) - m_\Psi : \nu \in \Gamma^{\circ\,\tau} \text{ and } \Psi(\nu) = 0\Big\} \\ &\quad\le \varliminf_{\delta\searrow 0}\, \varliminf_{n\to\infty} \frac{1}{n}\log\Big[\lambda^n\Big(\mathbf{L}_n \in \Gamma\,\Big|\, \big|\Psi(\mathbf{L}_n)\big| \le \delta\Big)\Big] \\ &\quad\le \varlimsup_{\delta\searrow 0}\, \varlimsup_{n\to\infty} \frac{1}{n}\log\Big[\lambda^n\Big(\mathbf{L}_n \in \Gamma\,\Big|\, \big|\Psi(\mathbf{L}_n)\big| \le \delta\Big)\Big] \\ &\quad\le -\inf\Big\{\mathbf{H}_\Psi(\nu) - m_\Psi : \nu \in \overline{\Gamma}^{\tau} \text{ and } \Psi(\nu) \le 0\Big\}; \end{aligned} \tag{1.12}$$

and

$$\begin{aligned} &\varlimsup_{\delta\searrow 0}\, \varlimsup_{n\to\infty} \frac{1}{n}\log\Big[\lambda^n\Big(\mathbf{L}_n \notin G\,\Big|\, \big|\Psi(\mathbf{L}_n)\big| \le \delta\Big)\Big] \\ &\quad\le -\inf\Big\{\mathbf{H}_\Psi(\nu) - m_\Psi : \nu \notin G \text{ and } \Psi(\nu) \le 0\Big\} < 0. \end{aligned} \tag{1.13}$$

Proof: First observe that if we can prove (1.12) and (1.13) in the situation when (1.10) holds, then we can prove (1.8) and (1.9), even when (1.10) does not hold, by simply replacing the original function Ψ with the function $\Psi \vee 0$. For this reason, we will assume throughout that (1.10) holds and will concentrate on (1.12) and (1.13).

Because the level sets of $\mathbf{H}_\Psi$ are sequentially τ-compact and Ψ is lower semi-τ-continuous, $\mathfrak{G}(\Psi)$ is obviously a sequentially τ-compact set which, by (1.5), must be non-empty. For the same reason, if F is a τ-closed subset of $\mathbf{M}_1(E)$,

$$\begin{aligned} &m_\delta(F) \equiv \inf\Big\{\mathbf{H}_\Psi(\nu) : \nu \in F \text{ and } \Psi(\nu) \le \delta\Big\} \\ &\quad\nearrow m_\Psi(F) \equiv \inf\Big\{\mathbf{H}_\Psi(\nu) : \nu \in F \text{ and } \Psi(\nu) \le 0\Big\} \end{aligned} \quad \text{as} \quad \delta \searrow 0. \tag{1.14}$$

Finally, if $\{\nu : \Psi(\nu) \le 0 \text{ and } \mathbf{H}_\Psi(\nu) < \infty\}$ is a convex set and μ and ν are both elements of $\mathfrak{G}(\Psi)$, then, because $\mathbf{H}_\Psi$ is convex on $\{\mu : \Psi(\mu) \le 0\}$, $\frac{\mu+\nu}{2} \in \mathfrak{G}(\Psi)$. In particular, this means that

$$\infty > \mathbf{H}_\Psi\left(\frac{\mu+\nu}{2}\right) = \frac{\mathbf{H}_\Psi(\mu) + \mathbf{H}_\Psi(\nu)}{2},$$

which, because $t \in [0,\infty) \longmapsto t\log t$ is strictly convex, is possible only if $\mu = \nu$.

Now suppose that $G \in \mathcal{B}_{\mathbf{M}_1(E)}$ is a τ-open set and that $\mu \in G$ satisfies $\Psi(\mu) = 0$ and $\mathbf{H}_\Psi(\mu) < \infty$. Set $f = \frac{d\mu}{d\lambda}$ and $\overline{f} = \frac{d\mu}{d\overline{\lambda}}$, respectively, and, for $n \in \mathbb{Z}^+$, define $f_n = f^{\otimes n}$ and $\overline{f}_n = \overline{f}^{\otimes n}$ on E^n. Finally, for $n \in \mathbb{Z}^+$ and $\delta \in (0,1)$, set

$$\Gamma_n(\delta) = \Big\{\mathbf{x} \in E^n : f_n(\mathbf{x}) > 0,\ \mathbf{L}_n(\mathbf{x}) \in G, \text{ and } \big|\Psi\big(\mathbf{L}_n(\mathbf{x})\big)\big| \le \delta\Big\}.$$

Then

$$\lambda^n\Big(\mathbf{L}_n \in G \text{ and } \big|\Psi\big(\mathbf{L}_n\big)\big| \le \delta\Big) \ge \int_{\Gamma_n(\delta)} \frac{1}{f_n(\mathbf{x})}\,\mu^n(d\mathbf{x}).$$

Moreover, by (1.7) together with the Law of Large Numbers,

$$\mu^n\Big(\Gamma_n(\delta)\Big) \longrightarrow 1 \quad \text{as} \quad n \to \infty. \tag{1.15}$$

Hence, by JENSEN's inequality,

$$\begin{aligned}
&\varliminf_{n\to\infty} \frac{1}{n}\log\Big[\lambda^n\Big(\mathbf{L}_n \in G \text{ and } \big|\Psi(\mathbf{L}_n)\big| \le \delta\Big)\Big] \\
&\qquad \ge \varliminf_{n\to\infty} \frac{1}{n}\log\left[\frac{1}{\mu^n\big(\Gamma_n(\delta)\big)}\int_{\Gamma_n(\delta)} \frac{1}{f_n(\mathbf{x})}\,\mu^n(d\mathbf{x})\right] \\
&\qquad \ge -\varlimsup_{n\to\infty} \frac{1}{n\mu^n\big(\Gamma_n(\delta)\big)}\int_{\Gamma_n(\delta)} \log f_n(\mathbf{x})\,\mu^n(d\mathbf{x}).
\end{aligned}$$

Note that

$$\int_{\Gamma_n(\delta)} \log f_n(\mathbf{x})\,\mu^n(d\mathbf{x}) = n\mathbf{H}_\Psi(\mu) - \int_{\Gamma_n(\delta)^\complement} \log f_n(\mathbf{x})\,\mu^n(d\mathbf{x})$$

($\Gamma^\complement$ denotes the complement of Γ) and that

$$\begin{aligned}
&\int_{\Gamma_n(\delta)^\complement} \log f_n(\mathbf{x})\,\mu^n(d\mathbf{x}) \\
&= \int_{\Gamma_n(\delta)^\complement} \log \overline{f}_n(\mathbf{x})\,\mu^n(d\mathbf{x}) - \sum_{k=1}^{n}\int_{\Gamma_n(\delta)^\complement} \zeta(x_k)\,\mu^n(d\mathbf{x}) \\
&= \int_{\Gamma_n(\delta)^\complement} \overline{f}_n(\mathbf{x})\log \overline{f}_n(\mathbf{x})\,\overline{\lambda}^n(d\mathbf{x}) - n\int_{\Gamma_n(\delta)^\complement} \zeta(x_1)\,\mu^n(d\mathbf{x}) \\
&\ge -\frac{1}{e} - nM\mu^n\Big(\Gamma_n(\delta)^\complement\Big)^{1-\frac{1}{p}},
\end{aligned}$$

where, in the final step, we have used HÖLDER's inequality and (1.1). After combining these with (1.15), we arrive at the conclusion that

$$\varliminf_{n\to\infty} \frac{1}{n}\log\Big[\lambda^n\Big(\mathbf{L}_n \in G \text{ and } \big|\Psi\big(\mathbf{L}_n\big)\big| \le \delta\Big)\Big] \ge -\mathbf{H}_\Psi(\mu) \tag{1.16}$$

for every $\mu \in G$ satisfying $\Psi(\mu) = 0$ and every $\delta \in (0,1]$. In particular, because of (1.10), this proves that, for each $\delta \in (0,1)$, there is an $N_\delta \in \mathbb{Z}^+$ such that

$$\lambda^n\Big(\big|\Psi(\mathbf{L}_n)\big| \le \delta\Big) \ge \exp\left[-n\frac{m_\Psi}{1-\delta}\right] \quad \text{for} \quad n \ge N_\delta. \tag{1.17}$$

Next, suppose that $F \in \mathcal{B}_{\mathbf{M}_1(E)}$ is a non-empty, τ-closed subset of $\mathbf{M}_1(E)$, and, for $\delta \in (0,1)$, set

$$\Theta_n(\delta) = \Big\{\mathbf{x} \in E^n : \mathbf{L}_n(\mathbf{x}) \in F \text{ and } \Psi\big(\mathbf{L}_n\big) \le \delta\Big\}.$$

Then

$$\begin{aligned}\lambda^n\Big(\Theta_n(\delta)\Big) &= \int_{\Theta_n(\delta)} \exp\Big[n\langle\zeta, \mathbf{L}_n(\mathbf{x})\rangle\Big]\,\overline{\lambda}^n(d\mathbf{x}) \\ &= \int_{E^n} \exp\Big[n\Phi_\delta\big(\mathbf{L}_n(\mathbf{x})\big)\Big]\,\overline{\lambda}^n(d\mathbf{x}),\end{aligned}$$

where $\Phi_\delta : \mathbf{M}_1(E) \longrightarrow [-\infty, M]$ is given by

$$\Phi_\delta(\nu) = \begin{cases} \langle\zeta,\nu\rangle & \text{if} \quad \nu \in F \text{ and } \Psi(\nu) \le \delta \\ -\infty & \text{otherwise.}\end{cases}$$

In particular, it is clear from this that

$$\lambda^n\Big(\Psi\big(\mathbf{L}_n\big) \le 1\Big) < \infty \quad \text{for all} \quad n \in \mathbb{Z}^+,$$

and therefore (1.11) holds with the N_δ in (1.17). Moreover, since Φ_δ is clearly bounded above, measurable, and (because of the uniform integrability guaranteed by (1.1)) upper semi-τ-continuous, the form of SANOV's Theorem in Theorem 3.2.21 together with the form of VARADHAN's Lemma in Lemma 2.1.8 of [**3**] apply and yield

$$\begin{aligned}&\varlimsup_{n\to\infty} \frac{1}{n}\log\left[\int_{E^n} \exp\Big[n\Phi_\delta\big(\mathbf{L}_n(\mathbf{x})\big)\Big]\,\overline{\lambda}^n(d\mathbf{x})\right] \\ &\quad \le \sup\Big\{\Phi_\delta(\nu) - \mathbf{H}\big(\nu|\overline{\lambda}\big) : \nu \in \mathbf{M}_1(E)\Big\} \\ &\quad = -\inf\Big\{\mathbf{H}\big(\nu|\overline{\lambda}\big) - \big\langle\zeta,\nu\big\rangle : \nu \in F \text{ and } \Psi(\nu) \le \delta\Big\} = -m_\delta(F).\end{aligned}$$

(Cf. (1.3) and (1.14).) In other words, we have now shown that

$$\varlimsup_{n\to\infty} \frac{1}{n}\log\Big[\lambda^n\Big(\mathbf{L}_n \in F \text{ and } \Psi\big(\mathbf{L}_n\big) \le \delta\Big)\Big] \le -m_\delta(F) \tag{1.18}$$

for every τ-closed $F \in \mathcal{B}_{\mathbf{M}_1(E)}$ and $\delta \in (0,1]$. In particular, for each $\delta \in (0,1)$, there is an $N'_\delta \in \mathbb{Z}^+$ such that

$$\lambda^n\Big(\Psi\big(\mathbf{L}_n\big) \le \delta\Big) \le \exp\left[-n\frac{m_\delta}{1+\delta}\right] \quad \text{for all} \quad n \ge N'_\delta, \tag{1.19}$$

where $m_\delta \equiv m_\delta(E)$.

The proof of (1.12) is now easy. Namely, if $G \in \mathcal{B}_{\mathbf{M}_1(E)}$ is τ-open, then, by (1.19),

$$\begin{aligned}
&\frac{1}{n}\log\Big[\lambda^n\Big(\mathbf{L}_n \in G\Big|\,|\Psi(\mathbf{L}_n)| \le \delta\Big)\Big] \\
&\quad \ge \frac{1}{n}\log\Big[\lambda^n\Big(\mathbf{L}_n \in G \text{ and } |\Psi(\mathbf{L}_n)| \le \delta\Big)\Big] - \frac{1}{n}\log\Big[\lambda^n\Big(\Psi(\mathbf{L}_n) \le \delta\Big)\Big] \\
&\qquad \ge \frac{1}{n}\log\Big[\lambda^n\Big(\mathbf{L}_n \in G \text{ and } |\Psi(\mathbf{L}_n)| \le \delta\Big)\Big] + \frac{m_\delta}{1+\delta}
\end{aligned}$$

for each $\delta \in (0,1)$ and all $n \ge N'_\delta$. Hence, by (1.16),

$$\varliminf_{n\to\infty} \frac{1}{n}\log\Big[\lambda^n\Big(\mathbf{L}_n \in G\Big|\,|\Psi(\mathbf{L}_n)| \le \delta\Big)\Big] \ge -\mathbf{H}_\Psi(\mu) + \frac{m_\delta}{1+\delta}$$

for every $\mu \in G$ with $\Psi(\nu) = 0$; and therefore the lower bound in (1.12) follows from (1.14) after one lets $\delta \searrow 0$. A similar argument, using (1.17), (1.18), and (1.14), proves the upper bound in (1.12). (The reader who is having trouble with the technical difficulties caused by the τ-topology may want to compare the situation here with the one treated in the final part of the proof of Theorem 3.2.21 in [**3**].)

□

The elements of the set $\mathfrak{G}(\Psi)$ described in (1.6) will be called *Gibbs states of the map* Ψ.

Remark 1.20. In their work, RUELLE and LANFORD avoid the direct consideration of infinite measures by confining their attention to regions of finite measure which grow in such a way that the *density of particles* tends to a prescribed limit; and their analysis has the distinct advantage that their conditioning need not be determined by functions of the empirical distribution alone. The advantage which we gain by restricting our attention to conditions determined by the empirical distribution is that SANOV's ideas are then directly applicable and lead to an immediate connection with entropy. In other words, we have made life easy for ourselves! Moreover, we are not even the first ones to have done so: S. ZABELL (via P. DIACONIS) has pointed out to us that the relevance of SANOV's theory to problems of this sort was observed and exploited by I. CSISZÁR [**2**] (cf. Remark 1.36 below).

Continuing in the setting of Theorem 1.4, let $\Lambda^\Psi_{n\delta}, \overline{\Lambda}^\Psi_{n,\delta} \in \mathbf{M}_1(E)$ be given by

$$(1.21) \quad \Lambda^\Psi_{n,\delta} = \lambda^n\Big(\,\cdot\,\Big|\Psi(\mathbf{L}_n) \le \delta\Big) \quad \text{and} \quad \overline{\Lambda}^\Psi_{n,\delta} = \lambda^n\Big(\,\cdot\,\Big|\,|\Psi(\mathbf{L}_n)| \le \delta\Big)$$

for $\delta > 0$ and $n \geq N_\delta$, and define $\mu^{\Psi,d}_{n,\delta}$ and $\overline{\mu}^{\Psi,d}_{n,\delta}$ be the distribution of $\mathbf{x} \in E^n \longmapsto \mathbf{x}_{(d)} = (x_1, \dots, x_d) \in E^d$ under $\Lambda^{\Psi}_{n,\delta}$ and $\overline{\Lambda}^{\Psi}_{n,\delta}$, respectively. It is then clear that

$$\int_E \phi \, d\mu^{\Psi,1}_{n,\delta} = \int_{E^n} \langle \phi, \mathbf{L}_n(\mathbf{x}) \rangle \, \Lambda^{\Psi}_{n,\delta}(d\mathbf{x})$$

and

$$\int_E \phi \, d\overline{\mu}^{\Psi,1}_{n,\delta} = \int_{E^n} \langle \phi, \mathbf{L}_n(\mathbf{x}) \rangle \, \overline{\Lambda}^{\Psi}_{n,\delta}(d\mathbf{x})$$

for every $\phi \in B(E; \mathbb{R})$. Hence,

$$\text{(1.22)} \qquad \begin{array}{ll} (1.9) \ \text{implies} & \mu^{\Psi,1}_{n,\delta} \longrightarrow \mathfrak{G}(\Psi) \\ (1.13) \ \text{implies} & \overline{\mu}^{\Psi,1}_{n,\delta} \longrightarrow \mathfrak{G}(\Psi), \end{array}$$

in the τ-topology as first $n \to \infty$ and then $\delta \searrow 0$. In fact, if

$$\text{(1.23)} \qquad \mathfrak{G}^{(d)}(\Psi) \equiv \left\{ \gamma^d : \gamma \in \mathfrak{G}(\Psi) \right\} \quad \text{for} \quad d \in \mathbb{Z}^+ \setminus \{1\},$$

then by a variation on the above line of reasoning, (1.9) and (1.13) imply, respectively, that $\mu^{\Psi,d}_{nd,\delta}$ and $\overline{\mu}^{\Psi,d}_{nd,\delta}$ tend in the τ-topology to $\mathfrak{G}^{(d)}(\Psi)$.

Unfortunately, in order to improve this last statement to get the convergence for the full sequence (i.e., not just the subsequence $\{nd\}_{n=1}^{\infty}$), it seems that one has to get the analogues of (1.9) and (1.13) for the d-dimensional empirical distribution functions

$$\text{(1.24)} \qquad \mathbf{x} \in E^{n+d-1} \longmapsto \mathbf{L}^{(d)}_n(\mathbf{x}) \equiv \frac{1}{n} \sum_{m=1}^{n} \delta_{(x_m, \dots, x_{m+d-1})} \in \mathbf{M}_1(E^d);$$

and this will require us to introduce some additional notation.

Given $d \in \mathbb{Z}^+ \setminus \{1\}$, $1 \leq k \leq \ell \leq d$, and $\nu \in \mathbf{M}_1(E^d)$, $\nu_{[k,\ell]} \in \mathbf{M}_1(E^{\ell-k+1})$ will be used to denote the marginal distribution of $\mathbf{x} \in E^d \longmapsto (x_k, \dots, x_\ell) \in E^{\ell-k+1}$ under ν, and $\nu_{[k]}$ will be used when $\ell = k$. Also, given $\nu \in \mathbf{M}_1(E^d)$, we will use $\mathbf{x} \in E^{d-1} \longmapsto \nu(x) \in \mathbf{M}_1(E)$ to stand for the ν-almost surely unique measurable map such that

$$\int_{E^d} \phi(x_1, \dots, x_{d-1}) \, \psi(x_d) \, \nu(d\mathbf{x}) = \int_{E^{d-1}} \phi(\mathbf{x}) \left(\int_E \psi(y) \, \nu(\mathbf{x}, dy) \right) \nu_{[1,d-1]}(d\mathbf{x})$$

for all $\phi \in B(E^{d-1};\mathbb{R})$ and $\psi \in B(E;\mathbb{R})$. (That is, in the language of probability theory, $\nu(x_1,\dots,x_{d-1})$ is the marginal distribution of x_d under the regular conditional distribution of ν given $(x_1,\dots,x_{d-1})$.) Finally, set

$$\mathbf{M}_1^{\mathbf{S}}(E^d) = \left\{\nu \in \mathbf{M}_1(E^d) : \nu_{[1,d-1]} = \nu_{[2,d]}\right\},$$

and define $\mathbf{H}_\Psi^{(d)} : \mathbf{M}_1(E^d) \longrightarrow (-\infty,\infty]$ so that

$$\mathbf{H}_\Psi^{(d)}(\nu) = \int_{E^{d-1}} \left(\int_E \frac{d\nu(\mathbf{x})}{d\lambda} \log \frac{d\nu(\mathbf{x})}{d\lambda}\, d\lambda\right) \nu_{[1,d-1]}(d\mathbf{x})$$

when $\nu \in \mathbf{M}_1^{\mathbf{S}}(E^d)$, $\nu << \lambda$, and $\Psi(\nu_{[1]}) \le 1$, and so that $\mathbf{H}_\Psi^{(d)}(\nu) = \infty$ for any other $\nu \in \mathbf{M}_1(E^d)$. Just as in the development of $\mathbf{H}_\Psi$, it is an easy matter to check that the integral involved in this definition is well-defined and that an alternative description of $\mathbf{H}_\Psi^{(d)}$ is

$$\text{(1.25)}\quad \mathbf{H}_\Psi^{(d)}(\nu) = \begin{cases} \mathbf{J}_{\overline{\lambda}}^{(d)}(\nu) - \langle \zeta, \nu_{[1]}\rangle & \text{if } \nu \in \mathbf{M}_1^{\mathbf{S}}(E^d) \text{ and } \Psi(\nu_{(1)}) \le 1 \\ \infty & \text{otherwise,} \end{cases}$$

where

$$\text{(1.26)}\qquad \mathbf{J}_{\overline{\lambda}}^{(d)}(\nu) \equiv \int_{E^{d-1}} \mathbf{H}\big(\nu(\mathbf{x})\big|\overline{\lambda}\big)\, \nu_{[1,d-1]}(d\mathbf{x}) \quad \text{for} \quad \nu \in \mathbf{M}_1^{\mathbf{S}}(E^d).$$

In particular, it follows from (1.25) that $\mathbf{H}_\Psi^{(d)}$ is lower semi-τ-continuous and bounded below. In addition, since (cf. Lemma 4.4.7 in [3]), $\mathbf{H}(\nu|\overline{\lambda}) \le d\,\mathbf{J}_{\overline{\lambda}}^{(d)}(\nu)$ for $\nu \in \mathbf{M}_1(E^d)$, the level sets of $\mathbf{H}_\Psi^{(d)}$ are sequentially τ-compact.

Theorem 1.27. *Let everything be as in Theorem 1.4, and, for a given $d \in \mathbb{Z}^+ \setminus \{1\}$, define $\mathbf{x} \in E^{n+d-1} \longmapsto \mathbf{L}_n^{(d)}(\mathbf{x}) \in \mathbf{M}_1(E^d)$, $\mathfrak{G}^{(d)}(\Psi)$, and $\mathbf{H}_\Psi^{(d)}$ as above. Then, for every τ-closed, measurable $F \subseteq \mathbf{M}_1(E^d)$, (cf. (1.5))*

$$\text{(1.28)}\qquad \begin{aligned} &\varlimsup_{\delta\searrow 0}\ \varlimsup_{n\to\infty} \frac{1}{n} \log\left[\lambda^{n+d-1}\left(\mathbf{L}_n^{(d)} \in F \,\middle|\, \Psi(\mathbf{L}_n) \le \delta\right)\right] \\ &\qquad \le -\inf\left\{\mathbf{H}_\Psi^{(d)}(\nu) - m_\Psi : \nu \in F \text{ and } \Psi(\nu_{[1]}) \le 0\right\}, \end{aligned}$$

and so, for every $G \in \mathbf{M}_1(E^d)$ which is a τ-open neighborhood of $\mathfrak{G}^{(d)}(\Psi)$,

$$\text{(1.29)}\qquad \begin{aligned} &\varlimsup_{\delta\searrow 0}\ \varlimsup_{n\to\infty} \frac{1}{n} \log\left[\lambda^{n}\left(\mathbf{L}_n^{(d)} \notin G \,\middle|\, \Psi(\mathbf{L}_n) \le \delta\right)\right] \\ &\qquad \le -\inf\left\{\mathbf{H}_\Psi^{(d)}(\nu) - m_\Psi : \nu \notin G \text{ and } \Psi(\nu_{[1]}) \le 0\right\} < 0. \end{aligned}$$

Moreover, when (1.10) *holds,* (1.28) *and* (1.29) *can be replaced by*

$$\tag{1.30} \begin{aligned}&\overline{\lim_{\delta\searrow 0}}\ \overline{\lim_{n\to\infty}}\ \frac{1}{n}\log\left[\lambda^{n+d-1}\left(\mathbf{L}_n^{(d)}\in F\,\Big|\,|\Psi(\mathbf{L}_n)|\le\delta\right)\right]\\ &\qquad\le -\inf\left\{\mathbf{H}_\Psi^{(d)}(\nu)-m_\Psi : \nu\in F \text{ and } \Psi(\nu_{[1]})\le 0\right\}.\end{aligned}$$

and

$$\tag{1.31} \begin{aligned}&\overline{\lim_{\delta\searrow 0}}\ \overline{\lim_{n\to\infty}}\ \frac{1}{n}\log\left[\lambda^{n}\left(\mathbf{L}_n^{(d)}\notin G\,\Big|\,|\Psi(\mathbf{L}_n)|\le\delta\right)\right]\\ &\qquad\le -\inf\left\{\mathbf{H}_\Psi^{(d)}(\nu)-m_\Psi : \nu\notin G \text{ and } \Psi(\nu_{[1]})\le 0\right\}<0,\end{aligned}$$

respectively.

Proof: By the same reasoning as we used at the start of the proof of Theorem 1.4, we may and will assume throughout that (1.10) holds and will restrict our attention to the (1.30) and (1.31). Furthermore, given (1.30), the only problem presented by (1.31) is whether

$$m_\Psi=\inf\left\{\mathbf{H}_\Psi^{(d)}(\nu) : \Psi(\nu_{[1]})\le 0\right\} \text{ and } \mathbf{H}_\Psi^{(d)}(\mu)=m_\Psi \implies \mu\in\mathfrak{G}^{(d)}(\Psi).$$

To see these, use the strict convexity of $t\in[0,\infty)\longmapsto t\log t$ to show that, for any $\mu\in\mathbf{M}_1^{\mathbf{S}}(E^d)$, $\mathbf{H}_\Psi^{(d)}(\mu)\ge\mathbf{H}_\Psi(\mu_{[1]})$ and that equality holds if and only if $\mu=\mu_{[1]}^d$. Thus, all that remains is to prove (1.30).

The strategy for proving (1.30) is essentially the same as the one which we adopted in the proof of the upper bound in (1.12). Thus, because of (1.17), everything comes down to showing that, for each $\delta\in(0,1)$,

$$\tag{1.32} \overline{\lim_{n\to\infty}}\ \frac{1}{n}\log\left[\lambda^{n+d-1}\left(\mathbf{L}_n^{(d)}\in F \text{ and } \Psi(\mathbf{L}_n)\le\delta\right)\right]\le -m_\delta^{(d)}(F),$$

where

$$\begin{aligned}m_\delta^{(d)}(F)&\equiv\inf\left\{\mathbf{H}_\Psi^{(d)}(\nu) : \nu\in F \text{ and } \Psi(\nu_{[1]})\le\delta\right\}\\ &\nearrow m_\Psi^{(d)}(F)\equiv\inf\left\{\mathbf{H}_\Psi^{(d)}(\nu) : \nu\in F \text{ and } \Psi(\nu_{[1]})\le 0\right\}\end{aligned}\quad\text{as}\quad\delta\searrow 0.$$

The proof of the second of these claims differs in no way from the derivation of (1.14), one simply uses the sequential τ-compactness of the $\mathbf{H}_\Psi^{(d)}$-level sets. Hence, we are left with the proof of (1.32), and this will require us to use a large deviation principle which is not covered by SANOV's Theorem.

Namely, what we will need to know is that, for any measurable, upper semi-τ-continuous, $\Phi : \mathbf{M}_1(E^d) \longrightarrow [-\infty,\infty)$ which is bounded above:

$$\overline{\lim_{n\to\infty}} \frac{1}{n} \log\left[\int_{E^{n+d-1}} \exp\left[n\Phi\left(\mathbf{L}_n^{(d)}(\mathbf{x})\right)\right] \overline{\lambda}^{n+d-1}(d\mathbf{x})\right] \le \sup\left\{\Phi(\nu) - \mathbf{J}_{\overline{\lambda}}^{(d)}(\nu) : \nu \in \mathbf{M}_1(E^d)\right\}, \tag{1.33}$$

where $\mathbf{J}_{\overline{\lambda}}^{(d)}$ is defined in (1.26). Before discussing the origins of (1.33), we will check that it provides a proof of (1.32). Indeed, define

$$\Psi^{(d)}(\nu) = \Psi\left(\frac{1}{d}\sum_{k=1}^{d} \nu_{[k]}\right) \quad \text{for} \quad \nu \in \mathbf{M}_1(E^d),$$

$$\zeta^{(d)}(\mathbf{x}) = \frac{1}{d}\sum_{k=1}^{d} \zeta(x_k) \quad \text{for} \quad \mathbf{x} \in E^d,$$

and

$$\Phi_\delta^{(d)}(\nu) = \begin{cases} \langle \zeta^{(d)}, \nu \rangle & \text{if } \Psi^{(d)}(\nu) \le \delta \\ -\infty & \text{otherwise.} \end{cases}$$

It is then an easy matter to check that (1.33) applies to $\Phi_\delta^{(d)}$ and that the probability being estimated in (1.32) is equal to

$$\int_{E^{n+d-1}} \exp\left[n\Phi_\delta^{(d)}\left(\mathbf{L}_n^{(d)}(\mathbf{x})\right)\right] \overline{\lambda}^{n+d-1}(d\mathbf{x}).$$

Hence, (1.32) follows from (1.33) together with (1.25).

In order to prove (1.33), we will use the DONSKER–VARADHAN large deviation theory for MARKOV chains as it is developed in Chapter IV of [**3**]. Thus, in the discussion which follows, all the references to material outside the present article will be to [**3**]; and what we will be doing is constructing a lexicon for translating the situation here into the situation treated there. To this end, set $\Sigma = E^d$, define the transition probability function Π on Σ by

$$\Pi(\sigma, \Gamma) = \int_E \mathbf{1}_\Gamma(x_2, \ldots, x_d, y)\, \overline{\lambda}(dy) \tag{1.34}$$

for $\sigma = (x_1, \ldots, x_d) \in \Sigma$ and $\Gamma \in \mathcal{B}_\Sigma$, let $\{P_\sigma : \sigma \in \Sigma\}$ be the corresponding MARKOV family of probability measures on $\Omega \equiv \Sigma^{\mathbb{N}}$, and set $P = \int_\Sigma P_\sigma \overline{\lambda}^d(d\sigma)$. It is then an easy matter to check that the distribution of $\mathbf{x} \in E^{n+d-1} \longmapsto \mathbf{L}_n^{(d)}(\mathbf{x}) \in \mathbf{M}_1(E^d)$ under $\overline{\lambda}^{n+d-1}$ coincides with the distribution of

$$\omega \in \Omega \longmapsto \mathbf{L}_n(\omega) \equiv \frac{1}{n}\sum_{m=1}^{n} \delta_{X_m(\omega)}$$

under P, where $X_m(\omega)$ is used to denote the mth-coordinate of $\omega \in \Omega$. Moreover, because the d-step transition $\Pi^d(\sigma,\,\cdot\,) \equiv \overline{\lambda}^d$, it is easy to verify that the all the conditions required in Section 4.1 (in particular, (**U**) at the bottom of page 100) are satisfied. Hence, Theorem 4.1.43 and Exercise 4.1.53 apply and lead to (1.33) with $\mathbf{J}_{\overline{\lambda}}^{(d)}$ replaced on the right hand side by the quantity $\mathbf{J}_\Pi$ described in (4.1.38). Finally, by Lemma 4.4.9, one has that, for Π's having form in (1.34), $\mathbf{J}_{\overline{\lambda}}^{(d)} = \mathbf{J}_\Pi$.

□

Given Theorem 1.27, in particular (1.29) and (1.31), we can proceed in exactly the same way as we did in the derivation of (1.22) to get the following.

Corollary 1.35. *If everything is as in* Theorem *1.4 and the measures*

$$\left\{\mu_{n,\delta}^{\Psi,d} : n \geq d \text{ and } \delta \in (0,1)\right\} \quad \text{and} \quad \left\{\overline{\mu}_{n,\delta}^{\Psi,d} : n \geq d \text{ and } \delta \in (0,1)\right\}$$

are defined in the way they were following (1.21), *then, for every* $\phi \in B(E^d;\mathbb{R})$,

$$\overline{\lim_{\delta\searrow 0}}\ \overline{\lim_{n\to\infty}}\ \inf_{\gamma\in\mathfrak{G}(\Psi)} \left|\langle \phi, \mu_{n,\delta}^{\Psi,d}\rangle - \langle \phi, \gamma^d\rangle\right| = 0,$$

and, when (1.10) *holds,*

$$\overline{\lim_{\delta\searrow 0}}\ \overline{\lim_{n\to\infty}}\ \inf_{\gamma\in\mathfrak{G}(\Psi)} \left|\langle \phi, \overline{\mu}_{n,\delta}^{\Psi,d}\rangle - \langle \phi, \gamma^d\rangle\right| = 0.$$

Remark 1.36. When λ is a probability measure and $\{\mu : \Psi(\mu)\}$ is a convex set with non-empty interior, CSISZÁR's results in [2] apply and yield a far sharper statement of Corollary 1.35.

2 Applications. Our goal in this section is to see what the results in Section 1 have to say about specific choices of the function Ψ, and we will begin with the classical case of an ideal gas. Thus, let λ be σ-finite measure on $(E, \mathcal{B}_E)$ and $U : E \longrightarrow [0,\infty)$ a $\mathcal{B}_E$-measurable function with the properties that

$$\lambda\left(U < 1\right) \wedge \lambda\left(U > 1\right) > 0 \tag{2.1}$$

and

$$(2.2)\quad \sigma \equiv \int_E \exp\left[-\alpha U^{\frac{1}{p}}\right] d\lambda < \infty \quad \text{for some } \alpha \in [0,\infty) \text{ and } p \in (1,\infty).$$

Define $\Psi : \mathbf{M}_1(E) \longrightarrow [0,\infty]$ by

$$(2.3)\qquad \Psi(\nu) = \begin{cases} \langle U,\nu\rangle - 1 & \text{when } U \text{ is unbounded} \\ \frac{|<U,\nu>-1|}{\|U\|_u} & \text{when } U \text{ is bounded} \end{cases}$$

for $\nu \in \mathbf{M}_1(E)$. It is then clear both that Ψ is $\mathcal{B}_{\mathbf{M}_1(E)}$-measurable and lower semi-τ-continuous and that $\{\nu : \Psi(\nu) \le 0\}$ is convex. In addition, (1.1) is satisfied when (cf. (2.2)) $\zeta \equiv \alpha U^{\frac{1}{p}} + \log \sigma$. Hence, the associated entropy function $\mathbf{H}_\Psi$ in (1.2) satisfies (1.3) and enjoys the properties described following (1.3). Also, by the Law of Large Numbers, if $\langle U,\mu\rangle < \infty$, then

$$\lim_{n\to\infty} \mu^n \left(\mathbf{x} \in E^n : \left| \frac{1}{n}\sum_{m=1}^n U(x_m) - \langle U,\mu\rangle \right| \ge \epsilon \right) = 0 \quad \text{for all} \quad \epsilon > 0;$$

and so (1.7) certainly holds. Finally, define $\gamma_\beta \in \mathbf{M}_1(E)$ by

$$(2.4)\quad \gamma_\beta(dx) = \frac{e^{-\beta U(x)}}{Z_\beta}\lambda(dx) \quad \text{for} \quad \beta \in \mathbb{R} \text{ with } Z_\beta \equiv \int_E e^{-\beta U}\, d\lambda < \infty.$$

Then,

$$(2.5)\quad \beta_\infty \equiv \inf\{\beta \in \mathbb{R} : Z_\beta < \infty\} \le 0 \text{ and } \beta_\infty = -\infty \text{ if } U \text{ is bounded},$$

$$(2.6)\qquad \beta \in (\beta_\infty,\infty) \longmapsto \langle U,\gamma_\beta\rangle \in (0,\infty) \text{ is strictly decreasing and continuous},$$

$$(2.7)\qquad \lim_{\beta\nearrow\infty} \langle U,\gamma_\beta\rangle < 1,$$

$$(2.8)\qquad \lim_{\beta\searrow\beta_\infty} \langle U,\gamma_\beta\rangle > 1 \quad \text{if} \quad \text{either } \lambda(E) = \infty \text{ or } \beta_\infty = -\infty,$$

and

$$(2.9)\qquad \mathbf{H}_\Psi(\gamma_\beta) = -\beta\langle U,\gamma_\beta\rangle - \log Z_\beta \quad \text{so long as} \quad Z_\beta < \infty.$$

Indeed, both (2.5) (remark that λ must be finite if U is bounded) and (2.9) are obvious, (2.6) comes from

$$\frac{d}{d\beta}\langle U,\gamma_\beta\rangle = \int_E \left(U - \langle U,\gamma_\beta\rangle\right)^2 d\gamma_\beta \quad \text{for} \quad \beta \in (\beta_\infty,\infty),$$

and both (2.7) and (2.8) are relatively straight-forward consequences of (2.1) together with (2.2).

With these preliminaries, we are a long way towards having proved the following theorem.

Theorem 2.10. *Let λ be a σ-finite measure on $(E, \mathcal{B}_E)$ and $U : E \longrightarrow [0,\infty)$ a $\mathcal{B}_E$-measurable function satisfying* (2.1) *and* (2.2); *and define $\Psi : \mathbf{M}_1(E) \longrightarrow [-1,\infty]$ as in* (2.3). *Then the results in Section 1 apply, and the set $\mathfrak{G}(\Psi)$ in* (1.6) *contains precisely one element γ_Ψ. In particular,* (1.8) *and* (1.9) *always hold, and, when $\langle U, \gamma_\Psi\rangle = 1$, so do* (1.12) *and* (1.13). *Finally, referring to* (2.4) *and* (2.5), *set*

$$\beta(U) = \inf\left\{\beta \in (\beta_\infty, \infty) : \langle U, \gamma_\beta\rangle \le 1\right\}.$$

Then $\beta(U) \in \mathbb{R}$, $Z_{\beta(U)} < \infty$, $\Psi(\gamma_{\beta(U)}) \le 0$, and

$$\text{(2.11)}\qquad \gamma_\Psi = \gamma_{\beta(U)} \quad \text{if and only if} \quad \beta(U)\Big(\langle U, \gamma_\Psi\rangle - \langle U, \gamma_{\beta(U)}\rangle\Big) \le 0.$$

In particular, the equality $\gamma_\Psi = \gamma_{\beta(U)}$ holds if either $\beta(U) \ge 0$ or $\beta(U) < 0$ and $\langle U, \gamma_\Psi\rangle = 1$; and, therefore, both it and $\langle U, \gamma_\Psi\rangle = 1$ hold when either $\lambda(E) = \infty$ or U is bounded.

Proof: To see that Section 1 applies, recall that we have already checked (1.7), and, when U is unbounded, (1.5) is an immediate consequence of (2.7) combined with (2.9) while, in the case when U is bounded, (1.5) follows from these two plus (2.8). Thus, $\mathfrak{G}(\Psi) \ne \emptyset$, and, because $\{\nu : \Psi(\nu) \le 0\}$ is convex, it must contain precisely one element. Hence, all that remains is to verify the assertions about $\beta(U)$ and the relationship between γ_Ψ and $\gamma_{\beta(U)}$.

We first note that, by (2.7), $\beta(U) < \infty$. Next, either $\beta_\infty > -\infty$, in which case $\beta(U)$ is certainly in $(-\infty,\infty)$, or $\beta_\infty = -\infty$, in which case (2.8) implies that $\beta(U) > -\infty$. To see that $Z_{\beta(U)} < \infty$, suppose the contrary, use (2.2) to see that $\lambda(U \le 2) < \infty$, and arrive at the contradiction:

$$\lim_{\beta \searrow \beta(U)} \langle U, \gamma_\beta\rangle \ge 2\left(1 - \lim_{\beta\searrow\beta(U)} \frac{\lambda(U \le 2)}{Z_\beta}\right) = 2.$$

Furthermore, we know that, by (2.8), $\langle U, \gamma_{\beta(U)}\rangle = 1$ when $\beta_\infty = -\infty$ and, by The Monotone Convergence Theorem, $\langle U, \gamma_{\beta(U)}\rangle \le 1$ in general. Hence, $\Psi(\gamma_{\beta(U)}) \le 0$.

Finally, because $\Psi(\gamma_{\beta(U)}) \le 0$, (2.9), combined with the same reasoning which led to (1.3), justifies

$$\begin{aligned}\mathbf{H}_\Psi(\gamma_{\beta(U)}) &\ge \mathbf{H}_\Psi(\gamma_\Psi)\\ &= \mathbf{H}(\gamma_\Psi | \gamma_{\beta(U)}) + \mathbf{H}_\Psi(\gamma_{\beta(U)}) - \beta(U)\Big(\langle U, \gamma_\Psi\rangle - \langle U, \gamma_{\beta(U)}\rangle\Big);\end{aligned}$$

and therefore,

$$\beta(U)\Big(\langle U,\gamma_{\Psi}\rangle - \langle U,\gamma_{\beta(U)}\rangle\Big) \geq \mathbf{H}\big(\gamma_{\Psi}\big|\gamma_{\beta(U)}\big).$$

Because $\mathbf{H}(\mu|\nu) = 0$ if and only if $\mu = \nu$, this completes the derivation of (2.11); and so, to complete the proof, it suffices to remark that $\langle U,\gamma_{\beta(U)}\rangle$ must be equal to 1 when $\beta(U) > 0$.

□

Remark 2.12. Observe that, when λ is finite but U is unbounded, $\Psi(\gamma_{\Psi})$ need not vanish. Indeed, simply take $E = (0,\infty)$ and consider the measure $\lambda(dx) = 2(1+x)^{-3}\mathbf{1}_{(0,\infty)}\,dx$ and the function $U(x) = \log(1+x)$. It is then easy to see that $\beta(U) = 0$ and, therefore, that $\Psi(\gamma_{\Psi}) = -\frac{1}{2}$.

The gas being modeled in Theorem 2.10 is ideal because the particles are *non-interacting*. A simple example of a system in which there is *interaction* can be constructed as follows. Let $U : E^2 \longrightarrow [0,\infty)$ be a measurable, symmetric function, and define $\Psi : \mathbf{M}_1(E) \longrightarrow [-1,\infty]$ by

$$\Psi(\nu) = \begin{cases} \langle U\nu,\nu\rangle - 1 & \text{if } U \text{ is unbounded} \\ \frac{|<U\nu,\nu>-1|}{\|U\|_u} & \text{if } U \text{ is bounded,} \end{cases} \tag{2.13}$$

where we have introduced the notation $U\nu$ to denote the function given by

$$U\nu(x) = \int_E U(x,y)\,\nu(dy) \quad \text{for} \quad \nu \in \mathbf{M}_1(E) \text{ and } x \in E.$$

We will assume that there exists a measurable function $\zeta : E \longrightarrow \mathbb{R}$ with the properties that, for some $a, b \in (0,\infty)$ and $p \in (1,\infty)$,

$$\begin{aligned} &|\zeta(x)|^p \leq a \inf_{y\in E} U(x,y) + b \quad \text{for all } x \in E \\ &\overline{\lambda}(dx) \equiv e^{-\zeta(x)}\,\lambda(dx) \quad \text{is a probability measure} \\ &\langle U\overline{\lambda},\overline{\lambda}\rangle < \infty. \end{aligned} \tag{2.14}$$

Clearly all of the conditions required in the discussion preceding Theorem 1.4 are satisfied by the function Ψ, and therefore the corresponding function $\mathbf{H}_{\Psi}$ in (1.2) satisfies (1.3) and enjoys the properties mentioned after that equation. In order to guarantee that (1.7) is satisfied, we will also make the assumption that

$$\int_E U(x,x)\,\mu(dx) < \infty \quad \text{whenever} \quad \mathbf{H}_{\Psi}(\mu) < \infty. \tag{2.15}$$

Theorem 2.16. *Let Ψ be given by (2.13) with a measurable, symmetric $U : E^2 \longrightarrow [0, \infty)$ satisfying the conditions in (2.14) and (2.15), and assume that there is a $\mu \in \mathbf{M}_1(E)$ for which $\Psi(\mu) \leq 0$ and $\mathbf{H}_\Psi(\mu) < \infty$. Then the results of Section 1 (i.e., Theorems 1.4 and 1.27 and Corollary 1.35) apply. Moreover, if $\gamma \in \mathfrak{G}(\Psi)$ satisfies either $\Psi(\gamma) < 0$ or $\Psi(\gamma) = 0$ and $\|U\gamma\|_{L^2(\gamma)} > 1$, then there exists a set $S(\gamma) \in \mathcal{B}_E$ and $\beta(\gamma) \in \mathbb{R}$ such that*

$$(2.17) \quad \gamma(dx) = \mathbf{1}_{S(\gamma)}(x) \exp\Big[\mathbf{H}_\Psi(\gamma) + \beta(\gamma)\Big(\langle U\gamma, \gamma\rangle - U\gamma(x)\Big)\Big]\, \lambda(dx).$$

In fact, if $\Psi(\gamma) < 0$, then $\beta(\gamma) = 0$.

Proof: In order to check that the results of Section 1 apply, it suffices to show that (1.7) is satisfied. To this end, first note that the classical TCHEBYCHEV proof of the weak Law of Large Numbers handles the case when U is a bounded function. To handle the general case, set $U_R = U \wedge R$ and $W_R = U - U_R$ for $R \in (0, \infty)$, and check that

$$\int_{E^n} \left(\frac{1}{n^2} \sum_{k,\ell=1}^{n} W_R\big(x_k, x_\ell\big)\right) \mu^n(d\mathbf{x}) = \left(1 - \frac{1}{n}\right) \langle W_R\mu, \mu\rangle + \frac{1}{n} \int_E W_R(x, x)\, \mu(dx).$$

Hence, the general result follows from the bounded case together with (2.15).

Turning to the characterization of $\gamma \in \mathfrak{G}(\Psi)$, set $f = \frac{d\gamma}{d\lambda}$, and take $S(\gamma) = \{f > 0\}$. To handle the case when $\Psi(\gamma) < 0$, let $\psi : E \longrightarrow R$ be a bounded, measurable function, and choose $\delta > 0$ so that, for all $t \in (-\delta, \delta)$,

$$\nu_t(dx) \equiv \frac{1 + t\psi(x)}{1 + t < \psi, \gamma >} \gamma(dx)$$

is a probability measure which satisfies $\langle U\nu_t, \nu_t\rangle < 1$. It is then an easy matter to check that, since $t \in (-\delta, \delta) \longmapsto \mathbf{H}_\Psi(\nu_t)$ is a smooth function with a minimum at $t = 0$,

$$0 = \int_{S(\gamma)} \psi\Big(\log f - \mathbf{H}_\Psi(\gamma)\Big)\, d\lambda;$$

and because this is true for every bounded, measurable ψ, (2.17) with $\beta(\gamma) = 0$ is immediate.

We now turn to the case when $\Psi(\gamma) = 0$ or, equivalently, $\langle U\gamma, \gamma\rangle = 1$. Under the condition that $\|U\gamma\|_{L^2(\gamma)} > 1$, it is clear that we can find an $M \in (0,\infty)$ with the property that $\int_E g\, U\gamma\, d\gamma > 1$, where $g \equiv M \wedge U\gamma$. Next, let a bounded, measurable $\psi : E \longrightarrow \mathbb{R}$ be given, set

$$f_{t,a,b}(x) = \big(1 + a + b\, g(x) + t\psi(x)\big)\, f(x) \quad \text{for} \quad (t,a,b) \in \mathbb{R}^3$$

and consider the mapping

$$F(t,a,b) \equiv \begin{bmatrix} t \\ \int_E f_{t,a,b}\, d\lambda \\ \frac{1}{2}\int_{E^2} f_{t,a,b}(x)\, U(x,y)\, f_{t,a,b}(y)\, d\lambda^2 \end{bmatrix}.$$

It is then an easy matter to see that F is smooth,

$$F(0,0,0) = \begin{bmatrix} 0 \\ 1 \\ 1 \end{bmatrix}, \quad \text{and} \quad \frac{\partial F}{\partial(t,a,b)}(0,0,0) = \begin{bmatrix} 1 & 0 \quad 0 \\ \langle \psi, \gamma\rangle & \mathbf{A} \\ \langle \psi U\gamma, \gamma\rangle & \end{bmatrix},$$

where $\mathbf{A}$ is an invertible 2×2 matrix which does not depend on ψ. Hence, by the Inverse Function Theorem, we can find a $\delta > 0$ and smooth maps $t \in (-\delta,\delta) \longmapsto a_\psi(t) \in \mathbb{R}$ and $t \in (-\delta,\delta) \longmapsto b_\psi(t) \in \mathbb{R}$ such that $a_\psi(0) = 0 = b_\psi(0)$ and, for each $t \in (-\delta,\delta)$,

$$\nu_t(dx) \equiv f_{t,a_\psi(t),b_\psi(t)}(x)\, \lambda(dx)$$

is a probability measure satisfying $\langle U\nu_t, \nu_t\rangle = 1$. In addition, there exist constants $c_1, \dots, c_4 \in \mathbb{R}$, which do not depend on ψ, such that

$$\dot a_\psi(0) = c_1\langle \psi, \gamma\rangle + c_2\langle \psi U\gamma, \gamma\rangle \quad \text{and} \quad \dot b_\psi(0) = c_3\langle \psi, \gamma\rangle + c_4\langle \psi U\gamma, \gamma\rangle.$$

Finally, since $t \in (-\delta,\delta) \longmapsto \mathbf{H}_\Psi(\nu_t)$ is a smooth function which has a minimum at $t = 0$, one now finds that there are numbers d_1, $d_2 \in \mathbb{R}$, which do not depend on ψ, such that

$$\int_E \psi\big(\log f - d_1 - d_2 U\gamma\big)\, d\gamma = 0;$$

from which (2.17) is clear.

□

Remark 2.18 . When it holds, the relation in (2.17) has a pleasing physical interpretation. Namely, it says that, at least on its support, the GIBBS state γ looks like the GIBBS state which one would predict for a system of non-interacting particles moving in the *mean-field* corresponding to γ itself. To see that, in general, (2.17) may fail to hold if $\Psi(\gamma) = 0$ but $\|U\gamma\|_{L^2(\gamma)} = 1$, consider the situation when $E = \mathbf{S}^1 = \mathbf{S}^1(1)$ and λ is the normalized HAAR measure on $\mathbf{S}^1$. Next, thinking of $\mathbf{S}^1$ as embedded in $\mathbb{C}$, define $\theta : \mathbf{S}^1 \longrightarrow [-\pi, \pi)$ to be the isomorphism determined by $x = e^{\sqrt{-1}\,\theta(x)}$ for each $x \in \mathbf{S}^1$. Finally, take

$$U(x, y) = 1 + \frac{\big(\theta(x) - 1\big)\big(\theta(y) - 1\big)}{25},$$

and define Ψ accordingly as in the second part of (2.13). It is then clear that $\langle U\mu, \mu\rangle = 1$ if and only if $\langle \theta, \mu\rangle = 1$. Hence, by Theorem 2.10, there is precisely one GIBBS state γ for Ψ, and

$$\gamma(dx) = \frac{1}{Z_\alpha} e^{-\alpha\theta(x)}\,\lambda(dx)$$

where $\alpha \in \mathbb{R}$ is determined by

$$Z_\alpha \equiv \int_{\mathbf{S}^1} e^{-\alpha\theta(x)}\,\lambda(dx) = \int_{\mathbf{S}^1} \theta(x)\, e^{-\alpha\theta(x)}\,\lambda(dx).$$

In particular, $\alpha \neq 0$ and so there can be no value of $\beta \in \mathbb{R}$ for which (2.17) holds.

Aside from the problem raised in Remark 2.18, Theorem 2.16 leaves open several questions. In the first place, at least when $\beta \geq 0$, (2.17) would be much improved if we knew that the set $S(\gamma)$ coincided with the set where $U\gamma$ is finite. Secondly, one like to know when $\mathfrak{G}(\Psi)$ contains only one element and, even if $\mathfrak{G}(\Psi)$ contains more than one element, when $\Psi(\gamma) = 0$ for all $\gamma \in \mathfrak{G}(\Psi)$. In order to address these questions, we will make use of the following.

Lemma 2.19. *For each* $\beta \in [0, \infty)$, *define*

$$\epsilon_\Psi(\beta) = \inf\left\{\mathbf{H}_\Psi(\nu) + \frac{\beta}{2}\langle U\nu, \nu\rangle : \nu \in \mathbf{M}_1(E)\right\}$$

and

$$\mathcal{E}_\Psi(\beta) = \left\{\mu \in \mathbf{M}_1(E) : \mathbf{H}_\Psi(\mu) + \frac{\beta}{2}\langle U\mu, \mu\rangle = \epsilon_\Psi(\beta)\right\}.$$

Then:

i) For every $\beta \in [0,\infty)$, $\mathcal{E}_\Psi(\beta) \neq \emptyset$ and $\bigcup_{\beta\geq 0}\mathcal{E}_\Psi(\beta)$ is sequentially τ-compact.

ii) For $\beta_1 < \beta_2$, $\mu_1 \in \mathcal{E}_\Psi(\beta_1)$, and $\mu_2 \in \mathcal{E}_\Psi(\beta_2)$, $\langle U\mu_2,\mu_2\rangle \leq \langle U\mu_1,\mu_1\rangle$.

iii) Suppose that $\{\beta_n\}_1^\infty \subseteq [0,\infty)$, $\mu_n \in \mathcal{E}(\beta_n)$ for each $n \in \mathbb{Z}^+$, $\beta_n \longrightarrow \beta$, and $\mu_n \longrightarrow \mu$ in the τ-topology. Then $\mu \in \mathcal{E}(\beta)$, and, if either $\beta \in (0,\infty)$ or U is bounded, then $\langle U\mu_n,\mu_n\rangle \longrightarrow \langle U\mu,\mu\rangle$.

iv) If $\mu \in \mathcal{E}_\Psi(\beta)$ and $\Psi(\mu) < 1$, then

$$\mu(dx) = \mathbf{1}_{\{U\mu(x)<\infty\}} \exp\Big[\mathbf{H}_\Psi(\mu) + \beta\langle U\mu,\mu\rangle - \beta U\mu(x)\Big]\,\lambda(dx). \tag{2.20}$$

Proof: Part **i)** is an easy consequence of the properties of $\mathbf{H}_\Psi$. To prove the monotonicity in **ii)**, simply note that

$$\begin{aligned}
\mathbf{H}_\Psi(\mu_1) &+ \frac{\beta_2}{2}\langle U\mu_1,\mu_1\rangle \\
&= \mathbf{H}_\Psi(\mu_1) + \frac{\beta_1}{2}\langle U\mu_1,\mu_1\rangle + \frac{\beta_2-\beta_1}{2}\langle U\mu_1,\mu_1\rangle \\
&\leq \mathbf{H}_\Psi(\mu_2) + \frac{\beta_1}{2}\langle U\mu_2,\mu_2\rangle + \frac{\beta_2-\beta_1}{2}\langle U\mu_1,\mu_1\rangle \\
&= \mathbf{H}_\Psi(\mu_2) + \frac{\beta_2}{2}\langle U\mu_2,\mu_2\rangle - \frac{\beta_2-\beta_1}{2}\Big(\langle U\mu_2,\mu_2\rangle - \langle U\mu_1,\mu_1\rangle\Big) \\
&\leq \mathbf{H}_\Psi(\mu_1) + \frac{\beta_2}{2}\langle U\mu_1,\mu_1\rangle - \frac{\beta_2-\beta_1}{2}\Big(\langle U\mu_2,\mu_2\rangle - \langle U\mu_1,\mu_1\rangle\Big).
\end{aligned}$$

As for part **iii)**, first observe that, when U is bounded,

$$\langle U\mu,\mu\rangle = \lim_{n\to\infty}\langle U\mu_n,\mu_n\rangle$$

and that, even when U is unbounded,

$$\langle U\mu,\mu\rangle \leq \underline{\lim}_{n\to\infty}\langle U\mu_n,\mu_n\rangle.$$

Thus, because $\mathbf{H}_\Psi$ is lower semi-τ-continuous, it is clear that $\mu \in \mathcal{E}_\Psi(\beta)$ in any case. In addition, if $\beta > 0$ and $\langle U\mu,\mu\rangle$ were strictly less than $\underline{\lim}_{n\to\infty}\langle U\mu_n,\mu_n\rangle$, one would derive the contradiction that

$$\mathbf{H}_\Psi(\mu) + \frac{\beta_n}{2}\langle U\mu,\mu\rangle < \mathbf{H}_\Psi(\mu_n) + \frac{\beta_n}{2}\langle U\mu_n,\mu_n\rangle$$

for sufficiently large n's. Finally, when U is bounded, there is no problem.

Turning to the proof of **iv**), set $f = \frac{d\mu}{d\lambda}$. Our first goal is to check that $\lambda\big(f = 0 \text{ and } U\mu < \infty\big) = 0$. Indeed, if this were not the case, we could choose $Z \in (0,\infty)$ so that the measure ν given by

$$\nu(dx) \equiv \frac{\mathbf{1}_{\{0\}}\big(f(x)\big)\, e^{-2U\mu(x)}}{Z}\, \lambda(dx)$$

is an element of $\mathbf{M}_1(E)$. Moreover, by an easy computation (using (2.14)) we would have that $\langle U\nu,\nu\rangle \vee \langle U\mu,\nu\rangle < \infty$. But this would mean that, for all sufficiently small $t \in (0,1)$, $\nu_t \equiv (1-t)\mu + t\nu$ satisfies $\Psi(\nu_t) < 1$ and, therefore, that

$$\begin{aligned} 0 &\le \frac{\mathbf{H}_\Psi(\nu_t) - \mathbf{H}_\Psi(\mu) + \frac{\beta}{2}\Big(\langle U\nu_t,\nu_t\rangle - \langle U\mu,\mu\rangle\Big)}{t} \\ &\le \mathbf{H}_\Psi(\nu) - \mathbf{H}_\Psi(\mu) + \log t + \frac{\beta}{2}\Big(\langle U\mu,\nu\rangle + t\langle U\nu,\nu\rangle\Big), \end{aligned}$$

which is obviously impossible. With this result, we have now proved that (2.20) holds on the set $\{f = 0\}$. To handle the set $\{f > 0\}$, we proceed in precisely the same way as we did in the proof of (2.17) when we were assuming that $\Psi(\gamma) < 0$.

□

Theorem 2.21. *Let everything be as in* Theorem 2.16*, only this time add the assumptions that*

$$\langle U\mu,\nu\rangle \le \frac{\langle U\mu,\mu\rangle + \langle U\nu,\nu\rangle}{2} \quad \text{if} \quad \mathbf{H}_\Psi(\mu) \vee \mathbf{H}_\Psi(\nu) < \infty, \tag{2.22}$$

$$\inf\Big\{\langle U\mu,\mu\rangle : \mathbf{H}_\Psi(\mu) < \infty\Big\} < 1, \tag{2.23}$$

and

$$\lambda(E) = \infty \quad \text{or} \quad U \text{ is bounded and } \int_{E^2} U\, d\lambda^2 \ge \lambda(E)^2. \tag{2.24}$$

Then $\mathfrak{G}(\Psi)$ contains exactly one element γ_Ψ. In fact, there is a unique $\beta(U) \in [0,\infty)$ such that

$$\gamma_\Psi(dx) = \frac{\mathbf{1}_{S(\Psi)}(x)}{Z(\Psi)} \exp\Big[-\beta(U)\, U\gamma_\Psi(x)\Big]\, \lambda(dx) \tag{2.25}$$

where

$$Z(\Psi) = \exp\Big[\beta(U)\, \langle U\gamma_\Psi,\gamma_\Psi\rangle + \mathbf{H}_\Psi(\gamma_\Psi)\Big]$$

and $S(\Psi) \equiv \{U\gamma_\Psi < \infty\}$. *Finally,* $\Psi(\gamma_\Psi) = 0$ *if either* U *is bounded or* $\beta(U) > 0$.

Proof: First note that, because of (2.22) and the strict convexity of $t \in [0,\infty) \longmapsto t\log t$, the set $\mathcal{E}_\beta(\beta)$ described in Lemma 2.19 contains exactly one element μ_β for each $\beta \in [0,\infty)$. Hence, by parts **i)** and **iii)** of Lemma 2.19, $\beta \in [0,\infty) \longmapsto \mu_\beta \in \mathbf{M}_1(E)$ is continuous as a mapping into the τ-topology and, so long as either $\beta \in (0,\infty)$ or U is bounded, $\lim_{\alpha\to\beta} \langle U\mu_\alpha, \mu_\alpha\rangle = \langle U\mu_\beta, \mu_\beta\rangle$. Next, using (2.23), choose $\mu\mathbf{M}_1(E)$ so that $\langle U\mu, \mu\rangle < 1$ and $\mathbf{H}_\Psi(\mu) < \infty$, and conclude from

$$\mathbf{H}_\Psi(\mu_\beta) + \frac{\beta}{2}\langle U\mu\beta, \mu_\beta\rangle \le \mathbf{H}_\Psi(\mu) + \frac{\beta}{2}\langle U\mu, \mu\rangle$$

that $\lim_{\beta\to\infty} \langle U\mu_\beta, \mu_\beta\rangle < 1$. Thus,

$$\beta(U) \equiv \inf\{\beta \in [0,\infty) : \langle U\mu_\beta, \mu_\beta\rangle \le 1\} < \infty;$$

and, by lower semi-continuity, $\Psi(\mu_{\beta(U)}) \le 0$. In fact, by (2.24) combined with **ii)** of Lemma 2.19 and the continuity already discussed, we have

$$\text{(2.26)} \qquad \Psi(\mu_{\beta(U)}) = 0 \quad \text{if either } U \text{ is bounded} \quad \text{or } \beta(U) > 0.$$

Finally, by combining (2.22) with (2.26) and the same reasoning as we used at the end of Theorem 2.10, we see that

$$\begin{aligned}\mathbf{H}_\Psi(\gamma_\Psi) &= \mathbf{H}(\gamma_\Psi|\mu_{\beta(U)}) + \mathbf{H}_\Psi(\mu_{\beta(U)}) \\ &\qquad + \beta(U)\left(\langle U\mu_{\beta(U)}, \mu_{\beta(U)}\rangle - \langle U\mu_{\beta(U)}, \gamma_\Psi\rangle\right) \\ &\ge \mathbf{H}(\gamma_\Psi|\gamma_{\beta(U)}) + \mathbf{H}_\Psi(\gamma_\Psi) + \frac{\beta(U)}{2}\left[\langle U\mu_{\beta(U)}, \mu_{\beta(U)}\rangle - \langle U\gamma_\Psi, \gamma_\Psi\rangle\right];\end{aligned}$$

which is possible only if $\gamma_\Psi = \mu_{\beta(U)}$. Hence, by (2.20) and (2.26), the proof is complete.

□

Remark 2.27. To see how uniqueness may break down when (2.22) does not hold, refer to the setting of Remark 2.18 and, for $\alpha \in (0, \pi^2/3]$, define $U_\beta : \mathbf{S}^1 \times \mathbf{S}^1 \longrightarrow [0,\infty)$ by

$$U_\alpha(x,y) = \frac{(\theta(x) - \theta(y))^2}{2\alpha}$$

and $\Psi_\alpha : \mathbf{M}_1(\mathbf{S}^1) \longrightarrow [0, \pi^2]$ by

$$\Psi_\alpha(\nu) = \frac{\alpha^2}{2\pi^2}\left|\iint_{\mathbf{S}^1\times\mathbf{S}^1} U_\alpha(x,y)\,\nu(dx)\nu(dy) - 1\right|.$$

It is then an easy matter to check that, for each $\alpha \in (0, \pi^2/3]$, all the conditions in Theorem 2.16 are satisfied. Now let $\mathfrak{G}(\Psi_\alpha)$ be the set of GIBBS' states for Ψ_α and, for $\eta \in [0, 2\pi)$, let $\mathbf{R}_\eta \in C(\mathbf{S}^1; \mathbf{S}^1)$ be rotation through angle η. It is then clear that $\mathfrak{G}(\Psi_\alpha)$ is $\mathbf{R}_\eta$-invariant in the sense that $\mu \circ \mathbf{R}_\eta^{-1} \in \mathfrak{G}(\Psi_\alpha)$ if $\mu \in \mathfrak{G}(\Psi_\alpha)$. In particular, given any $\mu \in \mathfrak{G}(\Psi_\alpha)$, use Theorem 2.10 to check that

$$\mu(dx) = \frac{1}{Z_{\beta_\alpha}} \exp\left[-\frac{\beta_\alpha}{\alpha}\Big(\theta(x) - \langle \theta, \mu\rangle\Big)^2\right] \lambda(dx),$$

where β_α is the unique $\beta \in [0, \infty)$ for which

$$\int_{\mathbf{S}^1} \frac{\theta(x)^2}{\alpha} \exp\left[-\frac{\beta\theta(x)^2}{\alpha}\right] \lambda(dx) = Z_\beta \equiv \int_{\mathbf{S}^1} \exp\left[-\frac{\beta\theta(x)^2}{\alpha}\right] \lambda(dx).$$

Conclude, in particular, that $\mathfrak{G}(\Psi_\alpha)$ contains only one element precisely when $\alpha = \frac{\pi^2}{3}$ (in which case $\beta_\alpha = 0$); and, when $\alpha \in (0, \pi^2/3)$, $\mathfrak{G}(\Psi_\alpha)$ has infinitely many elements.

Finally, notice that the statement of Theorem 2.21, in particular the conditions in (2.3) and (2.22), allows for the possibility that U explodes as one approaches the diagonal. For example, consider the situation when $E = \mathbb{R}^3 \times \mathbb{R}^3$ and U is given by

$$U(x, y) = \xi^2 + \eta^2 + \frac{\mathbf{1}_{\Delta^c}(p, q)}{|p - q|} \quad \text{for} \quad x = (p, \xi) \text{ and } y = (q, \eta),$$

where Δ is the diagonal set $\{(p, q) : p = q\}$. Next, let λ be a σ-finite measure on E, and use λ_1 and λ_2 to denote the marginal distributions of λ on, respectively, the first and second factors of E. Assuming that λ_1 is a finite measure for which $\iint \frac{1}{|p-q|} \lambda_1(dp)\lambda_1(dq) < \infty$ and that $\int e^{-|\xi|}\, d\lambda_2 < \infty$, it is straight-forward matter to check that the hypotheses of Theorem 2.21 are satisfied.

REFERENCES

[1] Borel, Émile, *Sur les principes de la théorie cinétique des gaz*, Ann. de l'École Norm. sup., 3e série **t.23** (1906), 9–33.

[2] Csiszár, I., *Sanov property, generalized I-projection and a conditional limit theorem*, Ann. Prob. **12 # 3** (1983), 768–793.

[3] Deuschel, J.-D. and Stroock, D. W., "Large Deviations," Academic Press Series in Pure & Appl. Math. 137, 1989.

[4] Diaconis, P. and Freedman, D., *A dozen de Finetti–style results in search of a theory*, Ann. Inst. Poincaré, Sup. au #2 . **23** (1987), 417–433.

[5] Kac, M., "Probability and Related Topics in Physical Sciences," Interscience, 1959.

[6] Khinchin, A., "Mathematical Foundations of Statistical Mechanics," Dover Press, 1949.

[7] Landford, O.E., *Entropy and equilibrium states in classical statistical mechanics*, in "Statistical Mechanics and Mathematical Problems," Edited by A. Lenard. Lecture Notes in Physics **20**, Springer, Berlin, 1973, pp. 1–113.

[8] Martin–Lof, Anders, "Statistical Mechanics and the Foundations of Thermodynamics," Lecture Notes in Physics #101, Springer–Verlag, 1979.

[9] McKean, H.P. Jr., *Geometry of Differential Space*, Ann. Prob. **1** (1973), 197–206.

[10] Mehler, F.G., *Über die Entwicklung einer Funktion von beliebig vielen Variabeln nach Laplaceschen Funktionen höherer Ordnung*, J. Reine u. Angewandte Mathematik (1866), 161–176.

[11] Ruelle, D., *Correlation functionals*, J. Math. Phys. **6** (1965), 201–220.

D. Stroock, M.I.T. 2-272, Cambridge, MA, 02139

O. Zeitouni, Electrical Engineering, Technion, Haifa, Israel

Power Counting Theorem in Euclidean Space

NORMA TERRIN AND MURAD S. TAQQU *

Consider integrals whose integrands are products of functions bounded near zero by powers of affine functionals and near infinity by different powers of the same functionals. The power counting theorem provides conditions for integrability.

1. Introduction and main result. For what values of $\alpha_0, \alpha_1, \alpha_2$ and α_3 does the integral

$$\int_{\mathbf{R}^3} \left| \frac{e^{i(x+y+z)} - 1}{i(x+y+z)} \right|^{\alpha_0} |x+1|^{\alpha_1} |y-2|^{\alpha_2} |z|^{\alpha_3} dx dy dz$$

converge?

The power counting theorem stated below gives the answer. It provides convergence conditions for a class of integrals on $\mathbf{R}^n$ whose integrands are products of functions bounded near zero by powers of linearly dependent affine functionals and near infinity by different powers of those functionals. The theorem is useful in studying asymptotic distributions of statistics of time series with long-range dependence, that is, time series whose covariance function decreases to zero like a power function (Terrin and Taqqu,

* This research was supported at Boston University by the National Science Foundation grant DMS-88-05627 and by the AFSOR grant 89-0115.

AMS 1980 subject classifications. Primary 26B99. Secondary: 60F17, 82A25

Keywords and phrases. Affine functionals, critical exponents, self-similar processes, long-range dependence

1989). It is related to Dyson's power counting theorem in mathematical physics (Dyson, 1949, Nakanishi, 1971), as extended by Lowenstein and Zimmermann (1975). Fox and Taqqu (1987) extended a basic lemma of Lowenstein and Zimmermann (1975) to obtain the condition for convergence of integrals on $[-t,t]^n$ when the integrands are products of powers of linearly dependent affine functionals. This paper generalizes their result from the $[-t,t]^n$ to $\mathbf{R}^n$. It also provides necessary and sufficient conditions for convergence.

One can use the generalized Hölder inequality, established in Avram and Taqqu (1989) to establish the sufficiency part, in the special case where the exponents are in $[0,1)$. In the present paper, however, the exponents can take any value in $(-\infty,\infty)$. This is a nontrivial extension because the positive and negative exponents can compensate each other. Moreover, since we integrate on $\mathbf{R}^n$, the integrand may diverge for large as well as small values of the argument.

The results obtained in this paper are used in Terrin and Taqqu (1989) and Terrin and Taqqu (1990) to establish non-central limit theorems for quadratic forms of stationary sequences with long-range dependence.

We start with some notation. For $i=1,\dots,m$, let $M_i(\mathbf{x})$ be a linear functional on $\mathbf{R}^n$. Let θ_i be real constants and set

$$T=\{M_i:\ i=1,\dots,m\}$$

and

$$T'=\{M_1+\theta_1,\dots,M_m+\theta_m\}.$$

T is a set of linear functionals and T' is a set of affine-linear functionals. For $0<a_i<b_i<\infty$, $c_i>0$ and real constants α_i and β_i define

$$P(\mathbf{x})=f_1(M_1(\mathbf{x})+\theta_1)\dots f_m(M_m(\mathbf{x})+\theta_m) \tag{1.1}$$

where $|f_i|$ is bounded above on (a_i,b_i) and

$$|f_i(y)|\le\begin{cases} c_i|y|^{\alpha_i} & \text{if } |y_i|<a_i,\\ c_i|y|^{\beta_i} & \text{if } |y_i|>b_i.\end{cases} \tag{1.2}$$

Our goal is to find conditions on the α_i's and β_i's such that

$$\int_{\mathbf{R}^n}P(\mathbf{x})d^nx<\infty, \tag{1.3}$$

and to show that those conditions are implied by (1.3) when (1.2) is replaced by

$$|f_i(y)| \geq \begin{cases} c_i|y|^{\alpha_i} & \text{if } |y_i| < a_i, \\ c_i|y|^{\beta_i} & \text{if } |y_i| > b_i. \end{cases} \tag{1.4}$$

Define

$$\begin{array}{llr} (i) & s_T(W) = \mathrm{span}(W) \cap T & \text{for } W \subset T, \\ (ii) & s_{T'}(W) = \mathrm{span}(W) \cap T' & \text{for } W \subset T', \\ (iii) & d_0(W) = r(W) + \sum_{s_{T'}(W)} \alpha_i & \text{for } W \subset T', \\ (iv) & d_\infty(W) = r(T) - r(W) + \sum_{T\setminus s_T(W)} \beta_i & \text{for } W \subset T, \end{array}$$

where span(W) denotes the linear span of W, $r(W)$ is the number of linearly independent elements of W, and summation over a subset of T is the summation over all indices of functionals in the subset.

We now state a first version of the power counting theorem.

Theorem 1.1. *If* $r(T) = n$,
(a) $d_0(W) > 0$ *for every nonempty subset* W *of* T' *with* $W = s_{T'}(W)$, *and*
(b) $d_\infty(W) < 0$ *for every proper subset* W *of* T, $W = s_T(W)$, *including the empty set, then*

$$\int_{\mathbf{R}^n} |f_1(M_1(\mathbf{x}) + \theta_1) \ldots f_m(M_m(\mathbf{x}) + \theta_m)| d^n x < \infty. \tag{1.5}$$

Remark 1.1. When integrating around zero, the presence of the additive constants θ_i may influence the convergence of the integral. However, the integral converges at infinity if and only if the integral without additive constants converges. Thus, condition (a), for convergence around zero, involves $T' = \{M_1 + \theta_1, \ldots, M_m + \theta_m\}$, while condition (b), for convergence around infinity, involves $T = \{M_1, \ldots, M_m\}$.

A set $W \subset T$ is said to be *padded* if for every linear functional M in W, M is also in $s_T(W \setminus \{M\})$. That is, M can be obtained as a linear combination of other elements in W.

There is a corresponding definition for affine-linear functionals. Let $L_i = M_i + \theta_i$, $i = 1, \ldots, m$. A set $W \subset T'$ is *padded* if for every L in W, L is also in $s_{T'}(W \setminus \{L\})$.

Corollary 1.1. *If $\alpha_i > -1$ in Theorem 1.1, then it suffices to verify (a) for sets W that are also padded. If $\beta_i \geq -1$, then it suffices to verify (b) for sets W that are also padded.*

Proof. Assume (a) for subsets of T' that are padded. Now suppose $W = s_{T'}(W)$ is not padded and partition W into W_1 and W_2 where $W_1 = \{L \in W : L \notin s_{T'}(W \backslash \{L\})\}$. If $W_2 \neq \emptyset$, then

$$d_0(W) = r(W) + \sum_W \alpha_i = r(W_2) + |W_1| + \sum_{W_2} \alpha_i + \sum_{W_1} \alpha_i$$

$$= d_0(W_2) + |W_1| + \sum_{W_1} \alpha_i \geq d_0(W_2) > 0,$$

since $W_2 = s_T(W_2)$ and W_2 is padded. If $W_2 = \emptyset$ (e.g. if W is a singleton) then $d_0(W) = |W_1| + \sum_{W_1} \alpha_i > 0$.

Now assume (b) for subsets of T that are also padded. Suppose $W = s_T(W)$ is not padded and partition W into W_1 and W_2 where $W_1 = \{M \in W : M \notin s_T(W \backslash \{M\})\}$. Then

$$\begin{aligned} d_\infty(W) &= r(T) - r(W_1 \cup W_2) + \sum_{T \backslash (W_1 \cup W_2)} \beta_i \\ &= r(T) - r(W_2) - |W_1| + \sum_{T \backslash W_2} \beta_i - \sum_{W_1} \beta_i \\ &= d_\infty(W_2) - (|W_1| + \sum_{W_1} \beta_i) \\ &\leq d_\infty(W_2) < 0. \quad \blacksquare \end{aligned}$$

Example 1.1.

$$(1.6) \qquad \int_{\mathbf{R}^n} \left| \frac{e^{i(x_1 + \ldots + x_n)} - 1}{i(x_1 + \ldots + x_n)} \right|^{\gamma_0} |x_1|^{-\gamma_1} \ldots |x_n|^{-\gamma_n} dx_1 \ldots dx_n.$$

In this case $T = T' = \{M_1 = x_1 + \ldots + x_n, M_2 = x_1, \ldots, M_{n+1} = x_n\}$, $(\alpha_1, \alpha_2, \ldots, \alpha_{n+1}) = (0, -\gamma_1, \ldots, -\gamma_n)$ and $(\beta_1, \beta_2, \ldots, \beta_{n+1})$ $= (-\gamma_0, -\gamma_1, \ldots, -\gamma_n)$. If we assume $\gamma_0 \leq 1$ and $\gamma_i < 1$, $i = 1, \ldots, n$, we may apply Corollary 1.1 The only sets $W = s_T(W)$ that are padded are $\emptyset$ and T. We have

$$d_0(T) = r(T) + \sum_T \alpha_i = n - \sum_{i=1}^n \gamma_i$$

and

$$d_\infty(\emptyset) = r(T) + \sum_T \beta_i = n - \sum_{i=0}^{n} \gamma_i.$$

Thus whenever $\gamma_0, \ldots, \gamma_n$ satisfy $n - \sum_{i=1}^n \gamma_i > 0$ and $n - \sum_{i=0}^n \gamma_i < 0$, the integral converges. For example, $\gamma_0 = 1$, $\gamma_i = \gamma$ with $1 - \frac{1}{n} < \gamma < 1$, $i = 1, \ldots, n$, is one solution.

Example 1.2.

$$\int_{\mathbf{R}^n} \left| \frac{e^{i(x_1+x_2)} - 1}{i(x_1 + x_2)} \right|^{\gamma_0} |x_1|^{-\gamma_1} |x_2|^{-\gamma_2} dx_1 dx_2 \tag{1.7}$$

is the same integral as in Example 1.1 with $n = 2$. Here, however, we let the γ_i's be arbitrary. Hence, in addition to the sets T and $\emptyset$, we must investigate all sets $W = s_T(W)$. We find

$$\begin{aligned} d_0(\{x_i\}) &= 1 - \gamma_i, \ i = 1, 2, \\ d_0(\{x_1 + x_2\}) &= 1 + 0, \\ d_\infty(\{x_1\}) &= 2 - 1 - \gamma_0 - \gamma_2 = 1 - (\gamma_0 + \gamma_2), \\ d_\infty(\{x_2\}) &= 2 - 1 - \gamma_0 - \gamma_1 = 1 - (\gamma_0 + \gamma_1), \\ d_\infty(\{x_1 + x_2\}) &= 2 - 1 - \gamma_1 - \gamma_2 = 1 - (\gamma_1 + \gamma_2). \end{aligned}$$

Hence (1.7) is finite if for $i = 1, 2$, $\gamma_i < 1$, $\gamma_0 + \gamma_i > 1$, and $\gamma_1 + \gamma_2 > 1$. For example, $\gamma_0 = 2$, $1/2 < \gamma_i < 1$, $i = 1, 2$, is one solution.

Example 1.3.

$$\int_{\mathbf{R}^2} f(x_1 + x_2 + 1) |x_1|^{-\gamma} |x_2|^{-\gamma} dx_1 dx_2, \tag{1.8}$$

where

$$f(u) = \begin{cases} |u|^{2\gamma - 2} & \text{if } |u| < 1, \\ |u|^{-1} & \text{if } |u| \geq 1, \end{cases}$$

and $1/2 < \gamma < 1$.

Our assumption on γ implies, as in Example 1.1, that condition (b) of Corollary 1.1 is satisfied, so we turn our attention to integration around zero. Since $(\alpha_1, \alpha_2, \alpha_3) = (2\gamma - 2, -\gamma, -\gamma)$, we have $\alpha_i > -1$, $i = 1, 2, 3$, and thus Corollary 1.1 is applicable. Because there are no padded sets, there is nothing to check. The integral is therefore finite.

Note that if $x+y+1$ were replaced by $x+y$, T' would be padded, and $d_0(T') = 2 + \sum_{i=1}^{3} \alpha_i = 0$. It is a consequence of Theorem 1.2 below that the integral would diverge.

Theorem 1.1 gives sufficient conditions for convergence. But are these conditions also necessary? That is, when (1.4) holds, does (1.5) imply (a) and (b)? A simple example shows that it does not: one has

$$\int_{-1}^{1} (x+1)^{-2/3}(x-1)^{-2/3}x^{-2/3}dx < \infty,$$

but if $W = \{x+1, x-1, x\}$, then $d_0(W) = 2 - 3(\frac{2}{3}) = 0$, and hence condition (a) is not met. However, a slight modification to (a) will make the converse of Theorem 1.1 hold. The modification involves the constants θ_i. We need first the following

Definition 1.1. A subset $\{M_{i_1} + \theta_{i_1}, \ldots, M_{i_k} + \theta_{i_k}\}$ of T' is *strongly independent* if $\{M_{i_1}, \ldots, M_{i_k}\}$ is linearly independent.

Note that strong independence implies linear independence but the converse is not true. For example, $\{x, x+1\}$ is linearly independent but not strongly independent.

Let

$$\hat{d}_0(W) = |W| + \sum_{s_{T'}(W)} \alpha_i. \tag{1.9}$$

The following power counting theorem provides conditions that are both necessary and sufficient for convergence.

Theorem 1.2. (Power counting theorem.) *Suppose* $r(T) = n$. *Then:*

(I) If f_i, $i = 1, \ldots, m$ *satisfies (1.2), the conditions*

(a') $\hat{d}_0(W) > 0$ *for every nonempty strongly independent subset* W *of* T', *and*

(b') $d_\infty(W) < 0$ *for every proper subset* W *of* T, $W = s_T(W)$, *including the empty set, imply*

$$\int_{\mathbf{R}^n} |f_1(M_1(\mathbf{x}) + \theta_1) \ldots f_m(M_m(\mathbf{x}) + \theta_m)| d^n x < \infty. \tag{1.10}$$

(II) Conversely, if f_i, $i = 1, \ldots, m$ satisfies (1.4), then (1.10) implies conditions (a') and (b').

Corollary 1.2. *If the integration in (1.10) is on $[-t,t]^n$, $t > 0$, instead of $\mathbf{R}^n$, then the integral converges if and only if (a') holds.*

Part I of Theorem 1.2 is proved in Section 2 and part II is proved in Section 3. Note that

$$(a) \Rightarrow (a'), \quad (b) \equiv (b') \tag{1.11}$$

but

$$(a') \not\Rightarrow (a). \tag{1.12}$$

One has $(a) \Rightarrow (a')$ because if W is strongly independent then W is also linearly independent, and hence

$$\hat{d}_0(W) = |W| + \sum_{s_{T'}(W)} \alpha_i = r(s_{T'}(W)) + \sum_{s_{T'}(W)} \alpha_i = d_0(s_{T'}(W)).$$

To show that $(a') \not\Rightarrow (a)$, consider $T' = \{x+1, x-1, x\}$ and $\alpha_1 = \alpha_2 = \alpha_3 = -2/3$. Here (a') is satisfied because all the strongly independent sets W are singletons and $\hat{d}_0(W) = 1 - 2/3 > 0$, but (a) is not satisfied because $d_0(T') = 2 - 3(\frac{2}{3}) = 0$.

Because of (1.11) it suffices to prove Theorem 1.2. This is done in the next two sections.

2. Proof of Part I of the Power Counting Theorem 1.2. Assume $r(T) = n$, fix a permutation $\sigma = (\sigma_1, \ldots, \sigma_m)$ of $\{1, \ldots, m\}$ and let

$$E_\sigma = \{\mathbf{x} \in \mathbf{R}^n : |M_{\sigma_1}(\mathbf{x})| \leq \cdots \leq |M_{\sigma_m}(\mathbf{x})|\}. \tag{2.1}$$

Construct a basis $B_\sigma = \{M_{\tau_1}, \ldots, M_{\tau_n}\}$ using the greedy algorithm: let $M_{\tau_1} = M_{\sigma_1}$ and let M_{τ_k} be the smallest element (in absolute value) in $T \backslash s_T\{M_{\tau_1}, \ldots, M_{\tau_{k-1}}\}$, $k = 2, \ldots, n$. The sets $T_1 = s_T\{M_{\tau_1}\}$, $T_k = s_T\{M_{\tau_1}, \ldots, M_{\tau_k}\} \backslash s_T\{M_{\tau_1}, \ldots, M_{\tau_{k-1}}\}$, $k = 2, \ldots, n$ form a partition of T. Define

$$q_k(\sigma) = \sum_{T_k} \beta_i, \quad k = 1, \ldots, n. \tag{2.2}$$

The permutation σ determines $T_1, \ldots, T_n$, and hence $q_1, \ldots, q_n$.

Lemma 2.1. *If $r(T) = n$, then $d_\infty(W) < 0$ for every proper subset W of T with $s_T(W) = W$ if and only if*

$$d_k(\sigma) = k + \sum_{i=n-k+1}^{n} q_i(\sigma) < 0, \quad k = 1, \ldots, n,$$

for every permutation $\sigma = (\sigma_1, \ldots, \sigma_m)$ of $\{1, \ldots, m\}$.

Proof. First we show sufficiency. Fix a permutation σ and let $W_k = T_1 \cup \ldots \cup T_{n-k}$, $k = 1, 2, \ldots, n-1$. We have $r(W_k) = n-k$, $s_T(W_k) = W_k$ and $T \backslash s_T(W_k) = T_{n-k+1} \cup \ldots \cup T_n$. Hence

$$\begin{aligned} 0 > d_\infty(W_k) &= n - (n-k) + \sum_{T \backslash s_T(W_k)} \beta_i \\ &= k + \sum_{T_{n-k+1}} \beta_i + \ldots + \sum_{T_n} \beta_i \\ &= k + \sum_{i=n-k+1}^{n} q_i(\sigma), \\ &= d_k(\sigma), \quad k = 1, 2, \ldots, n-1. \end{aligned}$$

When $k = n$,

$$0 > d_\infty(\emptyset) = n + \sum_T \beta_i = n + \sum_{i=1}^{n} q_i(\sigma) = d_n(\sigma).$$

To establish the necessity, let W be a proper subset of T with $s_T(W) = W$ and suppose $|W| = \mu$. Arrange the elements M_i of W in an arbitrary order. We get $W = \{M_{\sigma_1}, M_{\sigma_2}, \ldots, M_{\sigma_\mu}\}$, where $\sigma = (\sigma_1, \ldots, \sigma_m)$ is a permutation of $\{1, \ldots, m\}$. Let $T_1, \ldots, T_n$ be the partition of T corresponding to σ. Since by assumption W is proper and $W = s_T(W)$, either $W = T_1 \cup \ldots \cup T_{n-k}$ for some $k = 1, \ldots, n-1$, or $W = \emptyset$. If W is nonempty,

$$d_\infty(W) = n - (n-k) + \sum_{T \backslash W} \beta_i = k + \sum_{i=n-k+1}^{n} q_i(\sigma) = d_k(\sigma) < 0.$$

If $W = \emptyset$,

$$d_\infty(W) = n + \sum_T \beta_i = n + \sum_{i=1}^{n} q_i(\sigma) = d_n(\sigma) < 0. \quad \blacksquare$$

Lemma 2.2. *Let $\phi_1, \ldots, \phi_n$ be given real numbers. Then for all $t > 0$,*

$$\int_{t<|x_1|<\ldots<|x_n|} |x_1|^{\phi_1} \ldots |x_n|^{\phi_n} dx_1 \ldots dx_n < \infty, \tag{2.3}$$

if and only if

$$d_k = k + \sum_{i=n-k+1}^{n} \phi_i < 0, \quad k = 1, \ldots, n.$$

Proof. It suffices to consider the case $t = 1$. First we show that (2.4) implies (2.3), proceeding by induction on n. Clearly the lemma is true for $n = 1$. Suppose it holds for $n - 1$ and we are given $\phi_1, \ldots, \phi_n$ satisfying (2.4). The left side of (2.3) is equal to

$$\int_{1<|x_1|} |x_1|^{\phi_1} dx_1 \int_{|x_1|<\ldots<|x_n|} |x_2|^{\phi_2} \ldots |x_n|^{\phi_n} dx_2 \ldots dx_n.$$

Make the change of variables $u_i = x_i/|x_1|$, $i = 2, \ldots, n$. The integral becomes

$$\int_{1<|x_1|} |x_1|^{(n-1)+\sum_{i=1}^{n} \phi_i} dx_1 \int_{1<|u_2|<\ldots<|u_n|} |u_2|^{\phi_2} \ldots |u_n|^{\phi_n} du_2 \ldots du_n.$$

The integral on the left is finite since $(n-1) + \sum_{i=1}^{n} \phi_i + 1 = d_n < 0$, and the integral on the right is finite by the induction hypothesis.

To show that (2.3) implies (2.4) we also proceed by induction on n. Clearly the implication is true for $n = 1$. Assume it holds for $2, \ldots, n-1$, and suppose that $\phi_1, \ldots, \phi_n$ are such that

$$d_{k_0} = k_0 + \sum_{i=n-k_0+1}^{n} \phi_i \geq 0 \tag{2.5}$$

for some $k_0 \in \{1, \ldots, n\}$. We want to show that the left side of (2.3) diverges.

If $k_0 \geq 2$, then the left side of (2.3) is equal to

$$\int_{1<|x_1|<\ldots<|x_{n-k_0+1}|} |x_1|^{\phi_1} \ldots |x_{n-k_0}|^{\phi_{n-k_0}} \cdot |x_{n-k_0+1}|^{\phi_{n-k_0+1}} dx_1 \ldots dx_{n-k_0+1} \tag{2.6}$$

$$\cdot \int_{|x_{n-k_0+1}|<\ldots<|x_n|} |x_{n-k_0+2}|^{\phi_{n-k_0+2}} \ldots |x_n|^{\phi_n} dx_{n-k_0+2} \ldots dx_n.$$

With the change of variables $u_i = x_i/x_{n-k_0+1}$, $i = n-k_0+2, \ldots, n$, (2.6) can be written

$$\int_{1<|x_1|<\ldots<|x_{n-k_0+1}|} |x_1|^{\phi_1} \ldots |x_{n-k_0}|^{\phi_{n-k_0}}$$

$$\cdot |x_{n-k_0+1}|^{\sum_{i=n-k_0+1}^{n} \phi_i + (k_0-1)} dx_1 \ldots dx_{n-k_0+1}$$

$$\cdot \int_{1<|u_{n-k_0+2}|<\ldots<|u_n|} |u_{n-k_0+2}|^{\phi_{n-k_0+2}} \ldots |u_n|^{\phi_n} du_{n-k_0+2} \ldots du_n.$$

Since $1 + \sum_{i=n-k_0+1}^{n} \phi_i + (k_0 - 1) = d_{k_0} \geq 0$, the integral on the left diverges by the induction hypothesis.

If the largest k_0 such that (2.5) holds is $k_0 = 1$, then we can write the left side of (2.3) as

$$\int_{1<|x_1|<\ldots<|x_{n-1}|} |x_1|^{\phi_1} \ldots |x_{n-1}|^{\phi_{n-1}} dx_1 \ldots dx_{n-1}$$

$$\cdot \int_{|x_{n-1}|<|x_n|} |x_n|^{\phi_n} dx_n$$

$$= \int_{1<|x_1|<\ldots<|x_{n-1}|} |x_1|^{\phi_1} \ldots |x_{n-1}|^{\phi_{n-1}+\phi_n+1} \int_{1<|u_n|} |u_n|^{\phi_n} du_n$$

by setting $u_n = x_n/x_{n-1}$. The integral on the right diverges since $1+\phi_n \geq 0$. ∎

Let $K = 2\max_{1\leq i\leq m} |\theta_i|$, let E_σ be as in (2.1) and construct a basis for T, $B_\sigma = \{M_{\tau_1}, \ldots, M_{\tau_n}\}$, as in the beginning of this section. Note that $|M_{\tau_1}| \leq |M_{\tau_2}| \leq \ldots \leq |M_{\tau_n}|$ for $\mathbf{x}$ in E_σ. We may write $E_\sigma = \bigcup_{r=0}^{n} E_{\sigma,r}$ where

$$E_{\sigma,r} = \{\mathbf{x} \in E_\sigma : |M_{\tau_1}| \leq \ldots \leq |M_{\tau_r}| \leq K \leq |M_{\tau_{r+1}}| \leq \ldots \leq |M_{\tau_n}|\},$$

$r = 0, \ldots, n$, with

$$E_{\sigma,0} = \{\mathbf{x} \in E_\sigma : |M_{\tau_j}| \geq K,\ j = 1, \ldots, n\}.$$

Lemma 2.3. *Suppose* $\mathbf{x} \in E_{\sigma,r}$, $r = 0, 1, \ldots, n$, *and let* $f_1, \ldots, f_m$ *be as in (1.2). Then there exist constants* c_i', $i = 1, \ldots, m$ *such that*

$$\begin{cases} |f_i(M_i + \theta_i)| \leq c_i'|M_i + \theta_i|^{\alpha_i} & \text{when } M_i \in T_1 \cup \ldots \cup T_r, \\ |f_i(M_i + \theta_i)| \leq c_i'|M_i + \theta_i|^{\beta_i} & \text{when } M_i \in T_{r+1} \cup \ldots \cup T_n. \end{cases}$$

(*Convention*: $T_1 \cup \ldots \cup T_r = \emptyset$ for $r = 0$ *and* $T_{r+1} \cup \ldots \cup T_n = \emptyset$ *for* $r = n$.)

Proof. Since f_i is bounded except at 0 and ∞, (1.2) implies that for *any* $a_i' > 0$ and $b_i' > 0$ there exists $c_i' > 0$ such that

$$|f_i(y)| \leq \begin{cases} c_i'|y|^{\alpha_i} & \text{if } |y| < a_i', \\ c_i'|y|^{\beta_i} & \text{if } |y| > b_i', \ i = 1, \ldots, m. \end{cases}$$

If $M_i \in T_1 \cup \ldots \cup T_r$ then $M_i + \theta_i = a_{i1}M_{\tau_1} + \ldots + a_{ir}M_{\tau_r} + \theta_i$ and hence

$$|M_i + \theta_i| \leq (|a_{i1}| + \ldots + |a_{ir}|)K + |\theta_i| =: a_i'. \tag{2.8}$$

If $M_i \in T_{r+1} \cup \ldots \cup T_n$ then

$$|M_i| > |M_{\tau_{r+1}}| > K \geq |2\theta_i|. \tag{2.9}$$

Hence

$$|M_i + \theta_i| \geq \frac{1}{2}|M_i| > \frac{1}{2}K =: b_i'. \tag{2.10}$$

Substituting (2.8) and (2.10) in (2.7) establishes the lemma. ∎

Proof of Part I of Theorem 1.2. Since $\mathbf{R}^n = \bigcup_\sigma E_\sigma = \bigcup_\sigma \bigcup_{r=0}^n E_{\sigma,r}$, it suffices to show

$$\int_{E_{\sigma,r}} |f_1(L_1(\mathbf{x}) \ldots f_m(L_m(\mathbf{x}))| d^n x \tag{2.11}$$

is finite, where $L_i(\mathbf{x}) = M_i(\mathbf{x}) + \theta_i$, $i = 1, \ldots, m$. By Lemma (2.3), (2.11) is bounded by a constant multiple of

$$\int_{E_{\sigma,r}} \prod_{I_r} |L_i|^{\alpha_i} \prod_{J_r} |L_i|^{\beta_i} d^n x$$

where $I_r = \{i : M_i \in T_1 \cup \ldots \cup T_r\}$ and $J_r = \{i : M_i \in T_{r+1} \cup \ldots \cup T_n\}$. If $M_i \in T_s$, $s \geq r+1$, we have

$$\frac{1}{2}|M_i| \leq |M_i + \theta_i| \leq |M_i| + |\theta_i| \leq \frac{3}{2}|M_i| \tag{2.12}$$

by (2.9) and (2.10). By Lemma 4.1 of Fox and Taqqu (1987) there is a constant C_σ such that *

$$|M_{T_s}| \leq |M_i| \leq C_\sigma |M_{T_s}|. \tag{2.13}$$

Thus, by (2.12) and (2.13),

$$\frac{1}{2}|M_{T_s}| \leq |L_i| \leq \frac{3}{2}C_\sigma |M_{T_s}|,$$

and hence for any β_i (positive or negative) $|L_i|^{\beta_i} \leq c|M_{T_s}|^{\beta_i}$ for some $c > 0$ and all $\mathbf{x} \in E_{\sigma,r}$. As a result,

$$\prod_{T_s} |L_i|^{\beta_i} \leq c \prod_{T_s} |M_{T_s}|^{\beta_i} = c|M_{T_s}|^{q_s} \tag{2.14}$$

for some $c > 0$ and all $\mathbf{x} \in E_{\sigma,r}$. Hence

$$\prod_{J_r} |L_i|^{\beta_i} = \prod_{s=r+1}^{n} \prod_{T_s} |L_i|^{\beta_i} \leq c \prod_{s=r+1}^{n} |M_{T_s}|^{q_s} \tag{2.15}$$

for some $c > 0$ and all $\mathbf{x} \in E_{\sigma,r}$. Therefore (2.11) is bounded by a constant multiple of

$$\int_{E_{\sigma,r}} \prod_{I_r} |L_i|^{\alpha_i} \prod_{s=r+1}^{n} |M_{T_s}|^{q_s} d^n x.$$

Make the change of variables $y_i = M_{T_i}$, $i = 1, \ldots, n$. Since the transformation is linear, its Jacobian is constant and we have (2.11) bounded by a constant multiple of

$$\int_A \prod_{I_r} |\hat{L}_i(y_1, \ldots, y_r)|^{\alpha_i} \prod_{s=r+1}^{n} |y_s|^{q_s} dy_1 \ldots dy_n, \tag{2.16}$$

* The inequality in Lemma 4.1 of Fox and Taqqu (1987) involves L's, while (2.13) involves M's. The inequality is valid for the M's (just set the θ_i's in that lemma equal to zero.)

where $A = \{\mathbf{y} \in \mathbf{R}^n : |y_1| \le \dots \le |y_r| \le K \le |y_{r+1}| \le \dots \le |y_n|\}$ and $\hat{L}_i(y_1, \dots, y_r) = L_i(x_1, \dots, x_n)$. (Note that for $i \in I_r$, $\hat{L}_i$ depends only on $y_1, \dots, y_r$ because M_i is a linear combination of $M_{\tau_1}, \dots, M_{\tau_r}$). The integral (2.16) is equal to the product of integrals

$$\int_{|y_1|<\dots<|y_r|<K} \prod_{I_r} |\hat{L}_i(y_1, \dots, y_r)|^{\alpha_i} dy_1 \dots dy_r$$

$$\cdot \int_{K<|y_{r+1}|<\dots<|y_n|} \prod_{s=r+1}^{n} |y_s|^{q_s} dy_{r+1} \dots dy_n.$$

The second integral is finite by Lemmas 2.1 and 2.2. Let $\hat{T}' = \{\hat{L}_i : i \in I_r\}$. Since $\hat{d}_0(W) > 0$ for every strongly independent subset W of $\hat{T}'$, the first integral is finite by Theorem 3.1 of Fox and Taqqu (1987). ∎

3. Proof of Part II of the Power Counting Theorem 1.2. We divide the proof into two propositions.

Proposition 3.1. *If (1.4) holds, then (1.10) implies (b′).*

Proof. Assume (1.10). If (b′) does not hold, then by Lemma 2.1 there is some permutation σ and some $k \in \{1, \dots, n\}$ such that $d_k(\sigma) = k + \sum_{i=n-k+1} q_i(\sigma) \ge 0$, where q_i is as defined in (2.2). Let $\{M_{\tau_1}, \dots, M_{\tau_n}\}$ be the basis corresponding to σ, and

$$b = 2\max\{b_1, \dots, b_m,\ \theta_1, \dots, \theta_m\},$$

where b_i, $i = 1, \dots, m$ is as in (1.4). Set

$$E = \{\mathbf{x} : b < |M_{\tau_1}| < |M_{\tau_2}| < \dots < |M_{\tau_n}|\}.$$

Since $|M_i| > |M_{\tau_1}| > b > 2\theta_i$, we have $\frac{1}{2}|M_i| < |M_i + \theta_i| < \frac{3}{2}|M_i|$. Moreover, since (2.13) holds for $M_i \in T_s$, $s = 1, \dots, n$, $\frac{1}{2}|M_{\tau_s}| < |M_i + \theta_i| < \frac{3}{2}|M_{\tau_s}|$. We also have $|M_i + \theta_i| > \frac{1}{2}|M_i| > \frac{1}{2}b > b_i$. Hence by (1.4) there exist $c_1 > 0$ and $c_2 > 0$ such that

$$\int_E |f_1(M_1 + \theta_1) \dots f_m(M_m + \theta_m)| d^n x$$

$$
\begin{aligned}
&\geq c_1 \int_E |M_1+\theta_1|^{\beta_1} \dots |M_m+\theta_m|^{\beta_m} d^n x \\
&= c_1 \int_E \prod_{s=1}^{n} \prod_{T_s} |M_i+\theta_i|^{\beta_i} d^n x \\
&\geq c_2 \int_E \prod_{s=1}^{n} \prod_{T_s} |M_{\tau_s}|^{\beta_i} d^n x \\
&= c_2 \int_E \prod_{s=1}^{n} |M_{\tau_s}|^{q_s} d^n x \\
&= c_2 \int_{b<|y_1|<\dots<|y_n|} |y_1|^{q_1} \dots |y_n|^{q_n} dy_1 \dots dy_n = \infty
\end{aligned}
$$

by Lemma (2.2). ∎

To prove that (1.10) implies (a'), we need to establish a converse of Lemma 4.3 of Fox and Taqqu (1987).

Lemma 3.1. *Let $\phi_1, \dots, \phi_n$ be given real numbers. Then for all $t > 0$,*

$$
\int_{|x_1|\leq|x_2|\leq\dots\leq|x_n|\leq t} |x_1|^{\phi_1}|x_2|^{\phi_2} \dots |x_n|^{\phi_n} dx_1 \dots dx_n < \infty \tag{3.1}
$$

if and only if

$$
d_k = k + \sum_{j=1}^{k} \phi_j > 0 \quad \textit{for} \quad k = 1, \dots, n. \tag{3.2}
$$

Proof. It suffices to consider the case $t = 1$. That (3.2) implies (3.1) is the result of Lemma 4.3 of Fox and Taqqu (1987). To prove that (3.1) implies (3.2), one proceeds by induction on n, as in the proof of Lemma 2.2. Assuming $d_{k_0} = k_0 + \sum_{j=1}^{k} \phi_j \leq 0$ for some $k_0 \in \{1, \dots, n\}$, one makes the change of variables $u_i = x_i/x_{k_0}$, $i = 1, \dots, k_0 - 1$ if $k_0 \geq 2$ and $u_i = x_1/x_2$ if $k_0 = 1$. ∎

Proposition 3.2. *If (1.4) holds, then (1.10) implies (a').*

Set $L_i = M_i + \theta_i$, $i = 1, \dots, n$. In the proof of Proposition (3.2), a basis for $T' = \{L_1, \dots, L_m\}$ is required, so given a permutation $\sigma = (\sigma_1, \dots, \sigma_m)$ of $\{1, \dots, m\}$, we apply the greedy algorithm to T', obtaining a basis $B_\sigma = \{L_{\tau_1}, \dots, L_{\tau_r}\}$ and a partition $T_1, \dots, T_r$. Note that B_σ is different from the basis obtained in Section 2, since it is built from T' rather than T. Observe also that B_σ is linearly independent but need not be strongly independent, since it is possible that $r > n$. For example, if $T' = \{x+1, x-1, x\}$ and $\sigma = (1,2,3)$, then $B_\sigma = \{x+1, x-1\}$, so $r = 2$, while $n = 1$.

Proof of Proposition 3.2. Suppose ad absurdum that $\hat{d}_0(W) \leq 0$ for some strongly independent $W \subset T'$. Arrange the elements L_i of W in an arbitrary order. We get $W = \{L_{\sigma_1}, \dots, L_{\sigma_k}\}$, where $k = |W|$, and $\sigma = (\sigma_1, \dots, \sigma_m)$ is a permutation of $\{1, \dots, m\}$. Now let $B_\sigma = \{L_{\tau_1}, \dots, L_{\tau_r}\}$ be the basis for T' obtained with the greedy algorithm. Since W is strongly and hence linearly independent, we must have $L_{\tau_i} = L_{\sigma_i}$ for $i = 1, \dots, k$, and hence $W = \{L_{\tau_1}, \dots, L_{\tau_k}\}$. Set

$$E_t = \{\mathbf{x} \in [-t,t]^n : |L_{\tau_1}| \leq \dots \leq |L_{\tau_k}|\}.$$

By (1.4),

$$\int_{E_t} |f(L_1)\dots f(L_m)| d^n x \geq c_1 \int_{E_t} |L_1|^{\alpha_1} \dots |L_m|^{\alpha_m} d^n x$$

for some $c_1 > 0$. Furthermore, as an easy consequence of Lemma 4.1 of Fox and Taqqu (1987) we may reverse the inequality in Lemma 4.2 of Fox and Taqqu (1987) and obtain

$$\int_{E_t} |L_1|^{\alpha_1} \dots |L_m|^{\alpha_m} d^n x \geq c_1 \int_{E_t} |L_{\tau_1}|^{\Delta_1} \dots |L_{\tau_r}|^{\Delta_r} d^n x \tag{3.3}$$

where $\Delta_s = \sum_{T_s} \alpha_j$ and $c_1 > 0$. Make the change of variables $y_1 = L_{\tau_1}, \dots, y_k = L_{\tau_k}$. The right side of (3.3) diverges if

$$\int_{|y_1|<\dots<|y_k|<\epsilon} |y_1|^{\Delta_1} \dots |y_k|^{\Delta_k} dy_1 \dots dy_k$$

diverges for any $\epsilon > 0$. This last integral diverges by Lemma 3.1 since $k + \Delta_1 + \dots + \Delta_k = \hat{d}_0(W) \leq 0$. ∎

REFERENCES

[1] Avram, F. and Taqqu, M.S., " Hölder's inequality for functions of linearly dependent arguments," *SIAM J. Math. Anal.* **20** (1989), 1484-1489.

[2] Dyson, F.J., "The S matrix in quantum electrodynamics," *Phys. Rev.* **75** (1949), 1736.

[3] Fox, R. and Taqqu, M.S., "Central limit theorems for quadratic forms in random variables having long-range dependence," *Probab. Th. Rel. Fields* **74** (1987), 213-240.

[4] Lowenstein, J.H. and Zimmermann, W., "The power counting theorem for Feynman integrals with massless propegators," *Commun. Math. Phys.* **44** (1975), 73-86.

[5] Nakanishi, N., *Graph Theory and Feynman Integrals*, Gordon and Breach, New York, 1971.

[6] Terrin, N. and Taqqu, M.S., "Convergence in distribution of sums of bivariate Appell polynomials with long-range dependence." Preprint, 1989.

[7] Terrin, N. and Taqqu, M.S., "A noncentral limit theorem for quadratic forms of Gaussian stationary sequences." To appear in the *Journal of Theoretical Probability*, **3** (1990).

Norma Terrin
Department of Mathematics
Boston University
Boston, MA 02215

Murad S. Taqqu
Department of Mathematics
Boston University
Boston, MA 02215

Etude asymptotique des nombres de tours de plusieurs mouvements browniens complexes corrélés

Marc Yor

1. Introduction et énoncé des résultats.

(1.1) A la suite d'un échange de lettres portant sur les nombres de tours du mouvement brownien complexe, Mitchell Berger [1] a proposé d'étudier asymptotiquement les nombres de tours dans le cadre suivant, motivé par l'étude du vent solaire : "Consider, instead of one particle moving about several fixed points in the plane, letting several particles all move about each other. The particles are actually small disks, which move randomly on a plane. For solar applications, the disks can be regarded as the intersections of thin cylindrical tubes of magnetized gas with the solar surface..."

Simplifiant à l'extrême le cadre suggéré par M. Berger, je me suis intéressé au problème suivant : soient $Z^1, Z^2, \dots, Z^n, n$ mouvements browniens complexes indépendants issus de points $(z^i;\ 1 \le i \le n)$ distincts ; comment, du point de vue asymptotique, les processus $(Z^i;\ 1 \le i \le n)$ tournent-ils les uns autour des autres ?

De façon précise, pour $i < j$, soit $(\theta_{i,j}(t), t \ge 0)$ une détermination continue de l'argument autour de 0 de la trajectoire $(Z^i_s - Z^j_s,\ s \le t)$. Cette détermination est bien définie, car :

$$Z^i_s - Z^j_s = \sqrt{2}\, B^{i,j}_s, \tag{1.a}$$

où $(B^{i,j}_s, s \ge 0)$ est un mouvement brownien complexe issu de $\frac{z^i - z^j}{\sqrt{2}} \neq 0$ et donc, le processus $Z^i - Z^j$ ne visite presque sûrement pas 0.

Cet article apporte la réponse suivante à la question de M. Berger, dans la version simplifiée présentée ci-dessus ; on a :

$$\frac{2}{\log t}(\theta_{i,j}(t); 1 \le i < j \le n) \xrightarrow[t\to\infty]{(dist.)} (C_{i,j}; 1 \le i < j \le n) \tag{1.b}$$

où les variables $C_{i,j}$ sont des variables de Cauchy indépendantes, de paramètre 1.

Soulignons que, i et j étant fixés, la convergence en loi de $\frac{2}{\log t}\theta_{i,j}(t)$, lorsque $t \to \infty$, vers une variable de Cauchy standard, n'est autre, grâce à la représentation (1.a) du processus $Z^i - Z^j$, que le résultat maintenant classique de Spitzer [7] sur le nombre de tours asymptotique du mouvement brownien plan.

Le résultat (1.b) peut donc être interprété comme une extension multidimensionnelle du résultat de Spitzer, et il est remarquable que les variables $C_{i,j}$ soient indépendantes, bien que les mouvements browniens ($B^{i,j}$; $1 \leq i \leq j \leq n$) soient, dans leur ensemble, corrélés. De façon précise, remarquons que, pour i, j, k trois entiers différents, et $u, v \in \mathbb{R}^2 \simeq \mathbb{C}$, le processus :

(1.c) $$\left\{(u, B_t^{i,j})(v, B_t^{i,k}) - \frac{1}{2}(u,v)t\,;\; t \geq 0\right\}$$ [1]

est une martingale ; par la suite, nous dirons que les mouvements browniens $B^{i,j}$ et $B^{i,k}$ sont corrélés, et admettent $\frac{1}{2}\mathrm{Id}_{\mathbb{R}^2}$ pour matrice de corrélation.

(1.2) La remarque ci-dessus étant faite, il semble naturel de chercher à étendre le résultat (1.b) dans le cadre suivant : considérons, sur un espace de probabilité filtré $(\Omega, \mathcal{F}, (\mathcal{F}_t), P)$ un ensemble de p $(\mathcal{F}_t)$-mouvements browniens complexes $\{(B_t^k, t \geq 0);\ 1 \leq k \leq p\}$ issus respectivement de $z^k \neq 0$, et tels que : pour tous $k \neq \ell$, et tous $u, v \in \mathbb{R}^2$, il existe une matrice $A_{k,\ell}$ telle que :

(1.d) $$\{(u, B_t^k)(v, B_t^\ell) - (u, A_{k,\ell} v)t; t \geq 0\}$$

soit une $(\mathcal{F}_t)$-martingale. (Nous dirons alors que les mouvements browniens B^k et B^ℓ sont corrélés, au moyen de la matrice de corrélation $A_{k,\ell}$). On a alors l'extension suivante du résultat (1.b) :

Théorème 1. *On note $(\theta_t^k, t \geq 0)$ une détermination continue de l'argument de $(B_u^k, u \leq t)$ autour de* 0. *Alors, si :*

(1.e) *pour tous $k \neq \ell$, $A_{k,\ell}$ n'est pas une matrice orthogonale,*

c'est-à-dire : $A = A_{k,\ell}$ vérifie $A^* A \neq \mathrm{Id}$, on a :

(1.f) $$\frac{2}{\log t}(\theta_t^k; k \leq p) \xrightarrow[t \to \infty]{(dist.)} (C_k; k \leq p)$$

où les variables C_k sont des variables de Cauchy indépendantes, de paramètre 1.

Il est encore possible d'obtenir un résultat plus général que (1.f), en relaxant l'hypothèse (1.e), c'est-à-dire en permettant que certaines des matrices $A_{k,\ell}$ qui figurent en (1.d) soient des matrices orthogonales.

[1] On note ici, et dans toute la suite, (u,v) le produit scalaire usuel dans $\mathbb{R}^2$, et, plus généralement, dans $\mathbb{R}^d$.

A cet effet, examinons tout d'abord ce que signifie l'hypothèse (1.d) lorsque $A_{k,\ell}$ est une matrice orthogonale : en simplifiant les notations, considérons $(B_t, t \geq 0)$ et $(C_t, t \geq 0)$ deux mouvements browniens complexes admettant A, matrice orthogonale, pour matrice de corrélation, c'est-à-dire :

$$(1.d') \qquad \begin{array}{l} \text{pour tous } u, v \in \mathbb{R}^2, \\ \{(u, B_t)(v, C_t) - (u, Av)t \,;\, t \geq 0\} \text{ est une martingale.} \end{array}$$

En écrivant $v = A^*Av$, et en posant $\tilde{v} = Av$, $\tilde{C}_t = AC_t$ $(t \geq 0)$, la propriété (1.d′) peut être réécrite sous la forme équivalente :

$$(1.d'') \qquad \begin{array}{l} \text{pour tous } u, \tilde{v} \in \mathbb{R}^2, \\ \{(u, B_t)(\tilde{v}, \tilde{C}_t) - (u, \tilde{v})t \,;\, t \geq 0\} \text{ est une martingale.} \end{array}$$

Or, A étant une transformation orthogonale, $(\tilde{C}_t, t \geq 0)$ est un mouvement brownien plan ; en conséquence, si l'on note : $D_t = B_t - \tilde{C}_t$ $(t \geq 0)$, ce processus satisfait, à l'aide de (1.d″), la propriété suivante : pour tout $u \in \mathbb{R}^2$, $(u, D_t - D_0)^2$ est une martingale.

Il en résulte que le processus $(D_t, t \geq 0)$ est constant, et donc :

$$B_t - B_0 = AC_t - AC_0 \qquad (t \geq 0)\ P - \text{p.s.}$$

De plus, A étant une transformation orthogonale, l'accroissement de la détermination continue de l'angle du processus $(C_t, t \geq 0)$ entre deux instants est égal à celui correspondant à $(AC_t, t \geq 0)$, lorsque A est une matrice de rotation, et est changé en son opposé, lorsque A est la matrice d'une symétrie par rapport à une droite.

L'étude asymptotique des nombres de tours de $(B_t, t \geq 0)$ et $(C_t, t \geq 0)$ est donc ramenée à l'étude correspondante pour $(B_t, t \geq 0)$, les nombres de tours étant calculés relativement à deux points du plan.

Plus généralement, si en (1.d), toutes les matrices de corrélation $A_{k,\ell}$ sont des matrices orthogonales, l'étude asymptotique des nombres de tours des mouvements browniens B^k $(1 \leq k \leq p)$ autour de 0 est ramenée à l'étude correspondante pour un mouvement brownien $(B_t, t \geq 0)$, relativement à p points du plan $z^1, \ldots, z^p$, étude déjà faite par Pitman–Yor [5][2] et Le Gall–Yor [2], et dont nous rappelons maintenant le résultat principal : $(\theta_t^{z_i}, t \geq 0)$ désignant une détermination continue de l'argument de $(B_u, u \leq t)$ autour de z_i pour le processus B, on a :

$$(1.g) \qquad \frac{2}{\log t}(\theta_t^{z_i}; 1 \leq i \leq p) \xrightarrow[t \to \infty]{(dist.)} \xi_p \equiv (W_T + L_T(U)C_i; 1 \leq i \leq p)$$

où $C_1, \ldots, C_p$ sont p variables de Cauchy de paramètre 1,

---

[2]Pour un résumé succinct des résultats de [5], le lecteur pourra consulter Pitman–Yor [6].

W est un mouvement brownien réel issu de 0,

U est un mouvement brownien réfléchi issu de 0,

$T = \inf\{t: U_t = 1\}$, et $L_T(U)$ est le temps local en 0 de U jusqu'à l'instant T;

enfin, $C_1, \dots, C_p$, W et U sont indépendants.

Nous pouvons maintenant énoncer l'extension suivante du Théorème 1 ; cette extension est le résultat principal du présent article.

Théorème 2. *Soient $(B^1, \dots, B^p)$ p mouvements browniens plans corrélés au moyen des matrices de corrélation $A_{k,\ell}$ $(1 \le k < \ell \le p)$ telles que :*

(1.e) *pour tous k, ℓ, $A_{k,\ell}$ n'est pas une matrice orthogonale.*

Considérons de plus q points distincts $(z_1, \dots, z_q)$ tels que $B_0^i \neq z_j$, pour tous i, j. On a alors, en notant $\theta_t^i = (\theta_t^{i,z_j}; j \le q)$:

$$\text{(1.h)} \qquad \frac{2}{\log t}(\theta_t^i; 1 \le i \le p) \xrightarrow[t \to \infty]{(dist.)} (\xi_q^i; 1 \le i \le p)$$

le vecteur limite étant distribué comme suit:

- *les variables $(\xi_q^i; 1 \le i \le p)$ sont indépendantes :*
- *pour tout i, ξ_q^i est distribuée comme la variable ξ_q décrite en* (1.g).

(1.3) En conclusion de cette introduction, nous remarquons que l'étude asymptotique, faite par Le Gall–Yor [3] des nombres de tours d'un mouvement brownien $(B_t, t \ge 0)$ à valeurs dans $\mathbb{R}^3$, autour de p droites orientées $(D^1, \dots, D^p)$ de l'espace, telles que $B_0 \notin D^i$, pour tout i, est un cas particulier de l'étude proposée ci-dessus.

Pour simplifier la discussion, supposons que ces droites soient concourantes en 0. Le Gall–Yor [3] ont obtenu le résultat suivant :

$$\text{(1.i)} \qquad \frac{2}{\log t}(\theta_t^{D_i}; 1 \le i \le p) \xrightarrow[t \to \infty]{(dist.)} (C_i; 1 \le i \le p)$$

où les C_i sont p variables de Cauchy indépendantes, de paramètre 1, et $(\theta_t^{D_i}, t \ge 0)$ est une détermination continue de l'argument de $(B_u, u \le t)$ autour de D_i.

Le résultat (1.i) apparaît clairement comme un cas particulier de (1.f). C'est bien le cas, car si D est une droite de $\mathbb{R}^3$ passant par 0, et $\{\vec{u}_1; \vec{u}_2\}$ une base orthonormée du plan P passant par 0 et orthogonal à D, la détermination continue $(\theta_t^D, t \ge 0)$ n'est autre que la détermination continue de l'angle autour de 0 du mouvement brownien plan $\{(\vec{u}_1, B_t), (\vec{u}_2, B_t); t \ge 0\}$. Les différents mouvements browniens plans ainsi associés aux droites $(D^i; 1 \le i \le p)$ sont à l'évidence corrélés, et la matrice de corrélation de deux d'entre eux est une matrice orthogonale si, et seulement si, les droites orientées correspondantes D^i et D^j sont identiques ou opposées.

En définitive, le théorème 2 ci-dessus permet donc à la fois d'unifier les études de Pitman–Yor ([5], [6]) et Le Gall–Yor [2] d'une part, sur les nombres de tours asymptotiques d'un mouvement brownien plan autour d'un nombre fini de points, et, d'autre part, de Le Gall–Yor [3] sur les nombres de tours d'un mouvement brownien dans l'espace $\mathbb{R}^3$ autour de différentes droites, et de compléter ces deux études de manière naturelle pour des mouvements browniens plans corrélés de manière générale.

Soulignons encore que l'étude proposée par M. Berger [1] concernant la façon dont p particules browniennes tournent les unes autour des autres est, en un certain sens, "duale" de l'étude faite par Pitman–Yor [5] et Le Gall–Yor [2] concernant les nombres de tours d'une seule particule brownienne autour de p points du plan.

Pour être complet, signalons qu'un autre complément, dans une direction différente, de l'étude de Le Gall–Yor [3] sur les nombres de tours d'un mouvement brownien dans $\mathbb{R}^3$ autour de plusieurs droites est fourni par Le Gall–Yor [4] qui obtiennent des résultats asymptotiques sur les nombres de tours d'un mouvement brownien dans $\mathbb{R}^3$ autour de certaines courbes de l'espace.

2. Démonstration des Théorèmes 1 et 2.

(2.1) Comme le lecteur peut s'en douter, compte-tenu du fait, remarqué dans le sous-paragraphe **(1.3)**, que l'étude menée par Le Gall–Yor [3] des nombres de tours du mouvement brownien dans l'espace autour d'un nombre fini de droites est un cas particulier de l'étude entreprise ici, nous allons reprendre — pour la démonstration du Théorème 1 — certains des arguments développés en [3].

Pour faciliter la lecture, nous reprenons également les notations utilisées en [3], p. 623.

Ainsi, pour tout $i \le p$, $R_t^i = |B_t^i|$ $(t \ge 0)$ désigne la partie radiale de B^i ; il existe deux mouvements browniens réels ω^i et ρ^i tels que :

$$\theta_t^i = \omega^i(H_t^i); \log R_t^i = \rho^i(H_t^i), \text{ avec} : H_t^i = \int_0^t \frac{ds}{(R_s^i)^2}.$$

Introduisons également l'opérateur de scaling, de paramètre $c > 0$, agissant sur un processus générique $(U_t, t \ge 0)$: pour $c > 0$, on pose : $U_t^{(c)} = \frac{1}{c} U_{c^2 t}$ $(t \ge 0)$. De plus, définissons $T(U) = \inf\{t \ge 0 : U_t = 1\}$. Le résultat important suivant a été prouvé en [2] et [5] (la démonstration est également présentée, pour la commodité du lecteur, en [3]) :

$$\text{(2.a)} \qquad \frac{2}{\log t}\theta_t^i - \omega^{i,(c)}(T(\rho^{i,(c)})) \xrightarrow[t\to\infty]{(P)} 0, \text{ où } c = \frac{\log t}{2}.$$

Le résultat énoncé dans le théorème 1 découlera alors du résultat plus général suivant :

$$(2.b) \qquad \left(\rho^{i,(c)};\omega^{i,(c)};1\le i\le p\right)\xrightarrow[c\to\infty]{(d)}\left(\tilde\rho^{i};\tilde\omega^{i};1\le i\le p\right),$$

les $(2p)$ mouvements browniens réels

$$\left(\tilde\rho^{i},\tilde\omega^{i},1\le i\le p\right)$$

étant indépendants, et satisfaisant $\tilde\rho^i_0=\tilde\omega^i_0=0$. (d) désigne ici la convergence étroite des lois sur $C(\mathbb{R}_+,\mathbb{R}^{2p})$, muni de la topologie de la convergence compacte. Les variables de Cauchy indépendantes C_i qui figurent dans l'énoncé du Théorème 1 sont ici réalisées à l'aide de $\tilde\rho^i$ et $\tilde\omega^i$, de la façon suivante :

$$C_i=\tilde\omega^i(T(\tilde\rho^i)).$$

(2.2) Il s'agit maintenant de démontrer le résultat (2.b) sous l'hypothèse (1.e) faite dans l'énoncé du Théorème 1. Pour cela, nous nous appuierons essentiellement sur une version asymptotique du théorème de Knight sur les martingales continues orthogonales, version que nous présentons dans le Lemme 1 ci-dessous, qui reprend l'énoncé du Lemme 1 de [3].

Rappelons auparavant que, si $(M_t;t\ge 0)$ est une martingale locale continue telle que $M_0=0$, et $\langle M\rangle_\infty=\infty$, il existe un mouvement brownien réel $(\beta_t,t\ge 0)$, que nous dirons associé à (M_t), tel que $M_t=\beta_{\langle M\rangle_t}$ $(t\ge 0)$.

Lemme 1. *Soient $(M^1,\dots,M^k)$ k martingales locales continues, nulles en 0, définies sur un espace de probabilité filtré, et telles que :*

$$\langle M^1\rangle_\infty=\langle M^2\rangle_\infty=\dots=\langle M^k\rangle_\infty=\infty.$$

On note $\beta^1,\dots,\beta^k$ les mouvements browniens associés respectivement à ces martingales. Alors, s'il existe une fonction $u:]1,\infty[\to]0,\infty[$ telle que :

(a) *pour tout $i\le k$,* $\dfrac{1}{u(t)}\langle M^i\rangle_t\xrightarrow[t\to\infty]{(P)}\infty,$

(b) *pour tous i,j, avec $i\ne j$,* $\dfrac{1}{u(t)}\displaystyle\int_0^t|d\langle M^i,M^j\rangle_s|\xrightarrow[t\to\infty]{(P)}0,$

les mouvements browniens $(\beta^i;1\le i\le k)$ sont asymptotiquement indépendants, au sens suivant :

$$\left(\beta^{i,(c)};1\le i\le k\right)\xrightarrow[c\to\infty]{(d)}\gamma,$$

où γ désigne un mouvement brownien à valeurs dans $\mathbb{R}^k$, issu de 0.

Nous montrons maintenant l'assertion (3.b) en appliquant le Lemme 1 aux martingales locales $\theta^1, \ldots, \theta^p$, $\log R^1, \ldots, \log R^p$, avec $u(t) = (\log t)^{1+m}$, pour $0 < m < 1$.

A partir de maintenant, notre démonstration diffère sensiblement de celle faite en [3], p. 624–625, et s'appuie surtout sur le Lemme 5 ci-dessous. Remarquons tout d'abord que :

$$\frac{1}{u(t)}\langle \theta^i \rangle_t = \frac{1}{u(t)}\langle \log R^i \rangle_t = \frac{1}{u(t)} H_t^i \xrightarrow[t\to\infty]{(P)} \infty,$$

car $\frac{1}{(\log t)^2} H_t^i$ converge en loi vers une variable strictement positive (à savoir, avec les notations utilisées en **(2.1)**, $\frac{1}{4}T(\tilde{\rho}^i)$). Il nous reste à montrer que, pour $i \neq j$, les quantités :

$$\frac{1}{u(t)}\int_0^t |d\langle \theta^i, \theta^j \rangle_s|, \frac{1}{u(t)}\int_0^t |d\langle \theta^i, \log R^j \rangle_s|,$$
$$\frac{1}{u(t)}\int_0^t |d\langle \log R^i, \log R^j \rangle_s|$$

convergent en probabilité vers 0, lorsque $t \to \infty$. Or, on a :

$$\log R_t^i = \log R_0^i + \int_0^t \frac{(B_s^i, dB_s^i)}{(R_s^i)^2},$$

ainsi que :

$$\theta_t^i = \theta_0^i + \int_0^t \frac{(\hat{B}_s^i, dB_s^i)}{(R_s^i)^2}, \text{ où : } \hat{B}_s^i = (-\text{Im } B_s^i, \text{Re } B_s^i).$$

Remarquons maintenant que, si H et K sont deux processus prévisibles à valeurs dans $\mathbb{R}^2$ tels que: pour tout t, $\int_0^t |H_s|^2 ds < \infty$ et $\int_0^t |K_s|^2 ds < \infty$ P-p.s. on a :

$$\text{(2.c)} \qquad \langle \int_0^\bullet (H_s, dB_s^i), \int_0^\bullet (K_s, dB_s^j) \rangle_t = \int_0^t (H_s, A_{i,j} K_s) ds.$$

Cette formule (2.c) est obtenue aisément à partir de la définition en (1.d) de $A_{i,j}$ comme matrice de corrélation[3] en considérant le cas où H et K sont deux processus prévisibles simples, puis en prolongeant la formule par continuité.

On déduit alors de (2.c), en utilisant le fait que $A_{i,j}$ est une matrice de corrélation, et donc vérifie (3.b), la majoration suivante :

$$dV_s \leq \frac{ds}{R_s^i R_s^j},$$

[3] Nous anticipons ici légèrement sur la terminologie et les premiers résultats, élémentaires, du paragraphe 3 ci-dessous, auxquels le lecteur est invité à se reporter.

où dV_s désigne l'une quelconque des trois mesures aléatoires :

$$|d\langle\theta^i,\theta^j\rangle_s|,\ |d\langle\theta^i,\log R^j\rangle_s|,\ |d\langle\log R^i,\log R^j\rangle_s|.$$

Le résultat que nous cherchons à démontrer, à savoir :

$$\frac{1}{u(t)}\int_0^t dV_s \xrightarrow[t\to\infty]{(P)} 0$$

sera obtenu, dès que l'on aura prouvé que :

$$\text{(2.d)}\qquad \frac{1}{\log t}E\left[\int_1^t \frac{ds}{R_s^i R_s^j}\right] \text{ converge, lorsque } t\to\infty.$$

Un argument élémentaire permet de se ramener au cas où $B_0^i = B_0^j = 0$, et on obtient, en utilisant l'invariance par scaling du couple $(B_t^i, B_t^j; t\geq 0)$, et en notant : $G^i = B_1^i$ et $G^j = B_1^j$:

$$E\left[\int_1^t \frac{ds}{R_s^i R_s^j}\right] = \int_1^t \frac{ds}{s}E\left[\frac{1}{|G^i|\,|G^j|}\right] = (\log t)E\left[\frac{1}{|G^i|\,|G^j|}\right].$$

On a donc obtenu le résultat (2.d), car, d'après la fin du Lemme 5, appliqué avec $p=q=1$, et $d=2$, on a :

$$\text{(2.e)}\qquad E\left[\frac{1}{|G^i|\,|G^j|} < \infty\right].$$

(2.3) Il peut être intéressant de remarquer que le résultat technique obtenu en **(2.2)** ci-dessus, à savoir que les quantités :

$$\frac{1}{u(t)}\int_0^t |d\langle\theta^i,\theta^j\rangle_s|,\ \frac{1}{u(t)}\int_0^t |d\langle\theta^i,\log R^j\rangle_s|,$$
$$\frac{1}{u(t)}\int_0^t |d\langle\log R^i,\log R^j\rangle_s|$$

convergent en probabilité vers 0 dès que : $u(t) = (\log t)^{1+m}$, pour $m>0$, améliore le résultat correspondant de Le Gall–Yor [3] ; en effet, en [3], il est seulement montré que, lorsque θ^i désigne le processus du nombre de tours autour de la droite D^i, les p droites $D^1,\dots,D^p$ étant supposées concourantes en 0, les quantités ci-dessus convergent en probabilité vers 0 lorsque $u(t) = (\log t)^{1+m}$ pour $m > \frac{1}{2}$.

Cette restriction supplémentaire : $m>\frac{1}{2}$, qui s'avère donc être inutile, provenait du fait que, en [3], p. 634–625, on utilisait, pour la démonstration de ce résultat technique, la décomposition :

$$\text{(2.f)}\qquad \theta_t^i = \theta_t^i(\varepsilon) + \overline{\theta}_t^i(\varepsilon),\ \text{où } \theta_t^i(\varepsilon) = \int_0^t d\theta_s^i 1_{(B_s\in\Gamma_s^i)}$$

représente le processus des nombres de tours autour de la droite D^i ayant lieu à l'intérieur du cône de révolution Γ^i_ε d'axe D^i, et d'ouverture ε. Enfin, en développant en [3] la mesure aléatoire: $d\langle\theta^i,\theta^j\rangle_s$ à l'aide de (2.f), on utilisait finalement l'inégalité de Cauchy–Schwarz (ou Kunita–Watanabe) pour les crochets, ce qui amenait la restriction: $m > 1/2$.

(2.4) Enfin, la démonstration du Théorème 2 repose pour l'essentiel sur les mêmes arguments que ceux développés en (2.2) pour démontrer le Théorème 1, l'argument principal étant l'inégalité (2.e). La discussion est identique, mutatis mutandis, à celle menée en [3], au paragraphe (2.4), p. 625–626, auxquelles le lecteur pourra se reporter.

3. Matrices de corrélation et moments négatifs de variables gaussiennes corrélées.

(3.1) Soit d entier, $d \geq 1$. Dans toute la suite, la notation G_d (ou G'_d) désigne une variable gaussienne centrée, à valeurs dans $\mathbb{R}^d$, admettant pour matrice de covariance $\mathrm{Id}_{\mathbb{R}^d}$. Soient maintenant G_d et G'_d deux telles variables, telles que le couple $\{G_d; G'_d\}$ soit gaussien. Il existe une matrice A, que nous appelerons matrice de corrélation (sous-entendu: de G_d et G'_d) telle que:

$$\text{(3.a)} \qquad \text{pour tous } u, v \in \mathbb{R}^d, \; E[(u, G_d)(v, G'_d)] = (u, Av).$$

On caractérise tout d'abord les matrices de corrélation.

Lemme 2. 1) *A, matrice $d \times d$, est une matrice de corrélation, si, et seulement si:*

$$\text{(3.b)} \qquad \textit{pour tout } v \in \mathbb{R}^d, \; \|Av\| \leq \|v\|.$$

2) *Si A est une matrice de corrélation, et si la relation* (3.a) *est satisfaite, il existe une variable gaussienne ξ, centrée, à valeurs dans $\mathbb{R}^d$, de matrice de covariance: $I - A^*A$, indépendante de G_d, telle que:*

$$G'_d = A^* G_d + \xi.$$

Démonstration. 1) Si A est une matrice de corrélation, on déduit de (3.a), par application de l'inégalité de Cauchy–Schwarz, que:

$$\|Av\| \leq \|v\|, \text{ pour tout } v \in \mathbb{R}^d.$$

2) Supposons la condition (3.b) satisfaite, et considérons G_d donnée. Il existe alors, sur un espace de probabilité adéquat, une variable gaussienne ξ, à valeurs dans $\mathbb{R}^d$, centrée, indépendante de G_d, et de matrice de covariance $I - A^*A$. On définit alors: $G'_d = A^*G_d + \xi$, et il est immédiat que la relation (3.a) est satisfaite.

3) On vient également de montrer que la condition (3.b) est suffisante pour que A soit une matrice de corrélation. □

Nous étudions maintenant la condition (3.b) dans le cas particulier $d = 2$.

Lemme 3. $A = \begin{pmatrix} \alpha & \beta \\ \gamma & \delta \end{pmatrix}$ *satisfait la condition* (3.b) *si, et seulement si, les conditions* (3.c) *et* (3.d) *suivantes sont satisfaites :*

$$\text{(3.c)} \qquad -\text{dét}(I - A^* A) \equiv (\alpha\beta + \gamma\delta)^2 - (1 - \alpha^2 - \gamma^2)(1 - \beta^2 - \delta^2) \leq 0$$

et

$$\text{(3.d)} \qquad 1 - \alpha^2 - \gamma^2 \geq 0; \quad 1 - \beta^2 - \delta^2 \geq 0.$$

Démonstration. La formule donnant dét$(I - A^* A)$ est immédiate.

Notons : $I - A^* A = \begin{pmatrix} a & b \\ b & c \end{pmatrix}$.

La condition (3.b) équivaut à : $(u, (I - A^* A)u) \geq 0$, pour tout $u \in \mathbb{R}^2$. On a donc : $ax^2 + 2bxy + cy^2 \geq 0$, pour tous $x, y \in \mathbb{R}$. Ceci n'est possible que si, à la fois : $a \geq 0$, $c \geq 0$, et $b^2 - ac \leq 0$, ce qui n'est autre que les conditions (3.c) et (3.d). □

(3.2) Nous étudions maintenant la représentation d'une matrice de corrélation comme matrice de Gram–Schmidt. Introduisons tout d'abord la terminologie suivante : si r est un entier tel que $0 \leq r \leq d$, et A une matrice $d \times d$, on dira que A est de type r si le rang de la matrice $I - A^* A$ est r.

Nous pouvons maintenant énoncer le :

Lemme 4. 1) *Soit k entier, $k \geq d$, et $\{\vec{u}_1, \dots, \vec{u}_d\}$, $\{\vec{u}'_1, \dots, \vec{u}'_d\}$ deux suites composées de d vecteurs orthonormaux dans $\mathbb{R}^k$. Alors, la matrice :*

$$\text{(3.e)} \qquad A = \left((\vec{u}_i, \vec{u}_j{}'); i, j \leq d\right)$$

est une matrice de corrélation.

2) *Soit $d = 2$, et A matrice de corrélation 2×2.*

Alors, A admet une représentation de Gram–Schmidt (3.e) *avec $k = 2 + r$ (où $r = 0$, 1 ou 2), mais pas $k = 2 + r - 1$ si, et seulement si, A est de type r.*

Démonstration. 1) Soit G_k variable gaussienne, centrée, à valeurs dans $\mathbb{R}^k$, admettant $\mathrm{Id}_{\mathbb{R}^k}$ pour covariance. A l'évidence, la matrice A donnée par (3.e) est la matrice de corrélation du couple :

$$G_d^{\vec{u}} \equiv \left\{(\vec{u}_i, G_k); i \leq d\right\}, \quad G_d^{\vec{u}'} \equiv \left\{(\vec{u}'_i, G_k); i \leq d\right\}.$$

2) La démonstration peut naturellement être décomposée en l'étude des 3 cas : $r = 0, 1, 2$.

(i) $\underline{r = 0}$. A est de type 0, si, et seulement si, A est une matrice orthogonale.

La représentation de A sous la forme (3.e), avec $k = 2$, découle alors de ce que A est la matrice d'une rotation, composée éventuellement avec une symétrie par rapport à une droite. Inversement, si A admet la représentation (3.e) avec $k = 2$, le couple $\{\vec{u}_1', \vec{u}_2'\}$ peut être ramené au couple $\{\vec{u}_1, \vec{u}_2\}$ au moyen de la composition des deux transformations déjà citées, et A est une transformation orthogonale.

(ii) $\underline{r = 1}$. Ecrivons A sous la forme $\begin{pmatrix} \alpha & \beta \\ \gamma & \delta \end{pmatrix}$.

Grâce au Lemme 3, l'hypothèse $r = 1$ équivaut à :

$$\alpha\beta + \gamma\delta = \varepsilon(1 - (\alpha^2 + \gamma^2))^{1/2}(1 - (\beta^2 + \delta^2))^{1/2}$$

pour un certain $\varepsilon = +1$ ou -1, et, de plus, l'une au moins des deux quantités : $a = 1 - (\alpha^2 + \gamma^2)$ ou $c = 1 - (\beta^2 + \delta^2)$ n'est pas nulle.

Si $\{\vec{u}_1, \vec{u}_2, \vec{u}_3\}$ est une base orthonormale de $\mathbb{R}^3$, on construit, à partir de cette base, deux vecteurs orthonormaux de $\mathbb{R}^3$:

$$\begin{aligned} \vec{u}_1{}' &= \alpha\vec{u}_1 + \gamma\vec{u}_2 + \big(1 - (\alpha^2 + \gamma^2)\big)^{1/2}\vec{u}_3 \\ \vec{u}_2{}' &= \beta\vec{u}_1 + \delta\vec{u}_2 - \varepsilon\big(1 - (\beta^2 + \delta^2)\big)^{1/2}\vec{u}_3 \end{aligned} \tag{3.f}$$

On a ainsi, à l'évidence : $A = \big((\vec{u}_i, \vec{u}_j'); i, j \leq 2\big)$.

Inversement, si l'on peut écrire la matrice A sous la forme (3.e) avec $k = 3$, mais pas $k = 2$, les vecteurs $\vec{u}_1'$ et $\vec{u}_2'$ peuvent toujours être représentés sous la forme (3.f), l'une au moins des 2 quantités $a \equiv 1 - (\alpha^2 + \gamma^2)$ et $c \equiv 1 - (\beta^2 + \delta^2)$ n'étant pas nulle.

Grâce au Lemme 3, on voit alors que la matrice $I - A^*A$ a un déterminant nul, mais n'est pas identiquement nulle ; $I - A^*A$ est donc de rang $r = 1$.

(iii) $\underline{r = 2}$. D'après l'étude des cas précédents, il nous suffit maintenant de démontrer que toute matrice de corrélation 2×2 peut s'écrire sous la forme (3.e), avec $k = 4$.

Soit $\{\vec{u}_1, \vec{u}_2, \vec{u}_3, \vec{u}_4\}$ une base orthonormale de $\mathbb{R}^4$.

Ecrivons $A = \begin{pmatrix} \alpha & \beta \\ \gamma & \delta \end{pmatrix}$. On cherche donc 4 réels $\tilde{\alpha}, \tilde{\beta}, \tilde{\gamma}, \tilde{\delta}$ tels que les 2 vecteurs :

$$\vec{u}_1' = \alpha\vec{u}_1 + \gamma\vec{u}_2 + \tilde{\alpha}\vec{u}_3 + \tilde{\gamma}\vec{u}_4 \quad \text{et} \quad \vec{u}_2' = \beta\vec{u}_1 + \delta\vec{u}_2 + \tilde{\beta}\vec{u}_3 + \tilde{\delta}\vec{u}_4$$

soient orthonormaux dans $\mathbb{R}^4$, ce qui équivaut à :

$$\text{(3.g)}\qquad \begin{aligned} &\tilde{\alpha}^2+\tilde{\gamma}^2 = 1-(\alpha^2+\gamma^2) \equiv a^2; \quad \tilde{\beta}^2+\tilde{\delta}^2 = 1-(\beta^2+\delta^2) \equiv c^2; \\ &\tilde{\alpha}\tilde{\beta}+\tilde{\gamma}\tilde{\delta} = -(\alpha\beta+\gamma\delta) \end{aligned}$$

Le problème se ramène donc à la recherche de deux réels μ et ν tels que :

$$\tilde{\alpha}+i\tilde{\gamma} = a\,e^{i\mu}, \quad \tilde{\beta}+i\tilde{\delta} = c\,e^{i\nu},$$

et à ce que la dernière équation figurant en (3.g), qui s'écrit maintenant sous la forme :

$$\text{(3.g}'\text{)}\qquad ac\cos(\mu-\nu) = -(\alpha\beta+\gamma\delta)$$

soit satisfaite.
Or, d'après l'inégalité (3.c) du Lemme 3, il existe $\eta \in \mathbb{R}$ tel que :

$$-(\alpha\beta+\gamma\delta) = ac\cos\eta.$$

En conséquence, l'équation (3.g′) sera satisfaite dès que :

$$\cos(\mu-\nu) = \cos\eta,$$

qui admet toujours des solutions (μ,ν). □

(3.3) Nous étudions maintenant certaines propriétés d'intégrabilité du couple (G_d, G'_d) de variables conjointement gaussiennes, à valeurs dans $\mathbb{R}^d$, pour d entier quelconque, en fonction de leur matrice de corrélation A. On a le

Lemme 5. *Soient $p,q \geq 0$. On a alors : $E\left[\frac{1}{|G_d|^p|G'_d|}\right] < \infty$ si, et seulement si*

$$\text{(3.h)}\qquad p+q<d+r,\ p<d,\ q<d,\ \textit{où } r \textit{ est le rang de } I-A^*A.$$

En particulier, si A n'est pas orthogonale, on a :

$$E\left[\frac{1}{|G_d|^p|G'_d|^q}\right] < \infty \ \textit{dès que} : p < \frac{d}{2}+\frac{1}{2} \ \textit{et}\ q < \frac{d}{2}+\frac{1}{2}.$$

Démonstration: 1) Pour nous convaincre du résultat, considérons tout d'abord les cas extrêmes $r=0$ et $r=d$.

Dans le cas $r=0$, on a : $|G_d| = |G'_d|$, et le résultat du Lemme 5 est alors ramené à l'équivalence :

$$E\left[\frac{1}{|G_d|^{p+q}}\right] < \infty \text{ si, et seulement si} : p+q<d,$$

ce qui est immédiat par passage en coordonnées polaires.

Dans le cas $r = d$, il nous suffit, à l'aide de la représentation $G'_d = A^*G_d+\xi$ donnée dans la partie 2) du Lemme 2, de montrer que la fonction : $z \to E\left[\frac{1}{|z+\xi|^q}\right] < \infty$ est uniformément bornée pour $q < d$; d'autre part, elle est identiquement infinie pour $q \geq d$.

Plus généralement, ce résultat de bornitude est vrai pour toutes les fonctions: $Tf(z) = E[f(z+\xi)]$ lorsque f est localement intégrable par rapport à la mesure de Lebesgue, et bornée dans le complémentaire d'un voisinage de 0. Pour simplifier l'écriture, nous ferons la démonstration de cette propriété uniquement dans le cas où ξ admet $\mathrm{Id}_{\mathbb{R}^d}$ pour matrice de covariance. On a alors :

$$Tf(z) = \frac{1}{(2\pi)^{d/2}} \int_{\mathbb{R}^d} da \exp\left(-\frac{|a-z|^2}{2}\right) f(a),$$

égalité dont on déduit la majoration :

$$\begin{aligned}|Tf(z)| &\leq \frac{1}{(2\pi)^{d/2}} \int_{|a|\leq\rho} da|f(a)| + \sup_{(|a|\geq\rho)} |f(a)| \\ &< \infty, \text{ pour } \rho \text{ suffisamment grand.}\end{aligned}$$

2) Etudions maintenant le cas général.

Montrons tout d'abord que la condition (3.h) est nécessaire.

On se place dans une base orthonormée qui diagonalise $I - A^*A$, supposée de rang r. On note $m = \rho\theta$ le point générique de $\mathbb{R}^d \setminus \{0\}$, où $\rho = |m|$, et $\theta = \frac{m}{|m|}$, puis $A^*m = \rho A^*\theta \equiv \rho a(\theta)$; enfin, les réels strictement positifs μ_i sont les racines carrées positives des valeurs propres de $I - A^*A$. On a alors :

$$E\left[\frac{1}{|G_d|^p|G'_d|^q}\right] < \infty \quad \text{si, et seulement si :}$$

$$\begin{aligned}\Phi_{p,q} \equiv \int_{S_{d-1}} \sigma(d\theta) \int_0^\infty \frac{d\rho\,\rho^{d-1}}{\rho^p} \exp\left(-\frac{\rho^2}{2}\right) \int_{\mathbb{R}^r} dx \exp\left(-\frac{|x|^2}{2}\right) f(\rho,\theta,x) \\ < \infty,\end{aligned}$$

où :

$$\begin{aligned}f(\rho,\theta,x) = \big((\rho a_1(\theta) + \mu_1 x_1)^2 + \cdots + (\rho a_r(\theta) + \mu_r x_r)^2 \\ + \ (\rho a_{r+1}(\theta))^2 + \cdots + (\rho a_d(\theta))^2\big)^{-q/2}.\end{aligned}$$

En faisant le changement de variables $x = \rho y$, il vient :

$$\Phi_{p,q} = \int_{S_{d-1}} \sigma(d\theta) \int_0^\infty d\rho \frac{\exp(-\frac{\rho^2}{2})}{\rho^{1+(p+q)-(d+r)}} \int_{\mathbb{R}^r} dy \exp\left(-\frac{|y|^2\rho^2}{2}\right) f(1,\theta,y).$$

Il apparaît ainsi que la condition: $p+q<d+r$ est nécessaire pour que $\Phi_{p,q}<\infty$.

Cette condition étant supposée satisfaite, on obtient en explicitant l'intégrale en $d\rho$:

$$\Phi_{p,q}=c\int_{S_{d-1}}\sigma(d\theta)\int_{\mathbb{R}^r}dy(1+|y|^2)^{-(d+r-p-q)/2}f(1,\theta,y).$$

La contribution de l'intégrale en dy, pour y grand, est finie dès que:

$$\int^{\infty}\frac{\rho^{r-1}d\rho}{\rho^{d+r-(p+q)}\rho^{q}}<\infty$$

c'est-à-dire: $p<d$. Par symétrie, il est également nécessaire que: $q<d$.

3) Supposons maintenant la condition (3.h) satisfaite. On a alors: $\Phi_{p,q}<\infty$ si, et seulement si, la quantité suivante:

$$\int_{S_{d-1}}\sigma(d\theta)\int_{|y|\le A}dy\Big((a_1(\theta)+\mu_1y_1)^2+\cdots+(a_r(\theta)+\mu_ry_r)^2 \\ +a_{r+1}^2(\theta)+\cdots+a_d^2(\theta)\Big)^{-q/2}$$

est finie, pour tout $A>0$.

En faisant le changement de variables: $z_i=\mu_iy_i+a_i(\theta)$ $(i\le r)$, et en posant $b(\theta)=(a_{r+1}^2(\theta)+\cdots+a_d^2(\theta))^{1/2}$, on voit que l'intégrale ci-dessus est finie dès que:

$$\int_{S_{d-1}}\sigma(d\theta)\int_0^{A'}\rho^{r-1}d\rho\frac{1}{(\rho^2+b^2(\theta))^{q/2}}<\infty.$$

Cette intégrale ne faisant plus intervenir p, il est facile de voir qu'elle est finie, puisque:

$$E\left[\frac{1}{|G'_d|^q}\right]\equiv c\,\Phi_{o,q}<\infty \text{ dès que } q<d.$$

Le lemme 5 est maintenant entièrement démontré. □

Remerciements. Je remercie vivement Mitchell Berger, dont la question (voir (**1.1**)) m'a fourni le point de départ de ce travail, ainsi que pour une longue discussion sur des sujets connexes à l'Université de St Andrews (Ecosse) en Avril 1989.

Je remercie également Jim Pitman et H. Kesten dont les suggestions m'ont permis d'améliorer la présentation d'origine de ce travail.

REFERENCES

[1] M. Berger, Communication personnelle (Avril 1989).

[2] J.F. Le Gall and M. Yor, *Etude asymptotique de certains mouvements browniens complexes avec drift,* Prob. Th. Rel. Fields **71** (1986), 183–229 .

[3] J.F. Le Gall and M. Yor, *Etude asymptotique des enlacements du mouvement brownien autour des droites de l'espace,* Prob. Th. Rel. Fields **74** (1987), 617–635.

[4] J.F. Le Gall and M. Yor, *Enlacements du mouvement brownien autour des courbes de l'espace,* Trans. Amer. Math. Soc. **317** (1990), 687–722.

[5] J.W. Pitman and M. Yor, *Asymptotic laws of planar Brownian motion,* Ann. Prob. **14** (1986), 733–779.

[6] J.W. Pitman and M. Yor, *The asymptotic joint distribution of windings of planar Brownian motion,* Bull. Amer. Math. Soc. **10**, Jan. 1984, 109–111.

[7] F. Spitzer, *Some theorems concerning 2-dimensional Brownian motion,* Trans. Amer. Math. Soc. **87** (1958), 187–197.

Université Pierre et Marie Curie
Laboratoire de Probabilités, Tour 56
4, place Jussieu – 3ème étage
75252 Paris Cedex 05, France

Progress in Probability

Editors

Progress in Probability is designed for the publication of workshops, seminars and conference proceedings on all aspects of probability theory and stochastic processes, as well as their connections with and applications to other areas such as mathematical statistics and statistical physics. It acts as a companion series to *Probability and Its Applications*, a context for research level monographs and advanced graduate texts.

We encourage preparation of manuscripts in some form of TeX for delivery in camera-ready copy, which leads to rapid publication, or in electronic form for interfacing with laser printers or typesetters.

Proposals should be sent directly to the editors or to:
Birkhäuser Boston, 675 Massachusetts Avenue, Cambridge, MA 02139, U.S.A.

1 ÇINLAR/CHUNG/GETOOR. Seminar on Stochastic Processes, 1981
2 KESTEN. Percolation Theory for Mathematicians
3 ASMUSSEN/HERING. Branching Processes
4 CHUNG/WILLIAMS. Introduction to Stochastic Integration
5 ÇINLAR/CHUNG/GETOOR. Seminar on Stochastic Processes, 1982
6 BLOOMFIELD/STEIGER. Least Absolute Deviation
7 ÇINLAR/CHUNG/GETOOR. Seminar on Stochastic Processes, 1983
8 BOUGEROL/LACROIX. Products of Random Matrices with Application to Schrödinger Operator
9 ÇINLAR/CHUNG/GETOOR. Seminar on Stochastic Processes, 1984
10 KIFER. Ergodic Theory of Random Transformations
11 EBERLEIN/TAQQU. Dependence in Probability and Statistics
12 ÇINLAR/CHUNG/GETOOR. Seminar on Stochastic Processes, 1985
13 ÇINLAR/CHUNG/GETOOR/GLOVER. Seminar on Stochastic Processes, 1986
14 DEVROYE. A Course in Density Estimation
15 ÇINLAR/CHUNG/GETOOR/GLOVER. Seminar on Stochastic Processes, 1987
16 KIFER. Random Perturbations of Dynamical Systems
17 ÇINLAR/CHUNG/GETOOR/GLOVER. Seminar on Stochastic Processes, 1988
18 ÇINLAR/CHUNG/GETOOR/FITZSIMMONS/WILLIAMS. Seminar on Stochastic Processes, 1989
19 ALEXANDER/WATKINS. Spatial Stochastic Processes: A Festschrift in Honor of Ted Harris on His 70th Birthday
20 HAAGERUP/HOFFMANN-JØRGENSEN/NIELSEN. Probability in Banach Spaces 6: Proceedings of the Sixth International Conference, Sandbjerg, Denmark 1986

21 EBERLEIN/KUELBS/MARCUS. Probability in Banach Spaces 7: Proceedings of the Seventh International Conference

22 PINSKY. Diffusion Processes and Related Problems in Analysis, Volume I: Diffusions in Analysis and Geometry

23 HAHN, MASON, WEINER. Sums, Trimmed Sums and Extremes

24 ÇINLAR ET AL. Seminar on Stochastic Processes, 1990

25 CAMBANIS, SAMORODNITSKY AND TAQQU. Stable Processes and Related Topics

26 CRUZEIRO/ZAMBRINI. Stochastic Analysis and Applications: Proceedings of the 1989 Lisbon Conference

27 PINSKY/VIHSTUTZ. Stochastic Flows, Vol. 2

28 DURRETT/KESTEN. Random Walks, Brownian Motion and Interacting Particle Systems

GPSR Compliance

The European Union's (EU) General Product Safety Regulation (GPSR) is a set of rules that requires consumer products to be safe and our obligations to ensure this.

If you have any concerns about our products, you can contact us on ProductSafety@springernature.com

In case Publisher is established outside the EU, the EU authorized representative is:

Springer Nature Customer Service Center GmbH
Europaplatz 3
69115 Heidelberg, Germany

Batch number: 10577880

Printed by Printforce, the Netherlands